Modern Trends in Applied Terrestrial Ecology

Modern Trends in Applied Terrestrial Ecology

Edited by:

R. S. Ambasht and
Navin K. Ambasht
Banaras Hindu University
Varanasi, India

Kluwer Academic/Plenum Publishers
New York, Boston, Dordrecht, London, Moscow

ISBN 0-306-47332-1

233 Spring Street, New York, NY 10013

http://www.wkap.nl/

10 9 8 7 6 5 4 3 2 1

Printed in the United States of America

CONTRIBUTORS LIST

1. **R. S. Ambasht**
Honorary Scientist
Department of Botany
Banaras Hindu University
Varanasi - 221 005, India
E-mail: rambasht@banaras.ernet.in

2. **Ernst E. Hildebrand**
Director Institut fur Bodenkunde und Waldernahrung
Albert Ludwigs Universitat Freiburg
Bertold str 17, 79089 Freiburg, Germany
E-mail: hildern@uni-freiburg.de

3. **H. Schak-Kirchner**
Institute for Soil Science and Forest Nutrition,
University of Freiburg, Bertold str 17, 79089
Freiburg, Germany

4. **Amy J. Symstad**
Illinois Natural History Survey
Lost Mound Field Station
3159 Crim Drive
Savanna, IL 61074, USA
E-mail: asymstad@inhs.uiuc.edu

5. **Youn Su Lee**
College of Agriculture and Plant Sciences
Kangwon National University
Chuncheon, 200-701, Republic of Korea
E-mail: younslee@kangwon.ac.kr

6. **Min Woong Lee**
Department of Agrobiology
Dongguk University, 26, Pil-dong, 3-ga,
Chung-gu, Seoul, 100-715, Korea
E-mail: mwlee@dgu.ac.kr

7. **Linda Fernandez**
Department of Environmental Sciences
University of California
Riverside, CA 92501, USA
E-mail: fernandez@ucr.edu

8. **J. Nigel B. Bell**
Centre for Environmental Technology
Imperial College of Science Technology and Medicine
Sillwood Park, Ascot, Berkshire SL5 7PY, London, UK
E-mail: n.bell@ic.ac.uk

9. **Fiona M. Marshall**
Imperial College at Silwood Park,
Ascot, Berkshire SL5 7PY, UK
E-mail: f.marshall@ic.ac.uk

10. **Navin K. Ambasht**
Ecology Research Laboratory
Department of Botany
Banaras Hindu University
Varanasi - 221 005, India
E-mail: asnambasht@yahoo.com

11. **John Parrotta**
National Program Leader, International Science Issues
USDA Forest Service, Research and Development
4th Floor, RPC
1601 N. KENT St
Arlington, VA 22209
Rio Piedras PR 00928-5000, USA
E-mail: jparrotta@fs.fed.us

12. **William H. McDowell**
Professor of Water Resources Management
Director, New Hampshire Water Resources Research Center
Department of Natural Resources, 219 James Hall
University of New Hampshire,
Durham, NH 03824, USA
E-mail: bill.mcdowell@unh.edu

13. **Jürgen Bender**
Institute of Agroecology
Federal Agricultural Research Centre (FAL)
Bundesallee 50 D-38116, Braunschweig, Germany
E-mail: Juergen.Bender@fal.de

14. **Hans Joachim Weigel**
Institute of Agroecology,
Federal Research Center for Agriculture (FAL)
Bundesalle 50, 38116 Braunschweig, Germany
E-mail: hans.wiegel@fal.ed

15. **Alyson K. Tobin**
Plant Science Laboratory
School of Biology, Sir Harold Mitchell Building
University of St. Andrews
St. Andrews KY16 9TH, UK
E-mail: at6@st-andrews.ac.uk

16. **Rainer W. Bussmann**
Department of Plant Physiology
University of Bayreuth
95440, Bayreuth, Germany
E-mail: ceja.andina@t-online.de

17. **Madhoolika Agrawal**
Department of Botany
Banaras Hindu University
Varanasi - 221 005, India
E-mail: madhoo@banaras.ernet.in

18. **Hansjorg Dietz**
Geobotanisches Institut ETH
Eidgenossische Technische Hochschule Zurich
Zurichberg str 38 CH 8044, Zurich, Switzerland
E-mail: dietz@geobot.umnw.ethz.ch

19. **Thomas Steinlein**
Leherstuhl fur Experimentelle Okologie und Okosystembiologie,
Universitat Bielefeld W4-107, Universitatsstr. 25
33615, Bielefeld, Germany
E-mail: Tom.Steinlein@biologie.uni_bielefeld.de

20. **Janusz Uchmanski**
Institute of Ecology, Polish Academy of Sciences
05092 Lomianki, Poland
E-mail: januch@polbox.com

21. **Gian Reto Walther**
Institute for Geobotany, University of Hannover
Nienburger Street 17, D-30167, Hannover, Germany
E-mail: Walther@geobotanik.uni-hannover.de

22. **Jorg Petersen**
Institute for Geobotany, University of Hannover
Nienburger Street 17, D-30167, Hannover, Germany
E-mail: petersen@geobotanik.uni-hannover.de

23. **Richard Pott**
Professor and Vice President
Institut fur Geobotanik
Nienburger str 17, D-30167
Hannover, Germany
E-mail: pott@mbox.geobotanik.uni-hannover.de

24. **Grizelle González**
IITF/USDA Forest Service
P.O. Box 25000
San Juan, PR 00928-5000, USA
E-mail: ggonzalez@fs.fed.us

25. **Michiro Fujihara**
Himeji Institute of Technology
Awaji Landscape Planning and Horticulture Academy
945-2 Nojimatokiwa, Hokindan-cho
Tsuna-gun, Hyogo 656-1726, Japan

26. **Yoshio Hada**
Faculty of Informatics
Department of Biosphere-Geosphere System Science
Okayama University of Science, 1-1 Ridai-cho
Okayama 700-005
Japan

27. **Gentaro Toyohara**
Miyajima Natural Botanical Garden
Department of Biological Sciences
Graduate School of Science
Hiroshima University
Mitsumaruko-yama, Miyajima-cho, Saiku-gun
Hiroshima 739-0543, Japan

PREFACE

Ecology and economics have Greek roots in *oikos* for "household", *logos* for "study", and *nomics* for "management". Thus, ecology and economics should have complemented one another for a proper growth and development without destruction, but, unfortunately, rapid industrialization, lure for fast financial gains, and commercialization activities have led to a widespread surge in pollution load, environmental degradation, habitat destruction, rapid loss of biodiversity, sudden rise in rate of extinction of many wildlife and wild relatives of domesticated animals and cultivated cereals and other plants, global climate changes creating global rise in temperature, and CO_2 levels and increased ultraviolet B at ground level. Although these threats to human health have led us to look to ecology for their solutions and guidance for sustainable development without destruction, the industrial and technology houses are looking for alternative methods of development and resource use methods. The two global conferences of the United Nations in 1972 and 1992, and international programs of Man and the Biosphere (MAB), International Biological Program (IBP), International Geosphere, Biosphere program (IGBP), and World Conservation Union (IUCN), of different commissions, United Nations Environmental Program (UNEP) efforts, Ramsar Conventions (for wetlands), and World Wide fund for Nature (WWF) (for nature in general and wildlife in particular) have focused attention of ecologists, naturalists, governments and Non-governmental organizations (NGOs) toward better conservation. Applied ecology has thus emerged as the most important unifying science concerned with lots of midterm corrections of old technologies and reformation of new ones to ensure development without destruction and with least pollution.

Ecological research results are fast pouring through numerous journals. In the mad race of paper publication, many substandard journals have mushroomed. Consequently, students concerned with application of ecological concepts, ideas, and ecosystem functions find it difficult and tiresome to reach the right literature in this chaotic and confused mass of published material. The senior editor of this book, based on his teaching experiments with batches of B.Sc. and M.Sc. students produced the *Textbook of Plant Ecology* in 1969, and the other editor (Navin K. Ambasht) joined the authorship from 10th revised edition (13th edition 1999, reprint 2001). Constant updating of the book helped the students to remain abreast. In order to put together different aspects of environment, another book, *Environment and Pollution*, was written in 1990, and its new fourth edition is in press (2002). The vastness of the present-day literature and quantum of increase year after year makes it difficult for an individual to do justice to providing correct information, and that, too, of different bioclimatic regions. It was felt that a truly international, multiauthored book would be the right choice. The editors floated the idea and discussed it with some of the world leaders in ecology. The response was encouraging. We invited specialists on selected topics to introduce from the beginning and then elaborate the contents to the present state of the art. We are extremely thankful to all authors who agreed and did full justice to the topics of applied terrestial ecology. We are sure that university students of botany, zoology, ecology, vegetation science, agriculture, forestry, soil science, wildlife management, ecological economics, population biology, ethnobotany, and pollution aspects will find it up to date and informative.

In terrestrial ecosystems, a good vegetal cover reflects high dominance and abundance, which, in economic terms refers to high levels of biomass productivity. Vegetal cover, besides including higher plants, should also include lower plants such as mosses, lichens, algae, and cyanobacteria. Dietz has discussed in this volume at macro and micro levels, spatial and temporal changes with respect to plant regeneration, competition, facilitation, seed predation, and microhabitat selection. He has described nondestructive tools of vegetal cover measurements, remote sensing, and computerized image analysis. The vegetation of any place constantly reacts with the environment and is in turn influenced and modified. This has led to the concept of succession. Walther et al. have highlighted the complexities of changes in vegetation structure and composition. Hitherto regarded as not so complex, directional, and therefore predictable, the authors have succeeded in showing the complexities and nonlinear nature of systems making predictions difficult or impossible. Ecomodeling for showing the limits of short- and long-term predictability would help policymakers in nature conservation. A specific site example of succession in pine forests in Japan is provided by Fujihara et al. Soils form the medium of anchorage and supply of water and nutrients to plants, as well as providing habitat for specialized categories of organisms. It is a very much alive subsystem of all terrestrial ecosystems, being the hub of decompositional activities and release of nutrients. González has explained nicely the soil organisms and decomposition processes for the benefit of not only soil scientists but also a vast array of other specialists who deal with cycling of materials in terrestrial systems. Soil porosity helps in providing a good balance of atmosphere, water, nutrients, symbionts, and pathogens for plants growing on them. Based on their research, Hildebrand and Kirchner have shown that soil deformation and compaction by heavy machinery in humid regions creates massive changes in porosity, aerobic biological activity, and many other desirable properties. Vehicle movements in forests, therefore, should be along prescribed lines and not diffused into tree covered zones. Ecology of soil-borne pathogens in crop fields are discussed by Lee and Lee.

Besides soil and vegetal cover, applied ecology also concerns prediction of effects of climate and land use change on "plant communities and ecosystem services." Symstad has discussed at length the above aspects of ecological plant classification systems linking functional response and functional effect groups. Still another way of looking at ecological dynamics is to reduce the groups to the level of individual and single populations. The well-known specialist Uchmanski has discussed individual-based approaches and they produce different results from those obtained by classical theoretical ecology methods. These methods have important applied implications in predicting or foreseeing the likely changes in populations and communities. The totality of variability at gene, species, and ecosystem levels is referred to as biodiversity. The United Nations Conference on Environment and Development (UNCED), in 1992, focused two main issues of greatest concern to humanity: (1) the rapid loss of biodiversity, and (2) global changes in environment. Ambasht and Ambasht have introduced the topic and reviewed modern literature on different definitions, myths and facts, causes of biodiversity losses and methods of biodiversity conservation. Biodiversity is important from a variety of ecological, social, ethical, and economic considerations. Linda Fernandez has conducted an interesting review of economics of biodiversity. Humans have increased the rate of extinction directly by the destruction of habitats, deforestation, introduction of exotic species that do not have predators, and pollution. Biodiversity losses reduce ecosystem productivity, stability, and sustainability; therefore, ecosystem conservation enhances the above features. Fernandez has discussed how economic value arises from the services that an ecosystem performs for society, both straightforward extractive resource flows, such as timber, and nonextractive services, such as flood protection and aquifier recharge, and nonphysical services, such as aesthetic and cultural values. A very significant part of biodiversity is concerned with traditional knowledge and plant use from an ethnobotanical angle. Bussmann has provided interesting information on traditional base knowledge about medicinal plants. An example from Southern Ecuador has been used to show the effects of an integrated approach to conserve biological and cultural diversity. Besides biodiversity, the UNCED showed great concern for the global, environmental problems, particularly the changes brought about by elevated levels of CO_2, UV-B, Ozone stress, etc. Bell and Marshall have shown how and to what extents ubiquitous air pollutants, SO_2, NO_x, and O_3 cause extensive loss of crop yield, particularly in the developing countries like India, Pakistan, Egypt, Mexico, and China. Based on their researches, considerable reduction in yield of *Phaseolus* beans, turnip, radish, wheat, rice, moong bean, palak, lettuce, etc are due to SO_2 and NO_2 around urban and periurban agriculture. Elevated CO_2 impacts on plants is separately discussed by Agrawal. Agrawal has shown that CO_2 rises in the future may increase biomass production, but with reduced protein content and antioxidant levels. The UV-B level is also getting enhanced at ground level due to reduction in the stratospheric ozone layer. This has adverse effects on crops. Tobin has done an interesting applied ecology review and has shown methods to minimize the loss particularly through increasing the UV-B tolerance in crops by conventional breeding methods. During the spring and summer months, rises in tropospheric ozone level is an alarming feature of modern global changes. Bender and Weigel reviewed this aspect for agricultural plants and semi-natural vegetation. O_3 can affect all aspects of plant growth and at all levels, from subcellular to whole ecosystems. He has discussed acute and chronic responses, mechanisms of action, and potential impacts on crop yields and natural vegetation.

Tropical forests are the hotspots for rich centers of biodiversity, biomass production, and CO_2 sink. They are also the most severely degraded systems needing urgent ecological attention. In the concluding part of this book, problems facing the tropical forests are addressed. Bussmann has compared the neotropical mountain forests bordering Ecuador and Peru with the paleotropical large mountain ecosystems of Mount Kenya in East Africa. Characteristically, plant biodiversity declines with rises in altitude. Fire is the dominant factor for regeneration in Kenya, while landslide is dominant in the Andes. Parrotta has taken a comprehensive look at the progressive degradation of tropical forest ecosystems, negatively impacted human societies, and economies at local, national, and regional scales. He has also shown how with modern knowledge, forest scientists can bring about reversal of forest degradation. The role of planted forests is highlighted to meet socio-economic and environmental goals. Finally, McDowell has shown that nutrient losses from tropical forests in humid climates are quite high; higher than temperate forest. Despite the scarcity of data, from his studies in Puerto Rican watersheds and theoretical analyses of impacts of increased nitrogen deposition in tropical landscapes, McDowell has predicted qualitative and quantitative differences in nutrient losses from tropical and temperate forests. Landscape management, use of vegetated buffer zones on stream banks, and use of innovative agricultural tillage and intercropping are suggested for improvement.

Acknowledgements:

During the course of preparation and publication, we received help from many people and we thank them all. We are particularly thankful to all the contributors who very readily agreed to take the trouble of writing these chapters with the purpose of introducing the topics and discussing the latest findings, ideas, and concepts from applied ecology view points. We are indeed very grateful to Andrea Macaluso, Editor, Anthony Fulgieri, Production editor, and other members of Kluwer Academic/Plenum Publishers, New York staff for taking great interest and more than normal pains in giving a proper shape to this volume and publishing it so nicely and in such a short time. We thank R. Prasad for his help during the preparatory phase. We also thank our family members, Annpurna, Pravin, Sandhya, Anupama, Prakriti, Sukriti, and Soumya for their help and forbearance. We are thankful to the Council of Scientific and Industrial Research and Indian National Science Academy, New Delhi for awarding research grants and to the authorities of the Banaras Hindu University for providing facilities.

R. S. Ambasht
Navin K. Ambasht
Banaras Hindu University, India

CONTENTS

1

The Influence of Compaction on Soil Structure and Functions in Forest Sites

E. E. Hildebrand and H. Schack-Kirchner

Introduction

Forest soils are severely taxed by forestry, as well as by human society in general. For instance, forest soils are expected to provide sustained tree nutrition, although nutritional requirements may drastically change with the age of the stand. A further expectation of rising importance is the continuous production of high-quality drinking water. The behavior of forest soils as chemically quasi-closed systems seems to conflict with the continuous need for a mobilizable nutrient pool. The regulation between both functions takes place at hierarchically structured inner surfaces of forest soils, which control the amount of dissolved components in the soil solution. Therefore, forest soils are like "intelligent" reaction vessels of high complexity.

Figure 1.1 shows the principle of regulation in a soil section of a few mm^3: Roots are ion sources, and ion sinks in the soil solution, which restores its ion status by dissolution (quasi-)equilibria with the soil solid phase. If these functions meet certain management requirements, a soil is designated "well buffered." We can consider the microsection in Figure 1.1 as a basic regulation complex of the soil. Although solid substances, liquids, and gaseous components are randomly distributed and mixed in soils, they are not a "pap" or a "paste," but complex formations consisting of such elementary cells. The spatial distribution of soil solid substances can be roughly described by means of soil structure classification. Unlike sediments, structure is therefore one of the most essential properties of soils (Richter, 1986).

Figure 1.1 clearly illustrates the strong dependence of ecological soil functions on structure. Growing roots simultaneously need water, nutrients, and oxygen in small soil

E. E. Hildebrand and H. Schack-Kirchner • Institute of Soil Science and Forest Nutrition, University of Freiburg, Bertoldstr. 17, 79089 Freiburg, Germany, email: Ernst.Hildebrand@bodenkunde.uni-freiburg.de

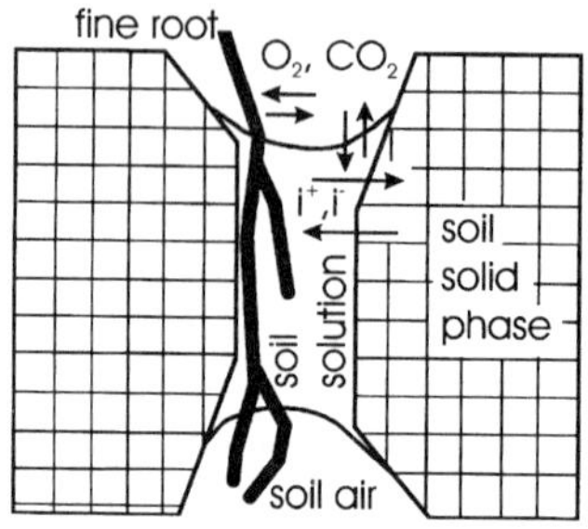

Figure 1.1. Basic regulation complex of a soil in the range of few mm^3 controlling the dissolution and matter transport in the soil solution.

volumes, which seems contradictory at first sight. Many tree species do not possess an aerenchyma in root tissues, which would allow direct use of the oxygen supply of the free atmosphere. Therefore, oxygen supply and carbon dioxide removal in the rhizosphere require the presence of a continuous, gas-filled pore space that is linked with the open atmosphere. This apparently conflicts with the need for sufficient plant-available water. Only close neighboring of oxygen and available water in small soil volumes allows tree roots to grow.

Soil Deformation by Heavy Machinery and Soil Structure

Soil structure is an important factor of simultaneous oxygen and water availability, as it provides hierarchically organized soil surfaces with frequent changes of interfaces between solid/water and solid/gas. Roots growing in well-structured soils inevitably find water and oxygen in close proximity. The important role of soil structure has caused, and still causes much scientific brainstorming in an attempt to describe soil structures by more or less simple and/or sophisticated physical or morphological parameters (Table 1.1). Such measurements are then used as independent variables to explain root or tree growth.

The diffusion coefficient may be seen as an integrating parameter of soil structure. Figure 1.2 shows, that a 10-year period of recovery is not sufficient for restoring the gas diffusion conditions of the control. Soil deformation always has long-lasting effects on gas turnover. Therefore, the gas exchange is considered to be the main link between soil structure and plant growth (Figure 1.3).

Gas Budget in the Rhizosphere of Forest Soils as a Link between Soil Structure and Root Growth

It is important to recognize that none of the parameters listed in Table 1.1 is of direct importance for root growth. Roots do not have any "interest" in water conductivity, pore-size distribution, and other useful inventories of soil physicists. They simply need water, nutrients, and oxygen, independent of the soil's physical status. Consideration of the shear strength might contradict this statement, but root growth is rarely mechanically restricted

Table 1.1. Technical and Ecological Evaluation of Some Soil Physical Measurements

Abbrev.	Parameter/method	Techn. effort	Sensitivity to soil deformation	Indicator quality for root growth	Literature (examples)
dB	Bulk density	Low	Medium	Indirect, poor	Standard method
pv	Total pore volume	Low	Medium	Indirect, poor	Standard method
pF	Pore size distribution	Low	Medium-high	Indirect, poor to medium	Standard method
kf	Sat. hyd. conductivity	Medium	High	Indirect, correlation with macropore gas budget	Standard method
ki	Air permeability	Medium	Very high	Indirect, correlation with macropore gas budget	Hansknecht et al. (1989)
ku	Unsat. hyd. conductivity	Very high, delicate	High, delicate	Parameter of water transport and water availability	Gluth (1987)
pq	Pore quotient	Medium	High	Indirect, effectiveness of transfer processes in macropores	Schack and Hildebrand (1987)
Ds	Diffusion coefficient	High	Very high	Parameter of O_2 supply and CO_2 disposal	Frede (1986)
	Hysteresis of water content	Delicate	Probably high	Indirect, parameter for pore continuity	Hildebrand and Wiebel (1986)
	Shear strength	Medium-high, delicate	Medium-high	Parameter for (important?) mechanical root growth resistance	Beekmann (1987)
	Seedling growth experiments	High	Very high	May serve as "biological penetrometer"	Hildebrand (1983), Corns (1988)
	Cross cuts	Very high	Out of competition	Visualization of micropedological environment of fine roots	Babel (1990)
	Scientific models	Very high	Out of competition	Isolation of key factors of root growth	Flühler (1973), Schack-Kirchner (1993)

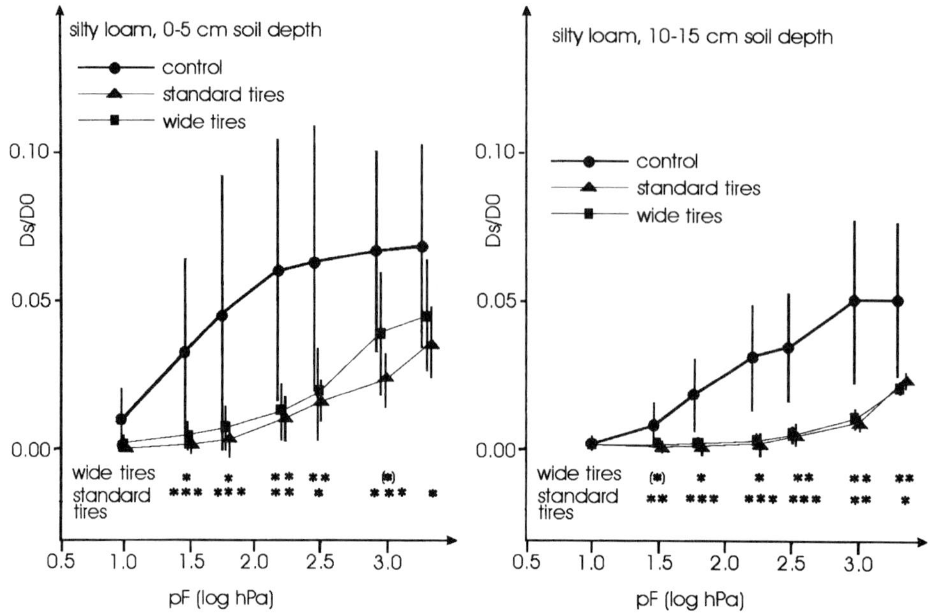

Figure 1.2. Apparent relative diffusion coefficients (mean values and standard deviations) in a 10-year-old skid trail in a beech stand (*Fagus sylvatica*) with a silty loam from loess in function of the water suction (Schack-Kirchner, 1994): right: 0–5 cm soil depth; left: 10–15 cm soil depth. The stars give the significance of the differences between treatment and control (nonparametric Wilcoxon test).

in forest soils. Moreover, it is not possible to construct penetrometers that emulate all forces that roots have to overcome (Helal, 1991).

Gas turnover, therefore, is a key process in the rhizosphere of forest soils. Results of Murach et al. (1993) clearly demonstrate this. With aeration treatment, the site-typical gradient of fine root frequency with depth could be inversed (Figure 1.4). Based on the working hypothesis that the highly sensitive monitor "soil atmosphere" contains key information for site ecology and ecosystem stability, we developed a device to measure gas concentration profiles in forest soils (O_2, CO_2, N_2, Ar, CH_4, C_2H_4, N_2O, sum of S-

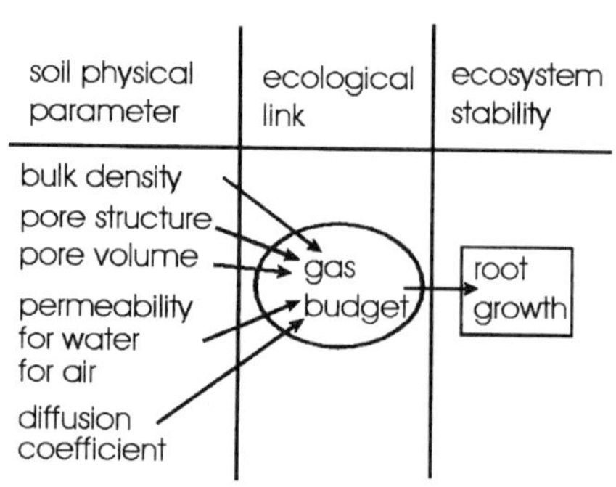

Figure 1.3. Gas budget as link between fragmenting soil physical parameters and root growth.

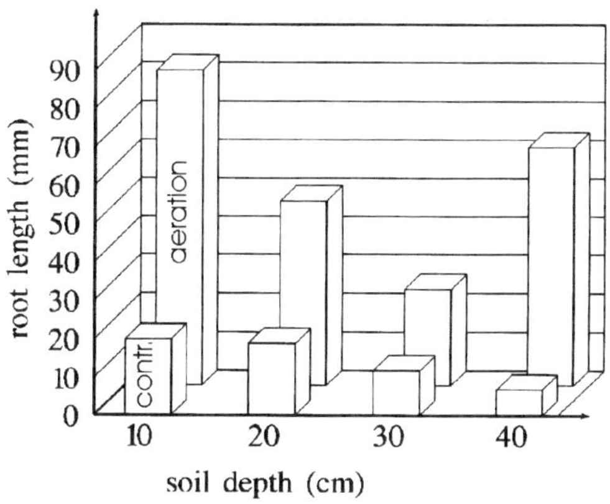

Figure 1.4. The significance of O_2 availability for root formation and distribution of Norway spruce (modified after Murach et al., 1993).

containing gases) with high spatial dissolution (1 cm vertical dissolution in the organic layers; Schack-Kircher, 1994).

A valid investigation of gas budgets in soils must be based on the correspondence of the measuring scale to the adjacent integration level on which the results are interpreted. Typical measuring and integration levels of scientists investigating gas transport, consumption, and production in soils are listed in Figure 1.5. The integration level of our measuring device consists of the gas-filled macropore space in the rhizosphere, designated as the "meso-scaled level" in Figure 1.5. The ecological relevance of this compartment is

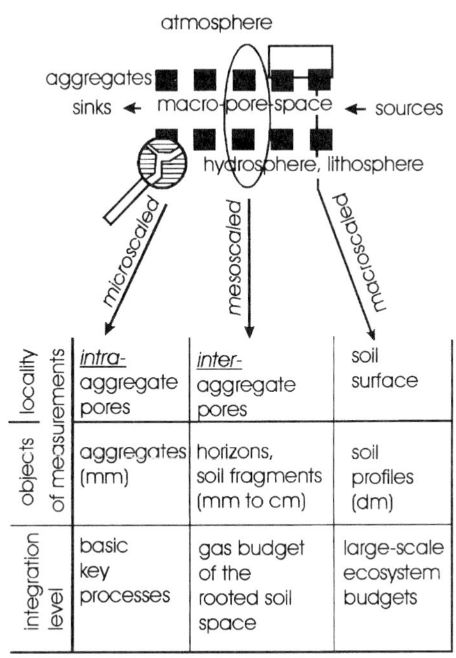

Figure 1.5. Integration levels of measuring gas concentrations in soils.

demonstrated by Babel (1990), who found by micropedological methods more than 80% of the fine roots of red spruce distributed in preformed macropores (>0.1 mm). In the case of beech, about 65% of the fine roots grow in preformed cavities. This also applies for mycorrhizal hyphae, which grow preferentially in macropores (Schack-Kirchner et al., 2000). It is important to note that a good aeration at the level of the air-filled macropores does not exclude anaerobic microsites within the soil aggregates. This fact seems to be of minor ecological importance, because tree roots and mycorrhizal hyphae seldom grow into aggregates. The extreme partial pressure gradients on the level of soil aggregates are, however, important to understand basic processes of gas transport in soils (Zausig and Horn, 1991).

Gas concentration profiles allow calculation of gas fluxes, assuming knowledge of diffusion coefficients and fulfilment of boundary conditions of diffusion transport equations. These assumptions are generally met when gas transport by convection is of minor importance. Simple diffusion laws (e.g., Fick's Laws) will fail if nonequimolar gas production and gas consumption occur. The comparison between measured soil gas concentrations and those predicted by the kinetic gas theory offers a simple tool to indicate whether diffusion laws alone are sufficient descriptors of the gas transport. Figure 1.6 includes examples of such comparisons.

A budget model of gas turnover, similar to budget models of solute transport with soil water, seems to be most suitable, because it recognizes the fact that oxygen supply and carbon dioxide discharge in the rhizosphere are key factors for root growth (Figure 1.1). Using physically defined transport laws (scientific model according to Hauhs, 1990), such a model could help to analyze ecological consequences of soil deformation, because soil atmosphere can be used as a highly sensitive and ecologically relevant monitor. The relations between gas concentrations presented in Figure 1.6 help to identify basic processes and use adequate transport models:

- *Case 1:* If predictions of basic diffusion transport laws are fulfilled (Figure 1.6, top), gas fluxes, production, and consumption can be calculated without restriction using numeric solutions of parametrized diffusion equations (i.e., Fick's Laws).
- *Case 2:* If relations between gas concentrations resemble those presented in Figure 1.6, bottom, a defined process identification by scientific models is more difficult. High spatial heterogeneity and fluctuations of gas concentrations are a result of overlaying processes. Scientific models can be technically adapted to some extent; however, if extreme heterogeneity is found, model resistance in the selected integration level may occur.

Schack-Kirchner (1993) found a considerable remote effect of surfacially increased diffusion resistance on the equilibrium oxygen concentration in the rhizosphere. Such a behavior is likely to cause flat rooting because of oxygen deficiency in the subsoil. Flat rooting leads to ecosystem lability, because external site factors (dryness, storm) are hardly compensated by soil regulatories. The final trigger (dryness, storm) for such cases of ecosystem breakdown is usually misinterpreted as the primary cause, which is a reduction of soil functions in deformed soils. This is why "resource-models" are pot sensitive enough to reveal hidden ecosystem defects, because such defects are disguised by compensatory mechanisms under "standard conditions" (Hauhs, 1990). One-dimensional approaches are adequate, assuming lack of lateral transport gradients. However, models based on this assumption do not reflect the real conditions of skid trails, which are deformed soil areas with extreme length/width relation. Lateral replenishment of oxygen

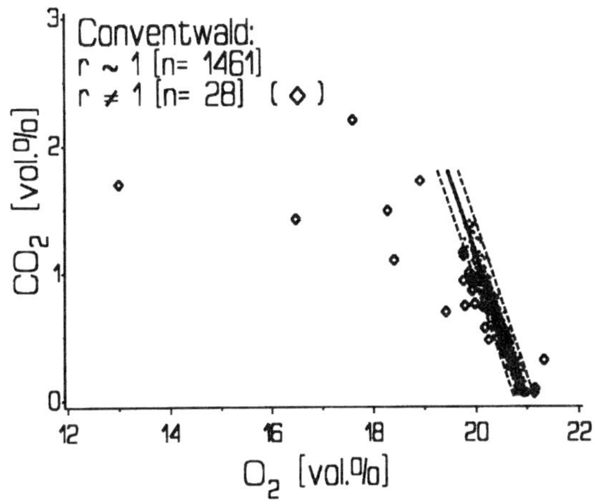

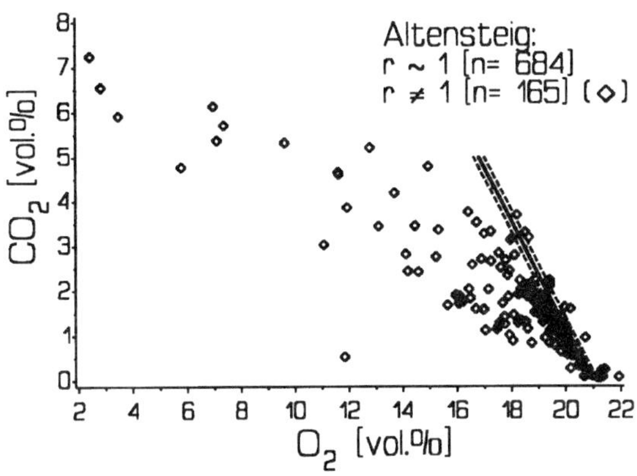

Figure 1.6. Top: The CO_2/O_2 relations in a well-aerated forest soil in the southern Black Forest, Germany (Ecosystem study "Conventwald" v. Wilpert et al., 2000) agree with predictions using diffusion transport laws, after Wood and Greenwood (1971). Bottom: Gas sinks or sources in an hydromorphic forest soil of the northern Black Forest (e.g., increased dissolution of CO_2 in the soil water at CO_2 partial pressures > 100 hPa) create deviations of the CO_2/O_2 relations predicted by the diffusion model of Wood and Greenwood (1971). Within the dashed lines, we assume a respiratory quotient $r = 1$ and a constant CO_2/O_2 gradient ratio of 1.2 ± analytical error (Schack-Kirchner, 1994).

concentrations in the subsoil must be considered. This requires a two-dimensional solution of the diffusion equations. The result of such a two-dimensional model is presented in Figure 1.7 (for details, see Schack-Kirchner, 1993).

It includes two 10-year-old skid trails in a beech stand on a silty loam near Freiburg (Hildebrand and Wiebel, 1986). The following model parameters were considered: depth function of diffusion coefficients; depth function of carbon content (both measured); C-dependent O_2 consumption (after Nadelhoffer, 1990); water status of 160 hPa; temperature in the topsoil 14°C, in the subsoil 10°C (selected boundary conditions). Figure 1.7 shows that the remote effect on horizontal and vertical oxygen impoverishment is higher in the

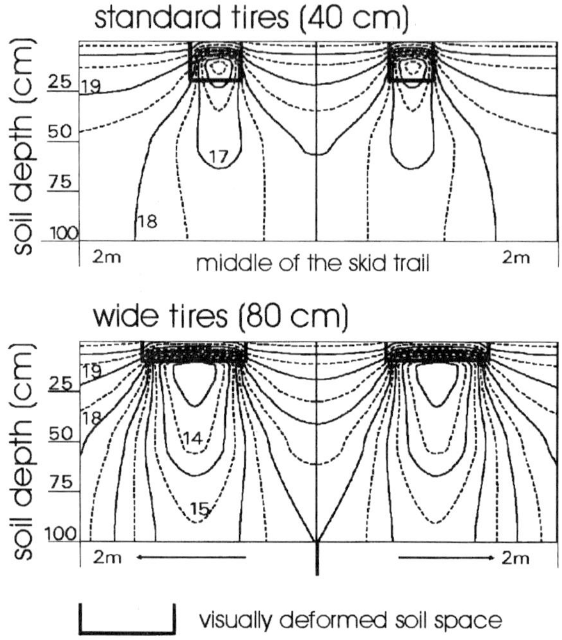

Figure 1.7. Result of a two-dimensional modeling of isolines of equilibrium oxygen concentrations in a silty loam below two 10-year-old skid trails (the skid trails are identical with those presented in Figure 1.2). Top: standard tires, 40 cm. Bottom: wide tires, 80 cm. The mechanically deformed soil space is marked.

case of the skid trail originating from wide tires regardless of whether the intensity of deformation is minor. On the first view, this seems to go against the common sense, but it simply shows *that the remote effect depends more on the size of the deformed soil area than on the intensity of deformation. It is important to note that such a trivial but ecologically important effect is frequently overlooked unless a process identification by model application is performed.* It is clear that model results based on mean values cannot reflect the reality in detail. We must concede that small but unrecognized heterogeneities (e.g., oxygen consumption by organic matter mixed into the soil) will have a strong influence on equilibrium oxygen concentrations in the soil. Although scientific models may simplify too much and create incorrect results for certain site conditions, they are still quite useful, because the scientist focuses on basic and powerful key parameters governing gas turnover in the soil.

In a field study, we tried to verify the interrelation between the size of a deformed soil area and its remote effect. Therefore, we compared the impact of two thinning concepts on the soil aeration status in a red spruce stand (30 years) on a clayey, loamy textured soil:

- Highly mechanized logging by harvester and forwarder, the skid trail was used once by a harvester (FMG Timberjack 1270), tire width 70 cm, and twice by a forwarder (Valmet 828, tire width 60 cm).
- Logging by horse after felling with chain saws; the investigation plot was passed 10 times by a horse (about 600 kg) skidding 1 to 4 logs of 0.08 m^3 volume. (We do not try to stop the march of time in tree logging; the horse was in this field study simply an easily available and cheap "biological walking machine".)

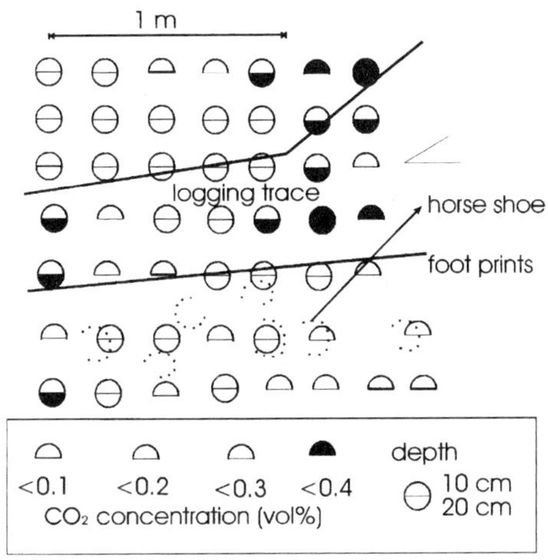

Figure 1.8. Results from gridwise determination of CO_2 concentrations in the soil air in 10- and 20-cm soil depth in skid trails and in the area logged by horse (after Schack-Kirchner, 1997).

If we consider the CO_2 concentration in the soil atmosphere as monitor for the quality of CO_2 removal, it becomes clears that horse prints have no ecological effect on the soil gas turnover, whereas in the center of the wheel track, the mean CO_2 concentrations are about eight times higher than outside of the track (see Figure 1.8). This finding tells us that the model results presented in Figure 1.7 indeed focused our view to an ecologically important relationship between the size of a deformed soil space and the remote effect on gas turnover.

Conclusions and Summary

Structure is an elementary soil property that causes the spatial proximity of soil/solid and soil solution/soil atmosphere interfaces. Therefore, growing roots can acquire water, nutrients, and oxygen at the same time, and nearly at the same place. Tree roots are especially dependent on this simultaneous, in certain aspects contradictory, availability of gases and liquids in the soil, because they cannot directly use the oxygen supply of the open atmosphere. Based on the ecological importance of soil structure, various efforts were made to describe structure with simple or complex soil physical parameters. Measurements of those are then used as independent variables to predict root and/or tree growth. This approach is also widely applied when investigating changes of soil functions by soil deformation. However, a great distortion is introduced with the use of linear and fragmented relations between soil physical parameters and growth response. *Roots do not "worry" about structure, but about the processes associated with structure.* For this reason it is necessary to look directly at the highly sensitive monitor "soil atmosphere," which may be seen as the "ecological link" between structure and root growth.

The effect of soil deformation on rhizosphere quality can be demonstrated using equilibrium models of oxygen availability fed with diffusion coefficients obtained from soil core measurements. In the case of skid trails, two-dimensional models must be applied, because their extreme length/width relation requires the inclusion of horizontal partial pressure gradients. The application of such a model to existing skid trails revealed the trivial, but widely neglected fact, that the deterioration of the equilibrium oxygen concentration in the rhizosphere depends on the size of the deformed soil compartment rather than on the intensity of soil deformation.

Scientific models sometimes simplify complex realities to the point of producing improper results. Nevertheless, they are still very useful. Bossel (1992) compares them with cartoons that focus the viewer's attention to hidden basic information. If we want to do more than accumulate case studies with linear, fragmented, and ecologically distorted cause/reaction relations, and if we really want to analyze how ecosystems are affected by soil deformation, we have no other choice than to work with scientific models. Yield, growth, and other processes that integrate over the whole range of site properties are variables of inadequate sluggishness. They cannot properly reveal ecosystem deficiencies, because they might be supported by compensatory "makeshift solutions."

References

Babel, U. (1990). Verteilung von Buchen- und Fichtenfeinwurzeln in Bezug zu Bodenhohlräumen und dichter Bodensubstanz. Allgemeine Forst-u. Jagd-Zeitung. 161: 109–112.

Beekmann, F. (1987). *Soil strength and forest operations*. Doctoral dissertation, Department of Forest Technique and Forest Products, Agricultural University, Wageningen, The Netherlands, 167 pp.

Bossel, H. (1992). Overview: Ecosystem and systems analysis. In: Ecologic and socioeconomic-system analysis and simulation. Institute for World Forestry and Ecology (Ed.), (pp. 51–76). Hamburg: Federal Research Centre for Forestry and Forest Products.

Corns, I. G. W. (1987). Compaction by forestry equipment and effects on coniferous seedling growth on four soils in the Alberta foothills. *Canadian Journal of Forest Research* 18: 75–84.

Flühler, H. (1972). Sauerstoffdiffusion im Boden: Mitteilungen der Eidgenoessischen Anstalt fuer das Forstl. Versuchswesen Bd. 49: 1–250.

Frede, H. G. (1986). Der Gasaustausch des Bodens. *Göttinger Bodenkundliche Berichte* 87: 1–130.

Gluth, G. (1987). Quantifizierung befahrungsbedingter Bodenveränderungen. *Holz-Zentralblatt* 112: 1618–1620.

Hansknecht, R., Hildebrand, E. E., and Hofmann, R. (1989). Vergleich des Verformungsverhaltens von Schlufflehm und lehmigem Sand nach praxisnaher Befahrung im Zustand der Frühjahrsfeuchte. Mitteilungen des Vereins Fuer Forstliche Standertskunde u. Forstplanzenzüchtung 34: 7–13.

Hauhs, M. (1990). Ecosystem modelling: Science of technology? *Journal of Hydrology* 116: 25–33.

Helal, M. (1991). Bodengefüge, Wurzelentwicklung und Wurzelfunktion. *Zeitschrift fuer Pflanzenernachrung. Bodenkde.* 154: 403–407.

Hildebrand, E. E. (1983): Der Einfluß der Bodenverdichtung auf die Bodenfunktionen im forstlichen Standort. *Forstwissenschaftliches Centralblatt* 102: 111–125.

Hildebrand, E. E., and Wiebel, M. (1986). Zur Bedeutung des Bodenwassergehaltes bei der Entstehung von Bodenschäden durch Befahrung. *Augemeine Forstzeitschrilt* 41: 617–622.

Murach, D., Ilse, L., Klaproth, F. U., and Wiedemann, H. (1993): Rhizotron-Experimente zur Wurzelverteilung der Fichte. *Forstarchiv* 64: 191–194.

Nadclhoffer, K. (1991): Microlysimeter for measuring nitrogen mineralization and microbial respiration in aerobic soil incubations. *Soil Science Society American Journal* 54: 411–415.

Richter, J. (1986). *Der Boden als Reaktor.* (Stuttgart: Ferdinand Enke Verlag. 1–239.

Schack, H., and Hildebrand, E. E. (1987). Einfluß der mechanischen Beanspruchung eines Sandbodens auf das Wurzelwachstum von Forstpflanzen. *Allgemeine Forst- und Jagdzeitung* 159: 27–34.

Schack-Kirchner, H. (1993). Modellierung des Bodengashaushaltes—Probleme und Lösungsansätze. Tagungsber. Arbeilsgemeinschaft Oekologie der Internationalen Biometrischen Gesellschaft 4: 50–64.

Schack-Kirchner, H. (1994) Struktur und Gashaushalt von Waldböden. Berichte des Forschungszentrums Waldökosysteme Reihe A, Bd 112, 145 p.

Schack-Kirchner, H. (1997). Soil air composition, gas transport and biologic activity in forest soils subjected to different harvesting methods. Proceedings of Symposium IUFRO XX World Congress, Projekt Group: Forest Operations and Environmental Protection. Tampere, Finland, pp. 115–122.
Schack-Kirchner, H., von Wilpert, K., and Hildebrand E. E. (2000) The Spatial distribution of soil hyphae in structured forest soils. *Plant and Soil* 224: 195–205.
von Wilpert, K., Zirlewagen, D., and Kohler, M. (2000) To what extent can silviculture enhance sustainability of forest sites under the immission regime in Central Europe? *Water, Air, and Soil Pollution* 112: 105–120.
Wood, J. T., and Greenwood, D. J. (1971). Distribution of carbon dioxide and oxygten in the gas phase of aerobic soils. *Journal of Soil Science* 22(3): 281–288.
Zausig, J., and Horn, R. (1991). Der Belüftungszustand eines Pelosol-Gleys als Funktion des Bodenwasserhaushaltes. *Mitteilnugen des Deutschen Bodenkundlichen Gesellschaft* 66: 263–266.

Acknowledgments

The research of gas budgets in forest soils was financed by the German Research Foundation (DFG).

2

An Overview of Ecological Plant Classification Systems

Linking Functional Response and Functional Effect Groups

Amy J. Symstad

Ecosystems and the communities of living things that comprise them are complex entities. Classification systems are one tool that ecologists use to understand these entities. By classifying organisms, communities, or ecosystems into distinct categories, an ecologist creates a conceptual model of the relationship between the entities being classified. Even though the entities in reality do not necessarily fit into discrete categories, but instead lie along a continuum, the criteria used to devise those categories serve as hypotheses that, when tested, lead to refinements of the model or, in some cases, its rejection. This process of determining the defining criteria of the categories, then evaluating their success in explaining observed patterns, leads to better understanding of the problem at hand. A classic example of this process comes from vegetation ecology.

The science of vegetation ecology began in earnest in the late 19th century, when some naturalists changed their focus from classifying plants into species to classifying plant communities according to the plant species within them (Mueller-Dombois and Ellenberg, 1974). One of the initial goals of this new science was to create a classification system similar to the taxonomic classification system of species. As the data accumulated, however, they revealed the continuous nature of plant communities. This made the discrete categories needed for a taxonomic-like vegetation classification systems unrealistic, and eventually, such systems were for the most part abandoned (Ricklefs, 1990). However, the exercise of trying to create such classification systems stimulated debate about the general nature of plant communities. In the United States, this debate was primarily between Clements and Gleason (e.g., Clements, 1916, 1936; Gleason, 1926), but it took place in other countries as well. The exchange of ideas surrounding this debate produced a better

Amy J. Symstad • Illinois Natural History Survey, Lost Mound Field Station, 3159 Crim Drive Savanna, IL 61074; email: asymstad@inhs.uiuc.edu understanding of the factors that determine the composition of plant communities through space and time.

Since that time, new ecological systems of classifying plants have emerged, based on the understanding that came from the debate. Two basic categories of these systems exist. The first has an objective not unlike that of early vegetation scientists mentioned earlier. Modern vegetation classification systems such as those being developed by the Ecological Society of America's Vegetation Classification Panel, aim to describe consistently the floristic composition of vegetation as it occurs in a landscape. These classification systems are developed regionally to map, describe, and inventory vegetative natural resources. The second type aims not to describe but to understand plant communities and their relationship to their environment. Thus, objectives of plant classification systems of this type include explaining the distribution of plant species in relation to climate, soil type, and disturbance, as well as understanding how plants affect ecosystem processes and properties.

Although the first type of ecological classification system has been deemed necessary for descriptive purposes, the second type is more responsible for stimulating further understanding of plant communities and ecosystems. The rest of this chapter concerns the second type. My goal is to provide an understanding of the variety of ecological plant classification systems of this type that exist. The remainder of the chapter is therefore organized into the following sections: purposes of plant classification systems, properties on which such systems are based, approaches to classifying plants, comparisons of results from different classification systems, and linkage of classification systems. An underlying theme is the examination of how different classification systems contribute to the understanding of plant community dynamics and effects on ecosystem functioning. This theme is particularly important because only through understanding will it be possible to achieve the basic goal of many modern plant classification systems—to predict how global change will affect future vegetation and ecosystem functioning.

Although global change affects all ecosystems, most examples in this chapter come from grasslands, the ecosystem with which I am most familiar. In addition, because they are the most commonly studied ecosystems for plant classification efforts, grasslands provide the best opportunity for comparing groups created by different classification systems.

For brevity, I henceforth refer to an ecological plant classification system as an EPCS.

Purposes of Ecological Plant Classification Systems

The general purpose of any EPCS is to simplify the complexity of a plant community, yet retain the information essential to achieving the specific goal (Noble and Gitay 1996). As illustrated in the next section, there are a large number of EPCSs and a sometimes confusing vocabulary that accompanies them. The main reason so many systems exist, in addition to researchers' individuality, is that there are many different goals, some quite specific, for classifying plants. Despite this, I suggest that there are three basic goals for classifying plants.

Goal 1: Understand How Species Are Distributed and Why They Are Distributed as They Are

Specific goals of this type can be rather simple, such as investigating the relationship between climatic variables and the distribution of individual species or predefined species groups (e.g., Paruelo and Lauenroth, 1996), or they can be much more complex. More complex goals in this category include discovering "emergent groups" of plants (Lavorel et al., 1997) and determining plant strategies.

In emergent groups, species share a suite of attributes.[1] These suites are usually derived via multivariate analysis of large data sets with many traits. Emergent groups may (1) reveal attributes particularly common in, and therefore perhaps adaptive to, a set of environmental conditions (e.g., Díaz and Cabido, 1997) and (2) when combined with species co-occurrence data, suggest assembly rules for a specific community type (Kindscher and Wells, 1995; Weiher et al., 1998) or (3) be compared to ecological groups derived via other means, in order to relate suites of attributes to specific plant behaviors (e.g., Lavorel, Rochette, et al., 1999). Although the multivariate techniques used to generate emergent groups do not reveal why certain suites of attributes are associated with particular environmental conditions, and therefore why species are distributed as they are, they are useful for generating hypotheses that can lead to answers to these questions.

A different EPCS set constructed largely to understand why plants occur where and when they do is one that seeks plant strategies, which are methods by which a species survives, and these methods are reflected in a suite of attributes assumed to be adaptive to certain environmental circumstances. The best known plant strategy system is the CSR scheme (Grime, 1974), which classifies plants based on three primary strategies—**C**ompetitive (efficient resource capture and utilization), **S**tress tolerant (able to tolerate continuously low resource levels), and **R**uderal (able to grow and reproduce in short intervals between disturbances). Since its initial proposal (Grime, 1974), much effort has been put into determining where individual plants fall within the CSR framework (e.g., Grime, 1977; Grime et al., 1988) and finding specific attributes that adequately characterize the primary strategies and provide measurable axes of differentiation (Hendry and Grime, 1993; Grime, Thompson, et al., 1997; Hodgson et al., 1999).

One of the main criticisms of the CSR scheme has been that its syndromes of attributes related to each strategy are not applicable outside of the plant community for which it was developed (Lavorel et al., 1997). This is a problem of most EPCSs. Recognizing this, Westoby (1998) suggested that three readily measured, objective plant attributes (specific leaf area, plant height, and seed mass) adequately capture and describe the variety of plant strategies that exist throughout the world. Its author believes that this "LHS" (Leaf–Height–Seed) strategy scheme should be useful not only to synthesize research investigating how plants interact with other plants and their environment, and therefore how they are distributed, but also to predict their response to global change (Westoby, 1998). The usefulness of this list of attributes has not yet been thoroughly evaluated.

Goal 2: Understand How Individual Species and Biodiversity Affect Ecosystem Functioning

EPCSs based on this goal are relatively new. They stem in large part from the current biodiversity crisis due to human effects on landscape structure, species redistribution,

---

[1] In this chapter, for the most part I follow the nomenclature convention of Lavorel et al. (1997), in which *trait* refers to a characteristic and *attributes* are the expressions of these traits. For example, seed size is a *trait* and 1 mm is an *attribute*. Others have used these and other terms somewhat interchangeably. The only time that I do not follow this convention is in Table 2.1, where Wilson (1999) uses *character* instead of *trait*.

climate, and nutrient cycling. It has been suggested that the loss of certain species, or biodiversity in general, may significantly affect the ecosystem services, such as food and fiber production or water purification, upon which humans depend (Daily et al., 1997; Schlapfer, 1999). Some experimental studies support this suggestion, in that they have found a significant, positive relationship between plant species richness and aboveground productivity and soil nutrient retention (e.g., Naeem et al., 1994; Tilman et al., 1996, 1997; Hector et al., 1999; Lepš et al., 2001). Many of these studies and others investigating the effects of biodiversity on ecosystem functioning classify plant species into "functional groups" based on differences in traits perceived or known to affect nutrient uptake, primary productivity, and other ecosystem properties (e.g., Tilman et al., 1997; Hector et al., 1999; Naeem et al., 1999; Wardle et al., 1999). Such classifications aid in understanding the mechanisms behind relationships, or the lack thereof, and between plant diversity and ecosystem functioning by clarifying which plant traits are responsible for the relationships found.

Goal 3: Predict How Species, Communities, Landscapes, and the Globe Will React to Environmental Changes, Especially Those Caused by Humans

A major impetus for the recent blossom of EPCS research has been climate models. These models have become more complex as their creators have attempted to incorporate two ecological facts. First, plants not only react to changes in climate, but they can also affect the climate. Second, not all plants react to climate changes or affect climate similarly. Consequently, climate modelers have asked ecologists to simplify the vast number of plant species into a small number of groups that represent the essence of vegetation change in response to changes in atmospheric CO_2 concentrations, temperature, and precipitation. This is necessary not only to provide information regarding vegetation feedback effects to the modelers, but also to figure out how agricultural crops will have to be adjusted, and what natural systems may be in danger if, as predicted, the earth warms.

Many of these specific goals, especially the former, have been very difficult to achieve (Körner, 1998). EPCSs that have met with more success are those seeking groups of species that respond to a particular perturbation in a similar way. Such systems are usually developed in relation to land use changes, such as abandoning cultivated fields or altering grazing regimes. They have been more successful than systems related to climate change for two reasons. First, environmental changes such as these are easily manipulated on scales relevant to the plant communities in question. Second, the magnitude and direction of land use changes expected to occur are relatively well known, especially compared to climate changes. In addition, ecological theories and EPCSs developed for the more pristine ecosystems on which ecologists have traditionally focused tend to be more relevant to land use changes than to climate change, because the disturbances involved are similar to those that are natural part of any ecosystem. Thus, even though Grime's CSR scheme was developed to understand how plant communities are structured, it can be used to predict changes in communities, because it incorporates species responses to disturbances (Grime, Hodgson, et al., 1997). Although most of the EPCSs developed to predict plant community response to perturbations aim to predict what these communities will look like in the future, results from this research could also be useful for ecologists trying to understand how to restore damaged ecosystems.

Properties on Which EPCS Are Based

EPCSs differ not only in their goals, but also in the properties on which they are based. Plant classification systems have been around for centuries, but they have proliferated recently. As a result, many names for ecological groups of organisms have been proposed, used, and confused. For example, the term *functional group* has a very specific meaning in some ecological literature (as discussed in detail below), but Naeem (Naeem and Li, 1997; Naeem, 1998) uses the term almost synonymously with *trophic level*. Also, in many published works, it is unclear how the authors decided to divide species into groups (Wilson, 1999). Because of this inconsistency in nomenclature, various attempts have been made to standardize the associated vocabulary and to classify the EPCSs, but none seem to have been successful yet (Semenova and van der Maarel, 2000). Noble and Gitay (1996) described five categories of EPCSs based on the properties used to establish groups within a system: taxonomy, structure, resource use, response to a defined perturbation, and role in ecosystem function. In this section, I describe in a broad sense how EPCSs are based on these five properties, as well as the vocabulary associated with EPCSs based on them. I will also refer to Wilson's (1999) detailed scheme of classifying EPCSs (Table 2.1).

Table 2.1. Wilson's Classification of Ecological Groups

I. *Beta guilds:* groups of species that occur in the same environmental conditions where environmental conditions are something for which organisms cannot compete (e.g., temperature, relative humidity). Species in the same beta guild tend to occur together in time or space because they require similar environments. Species in different Beta guilds tend to occur in different environments.
   A. Guilds based on observations of distribution:
      1. *Spatial distribution guilds:* based on instantaneous observations of species' spatial distributions (e.g., plants of dry calcareous grasslands).
      2. *Temporal distribution guilds:* based on observations of temporal variation of species occurrence or abundance (e.g., increase with rainfall).
   B. Guilds based on measurement of character:
      1. Descriptive (morphological, physiological) characters:
         a. *Subjective beta character guilds:* formed subjectively on just a few characters chosen *a priori* (e.g., Raunkiaer system).
         b. *Objective beta character guilds:* formed objectively by measuring a large number of characters, then using an objective numerical procedure to find groups that occur in different places or times (e.g., groups related to Grime's CSR theory).
      2. *Response guilds:* characters related to response to environmental factors (e.g., resprout after fire).

II. *Alpha guilds:* groups of species which are similar in the type of resources they use. Species in the same alpha guild tend not to occur together, at least at the small scale, because they competitively exclude each other.
   A. Rootian guilds defined by characters intended to represent resource use:
      1. *Subjective* a priori *Rootian guilds:* guilds defined subjectively and not tested (e.g., spring ephemerals).
      2. *Tested* a priori *Rootian guilds:* guilds defined subjectively but tested objectively, such as by looking for competitive exclusion through guild proportionality (e.g., Wilson and Roxburgh, 1994).
      3. *Objective character Rootian guilds:* guilds defined objectively (using numerical methods) based on character(s) intended to indicate similarity in resource use. Wilson (1999) states that this is difficult to do for plants, but relevant characters often relate to plant size and phenology. The character list for designating these guilds is often very similar to those used in forming objective beta character guilds
   B. *Intrinsic Guilds:* guilds defined by the behavior of the species (whether or not they co-occur).

Adapted from Wilson (1999; Table 2).

Taxonomy

One property used to group species in almost all EPCSs is taxonomy, a property used almost by default, because it is an established, readily available system of classifying plants. In addition, taxonomic groups sometimes share attributes that influence a plant's role in an ecosystem. For example, members of the mustard family (Brassicaceae) generally produce secondary chemicals that make them unpalatable to many herbivores. Thus, EPCSs are often based on taxonomy and perhaps a couple of other basic plant structural traits. In some cases, taxonomy does distinguish groups that seem to be ecologically relevant, such as grasses, forbs, and legumes (e.g., Hector et al., 1999), but many ecologists have recognized that anything but the broadest taxonomic groups are generally not relevant to how species function in an ecosystem. This is because plant taxonomic groups are based on traits, such as the morphology of reproductive organs, specifically selected for their minimal phenotypic variation (Grime, Hodgson, et al., 1997). In contrast, it is the traits that vary among individuals that are often most important to ecological plant functions, such as nutrient acquisition and shade tolerance (Duarte et al., 1995). Other reasons why taxonomically based plant classifications are often not useful for plant ecologists include the following:

- Taxonomic classifications are not readily transferred among ecosystems because of the ecosystems' different evolutionary pasts (Landsberg et al., 1999). The structural and functional aspects of plant composition in ecosystems in different parts of the world often converge, so that ecosystems are much more similar ecologically than taxonomically (MacArthur, 1972; Cody, 1975; Díaz et al., 1992).
- Taxonomic classifications describe the composition of a community but fail to improve the understanding of *why* species exist where they do, function as they do, or respond to a perturbation in a certain manner (Landsberg et al., 1999). In other words, they do not reveal the attributes of plants that explain their occurrence in a certain environment. Therefore, they have very little power to predict how species already in an ecosystem could react to a change in that environment, or which new species might invade.
- Populations of the same species can differ morphologically. Thus, two places with very similar composition may be quite different in their structure, and therefore function (Díaz et al., 1992).

Structure

The oldest plant classification systems are based on a plant trait that is relatively easy to assess: structure. Theophrastus (ca. 300 B.C.), a student of Aristotle, distinguished between trees, shrubs, and herbaceous plants. Centuries after Theophrastus, Raunkiaer (1934) made another great leap in plant classification. He developed a system based on the position of perennating organs in order to explain tolerance of physiological drought (especially that induced by cold temperatures) and therefore phytogeographic patterns. Some of the five main types of plants in Raunkiaer's system are used regularly in modern plant functional-type literature because of their relevance to disturbance response (e.g., McIntyre, Díaz, et al., 1999) and cold tolerance. Because this system is based on a limited number of subjectively selected traits, the Raunkiaer system, and others like it, fall under Wilson's (1999) "subjective beta character guild" class (Table 2.1). However, this small number of traits limits the usefulness of Raunkiaer's system for many applications and a variety of other structural traits often are considered when classifying plants for a specific purpose (Table 2.2). Noble and Gitay (1996) refer to the ecological groups based solely on such structural traits as "structural groups" or "life forms." In grasslands, the basic life forms consist of graminoids, forbs, subshrubs, shrubs, and trees, although the woody life forms can be minor components in many grassland ecosystems.

I believe it is safe to say that all ECPSs incorporate some structural traits. However, these traits generally are not sufficient because many traits other than a plant's structure and morphology influence how it acquires and uses resources and how it affects the surrounding environment, including other plants. Thus, other traits often are used for ECPSs based on specific goals.

Resource Use

Resources used by an organism, and the manner in which it uses them, are the third property on which EPCSs are often based. Compared to taxonomy and structure, this property is not very straightforward. In 1967, Root published a seminal paper in which he used the term *guild* to define "a group of species that exploit the same class of environmental resources in a similar way" (p. 335). Since then, the term has been used widely, often in a broader sense than Root's original definition. For example, Noble and Gitay (1996) refer to groups that use the same resource as *guilds*, then further (Gitay and Noble, 1997) refine the term into *structural guilds* and *functional guilds*. In their nomenclature, structural guilds comprise species that use the same resource but *not* in the same manner, whereas functional guilds comprise species that use the same resource in the *same* manner (*sensu* Root's definition). Wilson (1999) uses the term *guild* even more loosely, referring to any group of organisms as a guild and naming those groups that classify species according to similar resource use as *alpha guilds* (Table 2.1). He further divides his alpha guilds into those defined by co-occurrence patterns of the species (*intrinsic guilds*) or by characters that represent resource use (*Rootian guilds*), which can be obtained with a variety of methods (Table 2.1).

It is interesting to note here that Wilson (1999) argues that species within the same alpha guild tend not to occur together, because they competitively exclude each other. On the other hand, many recent publications claim that their method for grouping species of the same plant community together was because of similarities of the timing, location, or efficiency of the species' resource use (e.g., Hooper and Vitousek, 1997; Tilman et al., 1997; Symstad et al., 1998; Symstad and Tilman 2001). To resolve this apparent contradiction, one must simply acknowledge two things. First, plant communities arguably never reach a state of equilibrium in which competitive exclusion is complete. Thus, the existence of species depends not only on competitive interactions but also on regeneration properties (Grubb, 1977). Second, competitive interactions take place between individuals. For herbaceous plants, this is a scale much smaller than that of a whole community. Thus, species within the same alpha guild may co-occur in a community, but not necessarily right next to each other. For example, many grassland ecosystems exhibit high levels of apparent redundancy, in that species using the same resources at the same time from the same places co-occur (e.g., tall, warm-season grasses in the tallgrass prairie of North America). Some ecosystems seem to have very little redundancy within guilds, however, in that one or more guilds are heavily dominated by or consist of a single species (e.g., Sala et al., 1989; Lauenroth et al., 1997).

Table 2.2. Checklist of Traits That May Be Relevant to Ecosystem Function and Response to Climate Change and Disturbance

Traits	Relevant plant[a] or ecosystem[b,c] function
Morphological/Structural	
Life form* (tree, shrub, forb, graminoid)	
Position of dormant buds (belowground, surface, aboveground)	
HEIGHT (potential height at vegetative stage)	Competitive ability[a]
Lateral spread (plant diameter)	
Aboveground cover density (ACD: height at which an individual plant fills the three-dimensional volume defined by its height and lateral spread)	
Plasticity (e.g., of ACD)	
Canopy structure (e.g., flat-, erect-, partial-, protorosette)	Carbon and nitrogen cycling[b]; Aerodynamic conductance, interception, water relations, runoff, roughness/albedo, temperature buffering, soil stability[c]
Habit (e.g., mat, erect, scrambler, twiner, tufted)	
Stem density/secondary growth	Plant longevity[a], carbon storage[a,c]
Ramification (Díaz, Cabido, and Casanoves, 1999)	Structural complexity[c]
Depth of main root system (Skarpe, 1996; Walker et al., 1999)	Water budget[b]
Root distribution profile (Walker et al., 1999)	Water budget[b]
Root architecture (Díaz, Cabido, and Casanoves, 1999)	Water uptake[c]
Presence of reserve organs (Díaz et al., 1992)	Resilience[c]
SPECIFIC LEAF AREA (SLA: leaf area per unit leaf dry mass)	RGR, plasticity, stress tolerance, evergreenness, leaf longevity[a]; carbon and nitrogen cycling[b]
Leaf water content (Weiher et al., 1999)	Relative growth rate (RGR), plasticity, stress tolerance, evergreenness, leaf longevity[a]
Leaf length (Scholes et al., 1997)	
Leaf shape (Scholes et al., 1997)	
Leaf angle (Díaz et al., 1992)	
Leaf dissection (Díaz et al., 1992)	
Aboveground biomass (Díaz, Cabido, and Casanoves, 1999; Weiher et al., 1999)	Competitive ability, fecundity[a]; flammability[c]
Maximum total biomass (on per hectare basis, Walker et al., 1999)	Carbon and nitrogen cycling[b]
Life history	
Life span	Carbon and nitrogen cycling[b]; inertia[c]
Life cycle (e.g., biennial, short-lived perennial, perennial)	Plant longevity, space-holding ability, disturbance tolerance[a]; carbon and nitrogen cycling[b]
Time to first reproduction	
RGR (Grime, Thompson, et al., 1997)	Carbon and nitrogen cycling,[b] productivity[c]
Phenological	
Leaf phenology (e.g., deciduous, evergreen)	Carbon and nitrogen cycling[b]
Seasonality of growth	Carbon and nitrogen cycling[b]
Regeneration	
SEED MASS/size	Dispersal distance, longevity in seed bank, establishment success, fecundity[a]
Seed shape (Weiher et al., 1999)	Longevity in seed bank[a]
Seed dispersal vector	Dispersal distance, longevity in seed bank[a]; expansion over landscape[c]
Seed dormancy (e.g., transient, persistent)	Resilience[c]
Fecundity/seed number	Expansion over landscape[c] stress avoidance, disturbance[a]

Table 2.2. ***(continued)***

Traits	Relevant plant[a] or ecosystem[b,c] function
Onset of flowering (Weiher et al., 1999)	Avoidance
Length of flowering period	
Seasonality of germination	
Timing of seed release	
Recruitment frequency	
Vegetative reproduction (present/absent; obligate/facultative)	Space acquisition[a]
Resprouting ability	Disturbance tolerance[a]
Reproductive effort (flower mass/plant mass; Fernandez Ales et al., 1993)	
Maximum dispersal range (Platt and Weis, 1977)	
Pollination mode (Díaz, Cabido, and Casanoves, 1999)	Expansion over landscape[c]
Physiological	
Photosynthetic pathway (Skarpe, 1996; Walker, 1997)	
Heat resistance threshold (Walker, 1997)	
Frost resistance threshold (Walker, 1997)	
Optimum temperature for growth (Walker, 1997)	
Nutrient requirements/tolerances (Walker, 1997)	
Nitrogen-fixing capacity (Walker et al., 1999)	Carbon and nitrogen cycling,[b] rate of succession[c]
Dependence on mycorrhizal associations (Wilson and Hartnett, 1998)	Diversity, nutrient cycling, rate of succession[c]
Leaf nutrient content (Díaz, Cabido, and Casanoves, 1999; Walker et al., 1999)	Carbon and nitrogen cycling,[b] carrying capacity for herbivores[c]
Leaf turnover rate (Díaz, Cabido, and Casanoves, 1999)	Nutrient cycling, production efficiency[c]
Water-use efficiency (Walker et al., 1999)	Water budget[b]
Transpiration rate (Walker et al., 1999)	Water budget[b]
Minimum soil moisture tolerance (Platt and Weis, 1977)	
Grazing-related	
Secondary compounds (e.g., tannins, alkaloids, volatile oils, salt, resins, present/absent)	
Palatability (Scholes et al., 1997)	
Pubescence	
Prickliness	
Tensile strength of leaves	
Waxes	
Sclerophylly	
Silification	
Stem/leaf ratio	
Leaf thickness/succulence	
Leaf width	
Leaf surface area (Landsberg et al., 1999)	
Inflorescence prominence	
Position of active buds (distance from root to farthest meristem)	

(continued)

Table 2.2. (*continued*)

Traits	Relevant plant[a] or ecosystem[b,c] function
Maximum height relative to grazer	
Habit plasticity in response to defoliation	
Uprooting potential	
Regrowth potential (Landsberg et al., 1999)	

*Many authors (Lavorel et al., 1997; Landsberg et al., 1999, McIntyre, Lavorel, et al., 1999) recommend separating plants into life form before considering any other plant traits.
Note: Unless a reference is cited for the trait, it is found in Table 1 of McIntyre, Díaz, et al. (1999), compiled from many sources for disturbance response in herb-dominated vegetation. Grazing-related traits were separated to illustrate how some traits are relevant to ecosystems with specific disturbance regimes. Note, however, that some of these "response" traits (e.g., leaf pubescence, waxiness) may also influence ecosystem functions (e.g., water budget). The second column lists, for each trait, either the plant function(s) affected by the trait that Weiher et al. (1999)[a] consider key for predicting vegetation responses to disturbance or, the general ecosystem function that Walker et al. (1999)[b] and/or Díaz, Cabido, and Casanoves (1999)[c] describe as being particularly affected by the trait. The three traits in CAPITALS are those advocated by Westoby (1998) for a universal plant ecology strategy scheme.

EPCSs based on how species use resources (following Root's definition) are difficult to define because all plants use the same basic resources, but it is hard to tell how they use them. Thus, categories in EPCSs based on resource use are usually defined by traits such as root structure, mycorrhizal associations, and canopy structure, because it is fairly well understood how these control plants' use of nutrients, water, and light. As with taxonomy- and structure-based EPCSs, grouping plant species based on their resource use is generally not the ultimate goal of an EPCS. Instead, it is generally done in the context of ecosystem functioning.

Response to a Defined Perturbation

When I discuss the results of EPCSs in the comparison section below, many of the specific results come from those based on response to a defined perturbation. This bias occurs because it is one of four foci of Global Change and Terrestrial Ecosystems (GCTE; http://gcte.org/), a core project of the International Geosphere–Biosphere Program (e.g., Smith et al., 1997: Lavorel and Cramer, 1999). The mission of this focus is to find syndromes of plant attributes that can be used to predict how vegetation will respond to changes in land use practices, which are essentially perturbations (e.g., grazing, cultivation, fire). Although the ultimate goal of this type of EPCS is to predict the response of species to a certain perturbation based entirely on attribute syndromes, the development of such an EPCS must at some time include a direct observation of how some species actually do respond to that perturbation. This requirement for observation, as well as the organization afforded by GCTE, has produced a more consistent and thoroughly developed methodology for this type of EPCS than is associated with those based on other properties.

Despite this organization, the vocabulary associated with this type of EPCS is some of the most complicated and confused in the field. Although most researchers concerned with standardizing nomenclature have at some time referred to groups of plants that respond to a particular perturbation in a similar way as "response groups" (Noble and Gitay, 1996; Gitay and Noble, 1997; Lavorel et al., 1997; Wilson, 1999), such groups are often referred to as *plant functional types* (Smith et al., 1997). Gitay and Noble (1997) limit their definition of *response groups* to groups of organisms whose response to a perturbation is the same, but the mechanism by which they respond is not necessarily the same. If the response mechanism is the same, they use the term *functional groups*, which, like *guild*, is extremely nebulous. Its ambiguity, however, has stimulated a debate regarding how much response- and ecosystem function-based EPCSs overlap. This debate promises to increase the understanding of how plants interact with their environment.

Role in Ecosystem Function

Groups of species in response-based EPCSs are relatively straightforward to define—species can increase, decrease, or not change in response to a disturbance. Finding the attributes that can accurately predict these responses is more difficult. In contrast, the categories that comprise ecosystem function-based EPCSs are quite difficult to define. One reason for this difficulty is the inherent complexity of ecosystems. There are many ecosystem functions (e.g., primary production, nitrogen capture and retention, decomposition), each of which depends on different traits of the organisms and abiotic conditions of the ecosystem. Thus, groups of plants that affect one ecosystem function in a similar manner may not be grouped together when another ecosystem function is the focus of investigation (e.g., Solbrig, 1993; Vitousek and Hooper, 1993). On the other hand, because all ecosystem functions are tied together through energy flow, one would not expect an infinite number of functional group combinations. Second, it is very difficult to characterize species as to how they affect even a single ecosystem function because of the lack of data on which to base those characterizations (Grime, Hodgson, et al., 1997). This is not to mean that we do not understand anything about how plant attributes relate to ecosystem functioning—this is a basic goal in the field of ecophysiology. Indeed, Walker et al. (1999) list a number of traits they feel will be important to some key global change functions (Table 2.2). The problem lies in the fact that it is expensive, time-consuming, or impossible to measure many of these traits.

As with response-based EPCSs, the vocabulary associated with function-based EPCSs is confusing. *Functional group* (or type), a term used to refer to the groups produced by both of these EPCS types, can be defined either in the narrow or broad sense. Part of the ambiguity in this term stems from differences in opinion on how closely response- and function-based EPCSs coincide. In other words, do species that respond similarly to a perturbation via the same mechanism also play a similar role in ecosystem function? Noble and Gitay (1996) and Lavorel et al. (1997) suggest that the answer to this question is yes, whereas Walker et al. (1999) argue the opposite. Both sides of the debate have merit. A species' response to a disturbance often depends on how it uses resources, and resource use is tightly linked to ecosystem functioning. Thus, Lavorel et al. (1997, p. 475) state that functional groups (or types) can be defined in relation to "either the contribution of species to ecosystem processes or to the response of species to changes in environmental variables." On the other hand, Walker et al. (1999) suggest that dominant and minor species in a given ecosystem may be similar in their contribution to a specific function, but respond to environmental changes differently. Skarpe (1996) also states that species that affect an ecosystem function similarly will not necessarily respond to an environmental change in the same way. To overcome this ambiguity, Landsberg (1999) refers to groups derived for predicting species survival in response to a perturbation as "functional *response* groups," and to groups derived for predicting how species' responses affect ecosystem functions as "functional *effect* groups." Although Landsberg discusses functional effect groups only in the context of response to a perturbation, the term seems equally applicable to function-based EPCSs derived in a nondisturbance context. Thus, I follow this naming convention for the rest of this chapter, using *functional response group* to refer to a group of species that respond to a defined perturbation in a similar manner, with no implication of their role in ecosystem functioning, and *functional effect group* to refer to a group of species that play a similar role in a certain ecosystem function, with no implication of how they respond to a disturbance.

Curiously, although Wilson (1999) uses the term *functional type* in the title of his paper, he does not explicitly incorporate the functional ecosystem roles that plants play in his classification scheme. Function-based EPCSs could qualify as just about any of Wilson's guilds, but they are most similar to the Rootian guilds (Table 2.1).

Approaches to Classifying Plants

There are two basic approaches used to develop an EPCS. *A priori* approaches are based on some kind of model of how plant communities and ecosystems work. The degree to which these models are developed varies considerably. *A posteriori* approaches are used to search for patterns in data. These patterns, when used correctly, are used to generate hypotheses regarding the underlying causes of those patterns.

A Priori Approaches

The easiest, but least rigorous, *a priori* approach assumes that there are distinct groups of plants relevant to the processes/response of interest, then defines them based on professional experience with the ecosystem and the species. This professional experience generally involves experiments and observations that provide a general understanding of which processes and properties are important to the vegetation and ecosystem functions (Woodward and Cramer, 1996). In some cases, the person doing the classifying feels quite confident that the groups are "real," because the system in which he or she is working has a long history of focused ecological research. However, even the best of these "subjective" (Gitay and Noble, 1997) plant classification systems produce only candidate groups until they are tested (Vitousek and Hooper, 1993; Wilson, 1999), and testing sometimes reveals that some groups are too broadly defined (Symstad and Tilman, 2001).

A "deductive" (Gitay and Noble, 1997) approach for classifying plants is somewhat more rigorous than the subjective approach, because the plant groups are based on a set of axioms that describe an underlying model of how a system works. This approach is still somewhat subjective, however, in that it assumes that certain processes are more important than others in determining the structure and functioning of a certain ecosystem. For example, in their refinement of the Vital Attributes Model (Noble and Slatyer, 1980), Noble and Gitay (1996) determine that a primary axiom for determining vegetation dynamics in rain forests is that the persistence of a species in a given environment depends on two things—its method of persistence through a disturbance, and the conditions it requires for establishment. Based on this, they compiled an exhaustive list of all the possible plant types from the logical consequences of this axiom (e.g., persists by having a long-lived propagule pool that survives the disturbance, establishes only immediately after a disturbance) and the others of their ecosystem model. This list was then compressed into 13 plant types by combining types with the same dynamic behavior via different mechanisms (because they were not interested in mechanisms) and eliminating biologically impossible or rare combinations. The line between classification systems that are "subjective" versus "deductive" seems to depend largely on how clearly the classifier states the premises on which her or his groups are based. Because they start with the environment, then work down toward how the environment constrains the plant species, these classifications are similar to the environmentally based methods of Walker (1997) and Scholes et al. (1997).

A Posteriori Approaches

A posteriori, or *post hoc*, approaches for deriving EPCSs work from the bottom up, starting with the species and determining their relationship to the environment ("species-based" approaches; Scholes et al., 1997; Walker, 1997). These more objective plant classification systems are those that use "inductive" (Woodward and Cramer, 1996) or "data-defined" (Gitay and Noble, 1997) methods. The data by which these groups are defined are often the presence or abundance of species in different environmental conditions and, sometimes, measurements of a variety of traits of the species encountered in each sample. The data are then analyzed using a variety of clustering techniques to discern emergent groups.

In one of these methods, ecological groups are defined solely on whether or not they occur together, and therefore do not rely on attribute data (the "intrinsic guilds" in Table 2.1; Wilson, 1999). This method is often used in assembly rule literature and assumes that species that use resources similarly will not co-occur because of competitive exclusion. Groups of species that tend to be mutually exclusive within a community, and therefore belong to the same intrinsic guild, are deduced by comparing the actual distribution of species among samples to a null model distribution that assumes no interaction among species.

In most methods using the *a posteriori* approach, however, plant attributes are measured on a large number of species, on which groups are then based. A relatively simple inductive method is that used by Walker et al. (1999), in which they calculate the "ecological distance" (basically, the Euclidean distance squared) between all pairs of species based on a list of traits, then assume that species with small ecological distances fall into the same group. A more sophisticated method involves the use of multivariate techniques such as correspondence analysis. This method is used in determining functional response groups. Many varieties of this method exist, but the procedure generally consists of four steps (Landsberg et al., 1999; McIntyre, Lavorel, et al., 1999; Lavorel, 2002):

1. Select a set of traits to measure. How many and which traits to measure will be determined by ecological and practical factors. Ecologically speaking, the traits used depend on the disturbances of interest and expert knowledge of how the traits directly determine or are related to those disturbances. Practically speaking, most trait lists are limited by what can be found in the literature or easily measured in the field or on herbarium specimens. Using too many traits will make it difficult to find discrete groups of plants (Skarpe, 1996), but using too few will leave open the question of whether a critical trait was left out.
2. Use multivariate statistical methods (an example of which is thoroughly described in Lavorel et al., 1998) to relate plant *attributes* to certain environmental conditions (e.g., grazing intensity) or experimental treatments (e.g., grazed vs. ungrazed), and to determine whether there are syndromes of attributes (groups of attributes that occur together). This step produces attribute groups.

3. Use the same methods as in step 2 to relate plant *species* to the same conditions or treatments to produce species response groups.
4. Look for overlap between the groups produced by steps 2 and 3 to develop hypotheses to explain the groups discerned.

Usefulness of the Different Approaches

Whether an EPCS is derived using an *a priori* or an *a posteriori* approach, the procedure of determining the traits useful for categorizing species produces hypotheses that, when tested, illuminate the mechanisms behind plant community dynamics and effects on ecosystem functioning. However, each method used to classify plants into ecological groups has its problems. *A priori* approaches tend to be relatively subjective because of their implicit or explicit assumptions regarding the relative importance of the many factors that influence plant communities and ecosystems. *A posteriori* classifications tend to be more objective because of the numerical techniques on which they are based. They, too, suffer from some degree of subjectivity, however, in the selection of traits to measure (Woodward and Cramer, 1996) and in the interpretation of the analyzed data. One problem with the multivariate methods described earlier is that the groups defined are based on interpretations of graphs produced by canonical correspondence analysis or similar methods. One person looking at one of these graphs may see distinct groups, whereas another sees a continuum. The problem of creating groups in which only gradients exist is not unique to *a posteriori* approaches, however, because of the continuous nature of nature (Westoby and Leishman, 1997; Wilson, 1999).

In addition to varying in subjectivity, approaches differ in terms of their usefulness for different problems. Because *a posteriori* approaches rely on data to define groups, they generally offer less mechanistic understanding of the groups' structure than do deductive approaches. Indeed, Wilson and Gitay (1999) acknowledge they may never understand the attributes or mechanisms behind their intrinsic guilds. Thus, EPCSs based on *a priori* approaches may be more useful for predicting species and ecosystem responses in novel environmental circumstances (Woodward and Cramer, 1996). Methods using the *a posteriori* approach also often require a lot of effort to obtain the information about plant attributes on which many of them depend, so they may not be applicable to large-scale questions involving species in many ecosystems (Gitay and Noble, 1997).

Finally, it has been argued that the attribute syndromes used to define groups in EPCSs discerned via *a posteriori* approaches may exist in a plant community because of phylogenetic, not ecological, reasons (Kelly, 1996). Whether species share attributes for phylogenetic or ecological reasons is often difficult to discern, however, and because those attributes occur in the current set of environmental conditions, they should not be dismissed (Westoby et al., 1995; Díaz et al., 1998).

Because the *a priori* and *a posteriori* approaches each have their own pros and cons, some authors have suggested that the best approach is to iterate between the two (Scholes et al., 1997; Walker, 1997). Lavorel et al. (1998) claim that their multivariate method actually integrates the two approaches because it links environmental factors with species traits. The Integrated Screening Program (Hendry and Grime, 1993; Grime, Thompson, et al., 1997) is the most elaborate example of an integration of these two approaches, because it searches for plant functional types based on 67 traits but is based on a conceptual model of how vegetation is structured (the CSR system; Grime, 1977).

Whatever the approach used to classify plants into ecological groups, a few methodological caveats are warranted. First, it is best to start simple, elaborating only if necessary (Noble and Gitay, 1996). Second, many EPCSs assume that the observed vegetation is in equilibrium with the environment, but this usually is not the case (Skarpe, 1996; Walker, 1997). This lack of equilibrium reflects the long lag times often seen in vegetation science. Because of the varied response times of different plants to changes in the environment, such as those imposed in experiments testing possible future climate scenarios, functional response groups or functional effect groups deduced from such experiments may be misinterpreted (Skarpe, 1996). Finally, whether looking for emergent groups of attributes related to a climatic gradient or functional response groups of species with respect to a specific disturbance, it is important to consider whether recording just the presence of species (and therefore their attributes) versus species abundance will influence the interpretation of the results (Skarpe, 1996; Lavorel, Rochette, et al., 1999).

Emergent, Functional Response, and Functional Effect Groups: How Much Do They Overlap?

To summarize to this point, there are many reasons for developing an ecological plant classification system, many properties on which to base it, and many methods used to do so. From this, it is clear that these different combinations will yield different groups even within a single plant community (Körner 1993; Schulze and Mooney, 1993; Solbrig, 1993; Noble and Gitay, 1996; Skarpe, 1996; Woodward and Cramer, 1996; Gitay and Noble, 1997). Indeed, Chapin et al. (1996) warn that given the large number of goals set for EPCSs, it is very possible that they could yield more plant groups than there are plant species. However, some believe that the connectedness of everything in nature holds some promise that a single EPCS may be applicable to multiple goals (e.g., Chapin et al., 1996; Westoby, 1998). I believe that not only this is true, but that it is also necessary in order to progress toward the ultimate goal of most modem EPCSs to predict how human activities will affect plant communities, and the ecosystem services and functions that they influence. Thus, the rest of this chapter is dedicated to tying these different systems together.

The first step in doing this is to compare their outputs—the emergent, functional response, or functional effect groups defined by an EPCS. This can be done with two kinds of comparison. First, because the species composition of groups resulting from different studies is generally site- and study-specific, the only way to do a general comparison among EPCSs is to look at the traits that characterize them. Table 2.2 lists a large number of plant traits that have been used or suggested for use in developing EPCSs, as well as the relevance of each of these traits to different plant and ecosystem functions. Second, in some small number of cases, investigators used two different EPCSs as part of the same study; therefore, their results can be directly compared. In the following discussions of these two types of comparisons, I use data from grasslands only to illustrate the great variability in EPCSs even within a single ecosystem.

Emergent versus Functional Response Groups

How similar are the lists of traits that distinguish among emergent groups to those that distinguish among functional response groups? Comparing Tables 2.3 (emergent

Table 2.3. Emergent Groups of Plants or Attributes, Number and Type of Traits, and Why Emergent Groups Were Sought

Source and ecosystem	Purpose	Number and type of trait	Emergent groups
Lavorel, Rochette, et al., 1999; subhumid French Mediterranean old fields	Find general rules of plant behavior in response to agricultural land abandonment, comparing emergent groups to response groups.	5 Morphological	*3-y and 9-year-old fields:* Erect rosette,* low plasticity, moderate to large height; partial rosette, erect, low plasticity, moderate to large height; protorosette,* flat, plastic
		5 Regeneration	*3-year-old field:* Heavy seeds, moderate fecundity; heavy seeds, low fecundity; amemochorous, light seeds; zoochorous, light seeds; zoochorous, light seeds; anemochorous, very light seeds, high or very high fecundity *9-year-old field:* Barochorous, heavy seeds, low fecundity; zoochorous, heavy seeds, moderate fedundity; anemochorous perenneial; anemochorous, annual, winter–spring germination, high fecundity; anemochorous, fall–winter germination; zoochorous, light seeds
Lavorel et al. (1998); abandoned pasture, Brittany, France	Reduce large number of traits down to smaller number of "strategies" (attribute groups) for further use in finding response groups to various disturbances.	3 Morphological	Leafy, small, moderate lateral spread; leafy, large, large lateral spread; leafy, annual, long stems; rosette, compact, short; rosette, compact, medium; rosette, compact, tall; rosette, annual
		5 Regeneration	Heavy seeds; light, strongly dormant seeds, can reproduce vegetatively; light, moderately dormant, seeds, zoochorous, vegetative reproduction; very light seeds, anemochorous, annual; very light seeds, anemochorous, perennial
Ellery (1992) in Scholes et al. (1997); South African veld (rangeland)	Find clusters of grass species with similar combinations of attributes.	18	Tufted, moderately to strongly geophytic, obligately sexual, large-seeded, few-seeded, early- or prolonged-flowering (occur in wetter sites); rhizomatous or stoloniferous, strongly woody-based, small-seeded, many-seeded, late-flowering, C_4 nicotinamide-adenine-dinucleotide-malic enzyme (NAD-ME) photosynthesis (occur in drier sites)

Blackmore (1992) in Scholes et al. (1997), South African savanna	Find clusters of species with similar combinations of attributes	?	None found
Landsberg et al. (1999); grazed grasslands surrounding wooded groves, southwest Queensland, Australia	Find emergent groups of attributes to relate to grazing intensity; only grasses are included here.	5 Morphological	Small, sprawling, basal tussocks, with high SLA leaves; small, erect, high cover density, basal tussocks, with high SLA leaves; large, erect basal tussocks, with low SLA leaves; large, erect, intermediate-structure tussocks with moderate SLA leaves; moderate-sized, low cover density, intermediate tussocks; sprawling, low cover density, basal tussocks
		7 Grazing persistence	Low SLA, narrow leaves with inconspicuous trichomes, high uprooting potential, unprotected inflorescence; low SLA, narrow leaves with inconspicuous trichomes, low uprooting potential, stemmy; low SLA-leaves with conspicuous trichomes, low uprooting potential, stemmy, high plasticity; intermediate–high SLA, medium-width leaves with conspicuous trichomes, stemmy, low plasticity; intermediate-high SLA, medium-width leaves with inconspicuous trichromes, stemmy; intermediate-high SLA leaves with inconspicuous trichomes, low uprooting potential, high plasticity
		5 Regeneration	Long flowering period, numerous light seeds with no dispersal aids, low canopy seed bank; short flowering period, many seeds; short flowering period, heavy seeds, low canopy seed bank; long flowering period, many light seeds with dispersal aids; long flowering periods, rather many light seeds with parachute dispersal aids; long flowering period, very light seeds, high canopy seed bank

(*continued*)

Table 2.3. ***(continued)***

Source and ecosystem	Purpose	Number and type of trait	Emergent groups
Skarpe (1996), Kalahari savanna, south Africa	Find clusters of species with similar combinations of attributes that correspond to climate attributes.	6	Evergreen trees (sclerophyllous leaves and deep root systems); deciduous, fine-leaved trees (small leaflets and deep to moderately shallow roots—most *Acacia* species); deciduous broad-leaved trees; deciduous broad-leaved trees; deciduous, broad-leaved, low trees, (short, moderately shallow to moderately deep roots); perennial, narrow-leaved grasses (leaves often curl under water stress); perennial intermediate-leaved grasses; perennial broad-leaved grasses (leaves thin, soft); annual grasses; annual forbs; chamaephytes; geophytes
Kindscher and Wells (1995); tallgrass prairie, Kansas	Determine if guilds useful for understanding and interpreting the diversity of tallgrass prairie life forms exist.	28	C_4 (warm season) grasses; C_3 (cool season) grasses and sedges; annuals and biennial forbs; ephemeral spring forbs; spring forbs; summer/fall forbs; legumes; woody shrubs
Golluscio and Sala (1993), Patagonian steppe	Find groups of forbs that explain forbs' great diversity but small contribution to biomass.	6	Shallow-rooted mesophytes (annuals); shallow-rooted nonmesophytes; deep-rooted evergreen; deep-rooted deciduous
Díaz et al. (1992); submitted grasslands, central Argentina	Develop way of quantitatively describing plant community structure in terms of morphological characteristics of plants.	15	Annuals; prostrate stoloniferous plants; rosettes with underground reserve organs; graminoids of small–medium height; graminoids of medium–tall height; tussock grasses

*In this study, "erect rosettes," are essentially grasses and "protorosettes," have only cauline, not basal leaves.
Note: When the author did not separate traits into different categories, only the number of traits used is listed.

groups) and 2.4 (functional response groups) reveals six traits that have been both frequently measured and frequently cited as important distinguishing traits in studies seeking emergent and/or response groups: life form, plant height, canopy structure, life history, seed mass, and seed dispersal mode. Some of this overlap might be attributed to the fact that both types of groups were investigated in the same studies. However, this list of six traits corresponds fairly well with the results of two efforts to establish a set of core plant traits useful for a wide variety of purposes. The first of these efforts yielded 12 traits agreed upon by a diverse group of scientists as core common plant traits for predicting a species' response to disturbance (Weiher et al., 1999; see Table 2.2). Westoby (1998) suggested that just three traits (specific leaf area, plant height, and seed mass) are needed to represent adequately, but simply, the range of ecological strategies that occur among plants. Of the six traits common in Tables 2.3 and 2.4, two (plant height and seed mass) are on both the Westoby and Weiher et al. lists, and two others (life history and seed dispersal mode) appear on the longer list compiled by Weiher et al. One commonly measured and important trait notably lacking from both the Westoby and Weiher et al. lists is life form, which has been described as the trait that overrides almost all others in distinguishing response groups (Lavorel et al., 1997). This may be because stem density (Weiher et al., 1999) and plant height (Westoby, 1998; Weiber et al., 1999) include some of the same information and can be quantified. On the other hand, specific leaf area (SLA) has rarely been measured but is considered very useful by both Westoby and Weiher et al. because of its demonstrated relationship to leaf life span, leaf nitrogen, and relative growth rate (e.g., Grime et al., 1988; Reich et al., 1997), hard-to-measure traits that affect how quickly a plant becomes established, its plasticity of growth, and its palatability (Table 2.2). This trait will probably be used increasingly in future studies. This comparison across studies suggests that, in general, the traits useful in distinguishing emergent groups and functional response groups are similar.

Comparing the results of emergent and functional response EPCSs within studies yields a different conclusion. Even within the same plant community and investigation, attribute syndromes characterizing emergent groups do not always agree with the attributes that characterize response groups. This kind of comparison has been done, or can be done, with three of the studies in Tables 2.3 and 2.4 (Landsberg et al., 1999; Lavorel et al., 1998; Lavorel, Rochette, et al., 1999). For natural and artificial soil disturbances in an abandoned pasture in Brittany, Lavorel et al. (1998) found that their emergent and functional response groups based on morphological traits were quite similar, separating most strongly by canopy structure, then by plant height and volume. For regeneration traits, they found that the similarity was not as strong; emergent groups separated by seed mass, whereas functional response groups separated by seed dormancy (Tables 2.3 and 2.4). Similarly, in the second study (Lavorel, Rochette, et al., 1999), emergent and functional response group attributes were identical for morphological attributes but not for regeneration traits. Seed mass and dispersal agent separated emergent groups, whereas germination seasonality separated functional response groups (Tables 2.3 and 2.4). For grasses in an Australian rangeland (Landsberg et al., 1999), emergent and functional response group traits, whether morphological or regenerational, were not entirely consistent, with canopy structure plasticity, regrowth potential, and germination season separating functional response groups but not emergent groups, and SLA, fecundity, length of flowering period, and seed size separating emergent but not functional response groups. However, plant size, growth habit, and seed dispersal agent were important for distinguishing between both emergent and functional response groups (Landsberg et al., 1999; Tables 2.3 and 2.4).

Table 2.4. Plant Attributes Related to Grazing, Plowing, or General Disturbance

Ecosystem (source)	Disturbance	Plant trait	Attribute favored	Attribute disfavored	Notes
Mediterranean shrubland (Hadar et al., 1999)	Late-season grazing	Height Flowering phenology Life form	Small Early-flowering Geophytes (long-lived, slow-growing, low reliance on seeds)	Tall Late-flowering Grasses (tall, late-flowering)	Traits derived from responses of *a priori* groups to treatments
	shrub-clearing	None effected			
Mediterranean grassland (Fernandez Ales et al., 1993)	Grazing (enclosures erected at beginning of study)	Height Life cycle	Tall species constant through time Shorter-lived species constant through time	Tall species increased through time Shorter-lived species decreased through time	Annual plants only. Principal components analysis yielded plant height and seed weight as the most important, independent traits of nine tested. Authors' experience recommended adding life cycle to the trait list.
	Plowing	Seed weight Life cycle Height	Smaller Shorter NS	Larger Longer NS	
Montane grasslands, Argentina (Díaz et al., 1992)	Grazing	Height Life cycle Leaf size Habit	Short Annual Smaller Prostate or planophyllous	Tall Perennial Larger Vertical leaves	
Texas juniper/oak savanna (Fuhlendorf and Smeins, 1997)	Grazing intensity	Productivity Root system Height	Low Less extensive Short	High More extensive Medium	Traits were derived from groups of species that responded similarly. Only grasses were characterized.

Serengeti tall, mid-, and short grasslands (Belsky, 1992)	Soil disturbance	Life cycle	Annual	Perennial	Traits derived from reactions of *a priori* groups to treatments.
	Grazing (enclosures erected at beginning of experiment)	Height (of perennials)	Short	Tall	True only in short- and midgrass, not tallgrass sites of Serengeti
		Height	Short	Tall	Significant in later years of experiment when thick litter layer developed
		Method of clonal spread	Stolons	Rhizomes	Effect acted through tufted grasses' reliance on seeds for recruitment.
	Fire	Growth form	Tufted	Rhizomatous	No effects
Abandoned pasture, Brittany, France (Lavorel et al., 1998)	Soil scraping by wild pigs	Canopy structure	Leafy		"Attribute favored" indicates the attribute associated with each of the four treatments.
		Lateral spread	Large		
		Seed dormancy	None		
		Dispersal agent	Wind		
	General herbicide followed by:				
	Nothing	Canopy structure	Rosette		
		Lateral spread	Moderate		
		Seed dormancy	Moderate		
		Dispersal agent	Wind		
		Seed mass	Very light		
	Litter removal	Canopy structure	Compact rosette		
		Seed dormancy	Moderate		

(continued)

Table 2.4. (*continued*)

Ecosystem (source)	Disturbance	Plant trait	Attribute favored	Attribute disfavored	Notes
	Litter removal and soil scraping	Canopy structure	Leafy		
		Plant height	Short		
		Lateral spread	Moderate		
		Seed dormancy	Moderate		

Ecosystem (source)	Disturbance	Plant trait	Attribute favored		Attribute disfavored		Notes
			Grass	Forb	Grass	Forb	
Australian Mulga rangelands (grass-dominated area) (Landsberg et al., 1999)	Grazing intensity	Plant size	Small		Large		Of the 31 traits measured, only those significantly different among grazing levels are listed. "Attributes favored" are those associated with heavy grazing; "disfavored" are those associated with light grazing.
		Leaf size				Large	
		Habit	Sprawling tussock		Erect tussock		
		Canopy structure	Basal leaves		Leafy, branched		
		Canopy structure plasticity	High		Low		
		Regrowth potential	High		Low		
		Germination season	Cool		Warm		
		Dispersal agent	None	Animal	Wind		
		Fecundity		High		Low	
		Seed size		Medium		Small	
Mediterranean plowing pasture (Lavorel, McIntyre, et al., 1999)		Canopy structure	Flat rosette		Versatile		No other traits listed for grazing (below) responded significantly to plowing.
		Height	Medium		Short, tall		
		Seed mass	Light		Heavy		

	Grazing (enclosures erected at beginning of study)	Canopy structure		Leafy stems		Rosette	Authors conclude that plant volume, rather than height or lateral spread, would better reflect grazing tolerance.
		Height	Medium	Small	Tall	Tall	
		Lateral spread	Narrow	Narrow	Wide	Wide	
		Dormant bud position	*n.m.*		*n.m.*		
		Habit	*n.m.*	Flat	*n.m.*	Erect	
		ACD	*n.m.*		*n.m.*		
		Plasticity of ACD	*n.m.*	High	*n.m.*	Low	
		Seed weight	Light	Light	Heavy	Very light	
		Seed dormancy	Medium	High	Low	Low	
		Inflorescence position	Protected	*n.m.*	Exposed	*n.m.*	
		Seed dispersal agent	*n.m.*	Gravity	*n.m.*	Animal	

Ecosystem (source)	Disturbance	Trait	Disturbance specialist		Disturbance intolerant		Indifferent		Notes
			3-yr field	9-yr field	3-yr field	9-yr field	3-yr field	9-yr field	
Mediterranean old fields (Lavorel, Rochette, et al., 1999)	General herbicide, then soil scraped to 3 cm deep	Canopy structure	Partial rosette	Partial rosette	Erect rosette*	Erect rosette	Protorosette*	Protorosette	*In this study, "erect rosettes" are essentially grasses. "Protorosettes" have only cauline leaves.
		Lateral spread	Erect	Erect			Flat	Flat	
		Plasticity	Low	Low	Low	Low	High	High	
		Germination season	Winter–spring	Winter–spring or year-round	Autumn–winter	Autumn or autumn–winter			
		Seed mass (secondary to germination season)	Varied among blocks			Very light	Very light		

(*continued*)

Table 2.4. ***(continued)***

Ecosystem (source)	Disturbance	*Disturbance intensity* / *Resource supply*	High: Low	High: High	Medium: Low	Medium: High	Low: Low	Low: High	Notes
Agricultural landscape in southwestern Germany (Kleyer, 1999)	General agricultural (trampling, cutting, tearing, tilling, herbicides)	Dispersal					Various	Long-range	Attributes are derived from species distributions in agricultural landscape. Attributes in parentheses are those found in high disturbance conditions, when the disturbance was all aboveground.
		Seed weight	High						
		Age at first flowering		Some early	Late	Early			
		Life cycle	Annual (Perennial)	Annual					
		Seed number	Low	High		Low			
		Seed bank longevity	Short-term (Long-term)	Long-term		Transient	Various	Long-term	
		Shoot height	Short		Medium	Tall	Medium	Tall	
		Leaf distribution on stem	(Rosette or semirosette)		Along stem or semirosette				
		Leaf size			Small		Medium	Large	
		Leaf anatomy					Meso-or scleromorph		
		Lateral clonal expansion	(Medium)		Short		Long	Long	
		Vegetative regeneration (detachment of daughter plants)	(Rapid)	Some	Absent or low		Slow	Rapid	

Note: Information in this table comes from studies in which the authors described effects of a disturbance on plants with specific attributes. Unless otherwise noted, these effects were tested explicitly with multivariate methods in which attributes, not species or groups of species, were the response variables. In two studies (Landsberg et al., 1999; Lavorel, McIntyre, et al., 1999), these tests were done separately on grasses and forbs. All other studies combined all life forms in their analyses, "*n.m.*" indicates that a trait was not measured; NS or a blank cell indicates that the trait was not significantly related to a given disturbance treatment.

In their own comparison of emergent and functional response groups from these three studies, Lavorel and McIntyre (1999) concluded that emergent groups cannot always be expected to be useful surrogates for functional response groups. They suggest that congruence between the two types of groups will be best when the disturbance in question has been a major structuring force in the history of the plant community, assuming that emergent groups reflect dominant selection pressures on the plant species that make up the community. If the disturbance of interest has not been a part of the vegetation over evolutionary time, there still may be some overlap between the two types of groups if a disturbance that has similar selection pressures has shaped the community. For example, some attributes associated with drought-avoidance or tolerance are similar to those associated with grazing tolerance or avoidance (Lavorel and McIntyre, 1999). Thus, defining functional response groups using emergent groups instead of actually imposing the disturbance of interest may be difficult, because functional response groups are usually sought in order to predict the effect of a *change* in land use (i.e., a disturbance regime *different* from that with which the plant community evolved). This suggests that the use of EPCSs based on emergent groups will be useful more for understanding and explaining the current composition and structure of a plant community than for predicting its future.

Functional Response Groups versus Functional Response Groups

How applicable functional response groups generated by one study are to predicting the response of a slightly different community is also questionable. In their review summarizing work on functional response groups, Lavorel et al. (1997) concluded that (1) response groups generally separated out by life form, (2) annual or short-lived perennials are favored by disturbances that open gaps in a perennial-dominated matrix, and (3) small plants are favored by disturbances, presumably because of correlations between plant size and other traits. Some studies done since that review confirm at least some of these three suggestions (Fuhlendorf and Smeins, 1997; Hadar et al., 1999; Kleyer, 1999; Landsberg et al., 1999; Lavorel, McIntyre, et al., 1999; Table 2.4). Lavorel et al. (1997) admit that none of their conclusions were particularly earth-shattering. Nor were they particularly useful for predicting vegetation responses to land use change in anything more than a very general sense. The whole literature on the subject illustrates that traits that distinguish among response groups differ not only among disturbances (Belsky, 1992; Fernandez Ales et al., 1993; Lavorel, McIntyre, et al., 1999), but also among similar plant communities. A particularly striking example of the latter is one study in which the same disturbance was imposed on two old fields that were of different ages but otherwise similar (Lavorel, Rochette, et al., 1999). In it, morphological traits characteristic of specific responses were generally consistent in the young and old fields, but regeneration traits were not (Table 2.4). These results suggest that whether a given attribute is favored, disfavored, or not affected by a certain disturbance depends on the structure and composition of the community before it is disturbed (Díaz, Cabido, Zak, et al., 1996). In the case of the old fields, the greater canopy cover in older fields probably made a larger number of species disturbance-dependent for their existence, which was reflected by the different disturbance-favored attributes in the two fields (Lavorel, Rochette, et al., 1999).

Functional Response Groups versus Functional Effect Groups

Although there is not always congruence between emergent and functional response groups, or even among functional response groups from different situations, these differences are relatively minor compared to the differences between functional response and functional effect groups. These differences are attributable to two main factors: the traits used to characterize the groups, and the approach taken in developing the EPCS.

The list of traits used to characterize groups in response-based EPCSs overlap those of function-based EPCSs, but not completely. For example, size and longevity are plant traits that demonstrably affect both response to grazing (smaller, shorter-lived plants are favored) and ecosystem functions such as aboveground biomass and soil erosion (Landsberg, 1999). Furthermore, life form and life history (annual, biennial, perennial) figure prominently in both types of EPCS because they are strongly related to many physiological traits (Chapin, 1993; Box, 1996), at least when compared across many ecosystems. Both types of EPCS also often use canopy structure to distinguish among groups, though the level of detail used differs dramatically. Whereas ecosystem function-based EPCSs may separate forbs from graminoids, response-based EPCSs often separate forbs with a rosette structure from those with semirosette or leafy canopy structures, for example. Such discriminations have been shown to be necessary in response-based EPCSs because they influence factors such as how easily a plant is eaten by livestock (e.g., McIntyre, Lavorel, et al., 1999). Although this level of detail in canopy structure is rarely used to characterize functional effect groups, more refined effect-based EPCSs may find it useful because of its relationship to light and CO_2 use (Körner, 1993).

The main areas in which traits used for response- and effect-based EPCSs do not overlap seem to be in physiology and reproduction. The first is illustrated by Table 2.2, in which no physiological traits were listed by McIntyre, Lavorel, et al. (1999) as candidate traits for disturbance response categorizations. In contrast, many physiological traits are used when designating functional effect groups and trying to predict their feedback effects on climate change (Díaz, Cabido, and Casanoves, 1999; Walker et al., 1999). Because ecosystem functions of interest usually are related to nutrient cycling, physiological traits such as photosynthetic pathway, nitrogen-fixing capacity, and phenology are bases for ecosystem function-based EPCSs (Table 2.5). These traits directly or indirectly determine the quantity and type of nutrients a plant uses, the location and timing of nutrient extraction from and release to the environment, and the form in which they are released. These in turn affect productivity, decomposition, and other ecosystem functions.

On the other hand, functional effect groups are rarely based on regeneration traits, whereas these traits often figure prominently in response-based EPCSs. Seed mass, for example, occurs as a significant trait for distinguishing among groups in 6 of the 10 functional response group studies listed in Table 2.4 but in none of the 13 functional effect group studies in Table 2.5. Also, even though phenology is often used in both response- and ecosystem function-based EPCSs, the emphasis is on season of seed germination in the former, but on phenology of vegetative growth in the latter. The lack of regeneration traits in ecosystem function-based EPCSs comes from the fact that vegetative traits affect the acquisition of resources needed for vegetative growth, and therefore the resource cycles that are part of ecosystem functioning. Reproductive traits affect regrowth, recolonization, and migration, processes occurring after disturbances that kill or reduce the vigor of existing plants, and therefore open up windows of opportunity for recruitment (Armbruster, 1995; Díaz et al., 1998). This difference in emphasis on regeneration versus vegetative

Table 2.5. Functional Effect Groups from Grassland Studies (Defined with the Subjective/Deductive Approach)

Source	Ecosystem	Ecosystem function	Traits on which groups are based	Groups	Species within group consistent?
Sala et al. (1989)	Patagonian steppe shrub/grassland	Water use	Growth form (rooting depth, phenology)	Forbs	No (Golluscio and Sala, 1993)
				Grasses	Not tested
				Shrubs	Not tested
Hobbs (1997)	California annual grasslands	Many, including response to disturbance	Growth form, life history, nitrogen-fixing ability, phenology	Perennial grasses Legumes Early-season annuals Late-season annuals Geophytes	Not always
Hooper and Vitousek (1997, 1998); Hooper (1998)	California serpentine grasslands	Productivity, nitrogen and phosphorus cycling	Growth form, life history, nitrogen-fixing ability, phenology	Early-season forbs Late-season forbs Nitrogen fixers Perennial bunchgrasses	Not tested
Sims et al. (1978)	Western North American grasslands	General	Growth form, life history, phenology	Grasses and sedges Annual Cool-season Warm-season Perennial Cool-season Warm-season Shrubs/low trees Cool-season Warm-season Forbs Annual Cool-season Warm-season Perennial Cool-season Warm-season Succulents	Not tested
Wilson and Hartnett (1998)	Tallgrass prairie (greenhouse)	Positive effect of mycorrhizal colonization on productivity	Growth form, life history, phylogeny, photosynthetic pathway	Perennial C_4 grasses	Mostly
				Perennial C_3 grasses	No
				Annual grasses	Yes
				Perennial forbs	Partly
				Composites	No
				Legumes	Mostly
				Other	Mostly
				Annual/biennial forbs	Mostly

(continued)

Table 2.5. ***(continued)***

Source	Ecosystem	Ecosystem function	Traits on which groups are based	Groups	Species within group consistent?
Symstad et al., (1998)	Tallgrass prairie (greenhouse)	Aboveground productivity, nitrogen retention	Growth form, photosynthetic pathway, N-fixing capacity (phylogeny)	C_4 grasses C_3 grasses Legumes Nonleguminous forbs	No
Tilman (1996); Tilman et al. (1997)	Prairie/old field	Drought resistance, productivity, nitrogen use	Growth form, photosynthetic pathway, N-fixing capacity (phylogeny)	C_4 grasses C_3 grasses Legumes Nonleguminous forbs Shrubs/trees	Not always
Hector et al. (1999)	European seminatural grasslands	Productivity	Growth form, N-fixing capacity (phylogeny)	Grasses Nitrogen-fixing legumes Herbs	Not always
Scholes et al. (1997)	South African grasslands and savannas	General	Phenology, life history, leaf shape, leaf thickness/succulence, wetness of habitat, secondary compounds, thorniness, growth rate, N-fixing ability	Thorn shrub Thorn tree Broad-leaved shrub Broad-leaved tree Evergreen shrub Evergreen tree Succulent shrub Phreatophyte Succulent tree Geoxylic suffrutex Wire grasses Tuft grasses Creeping grasses Ephemeral forbs N-fixing forbs Perennial non-N-fixers Geophytes	Not tested
Wardle et al. (1999)	Australian pasture (non-native)	Productivity, soil nutrients, decomposition	Subjective	C_4 grasses C_3 annual grasses All C_3 grasses Dicotyledonous weeds	Not clear
Wardle et al. (2000)	Greenhouse	Productivity, microbial respiration, decomposition, soil nutrient concentrations	Subjective	C_4 grass C_3 grass N-fixing forb Non-N-fixing forb	Not tested

traits should be reconsidered, however, because regeneration traits will at least partly determine which species are present in what abundance after a plant community is disturbed. This species composition in turn affects ecosystem functioning as the vegetation changes over time following a disturbance (Symstad and Tilman, 2001). In addition, transient or subordinate species that respond to a disturbance differently than do the function-driving dominant species could affect ecosystem functioning in the long run by influencing the recruitment of the dominant species (Grime, 1998).

Functional response and functional effect groups also differ substantially in the way in which they are derived. Ecosystem function-based EPCSs are overwhelmingly derived using *a priori* approaches (Table 2.5), whereas response-based EPCSs are most often derived via *a posteriori* approaches. This fundamentally different way of reaching a classification system both stems from and causes differences in the generality of the groups that are produced. Compared to traits associated with responses to specific disturbances, traits that affect ecosystem functioning are relatively well understood and consistent within and among ecosystem types (Díaz et al., 2001). From tallgrass prairie to desert scrub, fast-growing plants with nutrient-rich leaves are associated with rapid decomposition and high productivity (Díaz, Cabido, and Casanoves, 1999). In contrast, it is difficult to predict a whole plant community's response to a disturbance from the traits of the plants that comprise it, because indirect effects of the disturbance (selective grazing, modified competitive interactions) may be more important than the direct effects of the disturbance on individual plants (Díaz et al., 2001). This was illustrated by the different responses of the same species to the same disturbance in fields of slightly different age, discussed in the "Functional Response Groups versus Functional Response Groups" section (Lavorel, Rochette, et al., 1999).

Thus, whereas the consistency of the relationship between certain plant traits and ecosystem functions makes it possible for investigators to feel fairly confident that their subjectively defined functional effect groups have real ecological meaning, the idiosyncratic nature of plant community responses to disturbances have so far required site-specific, numerically derived trait lists for response-based classification systems. This lack of general principles guiding response-based EPCSs has recently been acknowledged (Lavorel and McIntyre, 1999) and criticized (Semenova and van der Maarel, 2000), with some researchers feeling more confident than others that the problem can be overcome. Ecosystem function-based EPCSs seem to suffer from the opposite problem, in that the confidence that their users have in them is not necessarily well founded, because few of the systems have been rigorously tested (Table 2.5). For examplc, very few systematic studies of how all plant species in a given ecosystem affect one or more ecosystem functions exist, although the Integrated Screening Program (Hendry and Grime, 1993; Grime, Thompson, et al., 1997) is one notable exception. As a result, many of the traits used to define functional effect groups are derived from those of dominant species whose effects on an ecosystem process are well known. Whether all species with these traits affect that ecosystem process in the same way is rarely tested.

Linking Functional Response and Functional Effect Groups

Although one plant classification system clearly will not apply to all situations or goals, some universal plant traits are key to how a plant behaves in any environment. This is the basis for plant ecological strategy schemes such as those proposed by Grime (1974) and Westoby (1998). By concentrating on these key traits, it may be possible to link functional response and functional effect groups. Linking these groups will facilitate not only predictions of how changing climate and land use practices will affect vegetation, ecosystems, and ecosystem services, but also understanding of how plant communities are assembled and are related to ecosystem functioning.

A Tallgrass Prairie Example

The first step in linking these two types of EPCS is to make explicit comparisons of the groups they produce for the same ecosystem. Very few of these comparisons have been done (Gitay and Noble, 1997). One recent exception is that of Landsberg (1999), in which the overlap between groups based on the same data but different paradigms (functional response versus functional effect) was explored. In this comparison, the author found little direct conflict in predictions of plant species' response to grazing (i.e., never did one system predict a species would increase, whereas the other predicted it would decrease), but there was also not much overlap. There were many cases in which one EPCS predicted a change in the abundance of a particular species, whereas the other EPCS predicted that the species' abundance would not change (Landsberg, 1999). In both of the studies compared in this paper (Landsberg et al., 1999; Walker et al., 1999), the groups were derived via *a posteriori* methods, although Walker et al.'s trait list was constructed according to the hypothesis that minor species possessing functional attribute syndromes (height, cover density, SLA, life span, and leaf coarseness) similar to dominant species are functionally equivalent (i.e., belong to the same functional effect group).

A different type of comparison of functional response and functional effect groups can be done by examining the results of experiments in which the response of vegetation to a disturbance has been analyzed by changes in the abundance of *a priori* groups defined mostly for ecosystem function purposes. There are two main questions in such a comparison: (1) Do the species in the *a priori* functional groups respond consistently to the disturbance? and (2) Are there mechanistic reasons why attributes used to define the functional groups would also distinguish among response groups? As an example, I draw on studies from the tallgrass prairie of North America, where fire and grazing are perturbations with which the vegetation has evolved.

If tallgrass prairie vegetation is separated simply into C_4 grasses, C_3 grasses, forbs, and woody species (*sensu* Paruelo and Lauenroth, 1996), the answer to the first question is generally "no." By analyzing his data in a hierarchical manner, Gibson (1988) showed that perennial and annual C_4 grasses, and even perennial C_4 grasses of different maximum heights, responded differently to prescribed fire (Table 2.6). The response of annuals versus perennials within the C_3 grass and forb groups also differed in this study (Table 2.6). Coppedge et al. (1998) also found differences in response to fire of species within the four general functional groups listed earlier. Because of the overwhelming dominance of three tall, perennial C_4 grasses in the tallgrass prairie, they separated these three species from *Schizachyrium scoparium*, a midsized, perennial, C_4 bunchgrass, and other perennial grasses (C_4 and C_3). They also separated legumes from other, nonleguminous forbs, a common distinction made in functional groupings (e.g., Tilman et al., 1997). These finer distinctions, like Gibson's (1988), showed that not all C_4 grasses respond similarly to the same perturbation (Table 2.6). Species within the general C_4 grass group also varied in their response to bison grazing in this study, as did forbs versus legumes (Table 2.6). Thus, clearly groups that have been commonly defined for functional purposes need some

Table 2.6. Example of a *A Priori* Groups' Response to Disturbance in Tallgrass Prairie

Source	Disturbance	Groups favored	Groups disfavored	Groups indifferent	Notes
Gibson (1988)	Fire	Grasses C_4 grasses Perennial C_4 grasses tall and short C_3 annual forbs C_4 grasses as a whole	C_3 grasses Perennial C_3 grasses Forbs Perennial forbs all C_3 species	C_3 annual grasses C_4 annual grasses Medium C_4 perennial grasses C_4 annual forbs Annuals as a whole Perennials as a whole Woody species	Response to fire was measured by abundance of groups and subgroups relative to the time since the last fire (4-yr burning cycle).
Coppedge et al. (1998)	Fire	Forbs (?), legumes	Tallgrasses (?), sedges, and rushes	*Schizachyrium*, perennial grasses, annual grasses	Response to fire was measured by abundance of groups and subgroups relative to the time since the last fire.
	Fire season			Perennial grasses, sedges, and rushes	
	Spring	Tallgrasses, *Schizachyrium*			Grazing intensity was monitored as animals moved freely about. "Groups favored" are those positively affected by higher intensity grazing.
	Summer	Annual grasses, forbs, legumes			
	Fall	None			
	Bison grazing intensity	*Schizachrium* (?), forbs (?)	Annual grasses (?)	Tallgrasses, perennial grasses, sedges and rushes, legumes	"(?)" following a group indicates that significance of effects were different in analyses using single vs. multiple disturbances and environmental factors.

refinement if they are to be linked to response groups. These examples and other work (e.g., Sims et al., 1978; Howe, 1994; Lauenroth et al., 1997; Symstad and Tilman, 2001) suggest that prime possibilities for new axes of refinement are phenology within the nongraminoid groups, and life history, growth habit, and reproductive method in all groups.

Such refinements would not only aid in linking response and ecosystem function, but would also likely improve the ecosystem function-based EPCSs themselves. This is where the second question in comparing response to functional groups comes in, because mechanistic understanding of one will inevitably increase mechanistic understanding of the other. For the tallgrass prairie functional effect groups, there are some parallels between functional response and functional effect group attributes because of the attribute syndrome associated with photosynthetic pathway in grasses. The CO_2-concentrating morphology and physiology of C_4 species makes them generally more efficient in water use than C_3 species (Pearcy and Ehleringer, 1984), an attribute that has obvious effects not only on the ecosystem function of water cycling, but also on response to drought, a disturbance. In addition, because the C_4 photosynthetic pathway is efficient at greater temperatures than the C_3 pathway (Hatch, 1987), C_3 grasses grow primarily in cooler conditions (spring and fall), whereas C_4 grasses grow most actively in the hottest part of summer. This relationship between photosynthetic pathway and phenology is useful not only for functional effect groupings (because it describes when a species is most actively using nutrients), but also for functional response groupings. Prescribed fires in tallgrass prairies are usually conducted in spring and fall, when C_4 grasses are dormant and C_3 grasses are actively growing. Thus, C_3 grasses are generally less resistant to this type of disturbance than are C_4 grasses. Finally, the C_4 photosynthetic pathway is physiologically tied to greater nitrogen-use efficiency (Brown, 1985; Marschner, 1995) and therefore affects tissue nitrogen concentration. This in turn is related to palatability to herbivores (and therefore response to grazing) and the ecosystem function of decomposition. Although these examples show that there are mechanistic reasons why response and functional groupings may be based on similar traits, there is still considerable work to be done. This is particularly necessary in the more broadly defined groups such as "forbs," which in prairies are overwhelmingly C_3, but vary widely in phenology, growth form, rooting structure, and nutrient use—attributes important to both ecosystem function and response to disturbances.

Using Each Other's Methods

As mentioned earlier, response- and ecosystem function-based EPCSs both suffer from their own shortcomings. Those seeking functional effect groups could probably benefit from using the analytical and experimental techniques that the response groupers have developed for characterizing and assessing the validity of their groups. For example, comparing emergent groups of attributes relevant to ecosystem functioning to the untested functional effect groups designated by deductive approaches could yield some interesting results. When this was done in the Mediterranean scrub vegetation of stabilized sand dunes in southwestern Spain, the effort revealed four ecological groups relating to water use and succession, in addition to the two that had been previously recognized (Díaz Barradas et al., 1999). On the other hand, Westoby and Leishman's (1997) analysis for emergent groups in Australian semiarid woodland vegetation yielded five distinct groups that were not that different from what experienced field botanists would have designated using subjective methods. However, they also found that some groups that most field botanists would have broken out did not come out as cohesive units in the data-derived method. Thus, although finding emergent groups may sometimes be an arduous task, it can be a step toward finding rigorous functional effect groups by challenging ecologists to explain why they differ from subjectively defined groups.

Function-based EPCSs could also benefit from experimental testing—a key part of most response-based EPCSs. One type of experiment would be to measure the effect of removing each species of the same candidate functional group, one at a time, on the ecosystem function of interest (Chapin et al., 1996). If the system responds similarly to the removal of each species, then the validity of the functional group is increased. In contrast, if the ecosystem function of interest remains constant when all species in a candidate functional group are removed simultaneously, as happened in the semiarid shrubland of the southwestern United States (Reynolds et al., 1997), the EPCS may have to be rethought. Accordingly, species removal experiments could be used to define functional effect groups by monitoring the response of remaining species to the removal. Assuming that competition between species is strongest within the same functional effect group because these species use resources similarly, those species that respond most positively to the removal of a certain species belong to the same functional effect group as the species removed. Obviously, just as is the case for functional response groups, all functional effect groups cannot be experimentally generated or confirmed. However, these kind of experiments are necessary to assign any kind of certainty to the validity of function-based EPCSs used today, as well as to really understand which plant attributes affect important ecosystem functions. They could also help establish whether the suite of attributes that make species within a functional group affect an ecosystem function similarly also make them respond similarly to an environmental change (Walker et al., 1999).

Response-based EPCSs, on the other hand, could benefit from more attempts to use *a priori* methods, such as those based on the Vital Attributes Model (Noble and Slatyer, 1980; Noble and Gitay, 1996) or trait–environment filtering effects on community assembly (Weiher and Keddy, 1999), to define groups. Clearly, some *a priori* decisions are made in choosing which plant traits to measure and analyze. However, the thought processes involved in deducing the exact mechanisms by which these traits could affect community response would strengthen the conceptual basis of response-based EPCSs and produce hypotheses testable across a wide range of ecosystems.

Conclusion

All ecological plant classification systems are to some degree arbitrary groups in a continuous world. However, they are necessary, arbitrary groups if ecologists have any hope of improving predictions of ecosystem changes. It must also be acknowledged that it will not be possible to tie all plant classification systems together; however, a greater effort to link results of one type to the other is needed. This can be done through syntheses like this one, but more importantly, through experiments explicitly designed to test for relationships between functional response and functional effect attributes. If accurate predictions can be made, then there is at least a chance that appropriate actions can be taken to maintain, restore, and enhance biotic communities, ecosystems, and ecosystem services.

Acknowledgments

I would like to thank Chris Whelan, John Taft, and Brenda Molano-Flores for their valuable comments on an earlier draft of this chapter.

References

Armbruster, W. S. (1995). The origins and detection of plant community structure: Reproductive versus vegetative processes. *Folia Geobotanica Phytotaxonomica* 30: 483–497.

Belsky, A. J. (1992). Effects of grazing, competition, disturbance and fire on species composition and diversity in grassland communities. *Journal of Vegetation Science* 3: 187–200.

Box, E. O. (1996). Plant functional types and climate at the global scale. *Journal of Vegetation Science* 7: 309–320.

Brown, R. H. (1985). Growth of C_3 and C_4 grasses under low N levels. *Crop Science* 25: 954–957.

Chapin, F. S., III. (1993). Functional role of growth forms in ecosystem and global processes. In: J. R. Ehleringer and C. B. Field (Eds.), *Scaling physiological processes: Leaf to globe* pp. 287–312. San Diegeo: Academic Press.

Chapin, F. S., III, Bret-Harte, M. S., Hobbie, S. E., and Zhong, H. (1996). Plant functional types as predictors of transient responses of arctic vegetation to global change. *Journal of Vegetation Science* 7: 347–358.

Clements, F. E. (1916). Plant succession: Analysis of the development of vegetation. *Carnegie Institute of Washington Publication* 242: 1–512.

Clements, F. E. (1936). Nature and structure of the climax. *Journal of Ecology* 24: 252–284.

Cody, M. L. (1975). Towards a theory of continental diversities: Bird distribution over Mediterranean habitat gradients. In: M. L. Cody and J. M. Diamond (Eds.), *Ecology and evolution of communities* (pp. 214–257). Cambridge, MA: Belknap Press/Harvard University Press.

Coppedge, B. R., Engle, D. M., Toepfer, C. S. and Shaw, J. H. (1998). Effects of seasonal fire, bison grazing and climatic variation on tallgrass prairie vegetation. *Plant Ecology* 139: 235–246.

Daily, G. C., Alexander, S., Ehrlich, P. R., Goulder, L., Lubchenco, J., Matson, P. A., Mooney, H. A., Postel, S., Schneider, S. H., Tilman, D., and Woodwell, G. M. (1997). Ecosystem services: Benefits supplied to human societies by natural ecosystems. *Issues in Ecology* 2: 1–16.

Díaz, S., Acosta, A., and Cabido, M. (1992). Morphological analysis of herbaceous communities under different grazing regimes. *Journal of Vegetation Science* 3: 689–696.

Díaz, S., Briske, D. D., and McIntyre, S. (2001). Range management and plant functional types. In: R. Grice and K. Hodkingson (Eds.), *Global rangelands: Progress and prospects* (pp. 81–100). Wallingford, UK: CAB International.

Díaz, S., and Cabido, M. (1997). Plant functional types and ecosystem function in relation to global change. *Journal of Vegetation Science* 8: 463–474.

Díaz, S., Cabido, M., and Casanoves, F. (1998). Plant functional traits and environmental filters at a regional scale. *Journal of Vegetation Science* 9: 113–122.

Díaz, S., Cabido, M., and Casanoves, F. (1999). Functional implications of trait–environment linkages in plant communities. In: E. Weiher and P. Keddy (Eds.), *Ecological assembly rules: Perspectives, advances, retreats* (pp. 338–362). Cambridge, UK: Cambridge University Press.

Díaz, S., Cabido, M., Zak, M., Carretero, E. M., and Aranibar, J. (1999). Plant functional traits, ecosystem structure and land-use history along a climatic gradient in central-western Argentina. *Journal of Vegetation Science* 10: 651–660.

Díaz Barradas, M. C., Zunzunegui, M., Tirado, R., AinLhout, F., and Novo, G. (1999). Plant functional types and ecosystem function in Mediterranean shrubland. *Journal of Vegetation Science* 10: 709–716.

Duarte, C. M., Sand-Jensen, K., Nielsen, S. L., Enriquez, S., and Agusti, S. (1995). Comparative functional plant ecology: Rationale and potentials. *Trens in Ecology and Evolution* 10: 418–421.

Fernandez Ales, R., Laffarga, J. M., and Ortega, F. (1993). Strategies in Mediterranean grassland annuals in relation to stress and disturbance. *Journal of Vegetation Science* 4: 313–322.

Fuhlendorf, S. D., and Smeins, F. E. (1997). Long-term vegetation dynamics mediated by herbivores, weather and fire in a Juniperus-Quercus savanna. *Journal of Vegetation Science* 8: 819–828.

Gibson, D. J. (1988). Regeneration and fluctuation of tallgrass prairie vegetation in response to burning frequency. *Bulletin of the Torrey Botanical Club* 115: 1–12.

Gitay, H., and Noble, I. R. (1997). What are functional types and how should we seek them? In: T. M. Smith, H. H. Shugart, and F. I. Woodward (Eds.), *Plant functional types: Their relevance to ecosystem properties and global change* (pp. 3–19). Cambridge, UK: Press Syndicate of the Cambridge University Press.

Gleason, H. A. (1926). The individualistic concept of the plant association. *Bulletin of the Torrey Botanical Club* 53: 7–26.

Golluscio, R. A., and Sala, O. E. (1993). Plant functional types and ecological strategies in Patagonian forbs. *Journal of Vegetation Science* 4: 839–846.

Grime, J. P. (1974). Vegetation classification by reference to strategies. *Nature* 250: 26–31.

Grime, J. P. (1977). Evidence for the existence of three primary strategies in plants and its relevance to ecological and evolutionary theory. *American Naturalist* 111: 1169–1194.

Grime, J. P. (1998). Benefits of plant diversity to ecosystems: Immediate, filter and founder effects. *Journal of Ecology* 86: 902–910.

Grime, J. P., Hodgson, J. G., and Hunt, R. (1988). *Comparative plant ecology: a functional approach to common British species*. London: Unwin Hyman.

Grime, J. P., Hodgson, J. G., Hunt, R., Thompson, K., Hendry, G. A. F., Campbell, B. D., Jalili, A., Hillier, S. H., Díaz, S., and Burke, M. J. W. (1997). Functional types: Testing the concept in Northern England. In: T. M. Smith, H. H. Shugart, and F. I. Woodward (Eds.), *Plant functional types: Their relevance to ecosystem properties and global change* (pp. 122–150). Cambridge, UK: Cambridge University Press.

Grime, J. P., Thompson, K., Hunt, R., Hodgson, J. G., Cornelissen, J. H. C., Rorison, I. H., Hendry, G. A. F., Ashenden, T. W., Askew, A. P., Band, S. R., Booth, R. E., Bossard, C. C., Campbell, B. D., Cooper, J. E. L., Davison, A. W., Gupta, P. L., Hall, W., Hand, D. W., Hannah, M. A., Hillier, S. H., Hodkinson, D. J., Jalili, A., Liu, Z., Mackey, J. M. L., Matthews, N., Mowforth, M. A., Neal, A. M., Reader, R. J., Reiling, K., Ross-Fraser, W., Spencer, R. E., Sutton, F., Tasker, D. E., Thorpe, P. C., and Whitehouse, J. (1997). Integrated screening validates primary axes of specialisation in plants. *Oikos* 79: 259–281.

Grubb, P. J. (1977). The maintenance of species-richness in plant communities: The importance of the regeneration niche. *Biological Reviews* 52: 107–145.

Hadar, L., Noy-Meir, I., and Perevoltsky, A. (1999). The effect of shrub clearing and grazing on the composition of a Mediterranean plant community: Functional groups versus species. *Journal of Vegetation Science* 10: 673–682.

Hatch, M. D. (1987). C_4 photosynthesis—a unique blend of modified biochemistry, anatomy and ultrastructure. *Biochimica et Biophysica Acta* 895: 81–106.

Hector, A., Schmid, B., Beierkuhnlein, C., Caldeira, M. C., Dimer, M., Dimitrakopoulos, P. G., Finn, J. A., Freitas, H., Giller, P. S., Good, J., Harris, R., Högber, P., Huss-Danell, K., Joshi, J., Jumpponen, A., Körner, C., Leadley, P. W., Loreau, M., Minns, A., Mulder, C. P. H., O'Donovan, G., Otway, S. J., Peira, J. S., Prinz, A., Read, D. J., Scherer-Lorenzen, M., Schulze, E.-D., Siamantziouras, A.-S.D., Spehn, E. M., Terry, A. C., Troumbis, A. Y., Woodward, F. I., Yachi, S., and Lawton, J. H. (1999). Plant diversity and productivity experiments in European grasslands. *Science* 286: 1123–1127.

Hendry, G. A. F., and Grime, J. P. (Eds.) (1993). *Methods in comparative plant ecology.* London: Chapman & Hall.

Hobbs, R. J. (1997). Can we use plant functional types to describe and predict responses to environmental change? In: T. M. Smith, H. H. Shugart, and F. I. Woodward (Eds.), *Plant functional types: Their relevance to ecosystem properties and global change* (pp. 66–90). Cambridge, UK: Cambridge University Press.

Hodgson, J. G., Wilson, P. J., Hunt, R., Grime, J. P., and Thompson, K. (1999). Alocating C-S-R plant functional types: A soft approach to a hard problem. *Oikos* 85: 282–294.

Hooper, D. U. (1998). The role of complementarity and competition in ecosystem responses to variation in plant diversity. *Ecology* 79: 704–719.

Hooper, D. U., and Vitousek, P. M. (1977). The effects of plant composition and diversity on ecosystem processes. *Science* 277: 1302–1305.

Hooper, D. U., and Vitousek, P. M. (1998). Effects of plant composition and diversity on nutrient cycling. *Ecological Monographs* 68: 121–149.

Howe, H. F. (1994). Response of early- and late-flowering plants to fire season in experimental prairies. *Ecological Applications* 4: 121–133.

Kelly, C. K. (1996). Identifying plant functional types using floristic data bases: Ecological correlates of plant range size. *Journal of Vegetation Science* 7: 417–424.

Kindscher, K., and Wells, P. V. (1995). Prairie plant guilds: A multivariate analysis of prairie species based on ecological and morphological traits. *Vegetatio* 117: 29–50.

Kleyer, M. (1999). Distribution of plant functional types along gradients of disturbance intensity and resource supply in an agricultural landscape. *Journal of Vegetation Science* 10: 697–708.

Körner, C. (1993). Scaling from species to vegetation: The usefulness of functional groups. In: E.-D. Schulze and H. A. Mooney (Eds.) *Biodiversity and ecosystem function* (pp. 117–140). Berlin: Springer Verlag.

Körner, C. (1998). Tropical forests in a CO_2-rich world. *Climatic Change* 39: 297–315.

Landsberg, J. (1999). Response and effect—different reasons for classifying plant functional types under grazing. In: D. Eldridge and D. Freudenberger (Eds.), *People and rangelands: Building the future. Proceedings of the VI International Rangeland Congress*, Aitkenvale, pp. 911–915.

Lauenroth, W. K., Coffin, D. P., Burke, I. C., and Virginia, R. A. (1997). Interactions between demographic and ecosystem processes in a semi-arid and an arid grassland: A challenge for plant functional types. In: T. M. Smith, H. H. Shugart, and F. I. Woodward (Eds.), *Plant functional types* (pp. 234–254). Cambridge, UK: Cambridge University Press.

Lavorel, S. (2002). Plant functional types. In: H. A. Mooney and J. Canadell (Eds.), *The Earth system: biological and ecological dimensions of global environmental change.* (Vol. 2, pp. 481–489). Chichester: John Wiley & Sons.

Lavorel, S., and Cramer, W. (1999). Plant functional types and disturbance dynamics. *Journal of Vegetation Science* 10: 603–730.

Lavorel, S., and McIntyre, S. (1999). Plant functional types: Is the real world too complex? In: D. Eldridge and D. Freudenberger (Eds.), *People and rangelands: Building the future: Proceedings of the VI International Rangelands Congress*, Aitkenvale, pp. 905–911.

Lavorel, S., McIntyre, S., and Grigulis, K. (1999). Plant response to disturbance in a Mediterranean grassland: How many functional groups? *Journal of Vegetation Science* 10: 661–672.

Lavorel, S., McIntyre, S., Landsberg, J., and Forbes, T. D. A. (1997). Plant functional classifications: From general functional groups to specific groups of response to disturbance. *Trends in Ecology and Evolution* 12: 474–478.

Lavorel, S., Rochette, C., and Lebreton, J. D. (1999). Functional groups for response to disturbance in Mediterranean old fields. *Oikos* 84: 480–498.

Lavorel, S., Touzard, B., Lebreton, J. D., and Clement B. (1998). Identifying functional groups for response to disturbance in an abandoned pasture. *Acta Oecologica International Journal of Ecology* 19: 227–240.

Lepš, J., Brown, V. K., Diaz Len, T. A., Gormsen, D., Hedlund, K., Kailová, J., Korthals, G. W., Mortimer, S. R., Rodriguez-Barrueco, C., Roy, J., Santa Regina, I., van Dijk, C., and van der Putten, W. H. (2001). Separating the chance effect from other diversity effects in the functioning of plant communities. *Oikos* 92: 123–134.

MacArther, R. H. (1972). *Geographical ecology.* Princeton, NJ: Princeton University Press.

Marschner, H. (1995). *Mineral nutrition of higher plants.* San Diego: Academic Press.

McIntyre, S., Díaz, S., Lavorel, S., and Cramer, W. (1999). Plant functional types and disturbance dynamics: Introduction. *Journal of Vegetation Science* 10: 604–608.

McIntyre, S., Lavorel, S., Landsberg, J., and Forbes, T. D. A. (1999). Disturbance response in vegetation: Towards a global perspective on functional traits. *Journal of Vegetation Science* 10: 621–630.

Mueller-Dombois, D., and Ellenberg, H. (1974). *Aims and methods of vegetation ecology.* New York: Wiley.

Naeem, S. (1998). Species redundancy and ecosystem reliability. *Conservation Biology* 12: 39–45.

Naeem, S., and Li, S. (1997). Biodiversity enhances ecosystem reliability. *Nature* 390: 508–509.

Naeem, S., Thompson, L. J., Lawler, S. P., Lawton, J. H., and Woodfin, R. M. (1994). Declining biodiversity can alter the performance of ecosytstems. *Nature* 368: 734–737.

Naeem, S., Tjossem, S. F., Byers, D., Bristow, C., and Li, S. B. (1999). Plant neighborhood diversity and production. *Ecoscience* 6: 355–365.

Noble, I. R., and Gitay, H. (1996). Functional classification for predicting the dynamics of landscapes. *Journal of Vegetation Science* 7: 329–336.

Noble, I. R., and Slatyer, R. O. (1980). Use of vital attributes to predict successional changes in plant communities subject to recurrent disturbances. *Vegetatio* 43: 5–21.

Paruelo, J. M., and Lauenroth, W. K. (1996). Relative abundance of plant functional types in grasslands and shrublands of North America. *Ecological Applications* 6: 1212–1224.

Pearcy, R. W., and Ehleringer, J. (1984). Comparative ecophysiology of C_3 and C_4 plants. *Plant, Cell and Environment* 7: 1–13.

Platt, W. J., and Weis, I. M. (1977). Resource partitioning and competition within a guild of fugitive prairie plants. *American Naturalist* 111: 479–513.

Raunkiaer, C. (1934). *The life forms of plants and statistical plant geography.* Oxford, UK: Oxford University Press.

Reich, P. B., Walters, M. B., and Ellsworth, D. S. (1997). From tropics to tundra: Global convergence in plant functioning. *Proceedings of the National Academy of Sciences* (USA) 94: 13730–13734.

Reynolds, J. F., Virginia, A., and Schlesinger, W. H. (1997). Defining functional types for models of desertification. In: T. M. Smith, H. H. Shugart, and F. I. Woodward (Eds.), *Plant functional types: Their relevance to ecosystem properties and global change* (pp. 195–216). Cambridge, UK: Cambridge University Press.

Ricklefs, R. E. (1990). *Ecology*. New York: Freeman.

Root, R. B. (1967). The niche exploitation pattern of the blue-gray gnatcatcher. *Ecological Monographs* 37: 317–350.

Sala, O. E., Golluscio, R. A., Lauenroth, W. K., and Soriano, A. (1989). Resource partitioning between shrubs and grasses in the Patagonian steppe. *Oecologia* 81: 501–505.

Schlapfer, F. (1999). Expert estimates about effects of biodiversity on ecosystem processes and services. *Oikos* 84: 346–352.

Scholes, R. J., Pickett, G., Ellery, W. N., and Blackmore, A. C. (1997). Plant functional types in African savannas and grasslands. In: T. M. Smith, H. H. Shugart, and F. I. Woodward (Eds.) *Plant functional types: Their relevance to ecosystem properties and global change* (pp. 255–268). Cambridge, UK: Cambridge University Press.

Schulze, E.-D., and Mooney, H. A. (1993). Ecosystem function of biodiversity: A summary. In: E.-D. Schulze and H. A. Mooney (Eds.), *Biodiversity and ecosystem function* (pp. 497–510). Berlin: Springer-Verlag.

Semenova, G. V., and van der Maarel, E. (2000). Plant functional types—a strategic perspective. *Journal of Vegetation Science* 11: 917–922.

Sims, P. L., Singh, J. S., and Lauenroth, W. K. (1978). The structure and function of ten western North American grasslands: I. Abiotic and vegetational characteristics. *Journal of Ecology* 66: pp. 251–285.

Skarpe, C. (1996). Plant functional types and climate in a southern African savanna. *Journal of Vegetation Science* 7: 397–404.

Smith, T. M., Shugart, H. H. and Woodward, F. I. (Eds). *Plant functional types: Their relevance to ecosystem properties and global change*. Cambridge, UK: Cambridge University Press.

Solbrig, O. T. (1993). Plant traits and adaptive strategies: Their role in ecosystem function. In: E.-D. Schulze and H. A. Mooney (Eds.), *Biodiversity and ecosystem function* (pp. 97–116). Berlin: Springer-Verlag.

Symstad, A. J., and Tilman, D. (2001). Diversity loss, recruitment limitation, and ecosystem functioning: Lessons learned from a removal experiment. *Oikos* 92: 424–435.

Symstad, A. J., Tilman, D., Willson, J., and Knops, J. M. H. (1998). Species loss and ecosystem functioning: Effects of species identity and community composition. *Oikos* 81: 389–397.

Tilman, D. (1996). Biodiversity: Population versus ecosystem stability. *Ecology* 77: 350–363.

Tilman, D., Knops, J., Wedin, D., Reich, P., Ritchie, M., and Siemann, E. (1997). The influence of functional diversity and composition on ecosystem processes. *Science* 277: 1300–1302.

Tilman, D., Wedin, D., and Knops, J. (1996). Productivity and sustainability influenced by biodiversity in grassland ecosystems. *Nature* 379: 718–720.

Vitousek, P. M., and Hooper, D. U. (1993). Biological diversity and terrestrial ecosystem biogeochemistry. In: E.-D. Schulze and H. A. Mooney (Eds.), *Biodiversity and ecosystem function* (pp. 3–14). Berlin: Springer-Verlag.

Walker, B., Kinzig, A., and Langridge, J. (1999). Plant attribute diversity, resilience, and ecosystem function: The nature and significance of dominant and minor species. *Ecosystems* 2: 95–113.

Walker, B. H. (1997). Functional types in non-equlibrium ecosystems. In: T. M. Smith, H. H. Shugart and F. I. Woodward (Eds.), *Plant functional types: Their relevance to ecosystem properties and global change* (pp. 91–103). Cambridge, UK: Cambridge University Press.

Wardle, D. A., Bonner, K. I., and Barker, G. M. (2000). Stability of ecosystem properties in response to above-ground functional group richness and composition. *Oikos* 89: 11–23.

Wardle, D. A., Bonner, K. I., Barker, G. M., Yeates, G. W., Nicholson, K. S., Bardgett, R. D., Watson, R. N., and Ghani, A. (1999). Plant removals in perennial grasslands: Vegetation dynamics, decomposers, soil biodiversity, and ecosystem properties. *Ecological Monographs* 69: 535–568.

Weiher, E., Clarke, G. D. P., and Keddy, P. A. (1998). Community assembly rules, morphological dispersion, and the coexistence of plant species. *Oikos* 81: 309–322.

Weiher, E., and Keddy, P. A. (1999). Assembly rules as general constraints on community composition. In: E. Weiher and P. A. Keddy (Eds.), *Assembly rules: Perspectives, advances, retreats* (pp. 251–271). Cambridge, UK: Cambridge University Press.

Weiher, E., van der Werf, A., Thompson, K., Roderick, M., Garnier, E., and Eriksson, O. (1999). Challenging Theophrastus: A common core list of plant traits for functional ecology. *Journal of Vegetation Science* 10: 609–620.

Westoby, M. (1998). A leaf–height–seed (LHS) plant ecology strategy scheme. *Plant and Soil* 199: 213–227.

Westoby, M., and Leishman, M. (1997). Categorizing plant species into functional types. In: T. M. Smith, H. H. Shugart, and F. I. Woodward (Eds.), *Plant functional types: Their relevance to ecosystem properties and global change* (pp. 104–121). Cambridge, UK: Cambridge University Press.

Westoby, M., Leishman, M. R., and Lord, J. M. (1995). On misinterpreting the "phylogenetic correction." *Journal of Ecology* 83: 531–534.

Wilson, G. W. T., and Hartnett, D. C. (1998). Interspecific variation in plant response to mycorrhizal colonization in tallgrass prairie. *American Journal of Botany* 85: 1732–1738.

Wilson, J. B. (1999). Guilds, functional types and ecological groups. *Oikos* 86: 507–522.

Wilson, J. B., and Gitay. H. (1999). Alternative classifications in the intrinsic guild structure of a New Zealand tussock grassland. *Oikos* 86: 566–572.

Wilson, J. B., and Roxburgh, S. H. (1994). A demonstration of guild-based assembly rules for a plant community, and determination of intrinsic guilds. *Oikos* 69: 267–276.

Woodward, F. I., and Cramer, W. (1996). Plant functional types and climatic changes: Introduction. *Journal of Vegetation Science* 7: 306–308.

3

Ecology of Soil-Borne Pathogens in Crop Fields

Youn Su Lee and Min Woong Lee

Introduction

Soil is a resourceful reservoir for many plant pathogens. Roots interact with soil-borne plant pathogens exhibiting temporal and spatial variations. If the pathogen becomes dominant, disease outbreak is the result. In general, it is well known that fungi constitute the largest number of plant pathogens. Soil-borne bacteria, nematodes, and viruses are also important in relation to plant health and productivity (Bruehl, 1975).

Among the diseases that affect crops there is seemingly no end. Roots of crops in the field are beset with problems described as damping-off, seedling blight, root rots, crown rots, foot rots, club roots, root browning, pink roots, root parasites, and wilts. Sometimes crops are killed before plants emerge from the soil; other times, the plants emerge, only to succumb as seedlings. In some cases, the struggle continues until plants either grow to be reasonably healthy or grow, but lead unthrifty or aborted lives. Plants sometimes succumb to disease as they become weakened or reach maturity. Root-infecting fungi are found the world over but perhaps do more damage in hot, dry regions. Fungi and bacteria cause most of soil-borne diseases of crops. However, many other factors may also affect the occurrence and development of root-infecting fungi (Bruehl, 1975).

Much research has concerned the preferred habitats of soil bacteria and the influence of available energy and nutritional sources on their behavior. With a few exceptions, the biological activities of most plant-pathogenic bacteria have escaped scrutiny in the rhizosplane, rhizosphere, and nonrhizosphere soils, and the extent to which they persist in soil is relatively unknown. Modifying cultural practices and observing the incidence of disease have developed the idea that plant-pathogenic bacteria vary in their capacity to survive in soil. Many management practices have been predicated on the known evidence that the pathogen did not persist long in soil. Grouping bacterial plant pathogens into

Youn Su Lee • Kangwon National University, Chuncheon, Korea.
Min Woong Lee • Dongguk University, Seoul, Korea.

ecological classes based on their ability to survive in soil is useful in the comparative study of the ecology and control of bacterial plant pathogens of crops.

Viruses rarely interact directly with plant roots; however, viruses are found in soils and cause disease situations in root tissue, and virus infection can alter the subsequent interaction between infected roots and other soil microbes. Very little is known about the role of viruses in natural ecosystems. Large numbers of virus-infected plants are annually harvested or die, and, as a result, virus-containing plant debris is constantly introduced into agricultural soils. However, viruses are obligate parasites and, under natural conditions, soil microbes and invertebrates degrade most of plant debris. Degradation of plant debris destroys many viruses. In addition to the natural destruction of the virus-infected debris, some plant viruses are not stable when exposed to the physicochemical forces and microbial enzymes they encounter when released from plant debris in soil. The inability of viruses to enter root tissues, in the absence of wounds or root-penetrating vectors, has been demonstrated repeatedly.

Plant-pathogenic nematodes are roundworms, usually small in size. They exist as migratory ectoparasites in the water film around and between soil particles, as migratory semiendoparasites that penetrate partway into underground plant parts, or as migratory or sedentary endoparasites in plant roots, stems, leaves, or flowers. Although there is a considerable amount of morphological diversity within the nematode group, all plant-pathogenic nematodes have a stylet or mouth spear, which is a hollow cylinder ranging in length from a minimum of about 10 to a maximum of about 140 μm. The stylet is used to rupture the eggshell during the hatching process and penetrate plant cell walls in the actual feeding processes.

Fungi

Almost all plant-pathogenic fungi spend part of their lives on their host plants and part in the soil, or in plant debris on the soil. Some fungi pass all of their lives on the host, and only the spores may land on the soil, where they remain inactive until they are again carried to a host on which they grow and multiply. Others, such as the apple scab fungus *Venturia*, must pass part of their lives on the host as parasites and part on dead tissues of the same host on the ground, as saprophytes, in order to complete their life cycle in nature. The latter group of fungi, however, remains continually associated with host tissues, whether living or dead, and in nature does not grow on any other kind of organic matter. A third group of fungi grow parasitically on the hosts, however, they continue to live, grow, and multiply on the dead tissues of the host after its death and may furthermore move out of the host debris into the soil or other decaying plant material on which they grow and multiply as strict saprophytes. The dead plant material that they colonize need not be related at all to the host they can parasitize. These fungi are usually soil pathogens, have a wide host range, and can survive in the soil for many years in the absence of their hosts. They, too, however, may need to infect a host from time to time, in order to increase their populations, because protracted and continuous growth of these fungi as saprophytes in the soil result in more or less rapid reduction in their numbers (Agrios, 1997).

During the parasitic phase, fungi assume various positions in relation to the plant cells and tissues. Some fungi, such as powdery mildews, grow outside the plant surface but send their feeding organs, haustoria, into epidermal cells of the plant. Some of the fungi, including *Venturia*, grow only between the cuticle and the epidermal cells. Others grow in the intercellular spaces between the cells and may send haustoria into the cells. Still others grow between and through the cells indiscriminately. Fungi that cause vascular wilts, such as *Fusarium*, grow inside the xylem vessels of infected plants, whereas the so-called endophytic fungi, growing within mostly symptomless grass plants, exist intercellularly in the various plant organs. Obligate parasites (biotrophs) can grow only in association with living cells, being unable to feed on dead cells. On the other hand, the mycelium of some nonobligate parasites never comes in contact with living plant cells, because their enzymes macerate and kill the plant cells ahead of the mycelium. In most cases, however, regardless of the position of the mycelium in the host, the reproductive bodies (spores) of the fungus are produced at or very near the surface of the host tissues to ensure their prompt and efficient dissemination (Agrios, 1997).

The survival and performance of most plant-pathogenic fungi depend greatly on the prevailing conditions of temperature and moisture, or the presence of water in their environment. Free mycelium survives only within a certain range of temperatures (-5 to $45^\circ C$) and in contact with moist surfaces, inside or outside the host. Most kinds of spores, however, can withstand broader ranges of both temperature and moisture, and carry the fungus through the low winter temperatures and the hot, dry, summer periods. Spores, however, also require favorable temperatures and moisture in order to germinate. Furthermore, fungi producing zoospores require free water for the production, movement, and germination of the zoospores (Agrios, 1997).

Plasmodiophoromycetes

Plasmodiophoromycete fungi are obligate parasites and they can survive in the soil as resting spores for many years. However, they are able to grow and multiply in only a few hosts. The plasmodium lives off the host cells it invades but does not kill these cells for a long time. On the contrary, in some diseases, many invaded and adjacent cells are stimulated by the pathogen to enlarge and divide, thus making available more nutrients for the pathogen. The pathogens spread from plant to plant by means of zoospores, by anything that moves soil or water containing spores, by infected transplants, and so on (Colhoun, 1958).

Polymyxa and *Spongospora* can also transmit destructive plant viruses. *Polymyxa graminis* is a vector of several viruses of grain crops and of peanuts, whereas *P. beta* is a vector of beet necrotic yellow vein virus. *Spongospora* is a vector of the potato mop-top virus. Club root pathogen *Plasmodiophora brassicae* is widely distributed all over the world. Club root can cause serious losses to susceptible varieties (Dobson and Gabrielson, 1983). Fields once infected with the club root pathogen remain so indefinitely and become unfit for cultivation of crucifers (Buczacki, 1983).

Chromistan Fungal-Like Oomycetes

Oomycetes have elongated mycelium containing cellulose and glucans but no crosswalls. They produce oospores as their resting spores and zoospores or zoosporangia as asexual spores. Important plant-pathogenic oomycetes belong to two orders, the Saprolegniales and Peronosporales. The Saprolegniales genus *Aphanomyces* is important as a plant pathogen, causing root rot diseases of many annual plants. *Pythium* and *Phytophthora* species of the order Peronosporales are important plant pathogens.

The plant diseases caused by oomycetes are basically of two types: (1) diseases that affect plant parts present in the soil or in contact with the soil (e.g., roots, lower stems, tubers, seeds, and fleshy fruits lying on the soil) caused by all the species of *Aphanomyces* and *Pythium*, and some species of *Phytophthora*; (2) diseases that affect only or primarily aboveground plant parts, particularly the leaves, young stems, and fruits, caused by some species of *Phytophthora*, by all of the downy mildew fungi, and by *Albugo*.

Pythium Species

Damping-off diseases of seedlings occur worldwide in valleys and forest soils, in tropical and temperate climates, and in every greenhouse. The disease affects seeds, seedlings, and roots of all plants. In all cases, however, the greatest damage is done to the seed and seedling roots during germination either before or after emergence. Losses vary considerably with soil moisture, temperature, and other factors. Quite frequently, seedlings in seedbeds are completely destroyed by damping off or die soon after they are transplanted. In many instances, poor germination of seeds or poor emergence of seedlings is the result of damping-off infections in the preemergence stage. Older plants are seldom killed when infected with the damping-off pathogen, but they develop root and stem lesions and root rots; their growth may be retarded considerably, and their yields may be reduced drastically. Some species of the damping-off fungus also attack the fleshy organs of plants, which rot in the field or in storage (Agrios, 1997).

Phytophthora Species

Species of *Phytophthora* cause a variety of diseases on many different types of plants, ranging from seedlings of annual vegetables or ornamentals to fully developed fruit and forest trees. Most species cause root rots, damping off of seedlings, and rots of lower stems, tubers, and corms similar to those caused by *Pythium* species. Others cause rots of buds or fruits, and still others cause blights of the foliage, young twigs, and fruit. Some species attack only one or two species of host plants (Ingram and Williams, 1991), but others may cause similar or different symptoms in many different kinds of host plants. The best-known species is *P. infestans*, the cause of late blight of potatos and tomatos, but several other species cause extremely destructive diseases on their hosts (Fry et al., 1992, 1993). *Phytophthora cactorum*, *P. cambivora*, *P. cinnamoni*, *P. citrophthora*, *P. fragariae*, *P. palmivora*, and *P. syringae* cause primarily root and lower stem rots, some twig blights and fruit rots of woody ornamental and fruit and forest trees, as well as vegetables and other herbaceous plants. Several other species, such as *P. capsici*, *P. crypogea*, *P. megasperma*, and *P. parasitica*, cause root, stem, and fruit rots of many vegetables, ornamentals, field crops, and also of some woody plants (Brasier, 1992; Hwang and Kim, 1995). Sixteen species of *Phytophthora* have been reported as pathogens of more than 46 plant species in Korea (Song et al., 1998). *Phytophthora* root rots of Chinese cabbage and spinach caused by *P. drechsleri* have recently been reported for the first time in Korea (Jee et al., 1999).

Ascomycetes and Imperfect Fungi

Several ascomycetes and imperfect fungi including *Fusarium* and *Verticillium* cause vascular wilts. Each cause diseases on several important crop, forest, and ornamental plants. Vascular wilts caused by *Fusarium* and *Vericillium* are widespread, very destructive, spectacular, and frightening plant diseases. They appear as more or less rapid wilting, browning, and dying of leaves and succulent shoots of plants, followed by death of the whole plant. Wilts occur as a result of the presence and activities of the pathogen in the xylem vessels of the plant. Entire plants may die within a matter of weeks, although in perennials, death may not occur until several months or years after infection. As long as the plant is alive, the wilt-causing fungi remain in the vascular tissues and a few surrounding cells. Only when the disease kills the infected plant do these fungi move into other tissues and sporulate at or near the surface of the dead plant (Nelson et al., 1981).

In general, the root and stem rot diseases caused by soil-borne ascomycetes and imperfect fungi first appear on the affected plant organs as water-soaked areas that later turn brown to black. In some diseases, lesion area is frequently covered by white fungal mycelium. The roots and stems are killed more or less rapidly, and the entire plant grows poorly or is killed. The fungi that cause these diseases are nonobligate parasites that live, grow, and multiply in the soil as soil inhabitants, usually in association with dead organic matter. High soil moisture and high relative humidity in the air favor these fungi. Most produce conidia and some produce ascospores occasionally or regularly. Several produce sclerotia. All of these fungi can overwinter as mycelium in infected plant tissues or debris, as sclerotia, or as spores. These stages also serve as inocula that can be spread and start new infections.

Several imperfect fungi that cause serious plant losses by attacking primarily the roots and lower stems of plants include *Rhizoctonia* (teleomorph: *Thanatephorus*) and *Sclerotium* (teleomorph: *Aethalium*) that attack primarily herbaceous plants. *Rhizoctonia* and *Sclerotium* are soil inhabitants and cause serious disease on many hosts by affecting the roots, stems, tubers, corms, and other plant parts that develop in or on the ground. These two fungi are known as sterile fungi due to their lack of sexual or asexual spores. The mycelium and texture of sclerotia distinguish these two species. *Rhizoctonia* species have characteristic mycelium and a uniform texture of sclerotia. However, *Sclerotium* species have internally differentiated sclerotia (Sneh et al., 1991; Punja, 1985).

Fusarium Species

Fusarium species cause vascular wilts, and root and stem rots on various host species worldwide, including vascular wilts of vegetable and flowers, herbaceous perennial ornamentals, plantation crops, and the mimosa tree (Booth, 1971). Most of the wilt-causing *Fusarium* belongs to the species *F. oxysporum*. Special forms or races of the fungus attack different host plants. The fungus that attacks tomato is designated as *F. oxysporum* f. sp. *lycopersici*; cucurbits, *F. oxysporum* f. sp. *conglutinans*; banana, *F. oxysporum* f. sp. *cubense*; cotton, *F. oxysporum* f. sp. *vasinfectum*; carnation, *F. oxysporum* f. sp. *dianthii*; and so on (Beckman, 1987; Booth, 1971; Nelson et al., 1981). *Fusarium* wilts are most severe in warm soil conditions and in greenhouses. Most fusarial wilts have disease cycles and development to that of the *Fusarium* wilt of tomato (Anonymous, 1991).

Instead of vascular wilts, several *Fusarium* species, but primarily *F. solani* and some *formae specialis* of *F. oxysporum*, cause rotting of seeds and seedlings (damping off); rotting of roots, lower stems, and crowns; and rots of corms, bulbs, and tubers. They affect many different kinds of vegetables, flowers, and field crops. These diseases occur worldwide and cause severe losses by reducing stands and the growth and yield of infected plants (Chambers and Corden, 1963). In Korea, *F. solani* and its *formae specialis*, especially, causes very severe disease problems in ginseng, pea, and soybeans (Jung et al., 1999; Lee, 1984). Various kinds of diseases caused by *Fusarium* species in Korea are well illustrated in a book published by Lee and Park (Lee and Park, 2001).

In root rot cases, taproots of young plants show a reddish discoloration that later becomes darker and larger. The discoloration may cover the taproot and the stem below the soil line, without a definite margin, or it may appear as streaks extending up to the soil line. Longitudinal cracks appear along the main root, whereas small lateral roots are killed. Plant growth is retarded, and in dry weather the leaves may turn yellow and even fall off. Sometimes infected plants develop secondary roots and rootlets just below the soil line that may be sufficient to carry the plant to maturity and production of a fairly good crop. In many cases, however, infected plants decline and die, with or without wilt symptom. In stem rots, infected plants wilt and die from rot of the stem at the base of the plant. Lesions develop on the stern at or below the soil line, and their edges are often pink or red. The lesions develop inward from the outside. In some plants, a brown discoloration extends into the stem above the ground. In older plants, roots have often rotted and sloughed off. *Fusarium* species also cause bulb, corm, and tuber rot in the field and in storage. The rot, which often starts, at wounds or through the cuts formed on such tissues during harvest, is generally dry and firm. The foliage turns yellow or brown and dies prematurely. Tubers usually develop small brown patches that soon enlarge, become sunken, and show concentric wrinkles that contain cavities lined with white mycelium. Parts of the tuber, or entire tubers, are eventually destroyed and become hard and mummified; in humid conditions, however, they are then invaded by soft rotting bacteria. Root and stem rots caused by *Fusarium* become more severe when plants exposed to the pathogen are stressed by low temperature, intermittent drought, or excessive soil water, herbicide, soil compaction, and subsurface tillage pans, which restrict root growth (Booth, 1971; Nelson et al., 1981).

Verticillium Species

Verticillium causes vascular wilts of more than 200 species of plants, including various vegetables, flowers, field crops, perennial ornamentals, and fruit and forest trees. *Verticillium* is also the main cause of potato early-dying disease (Powelson and Rowe, 1993). When *Verticillium* wilt first appears in a field, it is mild and local. In subsequent years, as the inoculum builds up and as new, more virulent strains of the fungus appear, the attacks become successively more severe and widespread, until the crop has to be discontinued or replaced with resistant varieties (Agrios, 1997; Pullman and Devay, 1982).

Two species, *V. albo-atrum* and *V. dahliae*, attack hundreds of kinds of plants, causing wilt and losses of varying severity. Both produce short-lived conidia. *V. dahliae* also produces microsclerotia, whereas *V. albo-atrum* produces dark, thick-walled mycelium but not microsclerotia. *V. albo-atrum* grows best at 20 to 25°C, whereas *V. dahliae* prefers slightly higher temperatures (25–28°C) and is somewhat more common in warmer regions (Pegg, 1974). Some *Verticillium* strains show host specialization, but most attack a wide range of host plants. *V. dahliae* overwinters in the soil as microsclerotia, which can live up to 15 years. Both species, however, can overwinter as mycelium within perennial hosts, in propagative organs, or in plant debris. *Verticillium* penetrates young roots or host plants directly or through wounds. The fungus is disseminated by contaminated seed, by vegetative cuttings and tubers, by wind, by surface water, and by soil, which may contain up to 100 or more microsclerotia per gram. Many fields have become contaminated with *Verticillium* for the first time by planting infected potato tubers or other crops. Solanaceous crops such as potato, eggplant, and tomato increase the fungus inoculum level in the soil. However, *Verticillium* is often found in uncultivated areas, indicating that the fungus is native to the soils and can attack susceptible crops as soon as they are planted (Jones and Crill, 1973).

Gibberella Species

Gibberella, one of many fungi causing blight of corn seedlings, may be carried on or in infected seed, or it may attack the seed and seeding from the soil. In either case, the germinating seed may be attacked and killed before the seedling emerges from the soil, or after emergence, in which case, the seedling may be killed or become dwarfed and chlorotic, and eventually die. Light-brown to dark-colored lesions are usually evident on the tap and lateral roots, and in the lower internode. Two species of *Gibberella*, *G. zeae*, and *G. moniliforme* (*fujikuroi*), are responsible for the symptom development on both corn and small grains. Both fungi produce ascospores in perithecia and *Fusarium*-type conidia. Perithecia are rather rare in *G. moniliforme*. The fungi overwinter as perithecia, mycelium, or chlamydospores in infected plant debris, particularly corn stalks. In the spring, during wet, warm conditions, ascospores are released and carried by wind to corn stalks or ears, which they penetrate directly or through wounds and cause infections. Conidia may be produced on infected corn debris, but they are more commonly produced on infected plant parts and serve as the secondary inoculum. The diseases are favored by dry weather early in the season and wet weather near or after silking. High plant density, high nitrogen, low potassium in the plant, and early maturity of hybrids make them more susceptible to the diseases (Anderson, 1948; Christensen and Wilcoxson, 1966; Dodd, 1980).

Sclerotinia Species

Genus *Scleotinia*, especially *S. sclerotiorum* and *S. minor*, cause destructive diseases of numerous succulent plants, particularly vegetables and flowers. Another species, *S. homeocarpa*, causes the destructive dollar spot disease of turf grasses. *Scleotinia* diseases occur worldwide and affect plants in all stages of growth, including seedlings, mature plants, and harvested products. *Sclerotinia* species, which overwinter as mycelium or sclerotia in dead or living plants, and in or on soil, germinate to produce apothecia or mycelium. Asci in apothecia discharge ascospores that land and germinate on host plants. Mycelium produced attacks lower stem, roots, and so on. It invades succulent tissues rapidly and the infected cells collapse. White, fluffy mycelium grows on surfaces of invaded tissue and continues to invade tissues, eventually forming sclerotia on the surface or in cavities of infected tissue (Anonymous, 1979; Dennis, 1983; Moline, 1984).

Rhizoctonia Species

Rhizoctonia diseases occur throughout the world, causing losses in almost all vegetables, flowers, several field crops, turf grasses, and even perennial ornamentals, shrubs, and trees (Sneh et al., 1991). Damping off is the most common symptom caused by *Rhizoctonia* on most plants it affects. It occurs primarily in cold, wet soils. A seedling stem canker, known as soreshin, is common and destructive in cotton, tobacco, and other seedlings that have escaped the damping-off or seedling blight phase of the disease. It develops under conditions that are not especially favorable to the disease. Root lesions form in seedlings and on partly grown or mature plants. Reddish-brown lesions usually appear first just below the soil line, but in cool, wet weather, the lesions enlarge in all directions and may increase in size and number to include the whole base of the plant and most of the roots (Burpee and Martin, 1992).

On low-lying plants, the infection spreads upward to the next leaves, until most or all leaves, and the head, may be invaded and rot, with mycelium and sclerotia permeating the tissues or nestled between the leaves. Brown patch on turf grasses is particularly severe during periods of hot and humid or wet weather, especially with heavy dew periods. In brown patch, *Rhizoctonia* usually kills only the leaf blades, and plants in the affected area begin to recover and grow again from the center outward, resulting in a doughnut-shaped diseased area. On fleshy, succulent stems and roots, and on tubers, bulbs, and corms, *Rhizoctonia* causes rooting of tissues. It also causes rot on fruits and pods lying on or near the soil. This rot develops most frequently in wet, cool weather and appears first in the field but may continue and spread to other fruits after harvest and (during transportation and storage (Parmeter, 1970).

Sclerotium Species

Sclerotium diseases occur primarily in warm climates and cause damping off of seedlings, stem canker, crown blight, and root, crown, bulb, tuber, and fruit rot. *Sclerotium* frequently causes severe losses of fleshy fruits and vegetables during shipment and storage. In Korea, *S. cepivorum* causes severe white rot diseases in major production areas producing garlic (Lee et al., 1998). On all infected tissues, and even on the nearby soil, the fungus produces numerous small, roundish sclerotia of uniform size that are white when immature, becoming dark brown to black on maturity. Each *Sclerotium* is differentiated into an outer, melanized rind, a middle cortex, and an innermost area of loosely arranged hyphae. *Sclerotium* species produce abundant white, fluffy, branched mycelium that forms numerous sclerotia but is usually sterile. The organism overwinters mainly as sclerotia. It is spread by moving water, infested soil, contaminated tools, infected transplant seedlings, infected vegetables and fruits, and, in some hosts, as sclerotia mixed with the seed. The fungus attacks tissues directly. However, the mass of mycelium it produces secretes oxalic acid and also pectinolytic, cellulolytic, and other enzymes, and kills and disintegrated tissues before it actually penetrates the host. Once established in the host plants, the fungus advances and produces mycelium and sclerotia quite rapidly, especially at high moisture and a temperature range of 30 to 35°C (Punja, 1985).

Bacteria

Almost all plant pathogenic bacteria develop mostly in the host plant as parasites on the plant surface, especially buds, as epiphytes, and partly in plant debris or in the soil as saprophytes. There are great differences among bacterial species, however, in their degree of development in one or the other environment (Agrios, 1997).

Bacterial pathogens such as *Agrobacterium tumefaciens*, which causes crown gall, *Pseudomonas solanacearum*, which causes the bacterial wilt of *Solanacearum* species, and particularly *Streptomyces scabies*, which causes the common scab of potato, are rather typical soil inhabitants. Such bacteria build up their populations within the host plants, but these populations only gradually decline when they are released into the soil. If susceptible hosts are grown in such soil in successive years, sufficiently high numbers of bacteria could be present to cause a net increase of bacterial populations in the soil from season to season. Most plant-pathogenic bacteria, however, can be considered soil invaders. Such bacteria enter the soil in host tissue and, because they have poor ability to compete as saprophytes, persist in the soil either as long as the host tissue resists decomposition by saprophytes or for varying durations afterward, depending on the bacterial species and soil temperature and moisture conditions (Bradbury, 1986; Starr et al., 1983).

When in soil, bacteria live mostly on plant material. Less often, they live freely or saprophytically, or in their natural bacterial ooze, which protects them from various adverse factors. Bacteria may also survive in or on seeds, other plant parts, or insects found in the soil. On plants, bacteria often survive epiphytically in buds, on wounds, in their exudate, or inside the various tissues or organs that they infect (Goto, 1992).

The relationship of bacterial plant pathogens to the soil can be discussed from various points of view. Based on the ability of plant-pathogenic bacteria to survive in the soil matrix in the absence of living host plants, they can be classified into soil inhabitants and soil invaders. Soil-inhabiting bacteria are indigenous to soil and include those that persist for many years after being introduced to soil via infested plants. Soil invaders are also introduced into the soil by infected crop plants, but survival of these bacteria is dependent to a large extent, on the persistence of host tissue; they tend to survive for only a short time as free cells in soil. A bacterial species may be a soil inhabitant in one environmental region and a soil invader in another. The combined characteristics of the bacterium and the soil in which it is located determine whether it will persist in a particular environment. Thus, some species including *Pseudomonas syringae* and *Erwinia carotovora* have been described both as soil-inhabiting and soil-invading organisms (Civerolo et al., 1987).

Soil Inhabitants

Probably the most truly soil-inhabiting plant-pathogenic bacterium is the actinomycete *Streptomyces scabies* sensu, the causal agent of common scab of *S. tuberosum*. This pathogen has been found in many arable soils and is possibly present in some nonagricultural soils as well. Also, opportunistic bacteria associated with soft rotting diseases are present in many soils. Among these are *P. marginalis* and pectolytic *Clostridium* spp. Additional species endemic in soil include *A. tumefaciens* and *P. solanacearum*. *A. tumefaciens* survived more than 8 months in pots of soil kept outdoors and may persist indefinitely. It has been reported to survive in soil in which susceptible crop plants had not been grown for 40 years. Certain strains of *P. solanacearum* are also able to survive for long periods in soil in the absence of host plants, but adequate evidence is not available to conclude that *P. solanacearum* is an endemic, ubiquitous inhabitant of soils prior to agricultural disturbance. However, it has been found in nonagricultural soils (Mount and Lacy, 1982).

Soil Invaders

Most plant-pathogenic bacteria probably can. best be regarded as soil-invading organisms. Even plant pathogens such as *Erwinia amylovora*, whose soil phase is not usually considered important because it normally does not serve as an inoculum source, are indeed also able to persist in soil for a limited period of time.

Soil invaders often enter the soil environment protected by host tissue, and because these bacteria have poor competitive saprophytic ability, they are able to survive in soil only on account of their position in the host tissue. *Clavibacter michiganense* pv. *sepedonicum* is quickly overgrown by secondary organisms. *Corynebacterium michiganense* pv. *insidiosum*, *C. fl.* pv. *flaccumfaciens*, and *C. michagnense* pv. *michiganense* all have soil survival periods closely related to persistence of host tissue. The corynebacteria survive in host tissue especially well under cold or dry conditions. Similarity *Xanthomonas* and some of the *Pseudomonas* species are relatively short-lived in soil. Their survival is also related to persistence of host tissue, and survival may be extended in tissue fairly resistant to decompostion. Thus, *X. campestris* pv. *campestris* survived in cabbage stalks in soil for as long as 2 years. Soil invaders may also be introduced into the soil as free cells by leakage from diseased plants or decomposing infected tissue. Populations of *Etwinia carotovora* in potato root zone soil increased as seed tubers decayed, but subsequently decreased to very low levels after the plants had been removed. The length of time soil invaders persist as free cells may be as short as a few clays or may end to several months and is especially dependent on soil conditions (Mount and Lacy, 1982).

Factors that Affect Living in Soil

One of the most important factors affecting survival of plant-pathogenic bacteria in soil is that of nutrient availability. Host-plant debris serves as a food base for many of the soil-invading bacteria, and their longevity is contingent on persistence of host debris. Subsequent to disintegration of host tissue, survival depends on innate characteristics of the bacteria themselves and on a large number of variable properties of the environments. Furthermore, the microbial flora and fauna of the soil have a profound effect on survival. Environmental factors, in addition to having a direct effect on survival of plant-pathogenic bacteria, also have an indirectly affect antagonistic and competing microorganisms (Mount and Lacy, 1982).

Viruses

Viruses rarely interact directly with plant roots; however, viruses are found in soils cause disease situations in root tissue, and virus infection can alter the subsequent interaction between infected roots and other soil microbes. Very little is known about the role of viruses in natural ecosystems.

Large numbers of virus-infected plants are annually harvested or die, and, as a result, virus-containing plant debris is constantly introduced into agricultural soils. However, viruses are obligate parasites and, under natural conditions, soil microbes and invertebrates degrade much plant debris. Degradation of plant debris destroys many viruses. In addition to the natural destruction of the virus-infected debris, some plant viruses are not stable when exposed to the physicochemical forces and microbial enzymes they encounter when released from plant debris in soil (Gray and Williams, 1971). Tobacco mosaic virus (TMV) is one of the exceptions in physicochemical viral stability and can be considered a problem in plant debris in the soil. As long as the plant parts harboring the virus are intact, the virus remains infective. Survival of more than 50 years has been noted in certain dried specimens (Agrios, 1997). Survival of infectious TMV for more than 22 months in soil-borne plant debris has been reported (Broadbent et al., 1965). Plant debris in the soil can also be an important source of virus inoculum for soil-inhabiting virus vectors. Tobacco necrosis virus released from debris may be adsorbed to the surface of motile, fungal zoospores and gain entry into healthy roots as the zoospore penetrates the root (Temmink et al., 1970). Roots of grape have been known to survive as soil debris for more than 4 years and supply grapevine fan leaf virus to nematode vectors (Raski et al., 1965). However, it was reported that air drying of soil rapidly inactivated all five types of viruses applied to soils by artificial methods or released by infected roots (Smith et al., 1969).

The inability of viruses to enter root tissues in the absence of wounds or root penetrating vectors has been demonstrated repeatedly (Cadman, 1963; Yarwood, 1960). Undamaged susceptible roots may quite literally be bathed in solutions containing high concentrations of infectious virus and remain uninfected.

With the exception of TMV, the common feature of soil-borne viruses is the relationship with soil-inhabiting biological vectors. Vectors include nematodes and fungi that attack and penetrate plant roots. Root-feeding aphids have also been implicated in virus transmission to roots (McLean, 1962). Considering the large number and diversity of soil-inhabiting invertebrates and microbes, the number and groups of vectors implicated in virus transmission to roots are small. Considerable progress in locating and identifying soil-inhabiting virus vectors has been made in the past.

Viruses transmitted by nematodes infect both woody and herbaceous plants, and can also be mechanically transmitted to susceptible hosts. Certain nematode-transmitted viruses can also be transmitted by pollen, seed, and insects to their plant hosts. At least 19 nematode species are known to transmit viruses to plants; however, vectors have been found only in the genera of *Xiphinema*, *Longidorus*, and *Trichodorus*, or closely related taxa. Transmission of viruses by nematodes to root cells seems to be a complicated process. In some cases, contamination of the nematode stylet, after feeding on virus-infected tissues, may be sufficient explanation for transmission. In many cases, it appears that there is an accumulation of virus in the animal. Accumulated viruses are apparently released gradually during feedings, and this behavior may account for the fact that individual nematodes remain infective for prolonged periods with only one acquisition feeding.

Fungi of three classes Chytridiomycetes, Plasmodiophoromycetes, and Oomycetes have been shown to be vectors of plant-pathogenic viruses. The common characteristic of these fungi is that they are all capable of producing motile, flagellated zoospores. Fungal transmission of viruses to plants was revealed in the late 1950s and early 1960s. The viruses transmitted to plants by fungi are heterogeneous and share no common morphological features. Some of the viruses have never been adequately described, whereas others are small polyhedral particles, and still some others are flexous rods. Teakle (1972) mentioned nine viruses identified as fungal transmitters. Among these are tobacco necrosis, lettuce big vein, and tobacco stunt viruses, transmitted by *O. brassicae*, potato virus X, transmitted by *Synchytrium endobioticum*, and wheat mosaic virus, transmitted by *Polymyxa graminis*. Additionally, potato virus X and tobacco necrosis virus are readily transmitted by mechanical means, and both viruses have other important methods of dissemination.

Nematodes

Almost all plant-pathogenic nematodes live part of their lives in the soil. Many live freely in the soil, feeding superficially on roots and underground stems, but even in the specialized sedentary parasites, the eggs, the preparasitic juvenile stages, and the males are found in the soil for all or part of their lives. Soil temperature, moisture, and aeration affect survival and movement of nematodes in the soil. Nematodes occur in greatest abundance in the top 15–30 cm of soil. Distribution of nematodes in cultivated soil is usually irregular and is greatest in or around the roots of susceptible plants, which they follow sometimes to considerable depths. The greater concentration of nematodes in the region of host plant roots is due primarily to their more rapid reproduction on the food supply available and also to attraction of nematodes by substances released into the rhizosphere. Most nematode eggs hatch freely in water in the absence of any special stimulus. However, so-called hatching-factor effects of substances originating from the roots that diffuse into the surrounding soil markedly stimulate the egg hatching of certain species (Agrios, 1997).

Nematodes spread through the soil slowly under their own power. The overall distance traveled by a nematode probably does not exceed a few meters per season. Nematodes move faster in the soil when the pores are lined with a thin film of water than when the soil is waterlogged. In addition to their own movement, however, nematodes can be easily spread by anything that moves and can carry particles of soil. Farm equipment, irrigation, flood or drainage water, animal feet, birds, and dust storms spread nematodes in focal areas, whereas nematodes are spread over long distances primarily with farm produce and nursery plants. A few nematodes that attack the aboveground parts of plants not only spread through the soil, but are also splashed to the plants by falling rain or overhead watering. Some species ascend wet plant stem or leaf surfaces on their own power. Contact of infected plant parts with adjacent healthy plants is one way of spreading (Dropkin, 1980).

In terms of habitat, pathogenic nematodes are either ectoparasites, species that do not normally enter root tissue but feed only from the outside on the cells near the root surfaces, or endoparasites, species that enter the host and feed from within. Both can be either migratory, living freely in the soil, and feed on plants without becoming attached, or move around inside the plant, or sedentary, living only within a root without moving.

Root-Knot Nematodes

Root-knot nematodes occur throughout the world but are found more frequently and in greater numbers in areas with warm or hot climates and short or mild winters. Root-knot nematodes are also found in greenhouses everywhere, when nonsterilized soil is used. They attack more than 2,000 species of plants, including almost all cultivated plants, and reduce world crop production by about 5%. Losses in individual fields, however, may be much higher (Baker et al., 1985).

Plants infected by this nematode lose vitality on root tips and from swellings of the roots. These effects not only deprive plants of nutrients but also disfigure and reduce the market value of many root crops. When the host plants are infected at the seedling stage, losses are heavy and may result in complete destruction of the crop. Infections of older plants may have only slight effects or may reduce yields considerably.

A life cycle is completed in 25 days at 27°C, but it takes longer at lower or higher temperatures. When the eggs hatch, the infective second-stage juveniles may migrate from within galls to adjacent parts of the root and cause new infections in the same root, or they may emerge from the root and infect other roots of the same plants or roots of other plants. Root-knot nematodes are usually congregated in the root zone from 10 to 25 cm below the surface, but galls have been found on peach and other roots only 2–2.5 cm deep. Root-knot nematodes have limited ability to move on their own power, but they can be spread by water or by soil clinging to farm equipment, or be otherwise transported into uninfested areas (Sasser and Carter, 1985).

Cyst Nematodes

Cyst nematodes cause a variety of plant diseases, mostly in the temperate regions of the world. Some species attack only a few plant species and are present over limited geographic areas, whereas others attack a large number of plant species and are widely distributed. The round cyst nematode *Globodera rostochiensis* is particularly severe on potato, tomato and eggplant. Other common cysts include *Heterodera avenae* on cereals; *H. glycines* on soybeans; *H. schachtii* on sugar beets, crucifers, and spinach; *H. tabacum* on tobacco, and *H. trifolii* on clover (Raski, 1950).

Soybean cyst nematode has been found in northeastern Asia, Japan, and Java, in most soybeans producing states of North America; and in Colombia and Brazil. It continues to spread slowly to new areas. Several other legumes, such as common bean and forage legumes, and a few nonleguminous plants, are also attacked by this nematode. Losses vary from slight to complete destruction of the crop. Sugar beet nematode occurs wherever sugar beets are grown in North America, Europe, the Middle East, and Australia, and is the most important nematode pest of sugar beet production. In warmer climates or late-planted crops, the yield is the greatest (Riggs and Wrather, 1992).

Citrus Nematode

Tylenchulus semipenetrans is present and common wherever citrus trees are grown. In some regions, in addition to citrus, the nematode also attacks grapevines, olive, lilac, and other plants. The life cycle of *T. semipenetrans* is completed within 6 to 14 weeks at 24°C. The male juveniles and adults do not feed and apparently do not play a role either in the disease or the reproduction of the nematode. The second-stage female juvenile, the only infective stage of the nematode, cannot develop without feeding, but it can survive for several years. In the soil, the citrus nematode occurs as deep as 4 meters (Inserra et al., 1994; Siddiqui, 1974; Van Gundy, 1985).

Lesion Nematode

The lesion nematode occurs in all parts of the world, where it attacks the roots of all kinds of plants, including cereals and other field crops, vegetables, fruit trees, and many ornamentals. Lesion nematodes reduce or inhibit root development by forming local lesions on young roots, which may then rot because of secondary fungi and bacteria. The life cycle of the various species of *Pratylenchus* is completed within 45 to 65 days. The nematodes overwinter in infected roots or in soil as eggs, juveniles, or adults, except for the egg-producing females, which seem to be unable to survive the winter. Adults and juveniles can infect and leave roots. The females, with or without fertilization, lay their eggs singly or in small groups inside infected roots. The eggs hatch in the roots or in the soil when released after root tissues break down. The emerging second-stage juvenile develops into the other juvenile stages and becomes an adult either in the soil or after it enters the root. When in the soil, the nematodes are susceptible to drying, and during periods of drought, they lie quiescent until the moisture increases and the plants resume growth (Loof, 1991; Zunke, 1991).

Burrowing Nematode

The burrowing nematode occurs widely in tropical and subtropical regions of the world and in greenhouses in Europe. *Radopholus similis* is the most important banana root pathogen in most banana-growing areas, where it causes the so-called banana root rot, blackhead toppling disease, or decline of banana. Host plants include avocado, tea, black pepper, coconut, coffee, corn, vegetables, grasses, weeds, and fruits. *Radopholus* spends its life and reproduces inside cavities in the root cortex, where it completes a life cycle in about 20 days. All juveniles and the females infect roots and can emerge from the roots and spread through the soil. Most of the spread of nematode from plant to plant, however, is through root contact or near contact. Long-distance spread of the nematode is primarily via infected plant materials. Fungi such as *Fusarium* and *Sclerotium* invade nematode-infected roots much more readily and further increase their rotting and destruction (Gowen and Queneherve, 1990; Williams and Siddiqui, 1973).

Stem and Bulb Nematode

The stem and bulb nematode *Ditylenchus* occurs worldwide but is particularly prevalent and destructive in areas with temperate climate. It is one of the most destructive plant-parasitic nematodes. It attacks a large number of host-plant parasitic nematodes. Host plants include onion, hyacinth, tulip, oat, alfalfa, and strawberry. Infective juveniles can withstand adverse conditions of freezing and extreme drying for long periods in fragments of plant tissue, in seeds, or in the soil. During favorable moisture and temperature, the preadult juveniles become active, enter the host, pass through the fourth molt, and become males and females. A complete cycle usually lasts about 19–25 days. Reproduction continues throughout the year. *D. dipsaci*, an internal parasite of bulbs, stems, and leaves, passes generation after generation in these tissues, escaping to the soil only when living conditions in the plant tissues become unfavorable. When heavily infected bulbs decay, juveniles move out and sometimes accumulate about the basal plates of dried bulbs as grayish-white, cottony masses called nematode wool, where they can remain alive for years (Hooper, 1972; Sturhan and Brzeski, 1991).

Interrelationships between Nematodes and Other Plant Pathogens

Although nematodes can cause diseases to plants by themselves, most live and operate in the soil, where they are constantly surrounded by fungi and bacteria, many of which can also cause plant diseases. In many cases, an association develops between nematodes and certain of the other pathogens. Nematodes become a part of an etiological complex resulting in a combined pathogenic potential that sometimes appears to be far greater than the sum of the damage that either pathogen can produce individually (Krupa and Dommercues, 1979).

Nematode–Fungus Disease Complexes

Several nematode–fungus disease complexes are known. *Fusarium* wilt of several plants increases in incidence and severity when the plants are also infected by the root-knot, lesion, sting, reniform, burrowing, or stunt nematodes. Similar effects have also been noted in disease complexes involving nematodes and *Verticillium* wilt, *Pythium* damping off, *Rhizoctonia* and *Phytophthora* root rots, and in some other instances. In none of these cases is the fungus transmitted by the nematode. However, plant varieties susceptible to the respective fungi are damaged even more when the plants are infected with the nematodes, the combined damage being considerably greater that the sum of the damages caused by each pathogen actin alone. Also, they apparently infect varieties ordinarily resistant to the fungi after previous infection by nematodes (Anonymous, 1972; Dropkin, 1980; Evans et al., 1993).

Nematode–Bacteria Disease Complexes

Relatively few cases are known. The root-knot nematode increases the frequency and severity of the bacterial wilt of tobacco caused by *Pseudomonas solancearum*, of the bacterial wilt of alfalfa caused by *Clavibacter michiganense* subsp. *insidiosum*, and of the bacterial scab of gladiolus caused by *Pseudomonas marginata*. In these cases, the role of the nematode seems to be that of providing the bacteria with an infection court and to assist bacterial infection by wounding the host. Interaction between some species of the seed-gall nematode *Anguina* and the phytopathogenic bacterium *Clavibacter toxicus* distorts or prevents normal formation of grass seed heads. The bacterium also produces corynetoxins, which are among the most potent toxins produced in nature and cause lethal neurological convulsions in most domestic animals fed infected grasses and seeds. The role of the nematode seems to be primarily that of a vector of the bacterium from plant to plant, and from year to year, and in facilitating entry of the bacterium into the host plant. It is not now known whether corynetoxins have any effect on the nematode (McKay and Ophel, 1993).

Nematode–Virus Disease Complexes

Several plant-pathogenic viruses, including grapevine fanleaf virus, tomato ringspot virus, raspberry ringspot virus, and tobacco rattle virus, are transmitted through the soil by means of nematode vectors. *Xiphinema*, *Longidorus*, and *Paratrichodorus* transmit only polyhedral viruses, which include most of the nematode-transmitted viruses, whereas *Trichodorus* and *Paratrichodorus* transmit two rod-shaped viruses, tobacco rattle virus and pea early browning virus. These can transmit some of the viruses after feeding on infected plants for as short a time as 1 hour, but the percentage of transmission increases with longer feedings, up to 4 days. Once they have acquired virus from an infected plant, the nematodes remain infective for periods of 2–4 months and sometimes even longer. All stages, juvenile and adult nematodes, can transmit viruses; however, the virus is not carried from one juvenile stage to another and to adults through molts, nor does the virus pass from adults, through eggs, to juveniles. Nematodes can only transmit certain viruses they ingest and carry within them to healthy plants, and this suggests a close biological association between the nematode vectors and the viruses they can transmit (Brown et al., 1995).

References

Agrios, G. N. (1997). Plant pathology (4th ed.). New York: Academic Press.

Anderson, A. L. (1948). The development of *Gibberella zeae* head blight of wheat. *Phytopathology* 38: 595–611.

Anonymous. (1972). *Commonwealth Institute of Helminthology descriptions of plant-parasitic nematodes*. Bucks, UK: Commonwealth Agricultural Bureau, Farnham Royal.

Anonymous. (1979). Symposium on *Sclerotina* (= *Whetzelinia*): Taxonomy, biology, and pathology. *Phytopathology* 69: 873–910.

Anonymous. (1991). Recent advances in *Fusarium* systematics: A symposium. *Phytopathology* 81: 1043–1067.

Baker, K. R., Carter, C. C., and Sasser, J. N. (1985). *An advanced treatise on Meloidogyne: Vol. 2. Methodology.* Raleigh, NC: Department of Plant Pathology, North Carolina State University, and U.S. Agency for International Development.

Beckman, C. H. (1987). *The nature of wilt disease of plants*. St. Paul, MN: APS Press.

Booth, C. (1971). *The genus Fusarium*. Kew, UK: Commonwealth Mycological Institute.

Bradbury, J. F. (1986). *Guide to plant pathogenic bacteria*. Kew, Surrey, UK: CAB International Mycological Institute.

Brasier, C. M. (1992). Evolutionary biology of *Phytophthora*. *Annual Review of Phytopathology* 30: 153–200.

Broadbent, L., Read, W. H., and Last, F. T. (1965). The epidemiology of tomato mosaic: X. Persistence of TMV-infected debris in soil, and the effects of soil sterilization. *Annals of Applied Biology* 55: 471–483.

Brown, D. J. F., Robertson, W. M., and Trudgill, D. L. (1995). Transmission of viruses by plant nematodes. *Annual Review of Phytopathology* 33: 223–249.

Bruehl, G. W. (1975). *Biology and control of soil-borne plant pathogens*. St. Paul, MN: APS Press.

Buczacki, S. T. (1983). *Plasmodiophora*: An inter-relationship between biological and practical problems. In: S. T. Buczacki, (Ed.), Zoosporic plant pathogens (pp. 161–191). London: Academic Press.

Burpee, L., and Martin, B. (1992). Biology of *Rhizoctonia* species associated with turf grasses. *Plant Disease* 76: 112–117.

Cadman, C. H. (1963). Biology of soil-borne viruses. *Annual Review of Phytopathology* 1: 143–172.

Chambers, L., and Corden, M. E. (1963). Semiography of *Fusarium* wilt of tomato. *Phytopathology* 53: 1006–1010.

Christensen, J. J., and Wilcoxson, R. D. (1966). Stalk rot of corn. Monograph No. 3. St. Paul, MN: APS Press.

Civerolo, E. L., Collmer, A., and Gillespie, A. G. (1987). *Plant pathogenic bacteria. Proceedings of the Sixth International Conference*. Boston: Martinus Nijhoff.

Colhoun, J. (1958). Clubroot disease of crucifers caused by *Plasmodiophora brassicae*. Commonwealth Mycological Institute. Phytopathological Paper No. 3, 1–108.

Dennis, C. (1983). *Post-harvest pathology of fruits and vegetables*. New York: Academic Press.

Dobson, R. L., and Gabrielson, R. L. (1983). Role of primary and secondary zoospores of *Plasmodiophora brassicae* in the development of clubroot in Chinese cabbage. *Phytophthora* 73: 559–561.

Dodd, J. L. (1980). The role of plant stresses in development of corn stalk rots. *Plant Disease* 64: 533–537.

Dropkin, V. H. (1980). *Introduction to plant nematology*. New York: Wiley.

Evans, K., Trudgill, D. L., and Webster, J. M. (1993). *Plant parasitic nematodes in temperate agriculture*. Wallingford, UK: CAB International.

Fry, W. E., Goodwin, S. B., Matuszak, J. M., Spielman, L. S., Milgroom, M. G., and Drenth, A. (1992). Population genetics and intercontinental migrations of *Phytophthora infestans*. *Annual Review of Phytopathology* 30: 107–129.

Fry, W. E., Goodwin, S. B., Dyer, A. T., Matuszak, J. M., Drenth, A., Tooley, P. W., Sujkowski, L. S., Koh, Y. J., Cohen, B. A., Spielman, L. J., Deahl, K. L., Inglis, D. A., and Sandlan, K. P. (1993). Historical and recent migrations of *Phytophthora infestans*: Chornology, pathways, and implications. *Plant Disease*: 77: 653–661.

Goto, M. (1992). *Fundamentals of bacterial plant pathology*. San Diego: Academic Press.

Gowen, S. R., and Queneherve, P. (1990). Nematode parasites of banana, plantains and abaca. In: M. Luc, R. A. Sikora, and J. Bridge (Eds.), *Plant parasitic nematodes in subtropical and tropical agriculture* (pp. 431–460). Kew, UK: CAB International.

Gray, T. R. G., and Williams, S. T. (1971). *Soil micro-organisms*. New York: Hafner.

Hooper, D. J. (1972). Commonwealth Institute of Helminthology Descriptions of Plant-Parasitic Nematodes, Set 1, No. 14, *Ditylenchus dipsaci*. St. Albans, UK.
Hwang, B. K., and Kim, C. H. (1995). *Phytophthora* blight of pepper and its control in Korea. *Plant Disease* 79: 221–227.
Ingram, D. S., and Williams, P. H. (1991). *Phytophthora infestans*: The cause of late blight of potato. *Advanced Plant Pathology* vol. 7.
Inserra, R. N., Voulas, N., and O'Bannon, J. H. (1994). *Citrus nematode biotypes and resistant citrus rootstocks in Florida*. Nematology Circular No. 205, Florida Department of Agriculture and Consumer Services, Division of Plant Industry.
Jee, H. J., Kim, W. G., and Cho, W. D. (1999). *Phytophthora* root rot of Chinese cabbage and spinach caused by *P. drechsleri* in Korea. *Plant Pathology Journal* 15(1): 28–33.
Jones, J. P., and Crill, P. (1973). The effect of *Verticillium* wilt on resistant, tolerant and susceptible tomato varieties. *Plant Disease Reporter* 57: 122–124.
Jung, Y. S., Kim, Y. T., Yoo, S. J., and Kim, H. G. (1999). Mycological characteristics of *Fusarium solani* f. sp. *pisi* isolated from pea, ginseng and soybean in Korea. *Plant Pathology Journal* 15(1): 44–47.
Krupa, S. V., and Dommercues, Y. R. (1979). *Ecology of root pathogens*. New York: Elsevier.
Lee, M. W. (1984). Disease problems in ginseng (*Panax ginseng*) caused by soil-borne microorganisms in Korea. Food and Fertilizer Technology Center (in Taiwan) Book Series 26: 120–136.
Lee, Y. H., Lee, W. H., and Lee, D. K. (1998). Screening for resistance of garlic cultivars to white rot caused by *Sclerotium cepivorum*. *Korean Journal of Plant Pathology* 14(6): 594–597.
Lee, Y. S., and Park, C. S. (2001). *Fusarium* species of Korea. Plant Pathogens of Korea No. 5. National Institute of Agricultural Science and Technology. 91 pp. Suwon, Korea.
Loof, P. A. A. (1991). The family Pratylenchidae. In: W. R. Nickle (Ed.), *Manual of agriculatural nematology* (pp. 336–421). New York: Dekker.
McKay, A. C., and Ophel, K. M. (1993). Toxicogenic *Clavibacter/Anguina* associations infecting grass seedheads. *Annual Review of Phytopathology* 31: 151–167.
Mclean, D. L. (1962). Transmission of lettuce mosaic virus by a new vector, *Pemphigus bursarius*, *Journal of Economic Entomology* 55: 580–583.
Moline, H. E. (1984). *Post-harvest pathology of fruits and vegetables: Postharvest losses in perishable crops*. University of California Publication Ne-87 (U. C. Bulletin No. 1914). University of California, Division of Agriculture and Natural Resources, Oakland, California.
Mount, M. S., and Lacy, G. H. (1982). *Phytopathogenic prokaryotes*, Vols. 1 and 2. New York: Academic Press.
Nelson, P. E., Toussoun, T. A., and Cook, R. J. (1981). Fusarium: *Disease, biology, and taxonomy*. University Park: Pennsylvania State University Press.
Parmeter, J. R., Jr. (1970). *Rhizoctonia solani*: *Biology and pathology*. Berkeley and Los Angeles: University of California Press.
Pegg, G. F. (1974). *Verticillium* diseases. *Review of Plant Pathology* 53: 157–182.
Powelson, M. L., and Rowe, R. C. (1993). Biology and management of early dying of potatoes. *Annual Review of Phytopathology* 31: 111–126.
Pullman, G. S., and DeVay, J. E. (1982). Epidemiology of *Verticillium* wilt of cotton: A relationship between inoculum density and disease progression. *Phytopathology* 72: 549–554.
Punja, Z. K. (1985). The biology, ecology, and control of *Sclerotium rolfsii*. *Annual Review of Phytopathology* 23: 97–127.
Raski, D. J. (1950). The life history and morphology of the sugar beets nematode *Heterodera schachtii*. *Phytopathology* 40: 135–152.
Raski, D. J., Hewitt, W. B., Goheen, A. C., Taylor, C. E., and Taylor, R. H. (1965). Survival of *Xiphinema index* and reservoirs of fan leaf virus in fallowed vineyard soil. *Nematologica* 11: 349–352.
Riggs, R. D., and Wrather, J. A. (1992). *Biology and management of the soybean cyst nematodes*. St. Paul, MN: APS Press.
Sasser, J. N., and Carter, C. C. (1985). *An advanced treatise on Meloidogyne*: *Vol. 1. Biology and control*. Raleigh, North Carolina State University, Graphics.
Siddiqi, M. R. (1974). *Tylenchulus semipenetrans*: Commonwealth Institute of Helminthology descriptions of plant-parasitic nematodes, Set 3, No. 34. St. Albans, UK.
Smith, P. R., Campbell, R. N., and Fry, P. R. (1969). Root discharge and soil survival of viruses. *Phytopathology* 59: 1678–1687.
Sneh, B., Burpee, L., and Ogoshi, A. (1991). *Identification of Rhizoctonia species*. St. Paul, MN: APS Press.

Song, J. H., Roh, S. H., Park, H. C., and Moon, B. J. (1998). Mycological characteristics of *Phytophthora nicotianae* var. *nicotianae* causing *Phytophthora* rot of strawberry and resistance of strawberry cultivars to the pathogen. *Korean Journal Plant Pathology* 14: 646–650.

Starr, M. P., Stolp, H., Trüper, H. G., Ballous, A., and Schlegel, H. G. (1983). Phytopathogenic bacteria. *The prokaryotes.* A handbook on habitats, isolation, and identification of bacteria. Berlin, Heidelberg and New York: Springer-Verlag.

Sturhan, D., and Brzeski, M. W. (1991). Stem and bulb nematodes, *Ditylenchus* spp. In: W. R. Nickle (Ed.), *Manual of agricultural nematology* (pp. 423–464). New York: Dekker.

Teakle, D. S. (1972). Transmission of plant virus by fungi. In: C. I. Kado and H. O. Agrawal (Eds.) *Principles and techniques in plant virology* (pp. 248–266). New York: Von Nostrand Reinhold.

Temmink, J. H. M., Campbell, R. N., and Smith, P. R. (1970). Specificity and site of *in vitro* acquisition of tobacco necrosis virus by oospores of *Olpidium brassicae. Journal of Genetic Virology* 9: 201–213.

Van Gundy, S. D. (1985). The life history of the citrus nematode *Tylenchulus semipenetrans. Nematologica* 3: 283–294.

Williams, K. J. O., and Siddiqui, M. R. (1973). Commonwealth Institute of Helminthology Descriptions of Plant-parasitic Nematodes, Set 2, No. 27, *Radopholus semilis*. St. Albans, UK.

Yarwood, C. E. (1960). Release and preservation of virus by roots. *Phytopathology* 50: 111–114.

Zunke, U. (1991). Observations on the invasion and endoparasitic behavior of the root lesion nematode *Pratylenchus penetrans. Journal of Nematology.* 22: 309–320.

4

Economics of Biodiversity

Linda Fernandez

Biodiversity, or biological diversity, is defined as variations in gene, species, and ecosystem habitat (terrestrial and aquatic) (Chapman, 1999). The extinction of plant and animal species results in loss of biodiversity. Humans have increased the rate of extinction through direct and indirect human activity encroaching on natural ecosystems. Direct activity consists of killing flora and fauna for consumption. Indirect activity consists of destruction of the habitat by (1) land conversion through tropical deforestation and draining of wetlands for agricultural use, (2) introduction of exotic species that do not have predators, and (3) pollution.

Ill-defined property rights and inefficient pricing of resources in which social benefits of biodiversity are not counted have fostered the direct and indirect activity. There is lack of recognition of value for humans to protect biodiversity. Ecosystem services have been "free" in the sense that their value has not been captured by the marketplace. An individual who either owns the area or does not is not thinking about the social benefits of preserving biodiversity if there is no avenue to reap the benefits.

Biodiversity is a public "free" good with nonexcludability and nonrivalry of local and global consumption. Public goods of ecosystems include functions performed by the ecosystem, including hydrological, biogeochemical, or nutrient cycles (Kormondy, 1996). For example, primary production, by providing food energy, drives all other functions of the ecosystem and all values that can arise from it (Mitsch and Gosselink, 1993). Similarly, the biodiversity contained in an ecosystem enhances its productivity, stability, and sustainability (Tilman, 1997).

Valuation

Economic value arises from the services that an ecosystem performs for society. These services may include straightforward extractive resource flows, such as timber and

Linda Fernandez • Department of Environmental Sciences, University of California, Riverside, CA 92501.

fish harvests, physical nonextractive services, such as flood protection and aquifer recharge, and nonphysical services, such as aesthetics or cultural value.

Many ecologists and resource economists have looked at the value of ecological services and function. For example, according to Daily (1997), who provides a detailed discussion of wetland ecological and human services, wetlands provide at least three major categories of ecological services:

1. *Geohydrological:* floodwater storage and conveyance, groundwater recharge, pollution assimilation, sediment trapping, and nutrient cycling.
2. *Habitat/production:* vegetation products; fishery, fur-bearer, bird habitat, and production.
3. *Ecosystem integrity:* natural open space, micro- and macroclimate regulation, carbon cycling, and biodiversity storehouse to provide resiliency.

Ecological services may support direct uses by humans as well as indirect passive uses. As such, ecological service value "flows" to these uses. For example, the floodwater buffering capacity of a wetland may provide valuable storm protection for off-site property owners downstream from the flood zone. In addition, the wetlands may provide pollution assimilation and sediment trapping functions that enhance water quality, supporting downstream water uses such as swimming or municipal drinking water supply. Furthermore, the habitat function of a wetland may support recreation such as hiking, bird watching or hunting both locally and elsewhere; for example, along a flyway. Finally, passive uses related to aesthetic appreciation or spiritual rejuvenation may occur and be highly valued by such users. In other words, for the wetland example, major categories of natural resource services to humans include but are not limited to the following:

1. *Recreational:* beach use/swimming, fishing, boating, wildlife viewing (including bird watching), and hunting.
2. *Commercial/public or private:* waterway navigation, drinking water, aquaculture, agricultural irrigation, commercial process water, property protection from storms, commercial and subsistence fishing and hunting, and timber harvest.
3. *Cultural/historical:* religious and cultural ritual uses, historical research.
4. *Health:* reduction in mortality and morbidity (e.g., through provision of clean water, air, and food).
5. *Scientific:* research promoting increased productivity in commercial and other activities, and the development of life-saving pharmaceuticals.
6. *Passive use:* species, habitat, and ecosystem protection for existence and for bequest value.

The first five categories of services represent *direct uses* by humans. The last category, passive or nonuse value, captures the notion that individuals may value natural resources independent of direct uses. Individuals may value the services provided to others or the protection of natural resources for their own sake (existence value), or as bequests (option value) to future generations. For example, the public may place value on the ability of an ecosystem to support endangered species, though the public never sees these species.

The variety of ecosystem services is illustrated in the rows of Table 4.1. For example, gas regulation is the regulation of the chemical composition of the atmosphere; climate regulation refers to temperature and precipitation regulation, and so on. Each of these services has a value to the human economy, as changes in the natural environment alter production processes, costs and amenities, as well as social welfare. The economic value of

Table 4.1. Methods for the Economic Valuation of Global Ecosystem Services

	Valuation technique					
Ecosystem service	Replacement cost	Avoided cost	Factor income	Travel cost	Hedonic pricing	Contingent valuation
Gas regulation		X	X			
Climate regulation		X	X		X	X
Disturbance regulation	X	X	X			
Water regulation	X	X	X	X		
Water supply	X	X	X	X	X	X
Erosion control	X	X	X	X		
Soil formation	X	X	X			
Nutrient cycling	X		X			
Waste treatment	X	X	X	X	X	X
Pollination	X		X			
Biological control	X	X	X	X		X
Refugia	X		X	X	X	X
Food production	X	X	X			
Raw materials	X	X	X			
Genetic resources	X	X	X	X	X	X
Cultural						X

Source of ecosystem services, de Groot (1992).

an ecosystem is related to its ability to provide goods or services of direct or indirect value to society. However, instances in which ecological values have been recognized in market transactions are the exception rather than the rule at present. Many ecosystem services have not qualified for market trading because they are not "private" in nature. For example, flood protection services of wetlands accumulate to everyone in the floodplain; no single recipient can be billed for the service provided.

How does one discover value inherent to wetlands or other habitats? Traditionally, land appraisals have been employed to evaluate land values in real estate transactions. Yet appraised values for wetlands usually consider only one component of value: the market value of an interest in wetlands. From an ecological perspective, however, there is the need to consider the total economic value of a wetland irrespective of interest, in order to begin to capture the service values identified previously.

Table 4.1 highlights six major ecosystem service valuation techniques in the columns when there are no adequate direct market valuations (prices determined via supply/demand mechanisms) to reference.

When there are no explicit markets for services, we must resort to more indirect means of assessing values. A variety of valuation techniques can be used to establish the WTP—willingness to pay or WTA—willingness to accept for these services (Farber et al., 1999). Table 4.1 highlights six major ecosystem service valuation techniques (in the columns) when there are no adequate market valuations:

1. *Avoided cost* (AC): services allow society to avoid costs that would have been incurred in the absence of those services; for example, wetlands flood control avoids property damages, or waste treatment perfomed by wetlands avoids health costs (Boardman et al., 1996).

2. *Factor income* (FI): services provide for the enhancement of incomes through use of the services as an input to produce a good sold in the market; for example, an acre of Chesapeake Bay wetlands is a factor input used to create or enhance jobs to produce and sell blue crabs for local income (Anderson and Rockel, 1991; Lynne et al., 1981).
3. *Travel cost* (TC): service demand may require travel, whose costs can reflect the implied value of the service; for example, hunting or recreation areas attract distant visitors whose value placed on that area must be at least what they were willing to pay to travel to it (Boardman et al., 1996).
4. *Hedonic pricing* (HP): service demand may be reflected in the incremental prices people will pay for associated goods; for example, housing prices at coastal locations exceed prices of inland homes in the same area (Boardman et al., 1996).
5. *Contingent valuation* (CV): service demand may be elicited by posing hypothetical scenarios that involve some valuation of alternatives; for example, asking people what they would be willing to pay for increased fish catch or deer bag (Boardman et al., 1996).
6. *Replacement cost* (RC): estimates of what it would cost to replace ecosystem services with man-made systems; for example, natural waste treatment can be replaced with costly treatment systems. The difficulty with replacement cost is that, in some circumstances, it may have no relationship to economic value; for example, the service whose cost of replacement is being evaluated may have no use to society, or its use has less value than the cost of replacement (Anderson and Rockel, 1991). Even with these limitations, RC can help establish useful valuation baselines.

Each of these methods has its strengths and weaknesses. Also, each ecosystem service has an appropriate set of valuation techniques that can be applicable to it. Some services may require that several techniques be used jointly. For example, the recreational value of an ecosystem includes not only the value that visiting recreationists place on the site (TC), but also the increased incomes associated with use of the ecosystem as a factor input to produce fish which are sold in the market for income (FI). It is important that use of alternative methods does not lead to double-counting. For example, one could estimate the value of an upstream wetland by either calculating the cost savings inherent in not needing to build levees (AC) or comparing the prices of otherwise similar lands that either are or are not protected from flooding by wetlands (HP). Because both techniques measure the same service, flood protection, only the most accurate estimate should be employed to avoid double counting value.

Valuation Techniques by Ecosystem Service

This section provides some guidance for determining the appropriate valuation techniques for individual ecosystem services presented in Table 4.1. Valuation techniques are presented by type of ecosystem service. Descriptions of ecosystem services are derived from de Groot (1992).

Gas Regulation

This service regulates atmospheric chemical composition, including CO_2, O_3, SO_x and NO_x levels, for example, which determine the chemical condition of our atmosphere.

Life exists within a narrow band of chemical balance, and alterations in that balance can have positive or negative impacts on natural and economic processes. The way in which forests sequester atmospheric carbon is a clear example of an ecosystem service. Such services may be evaluated using AC, RC, or FI approaches. For example, using trees to sequester carbon avoids the economic damages that excessive CO_2 levels can engender (AC); natural carbon sequestration avoids replacing that service with technology-based reductions in CO_2 emissions (RC). Or increases in crop yields resulting from alterations of atmospheric chemical composition can offset the need to use purchased fertilizer as an input to produce the crop, leading to increases in net income levels among affected farmers (FI).

Climate Regulation

This ecosystem service controls atmospheric temperature, precipitation, and humidity. Gas regulation plays a role in this service, but microclimate contributions to regional airsheds also play a role in determining climate at all scales. The economic value of this service stems from the role of climate in production and consumption of goods and services. For example, temperature and precipitation affect crop yields. Marginal increases in these services may avoid replacing the service through irrigation (RC); for example, they may increase overall crop production (FI). Reducing excessive heat conditions reduces the costs of heat-related illnesses (AC) or avoids the need to replace the service with air conditioning (RC). Regional climate conditions may be attractive enough to be reflected in increased property values (HP), and people may be directly solicited to determine their WTP for these preferred climate conditions (CV).

Natural Disturbance Regulation

This service refers to the ability of an ecosystem to moderate potentially disturbing natural events. For example, vegetative structures can alter potentially catastrophic effects of storms, floods, and droughts through surface resistance and attenuating storage capacity. Forests can moderate wind velocities, avoiding property damages from storms (AC), and act as sponges for storing water in storm events, reducing the needs to replace this service with downstream flood control projects (RC). Vegetative cover may moderate drought cycles and increase the level or stability of grazing incomes to ranchers and farmers (FI).

Water Regulation

This service relates to the function of the hydrologic component of ecosystems. For example, plants uptake water and evapotranspire, which facilities movement of water from soils into the atmosphere. Water and climate regulation are interrelated, as vegetation effects on temperature and humidity consequently affect precipitation. This service can be distinguished from disturbance regulation insofar as it refers to normal conditions rather than extreme events. Improvements in normal hydrologic services of ecosystems can improve soil moisture and reduce chronic drought conditions, affecting agriculture and silvaculture crop yields, which results in irrigation replacement cost savings (RC) or increased yields for which water regulation is an input to derive income from produced (FI). Improving forest cover can improve stream flow regulation, increasing commercial (FI) and recreational fisheries (TC).

Water Supply

This service refers to the storage and retention of water in vegetation, streams, lakes, and aquifers. It is related to the role of ecosystems in hydrologic cycles but focuses on storage capacity rather than the flow of water through the system. Its economic value stems from filtration and storage for potential future uses of water. Aquifers, for example, preserve a store of water that can be withdrawn when needed, hence avoiding replacement of this storage with more costly storage (RC). They provide a stock of usable water for agricultural production sold for income (FI). Adequate stores of water in lakes and streams improve their pollution dilution capacities, and help avoid health costs of pollution (AC) or the costs of more extensive waste treatment (RC). Lakes and streams provide a store of water necessary for recreational fishing (TC). Wild streams provide boating pleasure for which people would be willing to pay to maintain adequate water levels (CV). Lakes, streams and oceans have social amenity values for which people are willing to pay for through higher property values (HP).

Erosion Control

Ecosystem structures can reduce soil erosion. For example, tree roots enhance soil stability, and leaf litter can moderate raindrop impacts on forest floors. Preserving soils also preserves soil nutrients, which avoids costs of replacing these nutrients (RC) and also enhances agricultural and silvacultural incomes (FI). Reducing soil erosion in watersheds also increases impoundment lifetimes (AC), improves water quality and resulting recreational value (TC), reduces municipal water supply costs (AC), and improves the flood retention capacity of streams and lakes (AC).

Soil Formation

Soil is formed through the disintegration of rock over time and becomes fertile through the accretion of animal and plant organic matter, and via sediment replenishment. Soil becomes a habitat for microorganisms that are essential to nutrient cycles. Soil itself supports a variety of services, including (Daily, 1997): (1) disturbance regulation for hydrologic cycles, (2) physical support of plants, (3) storage and delivery system for nutrients, (4) treatment of organic wastes, and (5) regulation of atmospheric gases included in, for example, the carbon and nitrogen cycles. Soil formation is then valued in relation to all these interrelated services.

Nutrient Cycling

The flows of nutrients, such as nitrogen and phosphorous, are critical to maintaining plant and animal life. Alterations in the nutrient cycling services of ecosystems then alter the viability and productivity of plants and animals. Trees play critical roles in nutrient cycling by taking nutrients from soils, then replacing some of the nutrients with decaying organic matter from their own litter drop. If nutrient cycles are disrupted, lost soil nutrients must be replaced through fertilizer applications (RC). Nutrient cycle disruptions may affect incomes by reducing agricultural and silvicultural yields (FI).

Waste Treatment

Ecosystems perform natural waste treatment functions through dilution, assimilation, and chemical recomposition. In fact, pollution exists only when the naturally occurring waste treatment capacity of the ecosystem has been exceeded. For example, wetlands can be used to treat wastes from human and economic processes, thus avoiding replacement of those services with more costly treatment systems (RC), and avoiding adverse health effects (AC), as well as impacts on commercial fisheries (FI) and recreational uses (TC and CV). Alternative, man-made treatment systems are not only costly but also have social impacts reflected in changes to nearby property values (HP).

Pollination

This pervasive service is necessary for the maintenance of plants and associated animals. These natural services enhance commercial incomes in agriculture (FI), avoid the necessity of replacement with man-made pollination activity (RC), and are critical to maintaining the stock of genetic resources (RC), another ecosystem service.

Biological Control

This service refers to the ability of an ecosystem to control the sizes and distributions of biological populations through mechanisms such as predator–prey interactions and natural pest control mechanisms. The result is a balance of species appropriate to the ecosystem at hand. For example, excessive populations of deer can impose agricultural and silvicultural crop damage, imposing damage costs (AC), reducing incomes (FI), or requiring costly methods of artificial population control (RC). Underabundance of hunted species, such as elk or bear, reduces recreational hunting values of ecosystems (TC, FI, and CV).

Habitat, Refugia/Sanctuary

Ecosystems provide habitat for populations. Such habitat protects and preserves gene pools from which individuals may later migrate to repopulate exogenous areas. Furthermore, these habitats may offer sanctuary for species suffering from exogenous stress. This is a pervasive ecosystem service touching all biologically-based values to the human economy. Wetlands provide refugia (sanctuary) for commercial (FI) and recreational species (TC and CV). The recreational value is associated with viewing and hunting of wildlife. Wetlands refugia services can be replaced at a cost (RC) (e.g., via zoos or botanical gardens). People pay higher property prices to be near an ecosystem with abundant flora and fauna (HP). Because it is so pervasive, this service can be evaluated with most of the valuation techniques. Existence value can be measured through CV. The dollar amount can then be mutiplied by the number of acres that share the species. The Habitat Evaluation Procedure (HEP) has been used to identify these species, and the Habitat Suitability Index (HSI) actually measures the viability of a species, given information about the quality of the supporting habitat.

Food Production

This service refers to the obvious value of ecosystems as sources of food for humans and is attributable to other services listed earlier, such as nutrient cycling, pollination, and refugia. Man-made food production systems, such as hydroponic gardening and aquaculture, can replace natural systems but at an incrementally higher cost (RC). Reductions in this service capacity result in nutrition-related health costs (AC). Commercial use of ecosystems for food production creates incomes (FI).

Raw Materials

Like food production, extractive raw materials are a critical ecosystem service to economies. For example, the availability of timber resources avoids the need to find substitute building materials (RC). Available timber for shelter and heating permits low-cost housing (HP) and avoids health costs associated with inferior shelter and exposure (AC).

Genetic Resources

This biodiversity service has value insofar as natural systems maintain a stock of genetic information that can be used for a variety of economic purposes, such as medicine, pest control, development of ornamental and commercial plants, and so on. Nature-based medicines can help avoid health costs (AC), may be cheaper than man-made alternatives (RC), and their extraction and development can create incomes in markets (FI). The variety of genetic resources embodied in plants and animals (biodiversity) is an attractive curiosity to humans, who would pay to visit biologically diverse sites (TC) or live adjacent to such sites (HP). The stock of genetic resources has within it a multitude of unrevealed secrets for which society would pay to preserve the option to determine those secrets in the future [e.g., via bioprospecting (CV)].

Cultural

Nature is a basis for folklore and culture as humans have developed means of coping and interacting with nature. In other words, human culture is embedded in natural systems. These broad cultural values are intrinsic, often not explicit, and are particularly difficult to value. Hypothetical, or scenario-based valuation methodologies (CV) may be the only means of assessing these broad cultural values.

The term *cultural* is used to represent a broad array of services supportive of the cultural character of human communities. For example, ecosystems provide recreational nonuse, such as aesthetic value. Furthermore, there is a need to measure the value people place on the knowledge that a resource exists for its own sake (existence value) or for the sake of future generations (bequest value). By surveying people through CV, there is a means of eliciting values for both. There are historical and archeological use values, too; for example, an ecosystem could contain sites evidencing prior settlements or other human uses—sites of significant importance to society. There is a means of gauging such value through HP by measuring the incremental price someone is willing to pay to live near a historical site, or through the TC method, by the price someone pays to travel to visit the site. Last, ecosystems provide research and educational use; that is, they provide opportunities for scientific research and observations of nature and its processes. The value in this use may be measured through TC or CV methods.

Now that the ecosystem services and valuation techniques have been defined, ecosystems can be identified for their complete value, noting several components of this value (Chapman, 1999). Total environmental value (TNV) of an ecosystem is the summation of two components, nonhuman value (NHV) + total economic value (TEV). The following notation can be used in a summation equation:

$$\text{TEV} = \text{DUV} + \text{IUV} + \text{OV} + \text{DSV} + \text{BV} + \text{EV}$$

DUV = direct use value = consuming resource
IUV = indirect use value = production of good
OV = option value = future use
DSV = discovery value = undiscovered resources
BV = bequest value = preservation for future generations
EV = existence value = survival of species

$$\text{Use Value} = \text{DUV} + \text{IUV} + \text{OV} + \text{DSV}$$

$$\text{Passive Nonuse Value} = \text{BV} + \text{EV}$$

The following arrows depict connections underlying the ecosystem features and how policy and economics attempt to recognize these features.

Ecological Process → Ecosystem Function → Ecosystem Service →
% Capture → Value

Ecological Process → Ecosystem Function

Ecosystem functions are the result of interactions among characteristics, structure, and processes of the ecoystem (Turner et al., 2000). Use of a tool such as the hydrogeomorphic model (HGM) establishes a function profile of physical, chemical, and biological characteristics of ecosystem. In the following policy section, the HGM is discussed, with an example of its use in policies to protect wetlands.

Ecosystem Function → Ecosystem Service

Use of a tool such as the HEP relies on use of indicators for ecosystem health. Indicators of the health or integrity of ecological sites provide the only way by which the success of environmental policies can be measured (Haskell et al., 1992). In the following policy section, the HEP is discussed, with an example of wetlands policies.

Ecosystem Service → % Capture

Determine property rights or institutional arrangements to distinguish between public and private ownership of ecosystem area.

Ecosystem Service → % Capture → Value

Multiply units of service times percentage capture times a derived dollar value per unit of service from a valuation technique.

The economic value of an ecosystem is related to its ability to provide goods or services of direct or indirect value to society. However, instances in which "ecological" values have been recognized in market transactions are the exception rather than the rule at present. Most properties continue to reflect value based only on their ability to provide private goods such as timber or housing sites. Thus, land transactions typically occur in market settings between individual parties, and prices for such transactions reflect only values that can be "captured" through regular market exchanges.

Markets could provide capture mechanisms for ecological services if there were a better appreciation for, and means of measuring, the ways in which these services underpin more traditional economic activities. Market capture mechanisms work relatively well for "private" goods, where owners can deny access to the service if payments are not made and if making access available to one person essentially makes it unavailable to others. The private value of an ecosystem can be represented with the following formula:

$$\text{Private Value} = (\text{Social Value}) \times (\text{Private Capture Ratio}),$$

where social value is the full value to society of an ecosystem and the private capture ratio is the percentage of this value that can be apprehended by an individual resource owner. It may not be possible for private parties to capture all social values of owned ecosystems. The ability to capture depends on existing markets and institutions that may allow some capture. Marketable commodity values of ecosystems are capturable if ownership of those commodities is well established. Timber is a good example of value capturable through markets. Recreational values of properties may be marketable where entry can be restricted. By setting up a formal park with some fee for entry to recreate, the property owners can earn some recreational value in revenues. Likewise, as marketplace signals gradually define the value of ecosystem services such as atmospheric gas regulation (related to the regional balance of sulfur and nitrogen oxides, or the global balance of carbon dioxide), or water filtration and storage (related to wetlands), the capture certainty of these values grows significantly.

The principle that wetlands generate economic benefits and services shared by all of society underpins policies and programs aimed at conservation of biodiversity through ecosystem protection. Capturing this value in exchangeable or marketable currency and convincing society of the monetary currency equivalent of "natural currency" is a challenge. Because wetlands and other ecologically valuable lands are often under private ownership and control, private landowners may frame less than altruistic questions. In exchangeable currency, what is the value of a wetland or endangered species habitat? How much should landowners be compensated if they choose to retain and preserve wetlands? How should such compensation be arranged?

Property Rights Policies to Protect Biodiversity in Ecosystems

Because human activities place these services in jeopardy, policy intervention helps protect biodiversity to reap social benefits and pays for the right price to do so. One realm of policy for biodiversity protection relates to property rights. Property rights need to be established either over geographic domain or over access for use. Property rights for access take the form of designated time and range of spatial access, and should have a market price based on potential revenue stream from future discoveries and replication (use and discovery value). Public involvement in these rights through joint funding and information dissemination would lead to social ownership and benefits. Within public property (park), we therefore enforce and monitor nonconsumptive use (no removal of species)—only sightseeing, recreational use.

Another method of capture is through bargaining arrangements based on specified property rights for ecosystem services, where beneficiaries negotiate payments with service owners. For example, a community may be willing to pay for the groundwater recharge services of a wetland, even though arranging such payment across a large number of beneficiaries may be quite difficult. Or the owner of a golf course may be willing to pay adjacent or downstream wetland owners to treat pesticide runoff as an alternative to on-site treatment.

A new mechanism has evolved to capture the relationship between ecosystem services and regional or even global service users. For example, drawing on the need for institutions to protect themselves from unpredictable environmental impacts such as those related to adverse climatic conditions, catastrophe futures are now arising as an important vehicle for insuring against such events. "Securitizing environmental risk" relates to the ability of insurance providers to guarantee services in the face of significant potential impacts to a business enterprise.

There is a need to remove adverse subsidies that create a profitability gap in favor of activities that damage biodiversity. In this realm, it would be helpful to change the pattern of approving and financing land conversion in forms of direct payments, credit guarantees, tax exemptions, technical assistance, and the provision of below-cost infrastructure allowing for easy access to biodiversity-rich areas. Examples of policies that could be reversed for their negative impact to biodiversity include U.S. federal and state incentives to drain wetlands: (1) the mid-1800s Swamplands Act enabled states to eliminate 65 million acres of wetlands for agriculture; (2) the U.S. Dept. of Agriculture provided funds for draining wetlands for production of subsidized crops (with farm price and income support programs) through the 1980s; (3) the U.S. Army Corps of Engineers has rechanneled rivers for agricultural drainage; (4) the IRS has an Internal Revenue Code favoring wetland conversion investments through 1986.

As a result of these policies, 125 of the 215 million acres of wetlands in the continental United States have been lost.

There are examples of policies that act in the opposite manner, and instead focus on incentives to protect biodiversity.

Tax Programs

In Canada, a 1996 income tax act exempts from capital gains tax all donations of biodiverse lands made in perpetuity to government and charities. This encourages voluntary, nonregulatory stewardship measures by landowners for biodiversity conservation. In The Netherlands, since 1995, investments in green funds are tax-free, allowing investors in green projects to contract loans at reduced interest rates, usually 2% less than commercial rates.

Trust Funds

Most trust funds are endowments, in which trustee can spend the interest and dividends, but not the principal. This assures the continuity of funds for an indefinite period.

In Australia, the Revolving Fund for Nature finances purchases by the Trust for Nature Victoria of ecologically sensitive lands. Once land is purchased, the Trust draw up a legally binding covenant specifying restrictions and conservation activities required to maintain the habitat. The Trust resells the land to owners that use land in a manner dictated by the covenant. The returned capital is then used to purchase further lands for covenanting and resale.

Belize, Namibia, and Papua New Guinea have trusts. Funds come from a "conservation fee" charged to all arriving foreign visitors. For example, since 1996, Belize has a fee of $3.75, which raises approximately $500,000 per year. The funds are used for training park rangers, biodiversity monitoring, and environmental education.

The World Bank Trust Fund for Endangered Species has devoted $4.3 million of the fund that has stock and bond listings in the London Stock Exchange to help finance Uganda's mountain gorilla parks in perpetuity.

Debt for Nature Swaps

One way to relieve debt burden and simultaneously invest in resource conservation of biodiversity is through debt for nature swaps (DNS), which are contracts between two or more parties to enable conservation or the provision of environmental services. One party to the contract [government, nongovernmental organization (NGO) or bank] will pay off a portion of the loan, typically at a deep discount. Banks are willing to sell this debt at a paper loss, because they suspect they will not be repaid in full anyway. In exchange, another party to the DNS must agree to invest a certain amount of money into a resource conservation program in their country. Private DNS are negotiated by and use funds raised by private international conservation organizations such as Conservation International, the Nature Conservancy, and the World Wildlife Fund. Public DNS are negotiated by governments of developed countries, for example, Sweden, The Netherlands, and the United States.

The DNS tend to focus on increasing enforcement of preservation goals by prohibiting resource destruction and supporting sustained-yield resource development at existing preserves (Deacon and Murphy, 1997). The actual ownership of the natural resource does not change hands. Ownership of forests and other resources that are targeted in DNS is being established or strengthened where it formerly was weak or nonexistent because through the DNS, the benefits of enforcing ownership outweigh the costs.

DNS are one way in which residents of rich countries can express their demand for resource preservation in poor countries. An example is the impact they have had on overall debt: For 1992, DNS have reduced total debt in the Third World by around $400 million, well under 1% of total Third World debt of $1.3 trillion. About one-fourth of these swaps have been financed by private parties, primarily conservation NGOs. The rest of the efforts have been paid for by rich-country governments, mostly the United States.

Transferrable Development Rights

The premise behind transferrable development rights (TDRs) is that biodiversity is a land-use issue, because it depends on existence of habitats (Panayotou, 1992). Owners of land that is ideal for preservation can sell development rights to developers. The revenue from selling these rights compensates the owners for their inability to develop their land.

One example of TDRs in practice is in the New Jersey Pinelands, 1 million acres of marshes that support several endangered species. The Pinelands Development Commission created Pineland Development Credits (PDCs), a form of TDR. Landowners in environmentally sensitive marsh areas receive PDCs in exchange for limiting development at the rate of one PDC for every 39 acres of existing farmland, upland, or wetland. To create a demand for these credits, developers seeking to increase the standard density on land zoned for development are required to acquire one PDC for every four units of increase. The program involves a planner facilitating the market and designating sites for protection and development. The TDRs are a land management device that severs the development potential from the land and treats it as a separately marketable commodity. TDRs introduce divisibility and privatization of the good (habitat conservation). The Akamas Peninsula, Cyprus also uses TDRs.

TDRs are offered for sale both locally and internationally at an initial offer price, including the opportunity cost of the land (net present value of income from foregone development opportunity).

Ways of enhancing demand for TDRs through credits and offsets resemble the Global Environmental Facility that transfers TDRs out of critical environments in developing countries, with compensation by developed countries for the foregone development opportunities. Figure 4.1 indicates how developed countries can demand Q* of biodiversity by paying (through donation, conservation tax, contract with land manager like GEF). P* indicates the optimal price that takes into account the intersection of international demand with host country supply (marginal costs of preserving land and biodiversity Q*).

A Conservation Tax Can Generate Demand for TDRs

Tax revenues per year equal the country's total willingness to pay for habitat conservation, which equals the price of TDRs. Within the country, an individual can choose between paying conservation tax or purchasing a TDR. One's tax payment depends on the changes in the value of property and/or income according to tax base. A TDR purchase ends tax payment but incurs an opportunity cost of interest foregone on funds used to purchase the TDR minus (or plus) any appreciation (depreciation) of the value of the TDR that one holds.

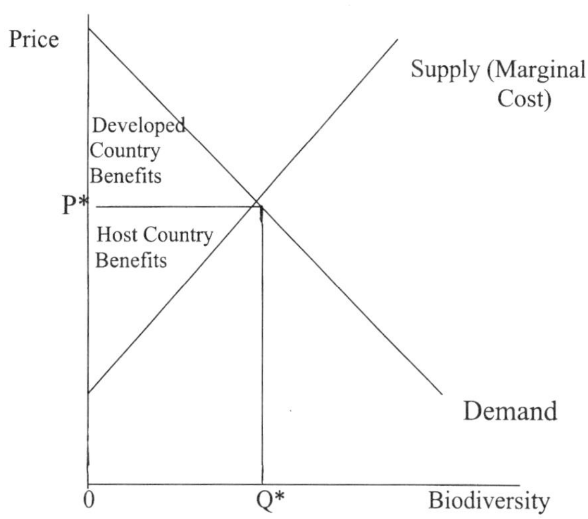

Figure 4.1. Demand and supply of biodiversity in an international context.

Owners of TDRs as shareholders in a habitat are entitled to dividends and/or earn capital gains on their appreciating TDRs, which will increase demand as increase species in the relevant habitat prove to be highly valuable.

Migratory Species Program Agreements between Several Countries

Biodiversity extends beyond borders. Several countries collaborate in formal financial and technical resource arrangements to maintain shared ecosystems. For example, the trilateral arrangement between Canada, Mexico, and the United States for migratory birds has an institution, the Commission on Environmental Cooperation, with three countries as decisionmakers and signatories to the North American Bird Conservation Initiative and North American Waterfowl Management Plan. Financial assistance, satellite tracking, breeding material, habitat preservation, and restoration and technical monitoring resources are shared along Pacific Flyway and other North American habitats for protection, restoration, and management (government, agencies, conservation groups, hunting associates, and landowners participate).

Breeding grounds in Canada and Alaska are maintained for nesting habitat, and the southern destinations in winter are maintained for food and growth habitat. Census data are jointly collected to monitor changes in population over time and space.

International Agreements

The Convention on Biological Diversity (1992) has three central features:

1. Nations are required to inventory and monitor their biological assets.
2. Sustainable conservation efforts (international funds from 0.7% of developed countries' annual gross domestic product (GDP), an increase from 0.4%, or from $55 billion to $98 billion).
3. Tropical nations have intellectual property rights to genetic material gathered from their land (forests) and regulation of trade of genetically engineered products.

The exception is when the nation agrees to a profit-sharing mechanism with a bioprospecting company. A total of 158 countries signed the Agreement (not including the United States).

Although the treaty is symbolic, it lacks the following:

1. Requirements for payment of the conservation funds.
2. The financial mechanisms to channel these funds.
3. Enforcement mechanisms to stop violations.
4. Difficulty in assessing compliance.

In order for international agreements to be meaningful, the four limitations of the Convention on Biological Diversity will have to be overcome.

Mitigation Banking Policy

In November 1995, national guidelines for wetlands mitigation banks (WMBs) were approved to help modify Section 404 of the U. S. Clean Water Act. WMBs are aimed at introducing economic incentives into wetlands protection and changing the timing of restoration and development activities. Unlike the case of issuing a permit at the same time a developer agrees to minimize impacts and restore degraded wetlands, the use of WMBs requires that restoration take place before a permit for development is issued. The WMBs work in the following manner: In order to obtain permission (a permit) to develop a wetlands area, a developer is required to have credits obtained from investment in the completed restoration of a WMB site in the same watershed. The number of credits that a developer obtains per dollars of investment depends on the success of wetlands restoration and this success is uncertain. If successful, the developer is able to "cash in" credits in exchange for permission to develop other wetlands. Therefore, a developer's demand for credits in the WMBs is derived from the demand for future development projects. The use of mitigation banks, to satisfy the "no-net-loss" policy of the United States Environmental Protection Agency (EPA), derives from the Fish and Wildlife Service (FWS) Mitigation Policy (46 FR 7644-7663, 1981), as well as Section 404(b)(1) of the Clean Water Act. The use is at the end of a series of required actions to protect wetland resources, including the process of "sequencing." This is required prior to the use of an approved bank as compensation for permitted wetland losses. Sequencing is a process whereby wetlands compensation is the last consideration only after efforts of impact avoidance and, to the greatest extent practicable, impact minimizations are exhausted.

The WMBs policy is an attempt to address the lag time and uncertainty in any restoration effort (Fernandez, 1999). It passes the uncertainty onto the developer in terms of requiring restoration in advance of development impacts to reduce net loss of wetlands. Regulatory approval of a development project depends on the level of restoration credits in the developer's WMB account. Credits are denominated in habitat units (Hus) and are a measure of habitat value. The number of HUs is the product of the number of species, or functions per acre of a wetland, times the number of acres. The number of species or functions per acre is a measure of natural processes present in wetlands (hydrology, chemical transformation, flora and fauna production).

The large size of a WMB site may offer greater likelihood of success by incorporating more ecosystem diversity than smaller, fragmented sites that developers may try to restore on their own. The larger site can be used by multiple developers, who are able to take advantage of economies of scale that do not occur on smaller, fragmented sites. There is evidence of a 3.1% decline in costs per acre for each 10% increase in project size. The multiple investors in the WMB can pool financial resources, planning, and scientific expertise that might be prohibitively expensive if financed individually by a developer on his or her own site.

WMBs sites are determined through a process referred to as "advanced identification" (ADID), which is a means of setting watershed conservation priorities by identifying wetlands functions and sites for protection and restoration. It is widely used in California, Florida, and Louisianna. ADID produces maps of viable sites, assesses functions and values of wetlands in a watershed, and designs a restoration program. It identifies the most degraded, least valuable wetlands for future development and selects other sites for restoration. In this manner, the process promotes the goal of maximizing public benefits from wetlands, while satisfying the constraint of no net loss of wetlands functions.

There are substantial costs involved in ADID. In addition to the hard costs of moving dirt, planting vegetation, and maintaining a new wetland over time, several soft costs are also involved in wetland creation, restoration, and enhancement. The most common include fees for environmental engineering, design and project oversight, and attorney and public agency permits and other legal compliance. Other soft costs include those associated with the lengthy conversion and monitoring/maintenance process. Some of these costs, such as the design of a restoration program for a particular area, are unavoidable. The fact that the public, rather than the WMB investor, bears these costs, means that ADID constitutes an implicit subsidy.

Any developer who decides to invest in a WMB will be facing some costs for restoration prior to receiving any benefits (wetlands credits). So the decision involves dynamic optimization of expected benefits minus costs subject to constraints on wetlands recovery. The reference to expected benefits is due to the growth or recovery function of wetland having some randomness associated with it. The decision of how much to try and restore, and when to stop, is influenced by several factors. Certainly, the degree of randomness in the growth function matters along with the interest rate, the cost parameters, and the value of the wetlands credits. How are those credits really measured? Well, so far, there is no agreement on the best way to assign a dollar value. It might be based on a per acre dollar real estate value. If there are available data on the services provided by the wetlands (particular species in the habitat, potential water quality benefits, etc.), there may be a way to generate a dollar value by conducting a variety of economic valuation techniques. For example, to assign value to the service of waste assimilation that the wetlands provide, a cost of replacement approach can be used; that is, if costs associated with any technology needed to perform the same service in the absence of the naturally occurring wetlands, they can be assigned to the value of the wetlands offering that service. For other services, such as habitat for a commercially harvested fish, use CV or TC methods for eliciting values from commercial fisherman would be appropriate.

If the WMB policy has any effect, it is in curbing further impacts to wetlands by stopping development before any restoration is attempted. The restoration efforts are quite experimental and no one agrees that recovering an ecosystem that supports a variety of species and functions is an obvious outcome of money invested in WMBs (Zedler, 1990). But it is an attempt, with effort to maximize success through comprehensive planning of the best site and plan of action to take in the restoration project.

Analysis of WMBs can be used to study policies such as the Endangered Species Act (ESA) of 1973. The act designates critical habitat and habitat conservation and recovery plans for sufficient interconnected habitat acres to support species listed as threatened and endangered. The ESA sets forth a strong national mandate to protect and manage endangered species and their habitats. In 1982, Congress amended the ESA to authorize the FWS to issue a permit for impacts to endangered species habitat, providing that a permit applicant prepare a "habitat conservation plan" (HCP). Ideally, the HCPs avoid and mitigate the "take" of the species to the extent practicable, and maintain the likelihood of the survival and recovery of the species—throughout a region or over a wide area. The "taking" of species by these actions is required to be compensated through the HCPs, using the WMB strategy of tying indicator species to compensation credits.

The result is a growing emphasis nationwide on HCPs that balance the conservation of listed species, with limited development in endangered species habitat. In recognition that a given parcel may encompass many endangered species and habitats distributed over a wide area, support for mitigation banks (similar to WMBs) as a compensatory mechanism for wildlife impacts is growing. Mitigation banks provide such opportunities in upland areas, as well as areas covered by the Clean Water Act. In effect, as with WMBs, the natural resource values of conservation banks are sold or traded as credits to those who must compensate for "incidental take" of endangered species. More aspects of the WMB policy could influence endangered species policies by insistence on changing the timing of compensation to increase recovery of species faster than they are impacted by new development of land. This may overcome the problems associated with uncertainty involved in setting the thresholds that dictate which species are listed or delisted as endangered species.

Bank and trade policies have created new assets for wetlands and other ecosystem owners. They provide a mechanism for privatizing or capturing social values, if the credits are clearly tied to those values. Both programs are still being developed in terms of comprehensive measures for credits, which have become the unit of currency in valuing these ecosystems; there is room for improving the currency to reflect social value in a meaningful way.

The U.S. Army Corps of Engineers (ACOE) and the EPA recognize mitigation banks as an acceptable mechanism of compensatory mitigation. A mitigation bank facilitates the trade of natural resource values to those who must compensate for resource impacts on land elsewhere. *Credits* and *debits* are the terms used to designate the units of trade (i.e., currency) in mitigation banking. Credits represent the accrual or attainment of aquatic functions at a bank; whereas debits represent the loss of aquatic functions at an impact or project site. Credits are debited from a bank when they are used to offset aquatic resource impacts (e.g., to satisfy a Section 10 or 404 permit requirement).

In order to realize the full economic potential of a bank, however, a bank developer must meet some specific requirements:

1. The bank should be in the same U.S. Geological Survey (USGS) cataloging unit, as defined by the Hydrologic Unit Map of the United States (1980), or an adjacent cataloging unit within the same river watershed as the impacted site.
2. The bank should be ecologically preferable to practicable on- and off-site individual mitigation options, as defined by federal wetland regulations.
3. The banking instrument, should be approved by a process that includes public review and comment (1972, c. 711, Section 62.1-13.3; 1982, c. 300; 1992, c. 836; 1996, c. 736.)

Banks can be on public or private lands. Cooperative arrangements between public and private entities to use public lands for mitigation banks may be acceptable. In some circumstances, it may be appropriate to site banks on federal, state, tribal or locally owned resource management areas (e.g., wildlife management areas, national or state forests, public parks, and recreation areas). The siting of banks on such lands may be acceptable if the internal policies of the public agency allow use of its land for such purposes and the public agency grants approval. Mitigation credits generated by this type of bank would supplement the public program(s) already-planned or in place. Baseline values represented by existing or already planned public programs, including preservation value, however, do not count toward bank credits.

In general, banks that involve complex hydraulic engineering features and/or questionable water sources (e.g., pumped) are more costly to develop, operate, and maintain. They also have a higher risk of failure than banks designed to function with little or no human intervention. Long-term maintenance requirements may be necessary and appropriate in some cases (e.g., to maintain fire-dependent plant communities in the absence of natural fire, to control invasive exotic plant species, etc.).

When uncertainties surrounding the technical feasibility of a proposed mitigation technique exist, appropriate arrangements (e.g., financial assurances, contingency plans, and additional monitoring requirements) should be in place to increase the likelihood of success. Such arrangements may be phased out or reduced when the attainment of prescribed performance standards is demonstrated.

Proposed uses of a mitigation bank are most appropriately made within the context of a comprehensive watershed plan. Such watershed planning efforts often identify categories of activities having minimal adverse effects on the aquatic ecosystem allowable under a general permit. In order to reduce the potential cumulative effects of such activities, it may be appropriate to offset these types of impacts by a mitigation bank established in conjunction with a watershed plan.

Credit may be given when existing wetlands are preserved in conjunction with restoration, creation, or enhancement activities, and when it is demonstrated that the preservation will augment the functions of the restored, created, or enhanced aquatic resource. The total number of credits available from the bank may reflect such augmentation. Preservation is implemented through appropriate legal mechanisms (e.g., transfer of deed, deed restrictions, conservation easement) to protect wetlands, accompanied by changes in land use or other physical changes (e.g., installation of restrictive fencing, etc.).

Credit pricing reflects both regulatory and financial considerations, such as bank supply/demand, cost of land, and bank development costs. Considerations include the following:

- Fair market value or current value
- Site acquisition transaction costs
- Planning and permitting costs
- Engineering design costs
- Equipment costs
- Cost of water for enhancement
- Cost of fencing and other physical habitat protection
- Administration, operation and maintenance costs
- Taxes, insurance, and finance charges
- Adjustments for inflation
- Long-term monitoring and reporting costs, and so on

The operation of successful, commercial wetlands credit markets requires the interaction of three agents: credit suppliers, permit applicants, and regulators. The objective of credit suppliers is to maximize the rate of return on investments in wetland restoration, to recoup the costs identified in the list. They try to minimize their cost of producing mitigation credits to maximize their own return on investment.

The objective of permit applicants is to maximize the rate of return on investments in development projects that impact wetlands. To the extent that permit applicants are required by regulators to provide mitigation, they try to minimize the cost of this requirement to maximize development returns.

The objective of regulators is to protect the wetland functions in a watershed on behalf of society as a whole. As discussed earlier, the Clean Water Act's Section 404 regulatory program reflects a policy goal of achieving no net loss in wetland function, to be followed by net gain, to meet this objective of protecting societal interests. These goals are the result of legal mandates that govern the administration of the regulatory program.

The differing objectives that credit suppliers permit applicants and regulators offer the potential for deal making, which is the essence of any incentive-based market. The challenge facing administrators is to establish and authorize a framework for the operation of credit market systems to meet the goals of all three interest groups. Credit suppliers and permit applicants each earn some profit while achieving the no net loss functional goal of regulators.

Government plays two roles in the supply and demand conditions that drive mitigation credit markets. First, credit markets could not exist in the absence of incentive-based government regulations that create demand for wetland development permits while granting permits conditionally on compensatory mitigation; in this vein, the regulators create "quasi-markets," markets that are not quite "free," but driven by incentive-focused guidelines attempting to link the marketplace to broad societal needs. Second, permit applicants are price-conscious but not necessarily quality-conscious when opting for price-optimum compensatory mitigation alternatives. It is the regulator, not the buyer of mitigation, who must impose "quality control" on the market through trading rules establishing how and when credits can be created and sold (Shabman et al., 1996).

The number of credits available for withdrawal (i.e., debiting) should generally be commensurate with the level of aquatic functions attained at a bank at the time of debiting. The level of function may be determined through the application of performance standards tailored to the specific restoration, creation, or enhancement activity at the bank site, or by an appropriate functional assessment methodology.

The success of a mitigation bank with regard to its capacity to establish a healthy and fully functional aquatic system relates to both the ecological and financial stability of the bank. Financial considerations are critical in early stages of bank development. It is generally appropriate in cases where there is adequate financial assurance, and where the likelihood of success of the bank is high, to allow limited debiting of a percentage of the total credits projected for the bank at maturity. Such determinations should take into consideration the initial capital costs needed to establish the bank and the likelihood of its success. However, it is the intent of this policy to ensure that those actions necessary for the long-term viability of a mitigation bank be accomplished prior to any debiting of the bank. In this regard, the following minimum requirements should be satisfied prior to debiting: Credits attributed to the preservation of existing aquatic resources may become available for debiting immediately upon implementation of appropriate legal protection accompanied by appropriate changes in land use or other physical changes, as necessary. Third parties may purchase credits; however, the wetland bank sponsor determines the cost of credits sold to a third party.

Ecosystem Assessment Techniques

Available techniques range from simple measurement of wetland acres to site-specific habitat evaluation methods, such as the U.S. Fish and Wildlife Service's HEP, and approaches to evaluate overall wetland functions. Other techniques include the wetland evaluation technique (WET) and hydrogeomorphic (HGM) procedures designed by several governmental agencies.

Habitat Evaluation Procedures

The U.S. Fish and Wildlife Service developed HEPs to assess the fish and wildlife habitat function of wetlands and other natural environments, and the nonmonetary value of common, surrogate fish and wildlife resources (Schamberger and Krohn, 1982). It involves use of a *habitat suitability index* (HSI) to describe the habitat requirements for particular species. In the case of wetlands, the target species would be those known to occur in wetland habitats. HEP, therefore, provides a species-specific approach based on the assumption that it is possible to describe numerically habitat quality and quantity.

The value of an HSI derives from an evaluation of the ability of essential habitat components to supply the life requisites of selected species of fish and wildlife. Evaluation involves using the same essential habitat components to compare existing and optimum habitat conditions for the species of interest. Optimum conditions are those associated with the highest potential densities of the species within a defined area. The HSI value obtained from this comparison thus becomes an index to carrying capacity for that species.

The index ranges from 0.0 to 1.0, and for operational purposes in HEP, each increment of change must be identical to any other. For example, a change in HSI from 0.1 to 0.2 must represent the same magnitude of change as a change from 0.2 to 0.3, and so forth. Therefore, HSI has a linear relationship to carrying capacity. This presents an operational restriction on the use of HSI in HEP. However, it is a restriction with which it is easy to comply; if the relationship between HSI and carrying capacity is unknown, it is assumed to be linear. If the relationship is nonlinear, conversion to a linear function is a corrective manipulation.

HEP attempts to incorporate concepts from both population and habitat ecological theories in evaluating habitat quality for specific species. The U.S. FWS develops HSI values for individual species using documented habitat suitability models that incorporate measurable essential habitat variables (e.g., percent canopy closure, etc.). HEP's unit of measurement is the *habitat unit* (HU). The arithmetic product of HSI values and area of available habitat derive the HUs for individual species. These values are useful in the HEP system for comparative purposes. No aggregation of species' HSI (or HUs) occurs. Many potential users tend to consider the HSI value as synonymous with the entire HEP system. This is not the case. HEP is like a bookkeeping ledger, because HEP and HSI both passively display and document values obtained from other sources. HEP, therefore, is a data management system; it is the data it manages (i.e., the index of quality and the quantity of available habitat) that is of interest in resource value assessment.

Wetland Evaluation Technique

The *wetland evaluation technique* (WET) developed by the Federal Highway Administration and later modified with support from the Army Corps Waterway Experiment Station is a method for determining bank credits by prioritizing functions and values based on local/regional importance and by quantifying the probability rating from a WET evaluation (Adamus et al., 1987). WET considers wetland functions to be the physical, chemical, and biological characteristics of a wetland, including the following characteristics:

- Groundwater recharge
- Groundwater discharge
- Flood flow alteration
- Sediment stabilization
- Sediment/toxicant retention
- Nutrient removal/transformation
- Production export
- Aquatic diversity/abundance
- Wildlife diversity/abundance

- Recreation
- Uniqueness/heritage

The WET is similar in nature to HEP but incorporates many wetland-specific factors not addressed by the HEP (Studt and Sokolove, 1996). WET, although having the ability to assess a full suite of wetland functions, produces only a qualitative assessment of the probability that a wetland performs the range of different functions (Scodari, 1997).

The WET evaluates these functions, and values, in terms of effectiveness, opportunity, social significance, and habitat suitability. Effectiveness assesses the capability of a wetland to perform a particular function. For example, a wetland that has no outlet has a high value for sediment retention, whereas a wetland downstream from a dam has a low value. Opportunity assesses the potential for a wetland to perform a specific function. For example, a wetland in a forested area having no potential sediment sources would have a low opportunity value for sediment retention. Social significance assesses the value of a wetland in terms of special designations, such as endangered species habitat. Social significance also includes economic considerations, such as potential value as a recreational area and strategic location, for example, the scarcity of the wetland type in a given region. The WET uses "predictors" that relate to the physical, chemical, and biological characteristics of the selected function. For example, whether a wetland has a constricted outlet could predict its effectiveness in storing floodwaters (Novitzki et al., 1997).

In applying the technique, first an investigator collects data on the target wetland, the area surrounding it, and the area downstream from it. Then he or she answers a series of questions about the wetland vegetation, the characteristics of the watershed, the wetland topography, and other features. By progressing next through a series of flowcharts (or an available computer software package), an evaluation can assign a probability rating to each of the functions, with the exception of recreation, listed in Table 4.1. Through this process, a habitat suitability rating for fish, waterfowl, and other wildlife is also defined (Adamus, 1988).

The probability rating developed at the end of the process is an estimate of the "likelihood" that a wetland will perform a function based on its characteristics. An investigator cannot estimate the degree or magnitude of the performance. Recreation is not evaluated, because scientific basis for a decision usually does not exist without an extensive, time-consuming data collection effort (Novitzki et al., 1997).

The major limitation of the WET methodology is its inability to be flexible to local or regional differences in wetlands. The need to apply the method to every wetland in every part of the United States is cumbersome. For example, most users are interested in a local area and prefer not to enter data repeatedly for local characteristics that are unlikely to change, as required in the WET approach. The WET originators intended to develop regional versions of the model to refine the thresholds of the ranking values, but this has not occurred.

Hydrogeomorphic Methods

In 1995, the U.S. ACOE introduced a new wetland assessment technique—the *hydrogeomorphic method* (HGM)—designed to overcome many of the shortcomings associated with HEP and WET (Smith et al., 1995). This method is designed to tailor wetland assessments to different wetland types that vary according to their position in the landscape, dominant water source, and movement of water through the system. This allows the assessment method to focus on the specific set of wetland functions associated with different wetland types. It provides a procedural framework for measuring the capacity of a wetland to perform these functions. The method relies on the use of "reference wetlands" to establish standards for the various functions associated with wetlands that are self-sustaining and show a high level of overall function. These standards define the basis for assessing and quantifying wetland functions and for estimating how they might consequently change development and restoration activity.

In the procedure, the investigator first classifies wetlands into regional subclasses based on hydrogeomorphic characteristics (geomorphic setting, water source, and hydrodynamics). This classification focuses attention on only those functions of wetlands in a class that are relevant to the targeted wetland and the landscape or ecosystem factors that are most likely to influence how wetlands in the class function. The investigator then develops a functional profile, or index, for each wetland subclass based on data collected from a regional reference standard. The reference standard reflects the range of conditions that occur within the particular wetland subclass. The investigator subsequently compares the functional capacity of a wetland under study to the functional index for the wetland subclass.

The HGM represents a considerable methodological advancement for wetland functional assessment. It promises to be particularly useful in the regulatory context for measuring gains and losses on a local scale resulting from permitted development projects and mitigation requirements. Because this method can be generalized to all wetland types, it could prove useful for evaluating changes in regional wetland functioning that results from the broader set of wetland protection and management programs.

Nevertheless, significant difficulties remain. HGM does not apply to entire, regional wetland systems. Moreover, the regional assessment of changes in wetland functioning resulting from wetland protection would be insignificant in itself to describe the effect on wetland goods and services. Wetland functional assessment produces measures of functional indices that are only suggestive of the capacity of wetlands to provide certain important outputs. The difficulty in fully defining and tracing the relationships between indices of wetland functioning and measures of the flows, and characteristics of wetland goods and services limits the ability to develop estimates of wetland protection benefits (Scodari, 1997).

Compensation Ratios

If the Section 404 permit requires compensation for adverse impacts caused by the permitted or authorized project, the type of mitigation needed to satisfy the requirements of the permit can affect the number of wetland acres required for compensation. The number of acres restored, created, or enhanced for each acre of wetlands impacted is the compensation ratio. For example, if the compensation ratio is 2 : 1, then two acres of wetlands must be restored, created, or enhanced for every one acre of wetland impacted by the permitted development project. In general, the minimum ratio of restoration or creation acres acceptable by regulators is 1 : 1. To account for uncertainties of speculative wetlands creation projects, the ratio is often higher, however, sometimes as high as 10 : 1. Because regulators usually favor restoration, the ratio of restored to impacted acres may be 2 : 1 for a restoration project. However, if the permit applicant wanted to use enhancement for mitigation instead of restoration, the ratio might increase to 3 : 1 or 4 : 1. Because preservation is the least favored method of mitigation, ratios for this type of mitigation are sometimes 10 : 1 or higher. These ratios are controversial and are often negotiated in different ways from region to region.

Many states have established their own guidelines for assigning credits and credit ratios. One interesting approach is that followed by Broward County, Florida. By ordinance, the County establishes a credit system based on a wetland benefit index (WBI). The level of various attributes of the site determines the WBI. These attributes include fish, wildlife, and habitat values/diversity, hydrophytic vegetation dominance, unpredictability of wetland mitigation system, surface water hydrology, surrounding landscape, the hydroperiod, soils, and plant cover. The total that an acre of land can earn is 1.0 points. A mitigation bank credit (MBC) is calculated using the following formula:

$$1.0 - (WBI \times HMC) = MBC,$$

where HMC equals the habitat mosaic credit (diversity of community types), which equals 0.9 for multiple habitat banks.

The Ordinance also contains a detailed formula for determining the mitigation ratio, which is a function of the quality of the impacted and replacement wetlands, whether the two sites are in the same basin, the type of wetland impacted, the type of mitigation technique used, and when the mitigation is performed relative to the impact. Specifically, the following formula calculates the ratio:

$$(I)(WBI^{I})(BRR)(MOD)(1 - AMBC) = (A)(MBC),$$

where I is acreage impacted, WBI^1 is the wetland benefit index of the impacted site, BRR is the base replacement ratio (determined by type of wetland impacted and whether the mitigation technique is creation or enhancement), MOD is the modifier (1.2 if mitigation is off-site, but in the same basin; 1.5 if mitigation is off-site, and in a different basin; and 0.95 if mitigation is performed and completed prior to impacts occurring), $AMBC$ is additional mitigation bank credits, which can be earned by providing a corridor to other environmentally significant protected lands, by providing critical type habitat for threatened or endangered species, by preserving additional lands, by constructing the bank prior to credit use, and by providing wetland benefits, as listed in the mitigation bank proposal, prior to credit use, and A is the acreage needed in the mitigation bank, and MBC, the mitigation bank credit (Crookshank, 1995).

Conservation Banking

A conservation bank is a publicly or privately owned piece of habitat managed for its natural resource values. As is the case of a defined WMB, a conservation bank sells credits to developers and others required by law to compensate for, or mitigate, detrimental environmental impacts of their activities. Credits can be "bundled" to purchase high-priority, ecologically important habitat for regional benefit that is generally beyond the reach of individual mitigation projects. The ESA provides the basis of support for a small number of conservation banks set up in various places throughout the United States for the purpose of conserving habitat for listed species through HCPs. A notable exception to the habitat approach is the broader ecosystem-based approach under testing in California and, to a limited extent, under way in the Northwest, Texas, and Florida (Marsh et al., 1996). In addition, a variation of the process is being used to address wildlife and other water-resource concerns in the San Francisco Bay–Delta and in south Florida (Marsh et al., 1996).

These credits fund habitat restoration at the bank site and provide a permanent endowment for operation of the bank in perpetuity as a wildlife preserve. The availability of such conservation banks also eases the regulatory process of permitting projects. These conservation banks function in the following ways:

- Allow the private and public sectors to harness market forces to improve significantly upon traditional ways of protecting and restoring wildlife habitat.
- Advance effective regional habitat conservation by encouraging the bundling of mitigation "credits" at sites recognized to be high priorities for habitat protection and restoration within watersheds and ecosystems.
- Provide a mechanism that assigns a monetary value to habitat, which in turn allows a landowner to obtain a financial return for conserving land rather than developing it.

The assignment of monetary value to habitat is an important aspect of a bank. The main function of all conservation or mitigation banks is to produce compensation credits that measure the value added to a given property. For example, a wetland conservation bank identifies and quantifies those services that a wetland provides, such as flood control and recreational opportunities, in terms of a currency for bank transactions. The currency could be equivalent to HUs, acres, or other metrics. Credit definition and valuation, however, remain the most complex issues in conservation banking.

The Natural Community Conservation Planning (NCCP) pilot projects in Southern California attempts to create a program for the entire range of an endangered ecosystem—the coastal sage scrub—and all the species that inhabit it, both rare and common (California Department of Fish and Game, 1996). The NCCPs, which are large-scale HCPs, are now being prepared for virtually all of the undeveloped urbanizing areas of Southern California. The NCCP planning area covers 6,000 square miles in five counties, and is expected to result in more than 300,000 acres of large habitat blocks preserved under permanent conservation management. Under formal agreement with the California Department of Fish and Game and the U.S. FWS, a conservation credit value originates. These credits are then available for sale to others, who are required to compensate for adverse environmental impacts in a development project or other activity.

Conservation banks, when used as a tool for helping to implement multispecies and -regional conservation plans, can provide greater flexibility and economic opportunity for the private landowner than the more traditional and restrictive WMBs. The ability to sell conservation credits provides landowners with an incentive to develop their lands for conservation rather than for other purposes. This type of incentive program will be critical to achieving long-term conservation goals and objectives throughout the state of California.

NCCP is a large step forward from traditional HCPs under Section 10 of the ESA. It focuses protection efforts on an entire ecosystem rather than exclusively on listed species and, as such, it covers both occupied and essential unoccupied habitat, and natural processes. The California state law that authorizes the program identifies a conservation standard of "no net loss of habitat value" for completed plans, a considerably higher benchmark than for losses allowed under the "jeopardy" standard by which HCPs are ultimately judged.

Mitigation and conservation banks are *ad hoc* arrangements between developers and regulators. Local and federal wetlands banking rules and guidelines dictate the use of a bank. Wetland bank owners provide mitigation mechanisms for the restoration, creation, enhancement, and preservation of habitat impacted by development. Through governmental dictum, banks can be privately or publicly owned. Owners may retain their credits for internal purposes or sell them in the market to others:

- Banks can be constructed for the sole use of one party (self-financed).
- Banks can be constructed by a group that then sells credits to the general public (market or fee based).
- Banks can be in the form of a trust that developers pay into, with proceeds used to carry out compensation activities later.

Conclusions

The preceding description of valuation techniques and policies to address ecosystem protection has focused on addressing the primary threats to biodiversity: lack of recognition of value and the need to formalize strategies to protect public goods embodied in biodiversity. In order to make significant strides in the continual challenge to protect biodiversity, it is imperative to integrate scientific information into the economic and policy arena. The examples of methods to assess wetlands ecosystems for indicator species and functions to value for policy demonstrate how critical scientific information helps to convey accurate measures of biodiversity. The examples of existing policies in various parts of the world to address habitat protection and focus on biodiversity are meant to serve as a building block to foster widespread action for biodiversity around the world.

References

Adamus, P. R. (1998). The FHWA/Adamus (WET) method for wetland functional asessment. In: D. D. Hook, W. H. McKee, Jr., H. K. Smith., J. Gregory, V. G. Burrell, Jr., M. R. Devoe, R. E. Sojka, S. Gilbert, R. Banks, L. H. Stolzy, C. Brooks, T. D. Mathews, and T. H. Shear (Eds.), *The ecology and management of wetlands: Vol. 2. Management, use and value of wetlands* (pp. 128–133). Portland, OR: Timber Press.

Adamus, P. R., Clairain, E. J., Smith, R. D., and Young, R. E. (1987). *Wetland evaluation technique (WET): Vol. II. Methodology.* Vicksburg, MS: U.S. Army Corps of Engineers Waterways Experiment Station.

Anderson, R., and Rockel, M. (1991). *Economic valuation of wetlands.* Discussion Paper No. 065. Washington, DC: American Petroleum Institute.

Boardman, A., Greenberg, D., Vining, A., and Weimer, D. (1996). *Cost–benefit analysis: Concepts and practice.* Upper Saddle River, NJ: Prentice-Hall.

California Department of Fish and Game Code. (1996). Section 2800-2825, Natural Community Conservation Planning Act.

Chapman, D. (1999). Environmental economics theory, application and policy. Reading, MA: Addison-Wesley, Longman.

Crookshank, S. L. (1995). Alternative wetland migration programs. Discussion Paper No. 77. Washington, D.C.: American Petroleum Institute.

Daily, G. C. (1997). *Nature's services.* Washington, D.C.: Island Press.

Deason, R., and Murphy, P. (1997). The structure of an environmental transaction: The debt for nature swap. *Land Economics* 73(1): 1–24.

de Groot, R. S. (1992). *Functions of nature: evaluation of nature in environmental planning, management and decisioin making.* Croningen: Wolters-Noordhoff.

Farber, S., Fernandez, L., Simpson, D., and Wilson, M. (1999, February). *Valuing ecosystem services: A framework for corporate stewardship*. Report prepared for EPRI under the auspices of the University of Pittsburgh, University of California–Santa Barbara, and the University of Wisconsin (Madison), respectively, under contract to Allegheny Power Company (TC9163-2), Palo Alto, CA.

Fernandez, L. (1999). An analysis of economic incentives in wetlands policies addressing biodiversity. *Science of the Total Environment*, 240: 107–122.

Haskell, B. D., Norton, B. G., and Costanza, R. (1992). What is ecosystem health and why should we worry about it? In: R. Costanza, B. Norfon, and B. Haskell (Eds.), *Ecosystem Health: new goals for environmental management*. Washington, D.C.: Island Press, pp. 35–49

Kormondy, E. J. (1996). *Concepts of ecology*. Upper Saddle River, NJ: Prentice-Hall.

Lynne, G., Conroy, P., and Prochaska, F. (1981). Economic valuation of marsh areas to marine production processes. *Journal of Environmental Economics and Management* (2): 175–186.

Marsh, L. L., Porter, D. R., and Salvesen, D. A. (Eds.) (1996). *Mitigation banking: Theory and practice*. Washington, DC: Island Press.

Mitsch, W. J., and Gosselink, J. G. (1993). *Wetlands* (2nd ed.). New York: Von Nostrand Reinhold.

Novitzki, R. P., Smith, R. D., and Fretwell, J. D. (1997). *Wetland functions, values, and assessment*. National Water Summary on Wetland Resources, USGS Water Supply Paper No. 2425. Retrieved online October 14, 1998 at http://h2o.usgs.gov/public/nwsum/WSP2425/functions.html.

Panayotou, T. (1992). The economics of environmental degradation: Problems, causes, and responses. In: A. Markandya and J. Richardson (Eds.), *Environmental economics: A reader*. New York: St. Martin's Press.

Schamberger, M. L., and Krohn, W. B. (1982). Status of the habitat evaluation procedure. *Transaction of the North American Wildlife and Natural Resources Conference* 47: 154–164.

Scodari, P. E. (1997). *Measuring the benefit of federal wetland programs*. Washington, D.C.: Environmental Law Institute.

Shabman, L., Scodari, P., and King, D. (1996). Wetland mitigation banking markets. In: L. Marsh, D. Porter, and D. Salvesen (Eds.), *Mitigation banking: Theory and practice* (pp. 109–138). Washington, DC: Island Press.

Smith, R. D., Ammann, A., Bartoldus, C., and Brinson, M. M. (1995). *An approach for assessing wetland functions using hydrdogeomorphic classification, reference wetlands, and functional indices*. Vicksburg, MS: Technical Report No. TRWRP-DE 10, U.S. Army Corps of Engineers Waterways Experiment Station.

Studt, J., and Sokolove, R. D. (1996). Federal wetland mitigation policies. In: L. L. Marsh, D. R. Porter, and D. A. Salveson (Eds.), *Mitigation banking: Theory and practice*. Washington, D.C.: Island Press, pp. 79–93

Tilman, D. (1997). Biodiversity and ecosystem functioning. In: Daily, G. (Ed.), *Nature's Services*. Washington, D.C.: Island Press, pp. 57–71.

Turner, R. K., van den Bergh, J., Soderquist, T., Barendregt, A., van der Staaten, J., Maltby, E., and van Ierland, E. (2000). Ecological–economic anlaysis of wetlands: Scientific integration for management and policy. *Ecological Economics* 35: 7–23.

U.S. Environmental Protection Agency. (1995). *Wetlands fact sheets*. Washington, DC: Author.

U.S. Environmental Protection Agency and U.S. Army Corps of Engineers (1990). Memorandum of agreement between the environmental protection, agency and the Department of the Army concerning the determination of mitigation under the Clean Water Act, section 404(b)(1) guidelines, 55 FR9210.

U.S. Fish and Wildlife Service. (1981). 46 Federal Register 7644-7663.

U.S. Fish and Wildlife Service. (1980). *Habitat evaluation procedures manual*. Washington, DC: Author.

Zedler, J. B. (1990). *A manual for assessing restored and natural coastal wetlands, with examples from southern California*. San Diego: Pacific Estuarine Research Lab, Biology Department, San Diego State University.

5

Effects of Air Pollution on Crops in Developing Countries

J. Nigel B. Bell and Fiona M. Marshall

Introduction

Agricultural practice in both the developed and developing world has always aimed to eliminate or minimize the numerous constraints on producing maximum yield of crops. These constraints may be abiotic, including nutrient deficiency, metal toxicity, salinity, drought, low and high temperatures, wind, and waterlogging. There are also numerous biotic constraints: invertebrate and vertebrate pests; fungal, viral, and bacterial pathogens; and trampling. Although vast sums of money are spent on overcoming these, there is another constraint that receives much less attention, but with evidence that it is potentially a widespread threat to crop production: air pollution. As early as the 17th century, the noted English diarist John Evelyn (1661) described graphically the effects of the smoke ridden air of London on ornamental plants and fruit trees growing in the city. Research into this phenomenon that commenced in the late 19th century was initially concerned with trees near smelters in both Europe and North America. In the final three decades of the last century, this research mushroomed, with extensive programs in many parts of the developed world. Much of this research was aimed at understanding the effects on crop yield and growth of the ubiquitous primary pollutants, sulphur dioxide (SO_2), and nitrogen oxides (NO_x), as well as the secondary pollutant ozone (O_3), indicating that all three pollutants were at different times and in different places having serious deleterious effects on crop productivity, with the two primary pollutants being associated mainly with industrial point sources and the urban fringe, whereas O_3 was widespread at phytotoxic levels in rural areas. Subsequently interest also developed into the potential for air pollution to have indirect adverse effects on crops by exacerbating the impacts of familiar

J. Nigel B. Bell and Fiona M. Marshall • Imperial College at Silwood Park, Ascot, Berkshire SL5 7PY, United Kingdom.

biotic stresses, such as pests and pathogens, and abiotic stresses, such as drought and frost.

In recent years, there has been a decline in interest in effects of air pollution on crops in the developed world, doubtless largely due to overproduction. The developing world is certainly not characterized by overproduction, with most countries striving desperately to increase yields of their staple crops to feed rapidly expanding populations. Air pollution has traditionally been viewed as a problem of "Western" countries, where the bulk of the industry and motor vehicles responsible for the major pollutants resides. However, emissions of some of these pollutants are being reduced as a result of introduction of stringent controls in recognition of their adverse effects on health, vegetation, aquatic ecosystems and materials, as well as a decline of polluting heavy industries. The latter are in fact transferring to the developing world, where rapid industrialization is taking place in many countries but with poor emission controls. At the same time motor traffic is growing at an enormous rate in the developing world, often using old and poorly maintained vehicles that play a major role in contributing to a deterioration in air quality. For example, the number of motor vehicles in China has increased by 11% annually over the last 30 years, and in India by 7% per annum over the past 10 years (Faiz and Sturm, 2000). Thus SO_2 and NO_x are increasing rapidly in many developing countries. We know far less about the third ubiquitous pollutant, O_3, because very little monitoring has been carried out in most countries, and most of this is restricted to the cities where concentrations are normally lower than in adjacent agricultural areas. However, it is quite clear that O_3 levels are elevated to potentially phytotoxic concentrations in developing countries (Chameides and Kasibhatia, 1994; Chameides et al., 1999). This is scarcely surprising in view of the conditions for the production of this secondary pollutant being amply fulfilled in such places. Ozone is a component pf photochemical smog, which was first detected in and around Los Angeles in the 1940s. After intensive research into this new smog phenomenon that not only had unpleasant effects on human health but also damaged vegetation, O_3 was identified as the main culprit in the latter case. Ozone is generated by photochemical reactions on NO_x and volatile organic compounds (VOCs). This requires high levels of emissions of these two categories of pollutants, and a climate with high temperatures, bright sunlight, and relatively still air, all pertinent to Los Angeles from the 1940s onwards (although recognition of the problem has resulted in drastic attempts to control emissions), and also to the present-day developing world.

Thus, we have every reason to believe that any problems identified due to air pollution in the developed world will be replicated in developing countries, where any effects on crop yield or quantity have potentially far more serious consequences for human welfare.

What studies have been carried out on the effects of these ubiquitous air pollutants on crop yield and quality in the developing world? The answer is almost zero. The exception is India, where most of the research has been devoted to point source field studies or controlled fumigations with concentrations above ambient levels. In recent years, interest has grown in air pollution in developing countries, doubtless stimulated by the international funding agencies and a growing environmental awareness in the local populations. However, effectively, all of this has been concerned with direct effects on human health. For example, a recent World Bank Study in four Bangladesh cities has estimated that 15,000 extra deaths per year occur due to air pollution, resulting in a loss of 0.7–2.0% of gross domestic product. If health is compromised by reduced food availability or quality caused by air pollution, then it could be argued that this represents a further serious indirect effect on the former.

This chapter provides an overview of current knowledge of air pollution impacts on crop yield and quality in developing countries. It concentrates entirely on the three ubiquitous pollutants already discussed, and also covers potential indirect effects via pests and pathogens.

What Do We Know About the Problem in the Developed World?

The answer is that we know a lot about some aspects, but little about others. In other words, despite the efforts of policymakers who have funded large programs in both North America and Europe, there are enormous gaps in knowledge. These include geographical issues, with European studies on O_3 being largely confined to central and northern Europe, whereas, paradoxically, the highest concentrations occur in the Mediterranean countries. In the case of these southern countries, the problem is exacerbated by the fact that, to a large extent, they grow different species and cultivars from the areas that have been more heavily researched; thus, studies in the latter may not be applicable to these locations. In the case of the developing world, this problem is even more acute, with doubtful relevance of "Western" research, as will be discussed later. Perhaps even more serious is the failure to address the real-world situation, where the presence of an ever-changing environment that includes stresses other than air pollution may change the magnitude of response. Thus, carefully controlled experiments in laboratory or quasi-laboratory conditions will not provide a realistic assessment of what happens in the field, unless many other factors are taken into account. Not only do environmental conditions such as soil and climatic factors influence the magnitude and, indeed, even the direction of pollution impacts (Bell, 1984), but as mentioned in the Introduction, pollution can also exacerbate the deleterious effects of pests and pathogens on crops.

Various attempts have been made over the last 20 years or so to quantify the economic losses to crops arising from air pollution, including in the United Kingdom, The Netherlands and the United States. The most ambitious and realistic of these was The National Crop Loss Assessment Network (NCLAN) program in the United States in the 1980s, aimed at producing a nationwide assessment in economic terms of the effects of ambient air pollution on the yields of the most widely grown crops. Experiments were carried out at five locations in different parts of the United States using a standard protocol. The crops were grown in sets of open-top chambers ventilated with charcoal-filtered or ambient air, or ambient air with different 7-hour concentrations of O_3 imposed on it. This generated dose–response relationships for seasonal, 7-hour O_3 concentrations and yield reductions of the crops concerned. Data from these functions were then integrated with monitored and modeled O_3 distribution across the United States to produce maps of yield loss. Economic models were then used to develop from the latter monetary losses for the country as a whole, indicating a loss of about $\$3 \times 10^9$ per year in the mid-1980s for the direct effects of O_3 (Adams et al., 1988). It should be noted that this is for O_3 alone, and that it excludes other high-value crops with more limited distribution. Furthermore, it ignores indirect abiotic or biotic stresses, whose individual effects on yield are much greater. It has in fact been argued that air pollution could have a greater effect on yield via exacerbating other stresses than by its direct impact (Bell et al., 1993).

Following NCLAN, a similar program instigated in Europe, funded by the European Commission, also demonstrated widespread losses to the limited number of crops studied, with O_3 being the main phytotoxic agent involved. Dose–response functions developed in the course of this program were utilized in the development of critical levels for O_3 under the auspices of the UN Economic Commission for Europe (UNECE). These critical levels are designed to provide protection against yield reduction for a range of pollutants being considered as thresholds in this respect (Table 5.1).

Trends in Pollutant Concentrations

SO_2 is emitted mainly from the combustion of coal and fuel oil, which have varying concentrations of sulphur(s), depending on source. Smelters are another important source, but more localized, and other industrial and domestic sources also play a role. In the developed world, road traffic is a relatively minor source in terms of national emissions due to low S content of fuel. However, some developing countries have much higher S levels in diesel fuel. For example, a recent study (Kitada and Asad, 1998) in Dhaka has shown that motor traffic contributes over 50% of the emissions, which results in mean concentrations around the city up to about 200 $\mu g\ m^{-3}$. Traffic also plays the major role in emissions of NO_x, with nitric oxide, the principal primary pollutant being rapidly oxidized to NO_2. In fact, all high-temperature combustion processes result in the emission of NO_x, mainly as a result of combination of atmospheric N_2 and O_2; thus, thermal power stations represent the other major source. For elevated O_3 levels to be produced above the maximum natural background of about 40 ppb, VOCs are necessary for photochemical reactions with NO_x and motor vehicles are a very important source, particularly those with the inefficient and poorly tuned engines so characteristic of developing countries.

As mentioned earlier, in the developed world, emissions of the key primary pollutants are generally falling, but the trend in most developing countries is strongly in the opposite direction. Thus, annual SO_2 emissions in Asia are predicted to increase from 34×10^6 tons in 1990, to 110×10^6 tons by 2020. In the case of NO_x, a global increase of from 40×10^6 tons in the mid-1980s to $55–66 \times 10^6$ tonnes per year by 2025 has been predicted (Lee et al., 1997), with substantially higher percentage increases in some developing countries, such as China. These increases in NO_x are predicted to cause widespread increases in O_3 levels across the developing world. Thus, Chameides and Kasibhatia (1994) estimated that an increase would take place in the area of cereal crops subjected to O_3 concentrations above the threshold for yield reductions from 9 to 35% of world area to 30–75%. More recently Chameides et al. (1999) modeled O_3 production in China and concluded that

Table 5.1. UNECE Critical Levels for Agricultural Crops (Ashmore and Wilson, 1994)

Pollutant	Concentration	Averaging time
SO_2	30 $\mu g\ m^{-3}$	Annual or winter mean
NO_x	30 $\mu g\ m^{-3}$	Annual or winter mean
O_3	AOT40 3000 ppb.h*	3-month growing season

*Cumulative dose in ppb.hours for daylight hours above a threshold of 40 ppb for 3-month growing season (Fuhrer et al., 1997).

concentrations were sufficiently high to reduce yield of winter wheat production over a large part of the country, notably in the Yangtze Delta, the southeast Yellow River Delta, the eastern Pearl River Delta, and Sichuan.

Direct Effects of Pollutants on Yield in Developing Countries

Sulphur Dioxide

Studies on the effects of ambient SO_2 concentrations on crop yield in developing countries are largely confined to transect studies away from point sources in India. In such investigations, plant material is usually grown in standard soils and containers, and exposed for all or part of the growing season along gradients of the pollutant. One important study of this type was reported by Singh et al. (1990), in which a local wheat cultivar was grown at different distances from a coal-fired power station in Uttar Pradesh. This generated a dose–response relationship, which was subsequently used by Ashmore and Marshall (1997) to estimate that farmers operating within 10 km of Indian power stations might experience yield losses of 10 to 60%, depending on the scale of emissions. Thus, it is apparent that location of industry in rural areas presents a serious threat to crop production on a local scale.

A rare example of a study in which open-top chambers have been employed in a developing country to study the effects of ambient SO_2 on crops, thereby using the main technique employed in developed world studies, was reported by Botha et al. (1990). They grew *Phaseolus vulgaris* in chambers ventilated with ambient or filtered air at a site near Cape Town. The results were somewhat inconclusive, with an 18% reduction in dry weight in ambient air in one experiment but no effect in another.

Despite evidence of increasing SO_2 emissions in developing countries, it would appear that so far as major agricultural areas are concerned, phytotoxic impacts are largely associated with point sources. However, this ignores the issue of urban and peri-urban agriculture, which is exposed in many places to high levels of both SO_2 and other pollutants. Such agricultural systems are almost entirely ignored by policymakers. Yet there is strong evidence that they play a very important role in the nutrition of the urban poor (United Nations Development Program, 1996). A recent research program on this matter carried out by the authors and collaborators in the Indian cities of Delhi and Varanasi have confirmed the importance of such agriculture both for the local economy and as a source of food security, direct income generation, and employment for the poor (Marshall et al., 2000). As part of this program, two summer and two winter crops were grown to maturity in standard soils and containers at sites in both cities, where air pollution levels varied from relatively clean to highly polluted. The summer crops were palak (*Beta vulgaris*) and moong bean (*Vigna radiata*), and the winter crops were wheat (*Triticum aestivum*) and mustard (*Brassica campestris*). Both moong bean and palak yield showed negative correlations ($p < 0.05$) with mean SO_2 concentration in Varanasi. The leaf dry weight ranged from 0.69 g-plant^{-1} at the site with the lowest SO_2 (40 $\mu g\ m^{-3}$) to 0.18 g-plant^{-1} at the site with highest SO_2 (110 $\mu g\ m^{-3}$) (M. Agrawal, personal communication cited in Marshall et al., 2000). In the winter experiment, concentrations ranged from 22 to 86 $\mu g\ m^{-3}$ in Delhi and 35–141 $\mu g\ m^{-3}$ in Varanasi. Both species' yield showed negative correlations with SO_2 concentrations at both sites, but this was only significant for wheat in Varanasi, where there was a 23% difference in grain yield between the sites with the highest and lowest SO_2 concentrations (C. K. Varshney and M Agrawal, personal communication).

Nitrogen Dioxide

Nitrogen dioxide has received much less attention than SO_2 but has the potential to be particularly important in and around urban areas. The study described earlier for SO_2 in Delhi and Varanasi also examined yield of the four crops in relation to ambient NO_2 concentrations, showing a negative correlation between moong bean yield and NO_2 concentrations ranging from 22 to 112 $\mu g\ m^{-3}$ in Varanasi (Marshall et al., 2000). In the winter experiments, the yields of both mustard and wheat in Delhi, and wheat only in Varanasi, were significantly negatively correlated ($p < 0.05$), with NO_2 concentrations ranging from 45 to 131 $\mu g\ m^{-3}$ (Marshall et al., 2000). A particularly serious issue was raised by the Varanasi study, where carbohydrate and energy, but not protein, contents showed significant negative relationships with both SO_2 and NO_2 in the two crops where such measurements were made viz. moong bean and wheat (M. Agrawal, personal communication). Thus, there is good evidence that not only yield but also crop quality is reduced by ambient urban and peri-urban air pollution. An important, potentially confounding factor in all these transect studies is the possibility of combined effects of pollutants. It has been known for some time that NO_2 and SO_2 can have additive and, more often, synergistic interactions (Ashenden and Mansfield, 1978), and that O_3 can contribute to this phenomenon (Bell, 1984). In some cases, antagonistic interactions have also been observed (Bell, 1984). Thus, all transect and filtration studies have to be viewed with caution due to the invariable presence of more than one phytotoxic pollutant. This is illustrated by the study of Singh et al. (1990), already discussed with respect to SO_2. In this case, the negative correlation between wheat yield and SO_2 was highly significant, but there was also a lower but significant correlation with NO_2.

Ozone

The effects of ambient O_3 have been studied in Mexico, Egypt, Pakistan, and India (reviewed in Ashmore and Marshall, 1999). Ozone is not a pollutant that it is appropriate to study using the transect approach, due to the absence of sharp concentration gradients. Indeed, O_3 invariably occurs at higher levels outside than within cities, where its precursors emanate, due to local destruction by NO at ground level. Thus, it is widespread over agricultural areas downwind of cities and presents a particular threat to rural crops. In the absence of transect studies, chamber filtration can be employed to determine impacts on yield—a costly and complex procedure. Luckily, there is another simple field technique that can be employed in lieu of the transect approach—the use of antiozonant chemicals applied to the foliage or as a soil drench. The earliest studies of this type utilized broad-spectrum fungicides such as benomyl, but in the late 1970s, Dupont de Nemours, Inc. developed a highly effective antiozonant, ethylenediurea (EDU), which can provide up to 100% protection against O_3-induced visible injury and yield reduction. Unfortunately, EDU is not available commercially and thus, it has to be manufactured specially for any experiment. Neither is it cheap, costing around £800 UK for 1 kg for the chemicals required for its manufacture; thus, it is suitable not as a commercial protectant, but rather as a valuable experimental tool.

Table 5.2 summarizes studies with EDU on crops in developing countries. Early reports on O_3-like symptoms were made by Bambawale (1986a, 1989), who observed them on local potato cultivars in the Indian Punjab. He demonstrated that these could be eliminated by application of EDU, this being the first demonstration of the phytoxicity of O_3 in the field in the Indian subcontinent. Bambawale (1986b) also demonstrated a similar phenomenon at another Indian northwestern plains location in Uttar Pradesh. Interestingly, these symptoms were recorded on the Indian side of the border with Pakistan, where our group subsequently carried out a series of experiments with open-top chambers and EDU out of the city of Lahore. An important experiment was conducted by Wahid et al. (2001) growing soybean, with and without EDU as a soil drench, at a site just outside Lahore and two other sites 35 km east and north of the city, respectively. Measurements of O_3 confirmed earlier studies, which showed that the pollutant increased continuously along a 39 km transect from the city center into the surrounding countryside (Shamsi et al., 2000). Substantial reductions in yield occurred in the non-EDU-treated plants compared with EDU-treated plants, with seed dry weight reduced by 53% at the site near Lahore, but by 65% and 74% at the two more rural locations, where O_3 levels were higher. This raises the very real possibility of O_3-induced yield reductions being widespread in the Punjab—the most important agricultural area of Pakistan.

The city in a developing country where O_3 has been recognized as a major problem for a long time is Mexico City, where topography, vehicle emissions characteristics and climatic factors all combine to produce exceptionally high levels of oxidants. Although most interest has been centered on effects on human health, some studies have investigated impacts on crops and trees, with visible symptoms of O_3 injury being reported on foliage. Laguette Ray et al. (1986) applied EDU to two cultivars of *Phaseolus* bean in Montecillos, outside Mexico City, and showed that yield was reduced in untreated plants by 29% in one cultivar but only 4% in the other. This demonstrates that cultivars of developing countries also have wide variations in sensitivity to O_3 within a single species, as demonstrated for many crops in the developed world.

Table 5.2. Effects of O_3 on Crops in Developing Countries as Demonstrated with EDU

Country	Crop	Effects on untreated cf. EDU-treated	Reference
India (Punjab)	*Solanum tuberosum* (potato)	17–30% leaf injury	Bambawale (1986a, 1986b, 1989)
India (Delhi)	*Lycopersicon esculentum* (tomato)	−26% shoot dry wt. −18% root dry wt.	Varshney and Rout (1998)
Egypt (Alexandria)	*Raphanus sativus* (radish)	−25% root dry wt.	Hassan et al. (1995)
	Brassica rapa (turnip)	NS	
Egypt (Abbis)	*Raphanus sativus* *Brassica rapa*	−37% root dry wt. −17% root dry wt.	Hassan et al. (1995)
Pakistan (Punjab)	*Glycine max* (soybean)	−53%, −65% −74% seed dry wt.	Wahid et al. (2001)
Mexico (near Mexico City)	*Phaseolus vulgaris* (bean)	−29% seed yield in one cv −7% in another cv	Laguette Rey et al. (1986)

NS, not significant at $p < 0.05$.

The first study of O_3 impacts on crops in the field in Africa was carried out by Hassan et al. (1985), who constructed two experimental plots in a suburban area of the Egyptian city of Alexandria and at the village of Abbis, 30 km to the south, in the Nile Delta agricultural region. A local cultivar of radish and of the less O_3-sensitive turnip were grown at each site, with and without EDU application. Lower O_3 levels were recorded in Alexandria than in Abbis, reflecting observations elsewhere of the pollutant being higher in rural areas. This was reflected in the relative effects of EDU at the two locations, in Alexandria, the dry weight of the edible root of radish was reduced by 25% but no significant effect was observed on turnip, yet in Abbis, reductions were 37% and 17%, respectively. This study, as in the case of the Lahore investigation for Pakistan, suggests that O_3-induced yield losses could be very high and widespread in the principal agricultural area of Egypt. Indeed, it can be postulated that the problem could he particularly severe in Egypt, where meteorological conditions are conducive to O_3 formation all year and extensive irrigation of crops in the Delta will result in high stomatal conductances and thus increased penetration of the pollutant into the foliage.

Further studies have been conducted into the effects of O_3 in the field in India. Thus, Varshney and Rout (1998) exposed tomato plants at three sites in Delhi, with and without EDU treatment. The results showed that shoot and root dry weight were reduced by 26% and 18%, respectively, in the untreated plants. A further unpublished study was carried out with moong bean (*Vigna radiata*) and palak (*Beta vulgaris*) grown in experimental plots up to maturity at nine sites across Delhi. Effects of EDU in increasing yield were again observed, and there was a significant relationship between crop dry weight and O_3 injury recorded at the sites on an O_3-sensitive indicator plant (*Nicotiana tabacum* cv. Bel-W3) (C. K. Varshney, personal communication, 1999).

Most of the information on effects of ambient O_3 in the developed world comes from open-top chamber filtration studies. However, the number of such studies conducted in the developing world is very limited. The most extensive investigation was conducted at an earlier date at the site on the edge of Lahore, where work with EDU subsequently showed a substantial O_3-induced reduction in yield of soybean. In this case, two local cultivars each of winter wheat and rice were grown in their successive growing seasons, repeated for two successive years. The crops were grown according to local agricultural practice in open-top chambers ventilated with ambient or charcoal-filtered air. The results are shown in Table 5.3. Remarkably consistent efforts were observed, with a reduction of 30–50% in yield of straw and grain of the two cultivars of each species in successive experiments. The charcoal filter will have removed other pollutants in addition to O_3, notably SO_2 and NO_2; thus their possible role in yield reduction must be considered. In the case of SO_2, concentrations at the site were negligible but there were elevated levels of NO_2. Consequently, parallel fumigation studies were conducted at Imperial College with the same cultivars of wheat fumigated with concentrations of O_3 and NO_2, either singly or in combination, similar to those recorded in the unfiltered chambers at Lahore (Maggs, 1996). These showed similar yield reductions to those reported for the Lahore chamber experiments in the case of O_3, but no effect of NO_2 either alone or in combination with O_3. Thus, there is extremely strong evidence that the yield reductions recorded at Lahore were caused by ambient O_3 alone, a matter further supported by the results of the subsequent experiment with EDU and soybean.

In many of these studies of impacts of O_3 in the field on crops in developing countries, exceptionally high yield reductions have been observed. The question is whether developing world cultivars are more sensitive to O_3 than their European/North American

Table 5.3. Mean Reductions in Yield Parameters for Wheat and Rice Grown in Unfiltered Air Compared with Filtered Air in Open-Top Chambers on the Outskirts of Lahore

Crop	Cultivar	Year	Straw dry wt.	Crop dry wt.
Oryza sativa	IRRI—6	1992	42%	37%
(rice)	Basmati—385		47%	42%
	IRRI—6	1993	37%	37%
	Basmati—385		51%	51%
Triticum aestivum	Pak—81	1991–1992	37%	47%
(wheat)	Chakwal—86	1992–1993	35%	35%
	Pak—81	1991–1992	19%	43%
	Chakwal—86	1992–1993	20%	33%

Data from Wahid et al. (1995a, 1995b); Maggs et al. (1995); Shamsi et al. (2000).

counterparts, or whether local environmental conditions or agronomic practices might render them particularly susceptible. No unequivocal answer can be given to these questions. In the latter case, it would be necessary to fumigate cultivars from both parts of the world under controlled conditions, manipulating climatic conditions and cultivation methods accordingly, and so far as we are aware, this has not been carried out. With respect to cultivar differences, there is some limited evidence that those from the developing world might be more sensitive. Thus, Wahid et al. (1995a, 1995b) considered carefully the response of their local cultivars of wheat and rice in relation to studies carried out in North America, Japan, and Europe using cultivars grown in these parts of the world. In the case of both species, the Pakistani cultivars appeared to be more sensitive than would be predicted on the basis of developed world studies, although differences in environmental conditions and agronomic practices could not be precluded from contributing to this observation. Further limited support for this theory has come from Hassan (1994), who simultaneously fumigated one Egyptian and three "Western" cultivars of both radish and turnip. For both species, the Egyptian cultivar was the most sensitive in terms of visible injury out of the four examined.

It is apparent that O_3 represents a major threat to crop production in many parts of the developing world. The studies outlined here provide experimental support for the predictions of Chameides et al. (1994, 1999) that phytotoxic O_3 concentrations are present in much of the developing world.

Potential Indirect Effects of Air Pollution on Crops in Developing Countries via Changes in Impacts of Biotic Factors

There is evidence from the developed world that the performance of both insect pests and fungal pathogens of crops are changed as a result of exposure to ambient levels of air pollution (Bell et al., 1993). Both field observations and controlled experiments have demonstrated that SO_2 and NO_2 exacerbate the impacts of both sucking and chewing insects, with the evidence pointing toward pollutant-induced changes in host chemistry being the causal mechanism, by rendering the plant as better nutritional value for the pests. In the case of O_3, the effects appear to be more complex, with negative impacts on the pest being observed in addition to stimulation in performance. Similarly, both negative and positive effects have been recorded for infection by fungal pathogens (Bell et al., 1993).

Although only a limited number of host–pest or pathogen systems have been examined in the developed world, literature on this issue is even more scant in developing countries. In the developed world, interest was initiated by observations that changes in intensity of infestation or infection occurred in polluted areas. Table 5.4 shows similar examples from the developing world, most of which are unpublished. Thus, a range of aphid and whitefly species appear to infest at higher than normal levels in an industrial area near Delhi, whereas sucking insect pest populations were higher on sorghum at a polluted compared with an unpolluted site near Ujjain in India. Also, at this latter location, it has been observed that pests of stored wheat grain were increased when the crops had been previously exposed to high SO_2 levels. So far as is known, this is the first indication anywhere that stored products could be affected in this way. In the Vaal Triangle—a heavily industrialized rural area of the South African Highveld—the maize stem borer has been reported to have a high level of infestation, while elsewhere in Africa, it has been noted that mealy bug infestation was particularly high at a location in Benin, where there was exposure to motor vehicle pollution.

Research aimed at elucidating causality of such observations is extremely limited. In Malaysia, bin Awang et al. (1991) fumigated cocoa plants with NO_2 and SO_2, and in combination, prior to being infested with cocoa mirids: All three pollutant treatments resulted in a significantly increased mean relative growth rate of the pest. Further research has been carried out in China, with Wu et al. (1996, 1997) showing that only moderate levels of SO_2 stimulated the growth and reproduction of the aphids, *Lipaphis erysimi* and *Myzus persicae*, feeding on turnip and rape, respectively. The potential significance of such impacts in the field in China has been demonstrated dramatically by Zheng et al. (1996), who exposed four local vegetable species (*Raphanus sativus*, *Brassica oleracea* var. *capitata*, *Lactuca sativa* var. *ungustana*, and *Brassica juncea* var. *involuta*) along a

Table 5.4. Field Observations of Changes in Pest and Pathogen Incidence in Relation to Air Pollution in Developing Countries

Location	Crops	Pest/pathogen and effect	Pollution source	Reference
India: Uttar Pradesh near Delhi (Noida)	Spinach, radish, pea, clover, wheat, sorghum; wild *Gardenia* and *Calotropis*	Increased aphid and whiefly infestation	Range of industries: SO_2, NO_x	M. Kasana (pers. comm., 1996)
India: Ujjain, Mahdya Pradesh	Wheat	Increase in pests of stored wheat grain from crop previously exposed to SO_2	General industrial	P.S. Dubey (pers. comm., 1997)
Ujjain	Sorghum	Mean sucking insect weight increase	General industrial	Rao et al. (1990)
South Africa: Vaal Triangle	Maize	Increase in maize stem borer	SO_2 from various industries	H. Van Hamburg (pers. comm., 1996)
Benin	Various spp.	Increase in mealy bug infestation	Motor vehicles (taxi-park)	S. McNeill (pers. comm., 1997)

gradient of air pollution in the industrial city of Chongqing. Not only were yields drastically reduced in relation to the prevailing SO_2 and NO_x levels but levels of aphid infestation also increased in parallel. More recently, our group has carried out a field study in India on the effects of SO_2 and NO_2 on two important crop pests. This was conducted in Varanasi at sites with different levels of pollution (Davies, 2000). The study provided strong evidence of increased performance of the mustard aphid (*Lipaphis erysimi*) on mustard and of the diamond-backed moth (*Plutella xylostella*) on cabbage in terms of a whole range of growth and reproductive parameters being increased at the more polluted sites, including reduction in larval period and time to adult emergence, and increase in reproductive period and fecundity (Davies, 2000). Thus, there is clear evidence that performance of major sucking and chewing insect pests of the developing world is enhanced by ambient pollution levels in agricultural areas of a major Indian city.

Research into the effects of air pollution on fungal pathogens in developing countries is even more limited than that on herbivorous insect pests. Again, much of the relevant research has been carried out in India, by a group at the Aligarh Muslim University (e.g., Khan and Khan, 1994), who also studied impacts on root-knot nematodes (e.g., Khan and Khan, 1997). A recent study by this group has demonstrated the potential complexity of such effects. Thus, Khan and Khan (1999) fumigated *Cucumis sativa* (cucumber) plants with 50, 100 and 200 ppb O_3 for 7 hours per day every 3 days over a 75-day period. The plants were infected with the important powdery mildew, *Spaerotheca fuliginea*, and the development of the disease was monitored. It was found that compared with control plants (17 ppb O_3) there was a significantly increased level of fungal colonization at 50 ppb O_3, but the reverse effect occurred at the two higher concentrations. This study has important implications in that the O_3 regime that caused a stimulation of pathogen performance reflects ambient conditions close to cities in the developing world.

Conclusions

All the research outlined in this chapter points toward ambient air pollution having major impacts on yield of crops in some parts of developing countries. In all areas studied, there is evidence for such effects, particularly within or in the proximity of large cities. Within urban areas, SO_2 and NO_2 appear to be the major problems, and this situation is likely to apply to the majority of cities in the developing world, particularly in the case of the latter pollutant, which is closely linked to emissions from the ever-growing volume of motor vehicles. Little is known about the distribution of elevated levels of O_3 in rural areas of developing countries, but experience from elsewhere and the studies in Egypt and Pakistan reported in this chapter indicate that phytotoxic concentrations may be widespread for at least 35 km away from cities. A particularly disturbing recent finding is the demonstration that India's ambient air pollution reduces crop quality. Thus, not only does the consumer have access to lower yields of staple foods, and thus, presumably, higher prices, but what is available is of a lower nutritional quality.

This strong evidence that air pollution impairs crop yield and quality in developing countries highlights the urgent requirement to quantify and elucidate the extent of this hitherto unexpected constraint on production. In particular, experiments need to be carried out to determine dose–response relationships between the main pollutants, both singly and in mixtures, and yield of local cultivars of key agricultural crops, grown under agronomic and climatic conditions prevailing in the areas of interest. This should be accompanied by extensive air pollution monitoring in both urban and rural areas, together with appropriate modeling, to assess the extent of the occurrence of phytotoxic concentrations. Thus, key information could be provided for a risk assessment for impacts on crops, which in itself can be used to formulate policies that will not only improve human health directly but also indirectly, via improved crop yield and quality.

References

Adams, R. M., Glyer, J. D., and McCarl, B. A. (1988). The NCLAN economic assessment: Approach, findings and conclusions. In: W. W. Heck, O. C. Taylor, and D. T. Tingey (Eds.), *Assessment of crop loss from air pollutants* (pp. 473–504). New York: Elsevier.

Ashenden, T. W., and Mansfield, T. A. (1978). Extreme pollution sensitivity of grasses when SO_2 and NO_2 are present in the atmosphere together. *Nature, London* 273: 116–127.

Ashmore, M. R., and Marshall, F. M. (1997). *The impacts and costs of air pollution on agriculture in developing countries*. Final technical report submitted to the Department for International Development, Environment Research Program, Project No. ERP6289. London: Imperial College of Science, Technology, and Medicine.

Ashmore, M. R., and Marshall, F. M. (1999). Ozone impacts on agriculture: An issue of global concern. *Advances in Botanical Research* 29: 32–52.

Ashmore, M. R., and Wilson, R. B. (Eds.). (1994). *Critical levels for air pollutants in Europe*. London: Department of the Environment.

Bambawale, O. M. (1986a). Evidence of ozone injury to a crop plant in India. *Atmospheric Environment* 20, 1501–1503.

Bambawale, O. M. (1986b). Detection and severity of ozone injury on potato crop in north western plains of India. *Proceedings of National Seminar on Environmental Pollution Control and Monitoring* CS10: 145–154.

Bambawale, O. M. (1989). Control of ozone injury on potato. *Indian Phytopathology* 42: 509–513.

Bell, J. N. B. (1984). Direct effects of air pollution on plants. In: C. Troyanowsky (Ed.), *Air pollution and plants* (pp. 116–127). Weinheim: VCH.

Bell, J. N. B., McNeill, S., Houlden, G., Brown, V. C., and Mansfield, P. J. (1993). Atmospheric change: Effect on plant pests and diseases. *Parsitology* 106: S11–S24.

bin Awang, M., Muhamad, R., and Hassan, S. A. (1991). The effects of sulphur dioxide and nitrogen dioxide singly or in combination on the growth performance of cocoa mirids (*Helopteltis theobromae*). *Pertanika* 14: 7–13.

Botha, A. T., Moore, L. D., and Visser, J. H. (1990). Gladiolus and bean plants as biomonitors of air pollution impacts in the Cape Town area. Proceedings of the 1st UAPPA Regional Conference on Air Pollution: Towards the 21st Century, Pretoria, Vol. 1, Paper No. 18.

Chameides, W. L., and Kasibhatia, P. S. (1994). Growth of continental scale metro-agro-plexes, regional ozone pollution and world food production. *Science* 264: 74–77.

Chameides, W. L., Li, X., Tang, X., Zhou, X., Luo, C., Kiang, C. S., St. John, J., Saylor, R. D., Liu, S. C., Lam, K. S., Wang, T., and Giorgi, F. (1999). Is ozone pollution affecting crop yields in China? *Geophysical Research Letters* 26: 867–870.

Davies, C. (2000). *Air pollution and agricultural insect pests in urban and periurban areas of India: A case study of Varanasi*. Ph.D. thesis, University of London.

Evelyn, J. (1661). *Fumifugium or the inconvenience of the aer and smoake of London dissipated: Together with some remedies humbly proposed*. London: W. Godbid.

Faiz, A., and Sturm, P. J. (2000). New directions: Air pollution and road traffic in developing countries. *Atmospheric Environment* 34: 4745–4746.

Fuhrer, J., Skärby, L., and Ashmore, M. R. (1997). Critical levels for ozone effects on vegetation in Europe. *Environmental Pollution* 97: 91–106.

Hassan, I. (1994). *Effect of ozone on Egyptian varieties of radish and turnip*. Ph.D. thesis, University of London.

Hassan, I. A., Ashmore, M. R., and Bell, J. N. B. (1995). Efffect of ozone on radish and turnip under Egyptian field conditions. *Environmental Pollution* 89: 107–114.

Khan, M. R., and Khan, M. W. (1994). Effect of ozone on conidea of *Sphaerotheca fuliginea*. *Indian Phytopathology* 47: 172–176.

Khan, M. R., and Khan, M. W. (1997). Effects of root-knot nematode, *Meloidogyne incognita*, on the sensitivity of tomato to sulfur dioxide and ozone. *Environmental and Experimental Botany* 38: 117–130.

Khan, M. R., and Khan, M. W. (1999). Effects of intermittent ozone exposures on powdery mildew of cucumber. *Environmental and Experimental Botany* 42: 163–171.

Kitada, T., and Azad, A. K. (1998). Study on the air pollution control system for Dhaka, Bangladesh. *Environmental Technology* 19: 443–459.

Laguette Rey, H. D., de Bauer, L. I., Shibata, J. K., and Mendoza, N. M. (1986). Impacto de los oxidante amtales en el cultivo de frigol, en Montecillos, estado de Mexico. *Centro de Fitopatologia* 66: 83–95.

Lee, D. S., Kohler, I., Grobler, E., Rohrer, F., Sausen, K., Gallardo-Klenner, L., Olivier, J. G. J., Dentener, F. J., and Bowman, A. F. (1997). Estimations of global NO_x emissions and their uncertainties. *Atmospheric Environment* 31: 1735–1749.

Maggs, R. (1996). *The effects of ozone and nitrogen dioxide on Pakistan wheat (Triticum aestivum) and rice (Oryza sativa) cultivars.* Ph.D. thesis, University of London.

Maggs, R., Wahid, A., Shamsi, S. R. A., and Ashmore, M. R. (1995). Effects of ambient air pollution on wheat and rice yield in Pakistan. *Water, Air, and Soil Pollution* 85: 1311–1316.

Marshall, F. M., te Lintelo, D. J. H., Wildig, Z., Stonehouse, J., Bell, J. N. B., Ashmore, M. R., and Batty, K. (2000). *The impacts and policy implications of air pollution on crops in developing countries.* Final Technical Report, Department for International Development, Environment Research Program R6992. London: Imperial College of Science, Technology and Medicine.

Rao, M. V., Gupta, G. K., and Dubey, P. S. (1990). Effects of relatively low ambient air pollution on the total sucking insect populations on *Sorghum vulgare. Tropical Ecology* 31: 66–72.

Shamsi, S. R. A., Ashmore, M. R., Bell, J. N. B., Maggs, R., Kafayat, U., and Wahid, A. (2000). The impacts of air pollution on crops in developing countries—a case study in Pakistan. In: *Environmental stress: Indication, mitigation, and eco-conservation.* M. Yunus, Ahmed, K. J., Singh, N., and Dekok, L. J. (Eds.), (pp. 63–71). Dordrecht: Environmental Academic Publishers.

Singh, J. S., Singh, K. P., and Agrawal, M. (1990). *Environmental degradation of Obra–Renukoot–Singrauli Areas and its impact on natural and derived ecosystems.* Project report submitted to Ministry of Environment and Forests, Government of India, 4/167/84/MAB/EN-2RE.

United Nations Development Program. (1996). *Urban agriculture: Food, jobs and sustainable cities.* New York: Author.

Varshney, C. K., and Rout, C. (1998). Ethylene diurea (EDU) protection against ozone injury in tomato plants at Delhi. *Bulletin of Environmental Contamination and Toxicology* 61: 188–193.

Wahid, A., Maggs, R., Shamsi, S. R. A., Bell, J. N. B., and Ashmore, M. R. (1995a). Air pollution and its impact on wheat yield in the Pakistan Punjab. *Environmental Pollution* 88: 147–154.

Wahid, A., Maggs, R., Shamsi, S. R. A., Bell, J. N. B., and Ashmore, M. R. (1995b). Effects of air pollution on rice yield in the Pakistan Punjab. *Environmental Pollution* 90: 323–329.

Wahid, A., Shamsi, S. R. A., Milne, E., Marshall, F. M., and Ashmore, M. R. (2001). Effects of oxidants on soybean growth and yield in the Pakistan Punjab. *Environmental Pollution* 113: 271–280.

Wu, K., Gong, P., and Li, X. (1996). Effects of SO_2-polluted rape plant on growth and reproduction of the turnip aphid, *Lipaphis erysimi* (Kaltenbach). *Journal of Environmental Sciences* 8: 414–418.

Wu, K., Gong, P., and Li, X. (1997). Effects of rape grown in SO_2-enriched atmospheres on performance of the aphid, *Myzus persicae*, (Ssulzer). *Entomologia Sinica* 4: 82–89.

Zheng, Y., Last, F. T., Xu, Y., and Meng, M. (1996). The effects of air pollution climate in Chonqing on four species of vegetable. *Chonqing Environmental Science* 18: 29–32 [in Chinese].

6

Applied Ecology of Biodiversity

Navin K. Ambasht and R. S. Ambasht

Introduction

Formation of new species and extinction of old ones have progressed hand in hand over long periods in the geological past in such a way that a rich variety of life forms has existed on terrestrial space and in the ocean waters. With the increase in richness of animals, and microbial life, each life form has acquired greater niche specialization and function. Competition, adjustment, symbiosis, food chains, and webs of eating and being eaten bring about some kind balance and dynamic equilibrium among the members of biotic communities and other ecosystem components. Kinds and sizes of populations of component species are controlled by the availability of space and food, prevailing biotic and abiotic stresses, catastrophic events, continued availability of propagules, and so on. Amplitude of fluctuations occur in seasonal cycles, particularly the duration of extremes of stressed situation, and competitive and symbiotic effects of native biotic associates. Populations of invader species have great bearing on the sustainability, or otherwise, of biodiversity. Phenomenon of succession (i.e., replacement of biotic communities at any place, particularly under anthropogenically disturbed situations) must also be considered while planning long-range strategies. To begin with, a variety of diaspores reach any virgin or freshly cleared habitat, but only some successfully establish; develop competition, symbiosis, and trophic links; and modify the microenvironmental complex in a manner that is no longer good for the existing communities, and allow new invasions and change in biotic communities structure. The successional phenomenon continues until a dynamic equilibrium of biotic and environmental complexes is reached and climax is attained. Even the climax communities are also not permanent and show aging over long periods of time. Evolution of new strains, varieties, ecotypes and species, if better equipped genetically and ecologically, cause replacement of weaker ones, and this may also lead to natural extinction. The normal status of biodiversity has not been adversely affected by natural speed of extinction. Anthropogenic disturbances are responsible for speeding the rate of

Navin K. Ambasht and R. S. Ambasht • Department of Botany, Banaras Hindu University, Varanasi, India; E-mail: rambasht@banaras.ernet.in.

extinction and retarding the speed of speciation in various ways. The emergence of humans as a dominant force, the rapid raise in their population and their invasion of all available, habitable space has disrupted natural balance and caused serious loss of biodiversity. Population size of thousands of species has been cut down to levels that threaten their future survival. Only a few species, such as human beings, have selected and cultivated crop species, domesticated animals, and those providing human food, such as goats, sheep, cattle, poultry, fish, shrimp, and so on, and have increased at the cost of other plants and animals. The worst part is that even the wild relatives of a few selectively cultivated cereals have not been protected, even though they are priceless gene banks. These and even hundreds of varieties so far cultivated have been discarded and replaced by genetically evolved, high-yield, disease-resistant strains. Humans used genes of wild relatives to improve the cultivated varieties, and in the future, when the stock of wild populations disappeared, then creation of new desired strains would be difficult. Loss of biodiversity is bound to limit seriously the choice and scope of producing new strains of cultivated plants and domesticated animals. With the advancement of ecology, the two most important problems identified by the United Nations Conference on Environment and Development (UNCED) held at Rio de Janeiro are those related to (1) biodiversity losses and (2) global changes (in climate).

Not only the quality of human life but also future survival and sustenance are at great risk, if an all-out effort is not made to tackle these two problems. Every day, we come across some new information on biodiversity loss or some other wild species that has become rare, threatened, or endangered. There is a general feeling that the weaker species suffer worst, but the strong species such as lions, tigers, and leopards are also endangered. These animals are being killed for their skins, bones, and furs. Many others are also being hunted for their flesh to obtain prized leathers, furs, beautiful feathers, horns, tusks, and so on. Despite stringent laws, many of the protected animals are being exterminated due to the extremely high cost of these material.

Expensive Shahtoosh shawls made of soft hairs of a Tibetan animal antlers were commonly available a few decades ago, but now the species has become endangered and is totally protected. Possession of an old Shahtoosh shawl is prohibited in India and punishable with fine and imprisonment, yet on payment of 40 or 50 thousand rupees one can easily obtain them in markets of Kashmir and Delhi, as sometimes reported in newspapers.

Newspapers frequently report confiscation of protected wildlife products. Thus, legislation and government vigilance do not effectively arrest biodiversity losses, because of the huge amount of money involved in illegal transactions. A peculiar example is the very rapid decline in population size of Indian vultures. These most efficient scavengers of dead animals, until a few years ago, could be seen in hundreds around the carcasses of dead animals. Vultures were always seen high in the sky and everywhere. They did not suffer shortages of food or from competition of any kind, yet during the past 10 years or so, vultures have become rare. Dead bodies keep rotting and stinking due to the absence of this efficient scavenger. No conclusive reason is available to explain this. Vultures are sturdy birds, capable of prolonged flight at high elevations. They occupy the top of food chain and are not killed and eaten by other animals or man. It is still not clear, how the population size of this species, free from competition, shortage of food, and territorial space, has so rapidly declined. One reason could be the destruction of the big trees of *Ficus religiosa* (Peepal), *F. benghalensis* (Bargad), and a few others on which these birds perched for night rest. They could not have adjusted to other trees for night rest. Another commonly cited reason is the possibility of spread of some viral disease in epidemic form. Scientific reasons have to be worked out soon on a priority basis. Meanwhile the Bombay Natural History Society has initiated a conservation program by producing their offspring in captivity and releasing them in nature.

Definitions

Ecologists use the term *diversity* and have developed methods to quantify it. "Species diversity is a statistical abstraction with two components" (i.e., richness and evenness or equitability). The product of richness and evenness is referred to as species diversity; the inverse of evenness is concentration. In a plant community with 100 species, the richness is 100. If all species were represented by an almost equal number of individuals, they would be referred to as most even, but if one or two species had very large population sizes, they would be called concentrated and dominant. Whittaker (1965) has differentiated three kinds of diversity depending upon the range of environment and community. An *alpha index of diversity* is concerned with a single community. But a study of more than one community in the same range of environment is called *beta index of diversity*.

A study of diversity extending to several communities in different ranges of environment is called a *gamma index of diversity*. Individuals of certain species outnumber others and exert a maximum controlling effect on the habitat; these are called *dominants*. Odum et al. (1969) have supplied the simplest quantification. They refer to diversity as the number of species encountered in the first 1,000 individuals (d = number of species per 1000 individuals in a community). Menhinick (1964) originated the formula:

$$d = \frac{S}{\sqrt{N}},$$

where d is diversity, S is the number of species, and N is the number of individuals. Margalef (1968) amended this formula:

$$d = \frac{S-1}{\log N}$$

Alpha diversity is also expressed as the number of species (or taxa) per unit space (Whittaker, 1972). Beta diversity is the change or replacement of species from one plot to another. For this, Jaccard's calculated index (J) of similarity can be a maximum of one, if there is a total similarity between the two plots in a range of environment, and a minimum of zero, if there is not a single common species between the two plots. Beta diversity can be expressed as $1 - J$ (i.e., inverse of similarity). Ashton (1998) has pointed out that the term *biodiversity* has been used in many diverse ways: Planners with backgrounds in social science use it to describe structural complexity or heterogeneity at the level of plant community; ecologists are interested in genetic diversity, within species, population and among species sharing a community. Ambasht and Ambasht (1998, 1999) have described and discussed the methodologies and scope of biodiversity.

Relative to Whittaker's (1960, 1972) alpha diversity, Huston (1994) states that it is a "within area" diversity (i.e., the number of species found to occur within an area of given size), in which only the presence, and not the number of individuals, is counted. In tropical rain forests the alpha diversity may be several times higher than in an equivalent area of temperate forests. Beta diversity refers to degree of species change along a habitat or physiographic gradient (i.e., it is essentially a measure of "between-area diversity.") It cannot be expressed as the number of species, but in terms of similarity or dissimilarity. Gamma diversity is also a measure of "between-area" diversity, in which a large region or landscape is considered (Cornell, 1985). Cody (1986, 1993) has considered it as a measure of turnover of species in different segments of a large habitat type. However, exact and agreed definitions are yet to come (Latham and Ricklefs, 1993).

Species richness of a site does not reflect the kind of taxa. For example, a place may be rich in birds and poor in butterflies, while in another habitat the reverse may be true. The relationship is not that simple. Schall and Pianka (1978) have found wide differences in associated animals for frog-rich areas in North America and Australia at similar latitude (north and south) and longitude (west and east). Different scientists who measure the presence may report higher values if their area of sampling is greater. Generally, the number of species doubles when the area is increased 16 times (United Nations Environment Program [UNEP], 1995).

UNEP (1995) has described *taxic diversity* as a measure of taxonomic dispersion in which monotypic taxa such as *Ginkgo biloba* are valued higher, because they represent a species, a genus, a family, and an order in the taxonomic classification. *Functional diversity* is another term now being used to give due weight or importance to functional attributes of a species, such that keystone species with stronger controlling ability are rated higher.

Biological diversity, or *biodiversity*, is the sum total of variability of life found at a single place. Life forms include plants and animals of all classes, sizes, and biomass, and microorganisms, including bacteria and viruses. Species is a well-defined taxonomic unit for plants and animals, but for microbial life, the concept of species may differ. Individuals of a species are similar in look and are interfertile. World Resource Institute (WRI), (World Conservation Union (IUCN), UNEP, (1992)) have defined biodiversity as the totality of genes, species, and ecosystems in a region. It is also defined as the genetic variability and diversity of life forms, such as plants, animals, and microbes.

Biodiversity is the ensemble interaction of genetic, species, and ecological diversity in a given place and time. Younes and diCastri (1996) have refined the definition as the ensemble and hierarchical interactions at the genetic, taxonomic, and ecological scales of organization and different levels of integration. (United Nations Education Scientific and Cultural Organization, 1994) has regarded *biodiversity* as an umbrella term for variability among living organisms from all sources, including terrestrial, marine and other aquatic ecosystems, and ecological complexes of which they are a part. Diversity can occur within species, and between species and ecosystems. Biodiversity is also a function of time (such as changes brought about by evolution) and space (i.e., biogeographical distribution).

Diversity within species is a common attribute of temporary changes found in ecad or ecophene populations and genetically fixed changes found in ecotypes. These changes within the framework of taxonomic species are brought about by environmental stresses or favorable situations found in the distribution area of concerned species.

Article 2 of the Convention on Biological Diversity (CBD) regards biodiversity as the variability among living organisms from all sources, including, among others, terrestrial, marine and other aquatic ecosystems, and the ecological complexes of which they are part. Diversity within species, between species, and of ecosystems is included. CBD was initiated by the UNEP in 1987. The convention was first adopted by 101 governments in 1992, in Nairobi, then came into force on December 29, 1993, and includes now about 120 governments. (International Union of Biological Sciences/Scientific Committee on

Problems of Environment/UNESCO) has published *From Genes to Ecosystems: A Research Agenda for Biodiversity*, edited by Solbrig (1991). UNEP (1995), in *Global Biodiversity Assessment*, has defined biodiversity as the "total diversity and variability of living things and of systems of which they are a part." Within the scope of biodiversity, *cultural diversity* is an important aspect that includes human influences for use and conservation practices. Many ethnobotanical aspects are also related to biodiversity. Traditional knowledge, customs, and uses have a great bearing on the present-day status of biodiversity (Gadgil and Berkes, 1991). Perception of biodiversity varies widely among different segments, such as biologists, sociologists, lawyers, naturalists, conservationists, ethnobiologists, and so on. The biodiversity issues have been a unifying force among people of various professions and pursuits. Inclusion of time frame has added the dimension of extinct species, as well as the creation of new ones by natural mutations, artificial breeding, and application of biotechnology.

The scope that biodiversity covers is compositional, functional, and cultural. Compositional components include the following:

1. Genetic diversity (i.e., variation of genes within species, such as the hundreds of traditional rice varieties found in India).
2. Species diversity (i.e., richness in the number of species).
3. Ecosystem diversity, in which sharp compartmentalization is not easily possible, because the expanse of communities and diversity refers to ecosystem functions rather than the species that perform it. Widely different taxa may perform similar functions. A cluster of such species forms a guild. Certain species that have overriding functions or controlling effects are called key-stone species. Cultural diversity includes human influences, customs, traditional practices, and so on. Shifting cultivation and a nomadic way of life are adjustments of human society to seasonal and longer ecosystem recovery cycles. WRI/IUCN/UNEP (1992) has regarded cultural diversity as helping people to adopt to changing conditions. "Cultural diversity is manifested by diversity in language, religious, beliefs, land management practices, art, music, social structure, crop selection, diet, and any number of other attributes of human society." (WRI/IUCN/UNEP, 1992, p. 3).

What We Know and Do Not Know about Biodiversity

Biodiversity is the totality of variability among living resources and the ecosystems of which they are a part. Variability can be conceived at various levels, such as genes, populations within species, species, and ecosystems. The commonest and easiest method of expression is the number of species at a place (alpha diversity). Man has made remarkable advancements in many fields of science and has explored space and learned about stars and planets, but when it comes to recording and scientifically describing different kinds of plants, animals, and microbes, there remains much more to be explored and recorded than hitherto known. Botanical and zoological naming and describing species' characteristics is a part of classification or taxonomy. A large number of species are described under folk classification and have vernacular names not yet listed in scientific floras or faunas. WRI/IUCN/UNEP (1992) has stated that although 1.4 million species are described, the total number, including the hitherto undescribed species may be nearly 10 million. About microbes, very little is known; there is a tendency to regard an estimate of

10 million as a lower figure. UNEP (1995) agrees on a working figure of 13 million as the most acceptable and of this, only one-eighth have so far been described. Wilson and Peter (1988) have estimated the number of plant species belonging to different taxa, such as blue-green algae (1,700), green algae (7,000, brown algae (1,500), red algae (4,000), Chrysophyta (12,500), ascomycetes (28,650), basidiomycetes (16,000), Bryophytes (16,600), Pteridophytes (11,294), gymnosperms (529), dicotyledons (170,000), and monocotyledons (50,000).

UNEP (1995) has estimated the numbers of currently accepted, scientifically described species for all categories of organisms. For green and red algae, the numbers are 10,000 and 5,000 respectively. The mosses and liverworts account for 16,000, and ferns, 10,000. Seed plants (i.e., gymnosperms and angiosperms together) have 240,000 species and fungi, 72,000. The Animalia have 1,320,000 species, of which Arthropoda alone accounts for 1,085,000. Chordata, which include amphibians, reptiles, fishes, birds, and mammals together have 45,000 recorded species. These figures are close approximations, with about a 20% degree of uncertainty. The total number of described species for all categories of organisms is 1,750,000, whereas a workable estimated figure of all species described and undescribed is 13,620,000, but the accuracy of these figures is rated as very poor. These figures were obtained by updating Hammond's (1992) species inventory. The ratio of described to expected undescribed species is very high in bacteria and viruses. Others with very high number of undescribed species are fungi, nematodes, protozoa, crustaceans, and insects. Highest shares of described versus total expected number of undescribed species are for vertebrates and land plants.

Information on the ratio between the described and undescribed number of species has great limitations. We do not know the within-population variability and population sizes of described species in almost all cases. We do not know the controlling functions of the members of a community. Grazing animals are known to increase the population of some grasses and forbs, and to decrease those of others. Key stone function of most species are also not known. The exact roles of biodiversity in global ecological balance, ecosystem functions, and human welfare are also not fully understood. But with the stark fact that the loss of biodiversity is rapidly taking place, we are left without the option of waiting until more information is generated but must begin globally and locally to conserve the biodiversity.

Myths and Facts

The subject of biodiversity, particularly its global ecological importance and economic value, has strongly emerged in social and academic worlds. In such a phenomenal rise in popularity, there are bound to be misconceptions about biodiversity and its management solutions. UNESCO (1994) has published certain myths, facts, and management actions in respect to biodiversity. The most important myth is that many biologists know well what kinds of biodiversity exist and what purposes they are serving. It is clear from the foregoing descriptions that we have not identified and described more than 1.5 million species out of the expected more than 10 to 13 million that exist today. Furthermore, among the 1.5 million species, we do not know the population sizes, niche functions, keystone roles, or even their rarity or threatened status. We should know these and the ecological services and economic products they provide. Therefore, improving knowledge on biodiversity is imperative for better appreciation and management. Another myth is that we can save all the threatened species from becoming extinct. The fact is that each species has a certain finite life span and through natural process will also become extinct. Presently, many more species are becoming extinct long before their naturally expected time. The wise course would be to attempt to minimize the rate of biodiversity loss. McKinney and Schoch (1998) have used the term *species triage* to refer to the priority a species deserves among many threatened ones for saving. *Triage* is a medical term used in operation theaters to give priority to such patients whose chances to survive are best, if attended to immediately. Some loss of biodiversity has been inevitable as the human population has heavily encroached on available terrestrial and wetlands. The best action in biodiversity conservation should be at ecosystem and landscape levels. This would help to protect the whole food chains and food webs, as well as competing and symbiotically adjusted species.

Another great myth or misconception is that prohibiting resource use is the best method of conservation. The concept of ecological conservation is based on sustainable use. The fact is that an element of local support and participation of inhabitants, and some minimal use must form an integral part of the biodiversity conservation plan. Sustainability must be ensured. In the present-day policies, in order to save gravely threatened species, a total ban on use or killing has been enforced. This has led to partial recovery of vanishing populations, but unauthorized poaching and illegal trade in wildlife products are responsible for poor results. A management plan involving the locals and deciding on minimal harvesting of the resource at a sustainable level would give the program better long-range success and reduce clandestine trade. Of course, in this kind of plan, the local human population has to be provided with new lifestyle alternative options, so that their dependence on protected areas are minimal or achieve a sustainable limit. The local population must not be employed as laborers but as true partners in both the profits and losses. For example, with respect to large lakes, excessive fishing for commercial purposes and hunting of game birds by contractors, decided in year-to-year auctions have led to gross loss of biodiversity and habitat productivity. If in the biodiversity conservation, the site is restored as a nice tourist center with aquatic sports and boating, the local people so far dependent on fish harvesting should have a share in the income from tourism, with involvement in managing the lake as responsible partners. Sustainable limits of fish harvesting and other resources must not be allowed to cross the limits of regeneration speed.

Most of the governments enact their biodiversity conservation plan by establishing a few biosphere reserves or national parks. This, again, is a myth. No protected area can remain unaffected by pollution of the air, water, and soil of the immediate surroundings and, to some extent, by pollution at far off places. The creation and management of biosphere reserves must be integrated with regional development. "Habitat protection is the single most effective means of protecting biodiversity at all levels of organisation" (UNESCO, 1994). Many people of the temperate regions have the misconception that their geographical region is not important from a biodiversity conservation viewpoint, because most biodiversity exists in the tropics. This is a myth. Temperate regions not only have a good share of biodiversity of flora and fauna characteristics but they are equally threatened. We are indeed not sure of the share of marine biodiversity with respect to tropical and temperate regions. Furthermore, the conservation of any flora or fauna affects for global benefit, and so does the loss. In the long range, we have to evolve global strategies, and all regional efforts should be a part of this global program. Such a perspective of biodiversity conservation came into force on December 29, 1993 (UNESCO, 1994). IUCN (1980) published the *World Conservation Strategy* and dealt with conservation use. The World Commission on Environment and Development (WCED) in 1987 published *Our Common Future*, and in 1991, the IUCN/UNEP/WWF (World Wildlife Fund for Nature) published *Strategy for Sustainable Living Use*. Partnership of local people and ensuring sustainability should be the guiding principles of biodiversity conservation. Sustainable use emphasizes that there is no decline in the biological diversity. This ensures the long-range needs of not only the present generation of people but also future generations.

Causes of Biodiversity Losses

A large number of factors and forces are responsible for the steep decline in the earth's biodiversity during the last century. Most of these are caused directly by anthropogenic activities of burgeoning populations, such as deforestation; conversion of wetlands into filled, flat surfaces; overexploitation of certain animals and plants for game, food, medicine, fur, skin, bone, horns, tusks, musk; habitat destruction narrowing the spectrum of a few selected cereal crops, discarding many coarse and less palatable (not necessarily less nutritive) species and traditional varieties; and the narrowing spectrum of traded animals and selective preference of some fishes, due to either discarding other fish species or competition of introduced ones. Human help, knowingly or unknowingly in migration of alien weeds is also responsible for eliminating some local weeds, with the aliens' success in gregariously covering the available space. This leads to extinction of local flora and fauna. Before taking up the details of causes, we briefly discuss *extinction*.

The total disappearance of certain species in a broad area but not totally from the earth is called *extirpation* (McKinney and Schoch, 1998). When the level of extinction is extremely widespread and the population size of remaining organisms is poor and without much effect on the ecosystem, it is called *ecological extinction*. Such species are called "living dead" or "living fossils." They need protection on priority basis. If a taxa totally disappears from earth, it is called *extinct*. Extinction of species has occurred naturally at a slow rate and at mass scale at long geological intervals during major global catastrophies. The last five major catastrophies occurred in Silurian (438 million years ago), Devonian (360), Permian (245), Jurassic (208) and Tertiary (65) times, causing massive extinction of both plants and animals. In recent geological periods megascale glaciation has been responsible for extinction.

Heywood and Watson (1995) have stated that against an estimated natural rate of mammalian extinction of one extinction per 200 years, the current rate must be roughly 100 times faster. Sepkoski and Raup (1986) have regarded the Triassic and Cretaceous episodes as most damaging, resulting into mass extinctions. The IUCN, in the *Red Data Book*, regard those taxa extinct that are no longer known to exist in wild. Convention on International Trade in Endangered Species (1979 conference) regards taxa as extinct that have not been found for at least 50 years. The World Conservation Monitoring Centre (WCMC; 1992 personal communication, cited by Heywood and Watson, 1995) recorded that, during the past four centuries, 484 animals (including 115 birds and 58 mammals) and 654 plants have become extinct, but the UNEP (1995) regarded this as an underestimate, especially for extinctions in the tropics. Furthermore, the estimate did not take into consideration all groups of animals and plants. Depending upon the status of decreasing number of individuals, Jain (1981) has followed IUCN in dividing taxa into (1) *Endangered*: "Taxa in danger of extinction and whose survival is unlikely if the causal factors continue operating are deemed to be in immediate danger of extinction"; (2) *Vulnerable*: Taxa believed likely to move into the endangered category in near future if the causal factors continue operating"; (3) *Rare*: Taxa with small world populations that are not at present endangered or vulnerable but are at risk. These taxa are usually localized within restricted geographical areas or habitats". The terms *vulnerable* and *rare* are sometimes confused. Jain (1981 p. 9) clarifies that rare species have small but relatively stable population, whereas vulnerable populations are on decline. Heywood and Watson (1995) have termed the categories (following World Conservation Monitoring Centre classification) *endangered*, *vulnerable*, *rare*, and *indeterminate*. The *indeterminate* category includes taxa whose status is uncertain with respect to extinction or the other three categories. The number of species of plants in the endangered category is 3,632, in the vulnerable category: 5,687, rare, 11,485, and indeterminate, 5,302 (i.e., a total of 26,106 according to the WCMC). Among animals, the total number of species in the four categories is 2,719 for vertebrates and 2,647 for invertebrates.

Several attempts have been made to determine the possible rates of extinction (Lovejoy, 1980; Raven, 1987; Reid, 1992). Reid (1992), using the species area curve method, has estimated the global loss of biodiversity to be about 25% at the species level between 1985 and 2015. This estimate is based on the extent of deforestation and loss of about 2,000 plant species per year in the tropics and subtropics.

There are a variety of causes for biodiversity losses causing shifts from vulnerable, to rare, to threatened status, leading ultimately to extinction. Loss of habitats, mostly through deforestation, mining, agriculture extension, and conversion of wetlands into uplands are important causes. With habitat destruction, the diversity of plants and animals is also destroyed. Even birds, which have the capability of moving away from stressed areas, may fail to establish in new habitats and are driven to a threatened state. Fragmentary habitat losses create discontinuity of growth of a continuous population structure. The gaps created may prevent natural flow of individuals and disruption of ecological life cycles of different species. In small, fragmented habitats, predators usually cause more impact on prey populations (Wilcove, 1985). Large animals require extensive and continuous natural habitats for their movement and to ensure abundant food supply. In fragmented patches, rate of recovery of vegetation may not match the greater demands of grazing animals. In the "Jhum," or slash-and-burn cycle of preparing forests for raising crops in the northeast Himalayas, the time period for recovery has been shortened to the extent that habitat restoration is not adequate before the next slash-and-burn treatment. This has made one of the world's rainiest places, once covered with very dense, multilayered rain forest, into a mosaic of eroded and rock-exposed patches. Tropical rain forests essentially have poor nutrient soil, because over the years nutrients are locked in the bodies of plants (i.e., nutrient cycling is intrabiotic). Once the forest is destroyed and the timber removed, the exposed recovery fails. In a tropical deciduous forest belt of India, Pandey and Shukla (1999) have studied the impact of different levels of disturbance on forest plant diversity on a 30-km-long transect of *Shorea robusta* plantation forest. A moderate level of disturbance has been found to result in higher diversity. They have identified "disturbance-friendly" species that overtake other plant populations in highly disturbed conditions. Pandey and Shukla (2001) have also found an increase in ramete production in *Clerodendron infortunatum*, *Mallotus philippensis*, and *Flacourtia indica* that accompanies the increase in disturbance, whereas certain other important species such as *Holarrhena antidysentrica* showed a decrease in ramete formation. Nonseed regeneration strategy in disturbed forest is highly important.

Forest management practices have far-reaching effects on biodiversity. A few high-yield or better timber-quality trees replace a large number of species occurring naturally in a forest. A vast variety of nontarget plants and animals perish when weedicides and pesticides are sprayed. Thinning practices not only involve removal of excess of seedlings and saplings of planted species but also removal of many "undesirable" species from forestry viewpoint (i.e., components of biodiversity). Structural as well as species diversity are greatly reduced in managed forests. Sometimes, a sudden realization of economic importance of so-far neglected component leads to overexploitation, and vulnerable species become endangered. Some years ago, we could locate in abundance plants of *Rauvolfia serpentina* and *Gloriosa superba* species in forests neighboring Varanasi, but now they are very rare as a result of overexploitation for medicinal purposes by pharmaceutical concerns. Such is the case with *Taxus buccata* in the outer Himalayas; taxol, a medicine for cancer, is obtained from this tree.

Some rare species of special interest are now threatened due to excessive collection for classwork by botany students during botanical excursions. Suppliers of scientific materials also collect certain species in abundance. *Gnetum* and *Psilotum* populations have suffered much. A large number of plant and animal components of ground flora and fauna also disappear with the destruction of forests. The WRI/IUCN/UNEP (1992) has observed that humans currently destroy 17 million hectares of forests globally every year. This is the most important anthropogenic cause of biodiversity loss. Checking deforestation is more important than planting trees from a biodiversity conservation viewpoint. Different aspects of biodiversity loss are described in the edited volume. *Biodiversity Loss* (Perring et al., 1995). Kormondy (1996) and Schulze and Mooney (1994) have also dealt with different aspects of biodiversity loss.

Another important cause of biodiversity loss is pollution and destruction of wetlands. Ambasht (1985) has shown that the magnitude of increased silt and soil nutrient load due to destruction of vegetation is 30 times greater in watersheds of the Chandraprabha River, a subtributary of River Ganga in India. Natural vegetal cover effectively retards from 90 to 97% of soil, nitrogen, and phosphorus runoff, and increases rain water infiltration. Similar checking of soil and nutrient input into 3,000 ha wetlands of Surhatal in India is also reported by Ambasht (1999). Kumar et al. (1996, 1997) have demonstrated the extent of reduction in soil, organic carbon, and nitrogen across an ecotone rich in biodiversity. Wetlands are the cradle of rich biodiversity and speciation, because they combine the best of terrestrial and aquatic habitats. Ambasht and Ambasht (1998) have studied diversity aspects of wetlands while reviewing the ecology of Indian wetlands. Seasonal wetlands, too, are the home of specialized plants. Species with xeric characteristics, such as *Alhagi camelorum*, easily perennate during prolonged inundation (Ambasht, 1958) on river banks, while many others need waterlogging to break the dormancy of their seeds, for example, *Polygonum plebejum* and *Chrozophora rottleri* (Lal and Ambasht, 1978). Ambasht et al. (1984) have found that plants with trailing habits fully cover the ground surface, such as *Phyla nodiflora*, on banks of the Gomati river in India prevent soil erosion above 95%, comparable to an efficient grass species such as *Cynodon dactylon*. Dugan (1990) has divided wetland loss causes into direct, indirect, and natural causes. Almost all types of wetlands, such as estuaries, open coasts, floodplains, freshwater marshes, lakes, peatlands, and swamp forests, severely suffer due to input of runoff from agricultural fields, deforested lands, and biocide sprays such as DDT in mosquito control programs. Solid wastes are most commonly dumped on wetlands around rural, urban, and industrial areas, thereby gradually filling the depressions for use as upland. Wetlands also lose considerable biodiversity when exploited for aquaculture and mariculture. It is fairly common to construct dikes, dams, and seawalls to control flood and storm water that have reduced the natural water supply to nearby wetlands. Besides these major, direct causes, the constructions cause direct sediments to find new routes along diverted watercourses and fill up flood plains and open coasts, as well as freshwater margins. Canals and construction of roads create new waterway flows and barriers, leading to sedimentation and upwelling of wetland sites, respectively. Among the natural causes of wetland losses, Dugan (1990) has listed drought, erosion, sea-level rise, and hurricanes and other storms as important causes. Overexploitation of wetland and water body plant and animal resources is responsible for the transition from once abundant to now vulnerable or threatened status of Gangetic dolphins, crocodiles, minks, many ducks, and certain turtles. According to McNeely et al. (1990), in past 400 years, 23 species of fishes have become extinct and 230 species have joined the threatened category. The biodiversity losses for amphibians and reptiles together are much worse, being 212 species extinct and 48,135 threatened. Indian wetlands are the rich storehouse of wild relatives of rice and jute (Tolba and Elkholy, 1992), and a very strong reason for wetland conservation is to sustain perpetuation of gene pools of wild relatives for all economically important species and varieties for future uses. Gene transfer from an Indian wild rice to high-yielding strains was used to increase production by inducing resistance to virus stunt disease. Ramsar convention is globally very effective in wetland biodiversity conservation. Mangrove wetlands are prized sites of biodiversity and house rare and threatened animals, such as the majestic, royal Bengal tiger. Mangrove forest trees yield construction timber, charcoal, tanins, and so on. Indonesia annually exported mangrove products worth 18 million $US and an equal quantity for domestic use (Hamilton and Snedeker, 1984). Not only the mangroves, but also huge swamp forests of *Taxodium* in the Amazonia, and *Dipterocarpus* in Indonesia, yield timber worth hundreds of millions of dollars. Biodiversity of these specialized habitats is also adversely affected by wetland degradation and destruction processes, mainly pollution and siltation. In checking soil runoff by vegetation on embankments into rivers and wetlands, responsible for upwelling of bottom and checking excessive inflow of nutrients causing eutrophication and ultimately biodiversity loss, Ambasht (1970), Ambasht and Srivastava (1994), and Kumar et al. (1992, 1996) have made quantitative measurements. The most effective species conserve above 90% of erosion and nutrient runoff. Lowrance et al. (1984), Cooper et al. (1986), and Verry and Timmons (1982) have regarded riparian vegetation as effective filters for excessive nutrients entering into rivers. Vegetation acts as filter strips (Schwer and Clausen, 1989) and removes toxic heavy metals (Lyngby and Brix, 1982). Biodiversity and sustained supply of biotic resources from wetlands and aquatic habitats are possible through ecomanagement and involving public participation.

One of the six main causes listed by WRI/IUCN/UNEP (1992) for biodiversity loss is the steadily narrowing spectrum of traded products from agriculture and forestry, and introduction of exotic species associated with agriculture, forestry, and fisheries (Heywood and Watson, 1995). There has been an increasing demand for food to meet requirements of a burgeoning human population. For this reason, emphasis on very large diversity of traditional varieties adjusted to different local environmental conditions and constraints with moderate productivity with very little external input, has switched to a very few, high-yielding varieties requiring heavy input of fertilizers, herbicides, pesticides, and repeated watering. This process helps to eliminate not only the really rich diversity of traditional crop varieties and valuable gene pools of their wild relatives, but also of the species and microbial soil populations killed by biocides. Excess nutrients and chemicals find their way from well-irrigated fields to groundwater, nearby wetlands, and rivers, along with runoff. Very similar is the fate of biodiversity from conversion of natural forests into even aged monocultures to obtain only selected timber varieties. Tropical moist forests may have about 500 tree species per hectare compared to about 40 tree species per hectare in temperate forests. Planktonic diversity in tropical oceans may be six times greater than that in temperate waters, with twice the fish diversity (McNeely et al., 1990). For Indian biodiversity, Ambasht et al. (1994) have reviewed the alarming decline of biodiversity and outlined conservation strategies. Most of the natural climax forest stands are rapidly being replaced by selected, high-quality timber trees, in which heavy external input is required to maintain the stands at a desirable stage of regeneration (arrested succession) and quality productivity. In the process of eliminating undesirable weedy species, there is a huge biodiversity loss of advantageous soil microbes, wild animal life, and plants used for fuel wood, forage, medicines, fruits, and leaf crops. In the natural forests on Vindhyan Hills in north and central India, only 25 to 30 years ago there were dense covers of mixed deciduous trees such as *Buchanania lanzan*, producing the nutritious dry fruits *chiraunji* used in Indian sweet dishes, then available abundantly from local tribals at Rs 10/kg and now at Rs 400/kg. This and other valuable trees, such as *Emblica officinalis*, yield fruits very rich in vitamin C, fibers, and a constituent of several Ayurvedic medicines, *Butea monosperma*, or flame of the forest, for bright-reddish, dense cover of flowers; harbors "lac," or sealing wax–producing insects, and *Diospyros melanoxylon*, yielding soft leaves worth several million rupees to wrap tobacco leaves in Indian cigarette industries (Biris), have been reduced. Wild bears, which preferentially fed on *Buchanania* seeds, have also become rare or threatened. Similar is the fate of numerous fish varieties in India. Wetland losses, degradation, and selective demand of a few preferred varieties of fish, particularly rohu (*Labio rohita*), have resulted in destruction of most other traditionally eaten varieties. The poor villagers, who could earlier catch enough small- and medium-size fish in village pools and ponds, suffer most, particularly in the rainy season. Through organized aquaculture, now rohu fishes from the state of Andhra Pradesh are being supplied to far-off places, up to 1,000–1,500 km away in refrigerated trucks. These are costwise cheaper than local fish of inferior table value. Likewise, the common, numerous varieties of chicken are now fast disappearing as cheaper and fleshier farm-raised broiler chickens and larger eggs are now abundantly available at cheaper prices than smaller, costlier eggs of traditional chickens. Populations of other popular, locally abundant table birds 30–40 years ago in North India are now rare. Birds not eaten or not killed for religious reasons have, however, survived. Indeed, rich diversity of ducks, cranes, flamingos, lapwings, peafowls, black and grey partridges, pheasants, and jungle fowl were abundant in wetlands and bushes. They are now protected and reduced in number of species and individuals. Jahangir, the Emperor of India some 400 years ago (1605–1627), was himself a great naturalist and made critical and interesting notes about hundreds of mammals, birds, and flowering plants (Salim Ali, 1927; Alvi and Rehman, 1968) of the then Indian Empire. From live plants and animals, he had paintings made by expert artists. His exact descriptions, local names, structure, size, and weight characteristics, and precise paintings of plants and animals have helped scientist to ascertain their correct botanical and zoological names. One of the prized color paintings, supposedly drawn from living specimen by Jehangir's artist is of the dodo, the now-extinct bird, along with loriquet, tragopan, ducks, and partridges; the paintings are reproduced by Alvi and Rehman (1968).

Another important reason for biodiversity loss is the lack of understanding about the long-range economic value, as well as the unseen ecological functions, of many of the plants and animals that today to the common man appear useless or insignificant. We just do not know necessary details about each and every recorded species, but we must not postpone their conservation until all information is gathered. By then much of the biodiversity would be lost forever. Heywood and Watson (1995) have stated that for the success of biodiversity management, protection, sustainable use, and benefit sharing, the policymakers and managers have to understand in correct perspective the problems and to select suitable tools and techniques to achieve the desired ends.

Many weeds in their native habitat do not become dominant, because as the ecosystems evolve, there are internal control mechanisms for population regulation. Some of these weeds, when transported to new habitats, soon drive out existing weed diversity and occupy all the available ground surface, and even the forest floors and cultivated fields. There are any number of examples available in which alien weeds are responsible for reducing the biodiversity of invaded places. *Parthenium hysterophorus* is one such example currently spreading to the whole of India.

There is a general lack of effective legislation and enforcement in most parts of the world. In overenthusiasm, as well as for quick popularity, numerous examples of meaningless or even wrong methods and ideas are spread to the public. The biodiversity and products of native are not only the property of human kind today but also of generations to come. The administrative setup allows unlawful but well organized and large-scale poaching and export of protected animal skins, horns, furs, elephant tusks, and so on. Illegally traded wild animals, according to the WWF estimate in 1990, yield 2–3 billion dollars per year globally. Many of the parts of very strong wild animals, such as rhino horns, bones, and the flesh of tigers and lions are used as aphrodisiacs and are therefore very expensive. Rhino horn price is 25,000 US$ per kg (McKinney and Schoch, 1998). Some rare forms of wildlife collected to be kept as pets are very expensive (e.g., the Amazon macaw for $30,000, and the mountain gorilla for $15,000. Elephant tusk cost $250 per pound. An ocelon coat costs $40,000. A grizzly bear claw necklace costs $2,500. These 1990 prices are from the U.S. Fish and Wildlife Service, compiled by McKinney and Schoch (1998). A small fraction of illegally traded material is confiscated, but instead of distributing it free to museums and educational laboratories, governments take pride in burning them, as if humans made them and therefore have the right to burn them. Burning is not a corrective measure. Why not let nature's finest confiscated creations be appropriately preserved for future generations that may not have the chance to see them, in the event they become extinct. Persons involved in such crimes against nature must be punished, and the recovered products of nature should not be destroyed. Bonfires of confiscated ivory and tiger bones, skins, and so on, by governments must be curbed. Overhunting of all wildlife and overexploitation of all valuable plants must be effectively stopped. Habitat destruction and drainage of toxic chemicals must be stopped. Control must be exercised over the invasion of new weeds.

Besides the previously described causes, there are a few secondary causes of biodiversity loss. When species that are a source of food to others, or that perform keystone functions, or on whom some other species are essentially dependent for their phenological events, such as breaking of seed dormancy, pollination, or dispersal of seeds, and so on, become locally extinct, they cause secondary biodiversity losses of dependent taxa. The loss of populations of their food plant, bamboo, is the cause of decline in population of panda bears in China. Dodo, the now-extinct bird, ate *Calveria* seeds, digested the outer part, and upon defecation, helped the scarified seed to germinate. In absence of the dodo, *Calveria* reproductive capacity is now much reduced due to inadequate scarification of its seed coat.

Endemism in Relation to Biodiversity

Many of the taxa with restricted areas of occurrence, unable to expand to wider areas, are more susceptible to extinction in the event of a local catastrophic event. They are called *endemic* and the phenomenon, *endemism*. Engler (1982) further broke down the traditional classification of endemism into (1) neoendemism and (2) palaeoendemism. Neoendemics are clusters of closely related species of recent origin, whereas paleoendemics are relics of a long evolutionary history. *Welwitschia mirabilis*, one of the most bizarre gymnosperms, is the only species of the order Welwitschiales restricted to the Namile Desert of Africa. *Ginkgo biloba*, a living fossil, well-distributed in Mesozoic Era, is now confined to China and Japan. It is a contracting or retrogressive endemic. Willis (1922) has discussed endemism in his age and area theory of geographical distribution and origin of species. Favarger and Constandriopoulos (1961) have divided endemics in four categories: (1) Palaeoendemics are ancient, high-ploidy-level plants whose diploid ancestors are already extinct; (2) schizoendemics are from divergence-vicariant species of equal ploidy; (3) patroendemics are young diploid species restricted to a region; and (4) apoendemics are widespread diploid species. The last three are neoendemics. Nelson et al. (1990) have shown that as a result of inadequate survey and ignorance about the full range of distribution, some pseudoendemic taxa are likely to be changed later to nonendemic status.

India is one of the 12 megabiodiversity regions of the world. Chatterjee (1939) estimated that 28% of the 3,169 species of Himalayan dicots are endemics. Wulff (1943) has shown that there are 200 endemic plant species in the European Alps. A high percentage of endemics exist in flora of islands such as Madagascar (66%), New Zealand (72%), and Hawaii (82%). Some species restricted to only a few thousand square kilometers are regarded as locally endemic (Terborgh and Winter, 1982; Gentry, 1986; Cowling and Holmes, 1992). Boykov (1983) developed a nomogram for determining average endemism expressed in terms of both percentage and absolute numbers. van Tol and Verdonk (1988) mentioned geographically isolated areas, such as certain islands, mountain peaks, ancient lakes, caves, thermal vents, hotsprings, and so on. In different geographic and climatic regions, endemics have some features that may be responsible for their extinction. In some endemic plants and animals, reported gigantism and dwarfism are often led in past to extinction. In birds, flightlessness is widespread among island endemics (Thaler, 1973; Lawler, 1982). Endemic species have narrow range of genetic variability compared to cosmopolitans (Moran and Hopper, 1983; Grant and Little, 1992; Sherman-Broyles et al., 1992).

Areas of abundant endemism are usually rich in biodiversity, such as in Mediterranean climates, shrublands, tropical forests, and coral reefs (Heywood and Watson, 1995). Endemics have suffered great biodiversity losses as a result of human settlers to new islands. About 50 to 100 endemic birds in Hawaiian islands up to late 18th century, and another 13 species up to the present time (Groombridge, 1993), have become extinct. Pimm (1995) has stated that about 63% of bird biodiversity in Pacific islands might have been lost due to human activities. Up to 90% of the extinction of avifauna have occurred on islands during the history of human settlements (Olsen, 1989). Endemic biodiversity richness is also a great asset for ecotourism. Native endemics need protection against invading species; hence utmost care is needed in the prevention of human agencies as carriers of alien species. Even biological control methods should not be allowed until adequate research in a confined situation has tested for safety against nontarget species. The IUCN *Red List* is an important document in endemism and biodiversity conservation.

Biodiversity Conservation

This could be divided into the following five principles:

1. Every species of plants and animals is a unique combination of genes that humankind cannot construct *de novo*. Therefore, we have to preserve species and prevent this extinction, so that future generations with greater and better scientific knowledge and biotechnology tools may study and make use of them.
2. In the long-range context, today's biodiversity conservation will yield economic benefits tomorrow. It should be regarded as an investment at local, regional, and global levels.
3. Economic development responsible for biodiversity loss has to be replaced by ecodevelopment that is biodiversity-friendly.
4. Biodiversity conservation planning should occur at all levels of human society. People's participation and sharing of costs and benefits are key to successful sustainability.
5. Human cultures, values, traditions, customs, ethnobotanical and ethnozoological knowledge and beliefs, nature-based ancient medications, and so on, are necessary parts of biodiversity and their conservation is of as much importance as that of wildlife. The essence of biodiversity conservation is to develop respect for all kinds of life, human societies, cultures, ecosystems and aesthetic values. WRI/IUCN/UNEP (1992) have elaborated the matter in 10 principles.

The term *conservation* should be considered in its broadest sense, including preservation, protection, and sustained use with public participation. The most extraordinarily complex ecosystem organization composed of myriad interactions should be conserved in a holistic manner. Human components of the locality, and their knowledge, requirements, and extent of participation should be considered.

Conservation as an ecological term refers not only to protection and preservation but also to sustainable regeneration of biotic and abiotic natural resources, so that there is an assured supply for human use. Despite conservation efforts, the exploitation of wildlife and degradation of habitats have taken place rapidly, and biodiversity has suffered heavily. Wilson (1992) advocated that United Nations Organization should observe the period from 1994 to 2003 as the International Biodiversity Decade and establishment of a Biodiversity Early Warning Network. We have described different kinds and causes of biodiversity losses; therefore, the conservation strategy also has to be a combination of methods and tools.

Every country has its own perspectives and policies, but for success, public education and awareness is the first step. Without this, government and others efforts go to waste. Another important aspect is utilization of traditional knowledge and experience not available in scientific literature. Ecologists, other biologists, foresters, engineers, agriculturists, aquatic scientists, anthropologists, and economists have to join hands and encourage biodiversity conservation involvement of local people as partners in gains and losses as necessary.

The biodiversity conservation methods are classified mainly into two categories: *in situ* and *ex situ*. The first involves tools and techniques to conserve species or other taxa, genetic varieties, ecosystems, habitats in nature or in the wild or the home of the taxa. *Ex situ* methods involve removal of organisms to be conserved from natural areas to protected situations.

In situ Conservation

Protection of areas of wilderness can be of different kinds and categories. The World Conservation Union (1994a, 1994b) has put them in six categories:

Strict Nature Reserve/Wilderness Area. This has two main types, first provide areas for research and monitoring, and second to protect wilderness for subsistence and recreation. The protected areas are ecologically representatives of the natural environment and free from disturbance.

National Parks. Outstanding natural or scenic areas of national and international importance, national parks are extensive areas of scientific and recreational value. Human activity and resource extraction are kept at a minimum level.

National Monuments/Natural Landmarks. These include small sites of special interest.

Habitat/Species Management Areas. To protect significant species or groups of species, or some interesting physical features, regulated harvesting of some resources is permitted.

Protected Landscapes and Seascapes. Mainly sites of national or international significance for public enjoyment, such as for recreation and tourism, these may include cultural and/or natural landscapes where traditional land use exists.

Managed, Resource-Protected Areas. Protected areas to maintain biodiversity and provide a continuous flow of natural products for human use. Traditional but sustainable methods are permitted.

In India, as elsewhere in the world, a large number of protected areas, biosphere reserves, national parks, wildlife sanctuaries, world heritage sites, and wetlands of national and international (Ramsar sites) importance are demarcated. There are 85 national parks and 448 wildlife sanctuaries in India (Personal communication, 1999, Department of Biotechnology, India). The Indian Council of Forestry Research has identified 309 forest preservation plots, of which 187 are in natural forests. These are *in situ* biodiversity conservation plots of ecodevelopment. The government takes care of the economic needs of the local people. Eight biosphere reserves are maintained at Man and Biosphere/UNESCO standards, including the over 2,200 km^2 Nanda Devi in the West Himalayas, and Nokrek and Manas in the East Himalayas. Sunderbans, the extensive coastal wetlands known for the royal Bengal tiger is another biosphere reserve. The famous rain forests of western Ghats and the Nilgiri Hills in south India are the other important reserve sites.

Kaziranga National Park, among the World Heritage Sites, is famous for one-horned rhinos, and Keoladeo Ghana National Park is a world-famous wetland for ducks, cranes, and other migratory birds. A great many wetlands have been declared to be of national importance, including six Ramsar sites, namely, Chilika Lake (114,000 ha), known for rich crops of fish and prawns; Harike (4,100 ha), Keoladeo-Ghana; 3,000 ha and Loktak (27,600 ha) Sambhar, and Wular (18,900 ha) Lakes. There are specific wildlife biodiversity conservation projects for tiger, Kashmir stags, elephant, and crocodile breeding. Twenty-three tiger reserves, accounting for 33,046 km^2 have helped to restore tiger population from a very critical stage since Project Tiger was launched in 1973. Projects for tiger, elephant, and rhino have helped to restore ecological integrity of not only habitats and populations of conserved species but also associated wildlife. In India, 26 centers for endemic plants have been recognized as needing urgent steps to conserve the rare endemic flora. These areas are on the mountains of Ladakh, Kumaun, Garhwal, Siwaliks, Sikkim, Arunachal, Lushai Hills, and Khasi Hills. In central India are Chhotanagpur, Pachmarhi, Simlipal, Bastar, and in south and west India are the Araku Valley, Cuddappa Hills, Saurashtra Kutch, Mahabaleshwar, and Khandala of Western Ghats, Nilgiris, Silent Valley, Palni, and the islands of Andaman and Great Nicobar.

Among important biosphere reserves in India are Nanda Devi, Manas, Kaziranga, Sundarbans, and Nilgiri. Among the National Parks are Dachigam, Corbett, Dudhwa, Palamau, Gir, Kanha, Periyar, and Bandipur. Some of the gravely threatened or endangered amphibians and reptiles are the monitor lizard, Ridley turtle, Ganges turtle, gharial and crocodile, and python. Among birds are Andaman teal, Assam partridge, cranes, pheasants, Great Indian bustard, mountain quail, monal, and hawks. Endangered mammals include the black buck, blue whale, capped langur, cheetah, chinkara, clouded leopard, dolphin, Himalayan ibex, Kashmir stag, musk deer, one-horned rhinoceros, tiger, wild buffalo, Indian elephant, flying squirrel, and so on. Ambasht and Ambasht (1998) have compiled the names of some gravely endangered plants from India and elsewhere: *Podophyllum*, *Aconitum*, *Rauwolfia*, *Psilotum*, *Nepenthes*, and so on. *Rafflesia arnoldii*, with the largest flower in the plant kingdom, is becoming rarer. Only one plant of *Ramosmania heterophylla* is recorded for the Indian Ocean. Similarly, only one plant of *Diospyros hemiteles* is recorded for Mauritus Island. Only one live plant of *Dicliptera dadsonii* is recorded for Equador. A few of the other globally threatened plants include *Saintpaulia ionantha*, *Paphiopelidium druryi*, *Punica protopunica*, *Neoritchia storku*, *Microcycas calacoma*, *Rhododendron santpaui*, *Rhus hookeri*, *Saussurea lappa*, *Colchicum luteum*, *Cyathea gigemtea, Osmunda regalis*, and so on.

On a worldwide scale, the IUCN (1994b) have compiled a list of the areas of different protected biomes. Tropical humid forests, magnificent formations with rich biodiversity, cover 10,513,210 km^2 and have 5.1% of their area under protection (53,833,388 ha). Subtropical/temperate rainforests/woodlands have 9.3% protected area, with 1.3% of the lake systems protected (663,486 ha). Of the largest biomes of warm deserts/semideserts, 4.1% are protected, over 98,400,670 ha. Mountain biomes and island systems are also protected, 8% and 9.9%, respectively.

In situ conservation will be more successful when large numbers of areas in a biome are designated, instead of a single large reserve. The number and size have to be codetermined per biome, taking into account endangered biota positions and local human resident factors. Between the sites, corridors or interconnections are of great importance to allow free flow of biota from one site to another. Each site is divided into zones of no disturbance, research and restricted use zones, and peripheral zones for tourism, and controlled harvests, and resource use.

The *in-situ* biodiversity conservation programs must include research and monitoring points, educational values, management for biotic resources and bioregions. In such programs, effective corridor management is a must, as most commonly these areas are slowly taken over by farmers, resulting into isolation or fragmentation of conservation areas (Saunders et al., 1991; Kattan et al., 1994).

Protected landscapes constitute a small portion of land area. Therefore, biodiversity conservation has to be spread to otherwise unprotected areas in a number of ways. In tropical agroforestry and agroecosystems, the rich diversity of species may include more than 100 plant species per field (Altieri and Hecht, 1990), which are used for construction of houses, huts, agricultural implements and carts, firewood, medicines, forage for livestock, and food for human beings. Traditional tropical agriculturists conserve wide diversity of domestic animals. Centers of origin of major cultivated crop plants are now reasonably known. These centers are the home of not only commonly cultivated species but also many other useful species of the same genera and their wild relatives. For example, 50 varieties of potato are cultivated by farmers of Andes in South America (Brush et al., 1981). Hundreds of traditional rice varieties have been cultivated in India, Thailand, Indonesia, and Japan. Exchange of seeds in such a large area has increased not only their genetic diversity but also useful disease-resistant traits and the ability to grow in some narrow microclimates, especially from dry land to submerged rice fields. Gene flows between cultivated and wild relatives in corn plants grown in Mexico have resulted into increased genetic diversity in *Zea mays* (Brush, 1986). These traditional practices require low external input. When viewed from a purely economic standpoint, the high production in modern agriculture often is not matched by the cost of higher inputs of modern gadgets, irrigation water supply, fertilizers, chemical pesticides, and so on. Thus, *in situ* conservation of crop genetic resources is an important aspect of global biodiversity conservation. This helps to sustain availability of wild but useful genetic traits that may be used in future crop improvements. This strategy to preserve *in situ* repositories of crop germ plasm has been advocated by Altieri (1987; Altieri and Merrick, 1987). The Food and Agricultural Organization Commission on Plant Genetic Resources, a permanent intergovernmental body of 122 members, has reaffirmed the concept that the heritage of mankind is subject to sovereign rights of nations over their genetic resources. Farmers' rights have also been recognized.

For most kinds of ecosystems, plant and animals resources need both *in situ* and *ex situ* methods to conserve their genetic resources.

Ex Situ Conservation

Humans have built gardens and zoos for several centuries. For recreation, pleasure and economic, aesthetic, and educational values, kings, rulers, and landlords have collected beautiful plants, rare ones, medicinal plants, and other interesting material from far and wide, and raised them in gardens. Perpetuation and spread of collected plants and animals were managed by the keepers of these gardens and zoos. So *ex situ* conservation, without the application of the term, has been adopted for several centuries. Organized botanical gardens and zoos were established by colonial, national, and state governments. Some of the large, world-famous botanical gardens were established in the 18th and 19th centuries.

Ex situ conservation in modern times has become specialized and biotechnology based. Specialized collection and preservation of economically important plants and their wild relatives are now being managed by chains of Bureaus of Plant Genetic Resources, such as the one in India, at New Delhi. Similarly, for domestic animals, there is the National Bureau of Animal Genetic Resources, in Karnal. These specialized bureaus are for specific categories of plants or animals, such as the National Bureau of Fish Genetic Resources, at Lucknow. The centers for trees are called *arboreta*; for cacti, *cactaria*; for roses, *rosaria*; for orchids, *orchidaria*; and so on. For preservation of seeds under dormant but living conditions, seed repositories have been established. The Indian National Gene Bank in the National Bureau of Plant Genetic Resources, at New Delhi, has 145,000 accessions of seeds, and its tissue culture repository maintains 800 accessions, with 1,000 samples preserved in very cold condition in liquid nitrogen (cryopreservation) (personal communication 1995, Department of Biotechnology, New Delhi). The Gene Bank for Medicinal and Aromatic Plants is located at Lucknow. There, desired genetic material is grown in field gene banks. On a global scale, there are over 1,500 botanical gardens and arboreta (Heywood and Heywood, 1991). How much of plant biodiversity conserved by these gardens is *ex situ* is difficult to guess, as we do not have any agreed estimate of the total number of plants now living on earth. Heywood (1992) states that about 80,000 species of higher plants are grown in the botanical gardens, which may be 25% of the total. The world-famous Kew Botanical Gardens holds 38,000 species, of which about 10% are in the threatened category. This garden, on 125 hectares, has more biodiversity of higher plants than that found in many countries. Many of the botanical gardens have developed adjuncts for seed banks, tissue cultures, and cryogenic preservation centers. Another important tool of propagation is *captive breeding* of rare or threatened species. Due to rapid biodiversity losses, some species are lost in nature and survive only in conservation gardens. In order to coordinate the conservation activities of the botanical gardens at a global level, Botanic Gardens Conservation International was established in 1987, to organize workshops to evolve priorities and address important issues. For the protection of endangered North American flora, the Plant Conservation Center of the Missouri Botanical Garden handles various work. Besides conservation, botanical gardens play a significant role in education.

Zoos and aquaria are established for the collection, growth, display, and captive breeding of land and water animals, respectively. According to the IUCN (1994b), there are about 800 professionally-managed zoos all over the world, in which about 700,000 individuals belong to about 3,000 species of mammals, birds, reptiles, and amphibians. The IUCN's Species Survival Commission Captive Breeding Specialist Group provides expert knowledge on captive housing. Aquaria are less developed, and are valued more for their scientific and economic value. The UNEP (1995), in *Global Biodiversity Assessment*, has estimated the number accessions of breeds of domesticated plants and animals in *ex situ* gene banks. For agricultural crops, conservation of very important traits such as disease resistance, high yield, easy storage, and transshipment are necessary. Humans derive about 50% of their calorie requirements from wheat, rice, maize, and potato (Prescott-Allen and Prescott-Allen, 1990). Among the fresh fruits most commonly used are hundreds of varieties of apples, oranges, bananas and mangoes. The most pressing *ex situ* conservation issue is of germplasm of selected cereals, fruits, fibers, and fodder crops for all future developmental needs of plant resources. For future food security, it is safe to combine *in situ* conservation at sites rich in wild relatives, and *ex situ* conservation for collection and preservation of agriculturally relevant germplasm. Genetic vulnerability and erosion are prevented through identification of the geographical distribution, research, education, and extension of knowledge and tools to common agriculturists. An alternative technology of replacement of crops in events of crop failure due to natural calamities, outbreaks of diseases, invasion by pests, and so on, is a necessary safety valve. Excessive use of pesticides is the largest single cause of soil biodiversity loss in agriculture lands and runoff in wetlands, and indirectly reduces economic returns as a result of poor pollination, and mortality of livestock and honey bees. In the United States, such losses are estimated to be at least $3 billion annually (Pimentel et al., 1992). The ever-increasing cost of pesticides and their increasing toxic impacts on biodiversity are due to the evolution of more and more resistant strains of insects and pests. To what extent should this competition between humans and nature continue? Adverse side effects have now reached greater negative values than advantages in terms of yield. In this race, there has been heavy mortality of natural predators such as birds, snakes, lizards, and so on. Search and application of biological controls must be a part of all biodiversity programs in agricultural systems.

In forest biodiversity conservation, *ex situ* plantations conserve desired genetical strains in conservation gardens. Instead of pure monocultures, it is always ecologically advantageous to have some degree of mixed populations and good ground flora. Agroforestry is now practiced with considerable advantage for obtaining timber, fuel, fodder, building material, medicinal plants, and agricultural products. Agroforestry and forest gardens are of much value in forest biodiversity conservation.

In India, there are 33 government-managed botanical gardens, including the National Botanical Garden at Howrah, and the National Botanical Research Institute at Lucknow. University botanical gardens mostly supply teaching material, but presently the Ministry of Environment and Forests has provided grants to some of these gardens to conserve rare and threatened species *ex situ*. There are 275 zoos, deer parks, safari parks, aquaria (personal communication 1999, Department of Biotechnology). Captive breeding has been particularly successful for Gangetic gharials, crocodiles, turtles, and white tigers. In the Vindhayan Hills at Chakia (India), an exclosure with 15-foot-tall barbed wire fence was made for lions from Gir in Gujarat in the early 1950s (Chandraprabha Sanctuary and Bhagwan Das Van Vihar). Although the lions have not survived and obviously perished as a result of human encroachment and damage to fencing, due to considerable protection, the natural vegetation and deer population have been preserved. Protected forest sites have become the source of seed collection from robust individual trees for nursery use and distribution of seedlings. The UNEP (1995), in outlining specific measures for *ex situ* forest conservation protection areas, has also emphasized protection of riparian forests. People's participation and use on ecological lines are recommended. Pott (1992) has highlighted for environmental conservation in commercial forestry enterprises. Important aspects are hydrology and erosion of soil and nutrients, microclimate, fires, pathogens, regeneration of resources, and biodiversity in forests. Cultural aspects are also important factors.

Ex situ biodiversity conservation is of greatest importance in wetlands, especially for fish cultures, ducks, shrimp, and traditionally grown floating rice, and so on. Cultivated and wild relatives of *Trapa*, *Nymphea*, *Oryza*, *Nelumbium*, and *Euryale* need *ex situ* protection in India, where they are of great economic value but are endangered due to habitat losses of isolated pockets of wetlands, which are their natural homes.

Biodiversity conservation has to be a continuous process, with an ecologically holistic approach, people's participation, economic returns, educational and tourism values, and sustainable character.

Some Conventions of an International/Transboundary Nature

The Convention on Biological Diversity (CBD) was signed by 30 governments in Rio Earth Summit in 1993, and by 41 governments in December 1993, when the CBD came in force (JoTreweek, 1999). The convention articulates biodiversity-related rights and obligations of nations. Individual nations are obliged to do whatever is needed to conserve biodiversity, the loss of which is a global problem. There are 42 articles in the Convention. Signatory countries are expected to identify their biological capital, develop a national biological diversity Conservation strategy, and also to make use of biological resources sustainably (Burhenne-Guilmin and Glowka, 1994). For European wildlife and natural habitats, the Bern Convention in 1979 was installed. Emphasis was on the conservation of habitats of wild plants and animals, particularly for vulnerable or endangered species. Migratory species particularly birds move across so many political and geographical boundaries that their conservation becomes all the more obligatory for all the countries involved. For water fowl, the Ramsar Convention in 1971 initiated international cooperation, but now the Convention has enlarged its scope to wetland sites and all biological resources. CITES, or the Convention on International Trade in Endangered Species of wild fauna and flora, came into force in 1973. The United Nations Convention on Law of the Sea was begun in 1982. The CBD has been active in bringing various other concerned conventions together, and accelerating the pace of biodiversity around the world.

Wynne (1993), in the Agenda for Conservation Action in the United Kingdom, has outlined a series of principles for biodiversity conservation: Biodiversity conservation must be an integral part of all government programs, policies, and actions, so that the biological resources are wisely used and sustained. Environmental appraisals, pollution controls, achieving higher environmental standards, and avoidance of repetition of actions that led to ecological damages in past are other aspects of the guiding principles.

Acknowledgments

We are thankful to the Council of Scientific and Industrial Research and the Indian National Science Academy for financial support and to Mr. Anthony Fulgieri for editorial corrections.

References

Altieri, M. A. (1987). *Agroecology, the scientific basis of alternative agriculture.* Boulder, CO: Westview Press.

Altieri, M. A., and Hecht, S. B. (1990). *Agroecology and small farm development.* Boca Raton, FL: CRC Press.

Altieri, M. A. and Merrick L. C. (1987). In-situ conservation of crop genetic resources through maintenance of traditional farming systems. *Economic Botany* 41: 86–96.

Alvi, M. A., and Rehman, A. (1968). *Jahangir the naturalist.* New Delhi: Indian National Science Academy.

Ambasht, R. S. (1958). Studies on the underground parts of *Alhagi camelorum* Fisch. *Proceedings of the National Academy of Sciences*, 28: 106–107.

Ambasht, R. S. (1970). Conservation of soil through plant cover of certain alluvial slopes in India. *Proceedings of the IUCN XI Technical Meeting* (pp. 44–48). Morges: IUCN.

Ambasht, R. S. (1985). Primary productivity and soil and nutrient stability of an Indian hilly savanna lands. In: J. C. Tothill, and J. J. Mott (Eds.), *Ecology and management of world's savannas* (pp. 217–219). Canberra: Australian Academy of Science.

Ambasht, R. S. (1999). Impact analysis of anthropogenic activities on productivity, soil erosion eutrophication and sedimentation of Surhatal (Ballia) lake and surrounding wetlands and habitat restoration. Final Technical Report, Ministry of Environment and Forest, Government of India Project No. J-22012/9/91-w.

Ambasht, R. S., and Ambasht, N. K. (1998). Ecology of Indian wetlands. In: S. K. Majumdar, E. W. Miller, and F. J. Brenner (Eds.), *Ecology of wetlands and associated systems* (pp. 104–116). Pennsylvania Academy of Sciences.

Ambasht, R. S., and Ambasht, N. K. (1998b). Biodiversity with special reference to soil management. In: D. Bagyaraj, A. Verma, K. K. Khanna, and H. K. Kheri (Eds.), *Modern approaches and innovations in soil management* (pp. 241–250). Merrut: Rastogi Publications.

Ambasht, R. S., and Ambasht, N. K. (1999). *A Text Book of Plant Ecology* (13th ed.). New Delhi: CBS.

Ambasht, R. S., Singh, M. P., and Sharma, E. (1984). Soil water and nutrient conservation by certain riparian herbs. *Environmental Management* 18: 99–104.

Ambasht, R. S., and Srivastava, N. K. (1994). Restoration strategies for degrading Rihand River and reservoir ecosystems in India. In: W. J. Mitsch (Ed.), *Global wetlands* (pp. 483–492). Amsterdam: Elsevier.

Ambasht, R. S., Srivastava, A. K., and Ambasht, N. K. (1994). Conserving the biodiversity of India: An ecological approach. *Indian Forester* 120: 791–798.

Ashton, P. S. (1998). Plant biodiversity in tropical forest. In: R. S. Ambasht (Ed.), *Modern trends in ecology and environment* (pp. 241–252). Backhuys: Leiden.

Boykov, B. A. (1983). *Ecological dictionary.* Ata: Nauka Kazakhskoi.

Brush, S. B. (1986). Genetic diversity and conservation in traditional farming systems. *Ethnobotany* 6: 151–167.

Brush, S. B., Carney, H., and Huaman, Z. (1981). Dynamics of Andean potato agriculture. *Economic Botany* 35: 70–88.

Burhenne-Guilmin, F., and Glowka, L. (1994). An introduction to the Convention on Biological Diversity. In: A. K. Krattiger, J. A. McNeely, W. H. Lesser, K. K. Miller, Y. S. Hil, and R. Senanayake (Eds.), *Widening perspectives in biological diversity* (pp. 15–19). Gland: IUCN.

Chatterjee, D. (1939). Studies on the endemic flora of India and Burma. *Journal of the Asiatic Society of Bengal* 5: 19–67.

Cody, M. L. (1986). Diversity, rarity and conservation in Mediterranean climate regions. In: M. E. Soule (Ed.), *Conservation biology: The science of scarcity and diversity* (pp. 123–152). Sunderland, MA: Scnauer.

Cody, M. L. (1993). Biodiversity components within and between habitats in Australia. In: R. E. Ricklefs, and D. Schluter (Eds.), *Species diversity in ecological communities: Historical and geographical perspectives* (pp. 147–158). Chicago: University of Chicago Press.

Cooper, J. R., Gillian, J. W., and Jacobe, T. C. (1986). Riparian areas as non-point pollutants. In: D. L. Cornell (Ed.), *Watershed research perspectives* (pp. 166–192). Washington, DC: Smithsonian Institute Press.

Cornell, H. V. (1985). Local and regional richness of cynipine gall wasps on California oaks. *Ecology* 66: 1247–1260.

Cowling, R. M., and Holmes, P. M. (1992). Endemism and speciation in lowland flora from the Cape Floristic Region. *Biological Journal of the Linnean Society* 47: 367–383.

Dugan, P. J. (1990). *Wetland Conservation.* Gland: IUCN.

Engler, A. (1882). *Varsuch einer Entwicklungsgeschichte der Planzenwelt.* Lepzig: Engelmann.

Favarger, R. C., and Constandriopoulos, J. (1961). Essai sur l'endemisme. *Bulletin de la Societe Botanique Suisse* 71: 384–408.

Gadgil, M., and Berkes, F. (1991). Traditional resource management systems. *Resource Management and Optimization* 8: 127–141.

Gentry, A. H. (1986). Endemism in tropical vs temperate plant communities. In: M. Soule (Ed.), *Conservation biology: The science of scarcity and diversity* (pp. 153–181). Sundarlund, MA: Sinaner.

Grant, W. S., and Little, R. M. (1992). How sedentary are greywing francolin *Francolinus africanus? Evolution* 46: 1477–1491.

Groombridge, B. (1993). IUCN *Red List of threatened animals.* Gland: IUCN.

Hamilton, L. S., and Snedeker, S. C. (Ed.). (1984). *Handbook of mangrove area management.* Gland: IUCN.

Hammond, P. M. (1992). Species inventory. In: B. Groombridge (Ed.), *Global biodiversity status of earth's living resources* (pp. 17–39). London: Chapman and Hall.

Heywood, C., and Heywood, V. H. (1991). *International directory of botanical gardens.* Koenigstain: V. Koeltz Scientific Books.

Heywood, V. H. (1992). Conservation of germplasm of wild plant species. In: O. T., Sandlung, K. Hindar, and A. H. D. Brown (Eds.), *Conservation of biodiversity for sustainable development* (pp. 189–203). Oslo: Scandinavian University Press.

Heywood, W. H., and Watson, R. T. (1995). *Global biodiversity assessment.* Cambridge, UK: Cambridge University Press.

Huston, M. A. (1994). *Biological diversity: The coexistence of species on changing landscapes.* Cambridge, UK: Cambridge University Press.

IUCN. (1980). World Conservation Strategy: Living resource conservation for sustainable development. Gland: IUCN/UNEP/WWF.

IUCN. (1994a). Guidelines for protected area management categories. Cambridge and Gland: World Conservation Monitoring Centre.

IUCN. (1994b). *United Nations list of national parks and protected areas.* Cambridge and IUCN-Gland: World Conservation and Monitoring Centre.

IUCN. (1994c). *Report of the global biodiversity forum.* Gland.

Jain, S. K. (1981). *Conservation of the threatened plants in India.* Plant Conservation Bulletin, No. 1. Howrah: Botanical Survey of India.

Joanna Treweek, (1999). *Ecological impact assessment.* Oxford, UK: Blackwell Science.

Kattan, G. H., Alvarez-Lopez, H., and Giraldo, M. (1994). Forest fragmentation and bird extinctions: San Antonio eighty years later. *Conservation Biology* 8: 138–146.

Kormondy, E. J. (1996). *Concepts of ecology* (4th ed.). New York: Prentice-Hall.

Kumar, R., Ambasht, R. S., and Srivastava, N. K. (1992). Conservation efficiency of five common riparian weeds in movements of soil, water and phosphorus. *Journal of Applied Ecology* 29: 734–744.

Kumar, R., Ambasht, R. S., Srivastava, A. K., and Srivastava, N. K. (1996). The role of some riparian wetland plants in reducing erosion of organic and selected cations. *Ecological Engineering* 6: 227–239.

Kumar, R., Ambasht, R. S., Srivastava, A. K., Srivastava, N. K., and Sinha, A. (1997). Reduction of nitrogen losses through erosion by *Leonotis nepaetfoila* and *Sida acuta* in simulated rain intensities. *Ecological Engineering* 8: 233–239.

Lal, B., and Ambasht, R. S. (1978). Growth of *Chrozophora rottleri* A. Juss in relation to different watering levels. *Indian Journal of Ecology* 5: 172–180.

Latham, R. E., and Ricklefs, R. E. (1993). Global patterns of tree species richness in moist forest: Energy diversity theory does not account for variation in species richness. *Oikos* 67: 325–333.

Lawler, T. E. (1982). The evolution of body size in mammals: Evidence from insular populations in Mexico. *American Naturalist* 119: 54–77.

Lewis, J. K. (1970). Production in grassland ecosystem. In: R. L. Dix, and R. C. Beidleman (Eds.), *U.S./I.B.P Grassland Ecosystem: a supplement.* Co: Colorado State University.

Lovejoy, T. E. (1980). A projection of species extinctions. In: *The Global 2000 Report to the President: Entering the twenty-first century.* Council of Environmental Quality and the Department of State, USA 328-331. Washington, DC: Government Printing office.

Lowrance, R., Todd, R. Jr., Fail, J., Hendrickson, J., Leonard, R., and Asmussen, L. (1984). Riparian forests as nutrient filters in agricultural watersheds. *Bioscience* 34: 374–377.

Lyngby, J. E., and Brix, H. (1982). Seasonal and environmental variation in concentrations of calcium, copper, lead and zinc in eelgrass (*Zostera marina* L.) in Limfjorel, Denmark. *Aquatic Botany* 14: 59–74.

Magalef, R. (1968). London *Perspectives in ecological theory.* Chicago: University of Chicago Press.

McKinney, M. L., and Schoch, R. M. (1998). *Environmental sciences: Systems and solutions.* London: John and Bart.

McNeely, J., Miller, K. R., Reid, W. V., Mittermeier, R. A., and Werner, T. B. (1990). *Conserving the world's biological diversity*: WRI, IUCN, World Bank, WWF, Conservation International, Washington, DC and Gland: IUCN.

Menhinick, E. F. (1964). A comparison of some species diversity indices applied to samples of field insects. *Ecology* 45: 858–862.

Moran, F., and Hopper, S. D. (1983). Genetic diversity and the insular population structure of the rare granite rock species *Eucalyptus caesia* Benth. *Australian Journal of Botany* 31: 161–173.

Nelson, B. W., Ferreira, C. A. C., de Silva, M. F., and Kawasaki, M. L. (1990). Endemism centres: Refugia and botanical collection intensity in Brazillian Amazonia. *Nature* 345: 714–716.

Odum, H. T., Cantlon, J. E., and Kornicker, L. S. (1969). An organisational hierarchy postulate for the interpretation of species individuals distribution, species entropy and ecosystem evolution and the meaning of species variety index. *Ecology* 41: 395–399.

Olsen, S. L. (1989). Extinction on islands: Man as a catastrophe. In: D. Western, and M. C. Pearl (Eds.), *Conservation for the twenty-first century* (p. 50–53). Oxford, UK: Oxford University Press.

Pandey, S. K., and Shukla, R. P. (1999). Plant diversity and community patterns along the disturbance gradient in plantation forests of sal (*Shorea robusta*). *Current Science* 77: 814–818.

Pandey, S. K., and Shukla, R. P. (2001). Regeneration strategy of plant diversity in degraded sal forest. *Current Science* 81: 95–102.

Perring, S., Maler, K. G., Folke, C., Holling, C. S., and Jansson, B. O. (Eds.), (1995). *Biodiversity loss*. Cambridge, UK: Cambridge University Press.

Pimentel, D., Acquary, H., Biltonen, M., Rice, P., Silva, M., Nelson, J., Lipner, V., Giordano, S., Hortwitz, A., and D'Amore, M. (1992). Environmental and economic costs of pesticide use. *Bioscience* 42: 750b–760b.

Pimm, S. L. (1995). Seeds of our own destruction. *New Scientist* 146: 31–35.

Pott, R. (1992). Environmental conservation practices in a commercial forestry enterprise. *South African Forestry Journal* 16: 36–39.

Prescott-Allen, R., and Prescott-Allen, C. (1990). How many plants feed the world? *Conservation Biology* 4: 365–374.

Raven, P. H. (1987). The scope of plant conservation problem world-wide. In: D. Bramwell, O. Hamman, V. Heywood, and H. Synge (Eds.), *Botanic gardens and world conservation strategy*, (pp. 19–29). London: Academy Press.

Reid, W. V. (1992). How many species will there be? In: T. C. Whitmore, and J. A. Sayer (Eds.), *Tropical deforestation and species extinction* (pp. 55–74). London: Chapman and Hall.

Salim, Ali. (1927). The Mughal Emperor of India as naturalist and sportsmen. *Journal of the Bombay Natural History Society.*

Saunders, D. A., Hobbs, R. J., and Margules, C. R. (1991). Biological conservation of ecosystem fragmentation: A review. *Conservation Biology* 5: 18–32.

Schall, J. J., and Pianka, E. R. (1978). Geographical trends in the numbers of species. *Science* 201: 679–686.

Schulze, E. D., and Mooney, H. A. (Eds.). (1994). *Biodiversity and ecosystem functions*. NY: Springer-Verlag.

Schwer, C. B., and Clausen, J. C. (1989). Vegetation filter treatment of daily milkhouse waste water. *Journal of Environmental Quality* 18: 446–451.

Sepkoski, J. J., Jr., and Raup, D. M. (1986). Periodicity in marine extinction events. In: D. K. Elliott (Ed.), *Dynamics of extinction* (pp. 3–36). NY: John Wiley.

Sherman-Broyles, S. L., Gibson, J. P., Hamrick, J. L., Buchner, M. A., and Gibson, M. J. (1992). Comparisons of allozyme diversity among rare and widespread *Rhus* species. *Systematic Botany* 17: 551–559.

Solbrig, O. T., (Ed.), (1991). From genes to ecosystem: a research agenda for biodiversity. International Union for Biological Science/Scientific Committee on Problems of Environment/United Nations Educational, Scientific and Cultural Organization Cambridge, Massaschusetts.

Terborgh, J., and Winter, B. (1982). Evolutionary circumstances of species with small ranges. In: G. T. Prance (Ed.), *Biological diversification in the tropics* (pp. 587–600). New York: Columbia Unversity Press.

Thaler, L. (1973). Nanisme et gigantisme insulares. *La Recherche* 37: 741–750.

Tolba, M. K., and Elkholy, O. A. (1992). The world environment 1972–1992. London: Chapman and Hall.

United Nations Environmental Program. (1995). *Global biodiversity assessment*, W. H. Heywood (Ed.). Cambridge, UK: Cambridge University Press.

United Nations Educational Scientific and Cultural Organization. (1994). *Biodiversity science, conservation and sustainable use*. Paris: UNESCO.

van Tol, J., and Verdonk, M. J. (1988). The protection of dragonflies (*Odonata*) and their biotopes. Strassburgh: European Committee for the Conservation of Nature and Natural Resources.

Verry, E. S., and Timmons, D. R. (1982). Waterborne nutrient flow through an upland peatland watershed in Minnesota. *Ecology* 63: 1456–1467.

Whittaker, R. H. (1960). Vegetation of Siskiyon Mountains, Oregon and California. *Ecological Monograph* 30: 279–338.

Whittaker, R. H. (1965). Dominance and diversity in land plant communities. *Science* 147: 250–260.

Whittaker, R. H. (1970). *Communities and ecosystem*. NY: MacMillan.

Whittaker, R. H. (1972). Evolution and measurement of diversity. *Taxon* 21: 213–215.

Wilcove, D. S. (1985). Net predation in forest tracks in the decline of migratory songbirds. *Ecology* 66: 1211–1214.

Willis, J. C. (1922). *Age and area: A study of geographical distribution and origin of species*. London: Cambridge University Press.

Wilson, E. O. (1992). *The diversity of life*. Cambridge, MA: Belknap Press.

Wilson, E. O., and Peter, F. M. (Eds.). (1988). *Biodiversity*. Washington, DC: National Academy of Sciences Press.

World Research Institute/The World Conservation Union/United Nations Environment Program. (1992). *Global biodiversity strategy*, Paris: UNESCO.

World Commission on Environment and Development. (1987). *Our Common Future*. Oxford University Press, Oxford.

Wulff, E. V. (1943). *An introduction to historical plant geography*. MA: Waltham.

Wynne, G. (Ed.). (1993). *Biodiversity challenge: An agenda for conservation action in the U.K.* Sandy: Royal Society for the Protection of Birds.

Younes, T., and diCastri, F. (1996). Biodiversity the emergence of a new scientific field, its properties and constraints. In: P. S. Ramakrishnan, A. K. Das, and K. G. Saxena (Eds.), *Conserving biodiversity for sustainable development* (pp. 1–9). New Delhi: Indian National Science Academy.

7

Restoration and Management of Degraded Tropical Forest Landscapes

John A. Parrotta

Introduction

Tropical forest loss and degradation, as well as degradation of lands that formerly supported forest, are proceeding at unprecedented rates, eroding biological diversity and prospects for sustainable economic development of agricultural and forest resources. Between 1980 and 1990, an estimated 15.4 million ha yr^{-1} of tropical forests and woodlands were destroyed or seriously degraded, principally through agricultural expansion, uncontrolled livestock grazing, logging, and fuelwood collection (Food and Agriculture Organization, 1993). This is equivalent to annual forest cover losses of 0.8%, the greatest losses occurring in moist deciduous forests (6.1 million ha, or 1.00%, y^{-1}) and in tropical rain forests (4.6 million ha, or 0.6%, y^{-1}). Significant losses during this period also occurred in upland forest formations (2.5 million ha, or 1.1%, y^{-1}) as well as in deciduous to very dry forest zones (2.2 million ha, or 0.5%, y^{-1}).

In India, for example, where deforestation and forest degradation trends have been partially offset by ambitious industrial and nonindustrial plantation development efforts over the past 20 years, forest cover[1] between the early 1980s and mid-1990s has reportedly declined only slightly, from 19.49 to 19.27% of the country's total geographic area (Forest Survey of India, 1997). However, these figures include forest plantation areas (which have increased significantly during this period in some States) and more importantly do not take into account the continuing rapid transformation of closed forest (forest cover with a canopy density >40%) to open forest (canopy density 10–40%). Over a 6-year period,

[1]Lands with a tree cover having a canopy density greater than 10%.

John A. Parrotta • International Institute of Tropical Forestry, USDA Forest Service, Río Piedras, PR 00928-5000, USA. Present address: USDA Forest Service, Research and Development, RPC 4th floor, 1601 North Kent Street, Arlington, VA 22209, USA: e-mail: jparrotta@fs.fed.us between the late-1980s and mid-1990s, closed forest in India declined from approximately 385,000 to 367,000 km^2 (11.2% of the country's total land area), whereas open forest increased from 250,000 to 261,000 km^2. Furthermore, over large areas of the country's remaining natural forests, human and livestock pressures (specifically, grazing, periodic fire, and fuelwood collection) have precluded effective natural regeneration to the extent that even without further cutting of canopy trees, these forests are in decline, with their complex ecological interactions between and among floristic, faunal, microbiological and abiotic components becoming increasingly simplified, less tightly integrated, and less productive. In such forests, it is but a short step between the inevitable mortality of remnant trees and their reclassification as "wastelands."

Historically, the most common response to land degradation has been abandonment or reliance on natural forest succession to restore lost soil fertility, species richness, and biomass productivity. Periodic land abandonment has been the basis of apparently sustainable traditional shifting cultivation and livestock herding systems worldwide. In many tropical regions, however, fallow periods are often shortened or eliminated due to increased population pressures and agricultural intensification. Without adequate inputs such as plant nutrients, productivity and land utility commonly decline. As a result, extensive areas of former forest lands in varying stages of degradation are in need of management to improve their productive capacity to meet basic human needs, to enhance prospects for effective watershed management, and to contribute to biodiversity conservation by providing improved natural habitat for wildlife.

Successful strategies for rehabilitating degraded areas must be based on the needs and priorities of stakeholders and the associated costs and benefits. A large proportion of marginal lands currently classified as degraded can and should be rehabilitated primarily for food production. New cropping practices (including agroforestry technologies) and more efficient agricultural resource management systems can assist this process and ensure that agriculture remains sustainable and, in the process, lessens local pressures on natural forest ecosystems. There remain, however, significant formerly forested areas in many countries that cannot be economically rehabilitated in the near term for either agricultural or intensive commercial forestry production, and degraded lands of high potential value for conservation and watershed protection.

An unknown proportion of these lands would naturally revert fairly quickly to secondary forest if the pressures on them (i.e., biomass harvesting, grazing, fire) were lifted, as has occurred in many areas of South and Southeast Asia, where Joint Forest Management programs have shown encouraging rates of forest recovery following implementation of effective community-based forest protection measures (Poffenberger and McGean, 1996). Other, more severely degraded, landscapes require some form of human intervention, or management, to facilitate their recovery. On these sites, persistent physical, chemical, and biological barriers, or stresses, preclude or severely limit the rate of natural forest succession. These barriers typically include some combination of the following "symptoms": recurrent fires; soil compaction, waterlogging, salinization, or other soil physical and chemical limitations; erosion and soil nutrient limitations; absence of obligate fungal or bacterial root symbionts; seasonal drought; low seed or rootstock availability; lack of suitable microhabitats for seed germination and seedling establishment; seed and seedling predation; and severe competition with grasses or ferns.

At present, most tropical reforestation efforts focus on the development of forestry and agroforestry systems aimed at maximizing production of a very limited number of tree species of economic importance. More rarely, forest restoration projects involve planting a diversity of forest species, with the objective of re-creating to a greater or lesser extent the forest ecosystem believed to have formerly occupied the landscape. This approach is being adopted by several forestry companies in southeastern Brazil, where legally protected riparian areas within large-scale *Eucalyptus* plantations established on previously deforested lands are being replanted with native forest species, in an effort to help conserve the region's rich flora and fauna.

Properly designed and managed, each of these approaches has its merits, with the former more likely to provide economic goods and services for a limited number of beneficiaries, and the latter stressing recovery of ecological services (i.e., watershed stabilization, soil and biodiversity conservation) for the benefit of a larger segment of society. Needed for large-scale application in many regions today are a variety of forest rehabilitation and management systems that simultaneously accelerate the recovery of native forest ecosystems (for the general benefit of society) while providing economically and socially valued forest products for a diversity of local beneficiaries. These systems should be based on the best available silvicultural and ecological knowledge, and adapted to local environmental and socioeconomic circumstances to ensure their successful implementation. In this chapter I discuss the application of current silvicultural and ecological knowledge to the design and management of forestry-based approaches to tropical landscape rehabilitation.

Forest Plantations: A Tool for Ecological Rehabilitation of Degraded Landscapes

Forest plantations utilizing appropriate tree and shrub species can play an important role in harmonizing tropical forest ecosystem rehabilitation and socioeconomic development goals (cf. Parham et al., 1993; Brown and Lugo, 1994; Parrotta and Turnbull, 1997; Kikkawa et al., 1998; Lamb and Tomlinson, 1994; Lamb, 1998). Plantations of fast-growing native and exotic trees are assuming an increasingly significant role in landscape management and the rural economy in many tropical regions (Food and Agricultural Organization, 1993; National Research Council, 1993). Such plantations provide timber and fiber for industrial development and, in some cases, fuel, small timber, and nonwood forest products for local communities, and are being used increasingly for rehabilitating deforested watersheds and other degraded landscapes, particularly in densely populated tropical regions (Evans, 1992; Brown and Lugo, 1994; Brown et al., 1997). According to United Nations estimates, tree planting for nonindustrial uses (i.e., woodlots for fuel and locally used timber, agroforestry plantings, and plantations established primarily for rehabilitation of deforested watersheds) increased dramatically between 1981 to 1990. During this period, approximately 12 million ha, or 64% of all new plantations in tropical countries, were established for these purposes. In 1990, the total nonindustrial plantation area in the tropics was estimated at 28.2 million ha, of which 82% is found in the Asian and Pacific region, 12% in Latin America and the Caribbean, and 6% in Africa (Food and Agriculture Organization, 1993).

Native or exotic species adapted to the stressful conditions that characterize degraded sites can also reverse degradation processes by stabilizing soils through development of extensive root systems; increasing soil organic matter via increased litter and fine root production and decomposition; and moderating soil pH and improving soil fertility (Sanchez et al., 1985; Parrotta, 1992).

The promotion of short-rotation plantations as a solution to the chronic wood shortages faced by hundreds of millions of people in tropical regions (Food and Agriculture Organization, 1981) has nonetheless raised concerns about their sustainability. Of particular concern is the risk that frequent harvest-related nutrient losses could result in soil fertility and biomass productivity declines over successive rotations, particularly on the inherently infertile or otherwise degraded soils where such plantations are often established (Fölster and Khanna, 1997; Gonçalves et al., 1997). Another serious concern raised by critics of short-rotation tropical plantations and monospecific rehabilitation plantation programs revolves around the (common) use of exotic species and their effects on soil fertility and biodiversity. It is often alleged that exotic species, particularly *Eucalyptus* species, more rapidly deplete soil nutrients and water, and that they inhibit the development of native flora in their understories. While the available evidence rarely, if ever, supports these allegations (Poore and Fries, 1985; George et al., 1993; Lugo, 1992b, 1997; Lugo et al., 1993; Silva Junior et al., 1995; Bone et al., 1997; Fang and Peng, 1997; Geldenhuys, 1997; Harrington and Ewel, 1997; Loumeto and Huttel, 1997; Parrotta and Knowles, 1999), these concerns persist, particularly among many restoration ecologists and conservation biologists. The results of these and other studies strongly suggest that the risk of declining productivity in plantations and negative impacts on soil fertility, hydrologic stability, and biodiversity are linked to the degree of soil degradation (physical, chemical, and biological) and the choice of planted species.

If plantations are to make a lasting positive contribution to the rehabilitation of degraded lands, a better understanding is needed of their structural and functional characteristics, specifically, their impacts on primary productivity, soil physical, chemical, and biological properties; local hydrology; nutrient cycling processes; and natural succession and biodiversity development above- and belowground (Bruijnzeel, 1997; Lal, 1997; Nambiar and Brown, 1997; Ewel and Miremath, 1998). For a given site, the choice of plantation species, plantation design, and management may all influence medium- and long-term environmental effects (Lugo, 1992a). Recent research has shown that tree species vary greatly with respect to their nutrient uptake, storage, and recycling patterns (cf. Wang et al., 1991; Lugo, 1992a; Toky and Singh, 1993, 1995; Montagnini and Sancho, 1990; Cuevas and Lugo, 1998; Parrotta, 1999), capacity to stabilize soils, increase soil organic matter and soil fertility (Verma et al., 1982; Gill et al., 1987; Sharma and Gupta, 1989; Sankaran et al., 1993; Parrotta, 1999), and facilitate the regeneration of native forest species in their understories. Nitrogen-fixing trees, mainly leguminous species, have been widely extolled for their soil-improving characteristics, related to their production of nitrogen-rich, often rapidly decomposed, leaf litter (National Academy of Sciences, 1979; Werner and Muller, 1990; MacDicken, 1994). These species are commonly used in agroforestry systems worldwide to great benefit, and have been recommended for mixed-species plantings with other, perhaps more highly valued, tree species to help offset potential soil nitrogen deficiencies and increase overall biomass productivity (Binkley and Giardina, 1997).[2]

---

[2]Although there have been some documented cases of increased productivity in mixed-species plantations in both temperate and tropical regions, the collective results of such studies have been inconclusive and show that accurate species/site matching and choice of complementary species strongly influence mixed-species plantation productivity (Food and Agriculture Organization, 1992).

Numerous studies during the past 15 years have demonstrated that plantations can accelerate, or catalyze, forest succession in their understories on sites where persistent ecological barriers to succession would otherwise preclude recolonization by native forest species, particularly where intensive silvicultural management is neglected. Examples of the "catalytic effect" of forest plantings on degraded sites can be found in many tropical countries, such as India (Bhaskar and Dasappa, 1986; Kushalappa, 1986; Soni et al., 1989; George et al., 1993; Srivastava, 1994), China (Yu et al., 1994; Fang and Peng, 1997; Zuang, 1997), Malaysia (Mitra and Sheldon, 1993), Indonesia (Kuusipalo et al., 1995), Australia (Keenan et al., 1997; Tucker et al., 1997), Uganda (Fimbel and Fimbel, 1996), Malawi (Bone et al., 1997), Congo (Loumeto and Huttel, 1997), South Africa (Knight et al., 1987; Lübbe and Geldenhuys, 1991; Geldenhuys, 1993, 1996, 1997; van Wyk et al., 1995), Puerto Rico (Lugo, 1992b; Parrotta, 1992, 1993, 1995, 1999; Lugo et al., 1993), Costa Rica (Guariguata et al., 1995; Powers et al., 1997; Haggar et al., 1997), and Brazil (Silva Junior et al., 1995; (Parrotta et al., 1997a).

These studies have shown that forest plantations very often accelerate natural successional processes by overcoming the barriers to natural regeneration. This is due to their influence on understory microclimatic conditions, vegetation structural complexity, and development of litter and humus layers during the early years of plantation growth. These changes lead to increased seed inputs from neighboring native forests by seed-dispersing wildlife attracted to the plantations, suppression of grasses or other light-demanding species that normally prevent tree seed germination or seedling survival, and improved light, temperature, and moisture conditions for seedling growth. In the absence of silvicultural management aimed at eliminating woody understory regeneration, even monospecific plantations are replaced by a mixed forest comprised of the planted species, and an increasing number of early and late successional tree species and other floristic elements drawn from surrounding forest areas.

For situations where the facilitated natural recovery of mixed native forest ecosystems is a desired goal, the success of the "catalytic plantation" approach depends on a number of site factors and management decisions, including plantation species selection (Parrotta, Knowles, et al., 1997):

- Relative to initially similar, unplanted sites, plantations generally have a marked positive effect on native forest redevelopment (succession) on severely degraded sites (such as mined lands and severely eroded areas), and on sites dominated by grasses or ferns, which otherwise preclude colonization by forest species.
- The degree to which plantations accelerate colonization and establishment by native forest species increases with increased site degradation and from drier to wetter sites, and generally decreases with increasing distance from remnant native forest stands (i.e., seed sources).
- Choice of plantation species can significantly affect the rate and trajectory of understory regeneration, due to a combination of factors, including overstory species effects on understory microclimate, soil chemical and biological characteristics, nutrient cycling processes, and their value to seed-dispersing wildlife.
- Structural complexity of the planted forest is an important determinant of subsequent biodiversity enrichment, due to the importance of habitat heterogeneity for attracting seed-dispersing wildlife and microclimatic heterogeneity required for seed germination for a variety of species. This suggests that broadleaf species yield

generally better results than conifers, and that mixed-species plantings are preferable to monocultures, due in part to their increased structural complexity.
- Wildlife, especially bats and birds, are of fundamental importance as dispersers of seeds and soil microorganisms in most tropical regions. Their effectiveness in facilitating plantation-catalyzed biodiversity development on degraded sites depends on the distances they must travel between seed sources (remnant forests) and plantations, the attractiveness of the plantations to wildlife (ability of plantations to provide habitat and food), and the condition of the forests from which they are transporting seeds (Wunderle, 1997).
- Larger-seeded forest species are far less likely to colonize degraded sites than smaller-seeded species, due to seed dispersal limitations, and therefore require management interventions (e.g., enrichment planting) to facilitate their establishment, particularly where forest restoration is a major objective.

Given the diverse needs of families, communities, and society for forest products traditionally extracted from (shrinking) natural forest areas and the importance of species selection on the long-term ecological effects of plantations, greater attention needs to be paid in most countries to the utilization of a broader range of native forest species than are currently used in plantation development and rehabilitation programs. Some countries, such as India, are very fortunate in that the silviculture of hundreds of the more common native tree species found in the country has been well studied and documented over an extended period (cf. Troup [1921] and recent revised editions). This knowledge, as well as that related to improved nursery production, site preparation, soil amelioration, and planting practices developed in recent years in the context of agroforestry and "wasteland" reforestation (cf. Khosla and Khurana, 1987, 1993; Prasad, 1988; Puri and Khosla, 1993; Marcar and Khanna, 1997; Rai, 1999), needs to be more widely communicated and used by those engaged in reforestation projects.

Restoration Forestry for Tropical Landscape Rehabilitation

Understanding the Context and Defining Goals

The first prerequisite for successful application of silvicultural and ecological knowledge to restoring degraded landscapes is an understanding of the local socioeconomic and ecological context. Landscape degradation rarely occurs in a social or political vacuum, and its underlying causes are in most cases linked to the needs and aspirations of stakeholders—local communities, private and public landowners, land managers, or leaseholders.

Whether the proximate causes of landscape degradation are dispersed (e.g., overgrazing by livestock, fuelwood collection, expansion of subsistence agriculture into marginally productive sites) or more focused (e.g., overcutting in timber concessions, large-scale public works projects, surface mining activities), the identification of the primary "actors" and an understanding of their motivations are essential for the development of effective restoration strategies that will reduce or eliminate degradation pressures at their source. Such evaluations may yield unexpected insights and help generate local community support for restoration activities, particularly if these activities are likely to improve yields and access to water resources, fodder, wood and non-wood forest products, or help to conserve local wildlife populations and natural areas of special cultural and spiritual value.

Another essential ingredient is an understanding of the local ecological context. This requires an evaluation of landscape conditions within and around sites targeted for rehabilitation or restoration, specifically, an analysis of local land-use and land-tenure patterns, the location and ecological condition of natural forest ecosystems (if present on the landscape), evaluation of vegetative cover/composition and degree of soil degradation, and ongoing degradation agents (e.g., fire, wind, or water erosion, and grazing). This information is critical, for it helps to identify the natural and anthropogenic barriers to natural forest ecosystem recovery that must be overcome, and defines the range of realistic possibilities and likely outcomes (and rates) of landscape rehabilitation efforts. Information on soil conditions, for example, is of obvious importance in determining which tree species will survive and grow well at a given site. The presence of free-ranging livestock and periodic incidence of fire will either further restrict choice of tree species for planting or indicate the need for some form of protection of sites during the rehabilitation process. Landscape floristics and the abundance/diversity of wildlife (especially seed-dispersing fauna, such as frugivorous birds and bats) are of great importance to the planning and design of projects aiming to restore native forests to an area, as discussed earlier.

Thus, an understanding of the biophysical and social/economic/political contexts of local landscape degradation provides a basis for the formulation of broad restoration objectives and specific goals that have sufficient public support (i.e., are consonant with the often diverse needs of stakeholders at local or broader societal levels), and a high probability of attainment within the local ecological and socioeconomic constraints. In densely populated rural areas where people depend heavily on forests, either directly or indirectly, landscape rehabilitation and/or forest restoration objectives are likely to be diverse: improvement of agricultural productivity, increased hydrological stability within upland watersheds (decreased incidence of flooding and maintenance of adequate base flows during dry seasons), reduction of soil erosion, increased fuelwood, small timber and fodder production, sustainable productivity of nonwood forest products, and/or maintenance or expansion of diverse wildlife habitats within or surrounding protected areas.

Designing Appropriate Forest Rehabilitation and Restoration Strategies

Once the objectives and goals for restoration are formulated and agreed upon, site-specific reforestation treatments must be designed, utilizing the best available technical silvicultural and ecological knowledge consistent with the particular interests of primary stakeholders. In complex rural settings, large-scale landscape rehabilitation efforts will likely call for a diversity of approaches suited to particular stakeholder objectives and appropriate to specific biophysical site conditions. These may include the following:

- Agroforestry systems combining tree crops for timber, fuelwood, fodder production, and soil amelioration, with cultivation of food crops, high-value species such as medicinal plants, and/or fodder.
- Intensively managed single- or mixed species plantations primarily for timber or fuel production.
- Less intensively managed plantations designed to yield both economically valued wood and nonwood forest products.

- Lightly managed plantings involving a variable number of tree species intended to "catalyze" natural forest succession and ecosystem restoration primarily for environmental benefits (such as soil conservation and watershed stabilization) in the near term, but offering multiple economic, social, and ecological (conservation) benefits in the longer term.

For each of these (or additional) options, a guiding principle for reforestation design and management is to "work with nature" to the greatest extent possible to meet specific rehabilitation or restoration goals at minimum cost. In this context, working with nature means targeting management interventions to overcome the major biophysical barriers to forest ecosystem recovery (such as fire, overgrazing, excessively harsh microclimatic conditions, severe soil limitations) and relying on natural processes to continue this task. This may be accomplished through.

- Selection of tree species that are well-adapted to the often highly stressed site conditions, and those that yield locally valued products and can contribute to ecosystem rehabilitation by virtue of their architectural features, influence on soils and nutrient cycling processes, and (for plantings designed to catalyze native forest succession) that offer habitat diversity (and perhaps food) for seed-dispersing wildlife populations present on the landscape.
- Use of high-quality planting stock and the most effective and affordable site preparation and planting methods to ensure survival and rapid early growth under specific local conditions.
- On extremely degraded sites that require special soil amelioration measures, such as mine-sites and other areas with biologically impoverished or chemically imbalanced substrates, efficient use of topsoil from nearby forests,[3] organic or inorganic soil amendments to enhance early growth of planted trees and help "jump-start" recovery of belowground ecological processes.
- Ensuring that plantings are sufficiently dense to bring about a rapid, favorable, change in microclimatic conditions so as to favor redevelopment of soil microbial, fungal, and faunal communities that are essential rehabilitation agents.
- Providing adequate site protection to prevent plantation failure.

Adaptive Management and Monitoring

Even when the considerable knowledge and experience of silviculturalists, agro-foresters, ecologists, and local people is used to design landscape rehabilitation or restoration strategies, the outcome will rarely be highly predictable, due in part to the vagaries of nature and man. Given this degree of uncertainty, an adaptive management approach is highly recommended. Rather than rigidly predetermining postestablishment management practices based on previous experiences at different locations, it is far better to incorporate some form of monitoring program and to use the results of periodic assessments of project results as a basis for management decisions.

---

[3]In the case of surface-mined sites, collection of topsoil prior to mining and its almost immediate reapplication to previously mined areas as part of the reclamation process is standard mining practice in many countries, and has been shown to be a cost-effective and essential component for moist tropical forest restoration on bauxite-mined lands in the Brazilian Amazon region (Knowles and Parrotta, 1995; Parrotta, Knowles, and Wunderle et al., 1997; Parrotta and Knowles, 1999).

The particular biophysical criteria and indicators adopted should be linked to site-specific objectives and goals but may include evaluations of tree survival, growth, and biomass productivity, understory crop yields (or natural regeneration density and diversity), erosion rates, soil organic matter development or changes in soil physical and chemical properties, and frequency of fire or other major disturbances. Information derived from periodic monitoring should form the primary basis for management decisions regarding which species or species mixtures to include or eliminate from future planting programs; intensity of early tending, weeding, and later silvicultural operations; and the necessity or utility of enrichment plantings for early shade-demanding species of particular economic or ecological value. Although such (monitoring) evaluations may be carried out by research scientists with technical skill and access to sophisticated research equipment and facilities, such measures can and should be adapted (or specifically designed) for easy use by the local people who will be the primary beneficiaries of successful restoration activities.

Research Needs

A long tradition of high-quality silvicultural research in many tropical countries, combined with more recent but intensive research in agroforestry and forest ecology, provides us with a solid foundation for facing the mounting challenge of economically rehabilitating degraded tropical landscapes and restoring at least a part of the complex forest ecosystems that formerly occupied such sites. The science, or art, of restoration ecology is relatively young, especially so in most tropical countries, and although much has been learned in recent years, an intensified research effort is needed to fill in many significant gaps in our knowledge.

During the past century, and particularly during the past two decades, countless reforestation projects have been carried out on degraded lands by private landholders, communities, nongovernment organizations (NGOs), government agencies, and international development organizations. Only a tiny fraction of these projects have been studied to evaluate their socioeconomic or environmental impact, and even fewer have been reported in the international scientific literature. Of those that have been studied, even casually, one often encounters innovative silvicultural techniques, developed by people with little or no formal training, that have tremendous potential for more extensive application. Perhaps one of most cost-effective strategies for making rapid progress in this emerging field of "restoration forestry" is to revisit and rigorously evaluate the history and outcomes of these often-extensive degraded land reforestation projects.

Increased research effort is required to develop and field-test improved propagation and planting techniques for a broader range of native forest species than that typically used in most reforestation programs today. Although earlier silvicultural research provides us with a great deal of knowledge in this area, the challenge of establishing diverse forests on severely degraded sites may require a modification of nursery practices and outplanting methods for many species of potential utility for restoration forestry. Direct sowing, which has not received the research attention it may deserve, is a potentially useful technique for low-cost establishment of many species under difficult site conditions, and may also be useful for enrichment plantings in established stands to increase their long-term biodiversity value.

Regarding the use of forest plantations to simultaneously produce timber (and other forest products) and catalyze native forest rehabilitation for long-term environmental benefits, experimental research is required to evaluate systematically the effects of site preparation, plantation species selection (both single- and mixed-species alternatives), understory management practices, and stand manipulation techniques (e.g., precommercial thinnings, liberation thinnings, enrichment plantings) on the productivity of both the planted crop and its regenerating understory. Additional research is needed to assess the associated economic and social costs and benefits, and the environmental impacts of alternative design and management systems related to biodiversity conservation, soil fertility rehabilitation, watershed stabilization, and carbon sequestration. These questions need to be addressed in the context of local ecological conditions.

In the area of wildlife ecology, research is needed to understand better the dynamics of animal seed dispersal in degraded landscapes under a variety of ecological conditions. This information can be used to develop appropriate plantation designs to encourage seed transport from remnant forest stands, and to determine the range of distances between seed sources and rehabilitation sites over which seed dispersal by animals is likely to be effective. The ecological role of invertebrates as regulators of primary production, pollination agents, vertebrate food sources, and agents in litter decomposition and soil development processes are very poorly understood by most silviculturalists and plant ecologists in the context of degraded land rehabilitation and forest restoration, and merit further study (Majer, 1989; Mboukou-Kimbatsa et al., 1998; Coleman and Hendrix, 2000).

Finally, soil microorganisms (microfauna and microflora) merit considerably more research attention. These components of the soil biota are critical contributors to ecosystem restoration due to their critical roles in soil formation, decomposition of organic matter, regulation of nutrient availability, nitrogen fixation and facilitation of nutrient uptake by trees and other forest species (Killham, 1994).

Conclusions

Restoration forestry, as an approach to rehabilitation of degraded tropical lands, has the potential to increase productivity of rural landscapes while restoring ecosystem functions for multiple environmental benefits, such as improved watershed stability, soil conservation, and biodiversity conservation. The synthesis and application of existing knowledge in the fields of tropical forest, wildlife, and landscape ecology, silviculture, and agroforestry, can offer rural people and other "stakeholders" facing progressive degradation of forest, soil, and water resources in many tropical regions, viable and sustainable forestry-based alternatives to improve their livelihoods and reverse ecosystem degradation processes for the benefit of local communities and society at large. To be effective, restoration forestry solutions must be designed and managed to fit local ecological conditions and, perhaps more importantly, the short- and long-term needs and goals of communities that presently deal with the negative consequences of deforestation and land degradation.

Acknowledgments

This work was conducted in cooperation with the University of Puerto Rico.

References

Bhaskar, V., and Dasappa (1986). Ground flora in Eucalyptus plantations of different ages. In: J. K. Sharma, C. T. S. Nair, S. Kedharmath, and S. Kondas (Eds.), (pp. 213–224). Peechi, Kerala, India: Kerala Forest Research Institute.

Binkley, D., and Giardina, C. (1997). Nitrogen fixation in tropical forest plantations. In: E. K. S. Nambiar and A. G. Brown, (Eds.), *Management of soil, water and nutrients in tropical plantation forests* (pp. 419–442). ACIAR Monograph No. 43. Canberra: Australian Centre for International Agricultural Research.

Bone, R., Lawrence, M., and Magombo, Z. (1997). The effect of *Eucalyptus camaldulensis* (Dehn) plantation on native woodland recovery on Ulumba Mountain, southern Malawi. *Forest Ecology and Management* 99: 83–99.

Brown, A. G., Nambiar, E. K. S., and Cossalter, C. (1997). Plantations for the tropics—their role, extent and nature. In: E. K. S. Nambiar and A. G. Brown, (Eds.), *Management of soil, water and nutrients in tropical plantation forests* (pp. 1–23). ACIAR Monograph No. 43. Canberra: Australian Centre for International Agricultural Research.

Brown, S., and Lugo, A. E. (1994). Rehabilitation of tropical lands: A key to sustaining development. *Restoration Ecology* 2: 97–111.

Bruijnzeel, L. A. (1997). Hydrology of forest plantations in the tropics. In: E. K. S. Nambiar and A. G. Brown, (Eds.), *Management of soil, water and nutrients in tropical plantation forests* (pp. 125–167). ACIAR Monograph No. 43. Canberra: Australian Centre for International Agricultural Research.

Coleman, D. C., and Hendrix, P. F. (2000). *Invertebrates as webmasters in ecosystems*. Wallingford, UK: CABI.

Cuevas, E., and Lugo, A. E. (1998). Dynamics of organic matter and nutrient return from litter fall of ten tropical plantation species. *Forest Ecology and Management* 112: 263–279.

Evans, J. (1992). *Plantation forestry in the tropics* (2nd ed.). Oxford, UK: Clarendon Press.

Ewel, J. J., and Hiremath, A. J. (1998). Nutrient use efficiency and the management of degraded lands. In: P. Gopal, P. S. Pathak, and K. G. Saxena (Eds.), *Ecology today: An anthology of contemporary ecological research* (pp. 199–215). New Delhi: International Scientific Publications.

Fang, W., and Peng, S. L. (1997). Development of species diversity in the restoration process of establishing a tropical man-made forest ecosystem in China. *Forest Ecology and Management* 99: 185–196.

Fölster, H., and Khanna, P. K. (1997). Dynamics of nutrient supply in plantation soils. In: E. K. S. Nambiar and A. G. Brown, (Eds.), *Management of soil, water and nutrients in tropical plantation forests* (pp. 339–378). ACIAR Monograph No. 43. Canberra: Australian Centre for International Agricultural Research.

Food and Agriculture Organization. (1981). Map of the fuelwood situation in the developing countries. *Unasylva* 33(Supple.)

Food and Agriculture Organization. (1992). *Mixed and pure forest plantations in the tropics and subtropics*. FAO Forestry Paper No. 103. Rome: Food and Agriculture Organization of the United Nations.

Food and Agricultural Organization. (1993). *Forest Resources Assessment 1990—Tropical countries*. FAO Forestry Paper No. 112. Rome: Food and Agriculture Organization of the United Nations.

Fimbel, R. A., and Fimbel, C. C. (1996). The role of exotic conifer plantations in rehabilitating degraded tropical forest lands: A case study from the Kibale Forest in Uganda. *Forest Ecology and Management* 81: 215–226.

Forest Survey of India. (1997). *The state of the forest report—1997*. Dehra Dun: Ministry of Environment and Forests, Government of India.

Geldenhuys, C. J. (1993). Management of forestry plantations to become effective stepping stones and corridors for forest migration. In: Everard, D. A. (Ed.), *The relevance of island biogeography theory in commercial forestry* (pp. 102–118). Environmental Forum Report. Pretoria: Division of Forest Science Council for Scientific and Industrial Research.

Geldenhuys, C. J. (1996). The Blackwood Group System: Its relevance for sustainable forest management in the southern Cape. *South African Forestry Journal* 177: 1–21.

Geldenhuys, C. J. (1997). Native forest regeneration in pine and eucalypt plantations in Northern Province, South Africa. *Forest Ecology and Management* 99: 101–115.

George, S. J., Kumar, B. M., and Rajiv, G. R. (1993). Nature of secondary succession in the abandoned Eucalyptus plantations of Neyyar (Kerala) in peninsular India. *Journal of Tropical Forest Science* 5: 372–386.

Gill. H. S., Abrol, I. P. and Sharma, J. S. (1987). Nutrient recycling through litter production in young plantations of *Acacia nilotica* and *Eucalyptus tereticornis* in a highly alkaline soil. *Forest Ecology and Management* 22. 57–69.

Gonçalves, J. L. M., Barros, N. F., Nambiar, E. K. S., and Novais, R. F. (1997). Soil and stand management for short-rotation plantations. In: E. K. S. Nambiar and A. G. Brown, (Eds.), *Management of soil, water and nutrients in tropical plantation forests* (pp. 379–417). ACIAR Monograph No. 43. Canberra: Australian Centre for International Agricultural Research.

Guariguata, M. R., Rheingans, R., and Montagnini, F. (1995). Early woody invasion under tree plantations in Costa Rica: Implications for forest restoration. *Restoration Ecology* 3: 252–260.

Haggar, J., Wightman, K., and Fisher, R. (1997). The potential of plantations to foster woody regeneration within a deforested landscape in lowland Costa Rica. *Forest Ecology and Management* 99: 55–64.

Harrington, R. A., and Ewel, J. J. (1997). Invasibility of tree plantations by native and non-indigenous plant species in Hawaii. *Forest Ecology and Management* 99: 153–162.

Keenan, R., Lamb, D., Woldring, O., Irvine, T., and Jensen, R. (1997). Restoration of plant biodiversity beneath tropical tree plantations in Northern Australia. *Forest Ecology and Management* 99: 117–131.

Khosla, P. K., and Khurana, D. K. (Eds.). (1987). *Agroforestry for rural needs*, Vol. 1. Solan: Indian Society of Tree Scientists.

Khosla, P. K., and Khurana, D. K. (Eds.). (1993). *Agroforestry for rural needs*, Vol. 2. Solan: Indian Society of Tree Scientists.

Kikkawa, J., Dart, P., Doley, D., Ishii, K., Lamb, D., and Suzuki, K. (Eds.). (1998). *Overcoming impediments to reforestation: Tropical forest rehabilitation in the Asia-Pacific region*. Proceedings of the 6th International BIO-REFOR Workshop, Brisbane, Australia, December 2–5, 1997.

Killham, K. (1994). *Soil ecology*. Cambridge, UK: Cambridge University Press.

Knight, R. S., Geldenhuys, C. J., Masson, P. H., Jarman, M. L., and Cameron, M. J. (Eds.). (1987). *The role of aliens in forest edge dynamics: A workshop report*. Occasional Report No. 22, Ecosystem Programmes, Division of Forest Science and Technology, Council for Scientific and Industrial Research, Pretoria.

Knowles, O. H., and Parrotta, J. A. (1995). Amazonian forest restoration: An innovative system for native species selection based on phenological data and performance indices. *Commonwealth Forestry Review* 74: 230–243.

Kushalappa, K. A. (1986). Nutrient status in Eucalyptus hybrid monoculture. In: J. K. Sharma, C. T. S. Nair, S. Kedharmath, and S. Kondas (Eds.), *Eucalypts in India: Past, present and future* (pp. 213–224). Peechi, Kerala: Kerala Forest Research Institute.

Kuusipalo, J., Ådjers, G., Jafarsidik, Y., Antii, O., Tuomela, K., and Vuokko, R. (1995). Restoration of natural vegetation in degraded *Imperata cylindrica* grassland: Understory development in forest plantations. *Journal of Vegetation Science* 6: 205–210.

Lal, R. (1997). Soils of the tropics and their management for plantation forestry. In: E. K. S. Nambiar and A. G. Brown, (Eds.), *Management of soil, water and nutrients in tropical plantation forests* (pp. 97–123). ACIAR Monograph No. 43. Canberra: Australian Centre for International Agricultural Research.

Lamb, D. (1998). Large-scale ecological restoration of degraded tropical forest lands: The potential role of timber plantations. *Restoration Ecology* 6: 271–279.

Lamb, D., and Tomlinson, M. (1994). Forest rehabilitation in the Asia-Pacific region: Past lessons and present uncertainties. *Journal of Tropical Forest Science* 7: 157–170.

Loumeto, J. J., and Huttel, C. (1997). Understory vegetation in fast-growing tree plantations on savanna soils in Congo. *Forest Ecology and Management* 99: 65–81.

Lübbe, W. A., and Geldenhuys, C. J. (1991). Regeneration patterns in planted and natural forest stands near Knysa, Southern Cape. *South Africal Journal of Forestry* 159: 43–50.

Lugo, A. E. (1992a). Comparison of tropical tree plantations with secondary forests of similar age. *Ecological Monographs* 62: 1–41.

Lugo, A. E. (1992b). Tree plantations for rehabilitating damaged forest lands in the tropics: In: M. K. Wali (Ed.), *Ecosystem rehabilitation: Ecosystem analysis and synthesis* (pp. 247–255). The Hague: SPB Academic.

Lugo, A. E. (1997). The apparent paradox of reestablishing species richness on degraded lands with tree monocultures. *Forest Ecology and Management* 99: 9–19.

Lugo, A. E., Parrotta, J. A., and Brown, S. (1993). Loss of species caused by tropical deforestation and their recovery through management. *Ambio* 22: 106–109.

MacDicken, K. G. (1994). *Selection and management of nitrogen-fixing trees*. Winrock International Institute for Agricultural Development, Morrilton, Arkansas and the Food and Agriculture Organization of the United Nations, Bangkok.

Majer, J. D. (1989). *Animals in primary succession—the role of fauna in reclaimed lands*. Cambridge, UK: Cambridge University Press.

Marcar, N. E., and Khanna, P. K. (1997). Reforestation of salt-affected and acid soils. In: E. K. S. Nambiar and A. G. Brown, (Eds.), *Management of soil, water and nutrients in tropical forests* (pp. 481–525). ACIAR Monograph No. 43. Canberra: Australian Centre for International Agricultural Research.

Mboukou-Kimbatsa, I. M. C., Bernhard-Reversat, F., and Loumeto, J. J. (1998). Change in soil magrofauna and vegetation when fast-growing trees are planted on savanna soils. *Forest Ecology and Management* 110: 1–12.
Mitra, S. S., and Sheldon, F. H. (1993). Use of exotic tree plantations by Bornean lowland forest birds. *Auk* 110: 529–540.
Montagnini, F., and Sancho, F. (1994). Aboveground biomass and nutrients in young plantations of indigenous trees on infertile soils in Costa Rica: Implications for site nutrient conservation. *Journal of Sustainable Forestry* 1: 115–139.
Nambiar, E. K. S., and Brown, A. G. (1997). Towards sustained productivity of tropical plantations: Science and practice. In: E. K. S. Nambiar and A. G. Brown, (Eds.), *Management of soil, water and nutrients in tropical plantation forests* (pp. 527–557). ACIAR Monograph No. 43. Canberra: Australian Centre for International Agricultural Research.
National Academy of Sciences. (1979). *Tropical legumes: Resources for the future*. Washington, DC: National Academy Press.
National Research Council. (1993). *Sustainable agriculture and the environment in the humid tropics*. Washington, DC: National Academy Press.
Parham, W. E., Duran, P. J., and Hess, A. L. (Eds.). (1993). *Improving degraded lands: Promising experiences from South China*. Bishop Museum Bulletin in Botany No. 3. Honolulu: Bishop Museum Press.
Parrotta, J. A. (1992). The role of plantation forests in rehabilitating degraded ecosystems. *Agriculture Ecosystems and Environment* 41: 115–133.
Parrotta, J. A. (1993). Secondary forest regeneration on degraded tropical lands: The role of plantations as "foster ecosystems." In: H. Lieth, and M. Lohmann (Eds.), *Restoration of tropical forest ecosystems* (pp. 63–73). Dordrecht: Kluwer.
Parrotta, J. A. (1995). The influence of overstory composition on understory colonization by native species in plantations on a degraded tropical site. *Journal of Vegetation Science* 6: 627–636.
Parrotta, J. A. (1999). Productivity, nutrient cycling and succession in single- and mixed-species plantations of *Casuarina equisetifolia, Eucalyptus robusta* and *Leucaena leucocephala* in Puerto Rico. *Forest Ecology and Management* 124: 45–77.
Parrotta, J. A., and Knowles, O. H. (1999). Restoration of tropical moist forests on bauxite mined lands in the Brazlian Amazon. *Restoration Ecology* 7: 103–116.
Parrotta, J. A., Knowles, O. H., and Wunderle, J. M., Jr. (1997). Development of floristic diversity in 10-year-old restoration forests on a bauxite mined site in Amazonia. *Forest Ecology and Management* 99: 21–42.
Parrotta, J. A., and Turnbull, J. W. (Eds.) (1997). Catalyzing native forest regeneration on degraded tropical lands. *Forest Ecology and Management* 99: 1–290.
Parrotta, J. A., Turnbull, J., and Jones, N. (1997). Catalyzing native forest regeneration on degraded tropical lands. *Forest Ecology and Management* 99: 1–8.
Poffenberger, M., and McGean, B. (Eds.). (1996). *Village voices, village choices: Joint forest management in India*. New Delhi: Oxford University Press.
Poore, M. E. D., and Fries, C. (1985). *The ecological effects of Eucalyptus*. FAO Forestry Paper No. 59. Rome: Food and Agriculture Organization of the United Nations.
Powers, J. S., Haggar, J. P., and Fisher, R. F. (1997). The effect of overstory composition on understory woody regeneration and species richness in 7-year-old plantations in Costa Rica. *Forest Ecology and Management* 99: 43–54.
Prasad, R. (1988). *Technology of wasteland development*. New Delhi: Associated Publishing Co.
Puri, S., and Khosla, P. K. (Eds.). (1993). *Nursery technology for agroforestry: Applications in arid and semiarid regions*. New Delhi: Oxford and IBH.
Rai, S. N. (1999). *Nursery and planting techniques of forest trees in tropical South Asia*. Dharwad, Karnataka: Punarvasu.
Sánchez, P. A., Palm, C. A., Davey, C. B., Szott, L. T., and Russell, C. E. (1985). Tree crops as soil improvers in the humid tropics? In: M. G. R. Cannell and J. E. Jackson (Eds.), *Attributes of trees as crop plants* (pp. 327–358). Huntingdon, UK: Institute of Terrestrial Ecology, National Environmental Research Council, Abbots Ripton.
Sankaran, K. V., Balasundaran, M., Thomas, T. P., and Sujatha, M. P. (1993). *Litter dynamics, microbial associations and soil studies in Acacia aurivuliformis plantations in Kerala*. Kerala Forest Research Institute Report No. 91, Peechi, Kerala, India.
Sharma, B. D., and Gupta, I. C. (1989). Effect of trees cover on soil fertility in western Rajasthan *Indian Forester* 115: 348–354.
Silva Junior, M. C., Scarano, F. R., and Souza Cardel, F. (1995). Regeneration of an Atlantic forest formation in the understory of a *Eucalyptus grandis* plantation in southeastern Brazil. *Journal of Tropical Ecology* 11: 147–152.

Soni, P., Vasistha, H. B., and Kumar, O. (1989). Biological diversity in surface mined areas after reclamation. *Indian Forester* 115: 475–482.

Srivastava, R. K. (1994). Re-establishment of shoals in grassland (a reverse process). *Indian Forester* 120: 868–870.

Toky, O. P., and Singh, V. (1993). Litter dynamics in short-rotation high density tree plantations in an arid region of India. *Agriculture Ecosystems and Environments* 45: 129–145.

Toky, O. P., and Singh, V. (1995). Nutrient dynamics in *Leucaena, Acacia* and *Eucalyptus*, short rotation, high density ("energy") plantations in arid India. *Journal of Arid Environments* 31: 401–413.

Troup, R. S. (1921). *The silviculture of Indian trees*. Oxford, UK: Clarendon Press.

Tucker, N. I. J., and Murphy, T. M. (1997). The effects of ecological rehabilitation on vegetation recruitment: Some observations from the wet tropics of North Queensland. *Forest Ecology and Management* 99: 133–152.

van Wyck, G. F., Everard, D. A., and Geldenhuys, C. J. (1995). *Forest ecotone development and succession: Experimental results and guidelines for forest rehabilitation and protection*. Report FOR-DEA 867. Pretoria: Council for Scientific and Industrial Research.

Verma, S. C., Jain, R. K., Rao, M. V., Misra, P. N., and Murty, A. S. (1982). Influence of canopy on soil composition of man-made forest in alkali soil of Banthra (Lucknow). *Indian Forester* 108: 431–437.

Wang, D., Bormann, F. H., Lugo, A. E., and Bowden, R. D. (1991). Comparison of nutrient-use efficiency and biomass production in five tropical tree taxa. *Forest Ecology and Management* 46: 1–21.

Werner, D., and Muller, P. (1990). *Fast growing trees and nitrogen fixing trees*. Stuttgart: Verlag.

Wunderle, J. M., Jr. (1997). The role of animal seed dispersal in accelerating native forest regeneration on degraded tropical lands. *Forest Ecology and Management* 99: 223–235.

Yu, Z. Y., Wang, Z. H., and He, S. Y. (1994). Rehabilitation of eroded tropical coastal lands in Gaundong, China. *Journal of Tropical Forest Science* 7: 28–38.

Zuang, X. (1997). Rehabilitation and development of forest on degraded hills of Hong Kong. *Forest Ecology and Management* 99: 197–201.

8

Nutrient Export from Tropical Rain Forests

William H. McDowell

Introduction

Water flowing through terrestrial ecosystems links atmosphere, vegetation, soil, and surface waters. Because of this intimate connection between a river and its valley (as eloquently described by Hynes, 1975), disturbance of terrestrial ecosystems in a humid basin is likely to be reflected in elemental losses in stream and river water. Early work using the small watershed approach at Hubbard Brook in New Hampshire demonstrated the power of quantifying elemental inputs and outputs to characterize the ecosystem-level response to a terrestrial disturbance (Likens et al., 1978). In the years since publication of that seminal paper, the value of using stream and river chemistry as a measure of the severity of disturbance in a landscape has been widely accepted (Ranger and Turpault, 1999). Because streams usually integrate and synthesize the effects of disturbance better than direct sampling of terrestrial landscapes, changes in stream chemistry provide a powerful tool for understanding the effects of disturbance on tropical rain forests (e.g., Schaefer et al., 2000).

Despite its broad utility and acceptance, there are limitations and caveats associated with the use of stream chemistry as an integrator of terrestrial nutrient cycles. For some elements, particularly nitrogen, quantitatively significant transformations can occur at the stream–groundwater interface (Chestnut and McDowell, 2000) or in the stream channel itself (Peterson et al., 2001). Another caveat is that nutrients lost in stream water may not reflect the nutrient pools available to plants. When flow paths are deep enough to be below the rooting zone, nutrients released by weathering, for example, may be lost in stream flow without ever being available for plant uptake (McDowell, 1998). Regardless of these caveats, the use of stream chemistry to characterize terrestrial disturbances is a powerful tool with wide acceptance in the scientific community.

William H. McDowell • Department of Natural Resources, University of New Hampshire, Durham, NH 03824, e-mail: bill.mcdowell@unh.edu

Several practical management considerations drive the study of nutrient dynamics following disturbance in tropical rain forests. Excessive nutrient loss can lead to declines in site fertility. In the northeastern United States, for example, depletions of calcium from forest harvesting (Hornbeck et al., 1990) and acidic deposition (Likens et al., 1996) are significant management concerns. The response of aquatic ecosystems to increased nutrient losses is also of concern, especially when sediment-bound particulate nutrients are lost from a site, as commonly occurs after forest harvesting (Binkley and Brown, 1993). Ultimately, the integrity of marine ecosystems is affected by delivery of nutrients and sediments from the terrestrial landscape. The recently documented hypoxic "dead zone" in the Gulf of Mexico appears to be a direct result of excessive losses of nitrogen from the terrestrial landscape, causing eutrophication and depleted oxygen levels. In the tropics, apparent worldwide declines in coral health and abundance may be tied in part to excess nutrient losses from terrestrial ecosystems (Lugo et al., 2000; Nystrom et al., 2000). Although such wide-scale phenomena are difficult to link directly to processes occurring in individual basins, they provide a significant impetus for developing broad, synthetic understanding of controls on nutrient losses that can be applied in regional and global models of biogeochemical cycles (Howarth et al., 1996).

In this chapter, I address the broad topic of controls on nutrient losses from tropical rain forests, with an emphasis on the effects of disturbance, first characterizing nutrient fluxes from relatively undisturbed sites, then examining the impacts of natural and anthropogenic disturbances on these fluxes, and finishing with a brief examination of the practical implications of these findings for landscape management. My approach throughout is to use export of nutrients in streams and rivers as an integrative measure of nutrient availability and flux from tropical forests.

Nutrient Cycling in Tropical Rain Forests

Early work on tropical rain forests emphasized their fragility, their apparently tight nutrient cycles, and the significance of biomass as a dominant reservoir of nutrient capital in the ecosystem (see reviews by Whitmore, 1989; Proctor, 1987). These ideas led naturally to the concept that tropical rain forests would be particularly sensitive to the effects of disturbance, with catastrophic nutrient losses likely to result from large-scale disturbance. More recent work has shown that many of these early paradigms regarding nutrient cycling in tropical rain forests were incorrect, or were only appropriate for a particular site or forest type. The elegant work of Stark and Jordan (1978) highlighted the potential importance of litter–root connections in phosphorus and calcium cycling in a particularly nutrient-poor site, San Carlos de Rio Negro (Venezuelan Amazon). They showed that when a solution of radiotracer ^{45}Ca is applied to the forest floor, it is retained in roots and does not appear in soil solution. Similarly, no ^{32}P was recovered in soil solution in the 5 months after application. Other work at San Carlos has shown that the soils, colloquially described as "white sands," are very low in exchangeable cations (Medina and Cuevas, 1989). This soil chemistry is reflected in the streams that drain them, which are equally low in base cations (Furch, 1984). Work at a large variety of other sites, however, has now shown that tropical rain forest soils are not particularly nutrient-poor as a group (e.g., Whitmore, 1989; Silver et al., 1994; McDowell, 2001). Furthermore, soils are the major pool of most available nutrients, not the biomass, and litterfall represents a small internal nutrient flux relative to standing stocks (e.g., McDowell, 1998).

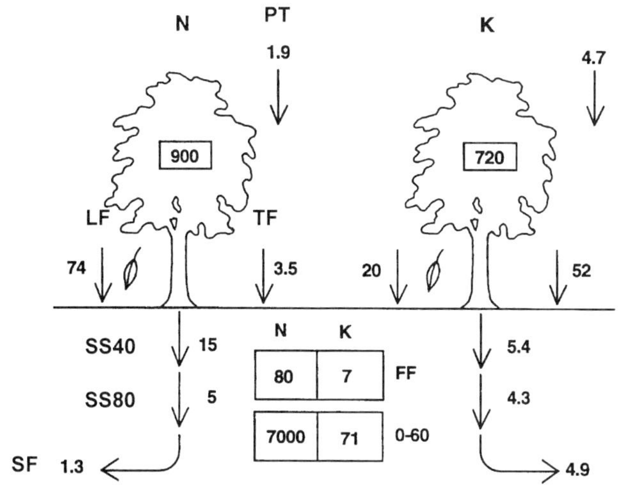

Figure 8.1. Standing stocks and internal fluxes of nitrogen and potassium in the tabonuco forest type, Luquillo Experimental Forest, Puerto Rico. Standing stocks (boxes) are kg/ha; fluxes (arrows) are kg/ha/yr. PT = precipitation; TF = throughfall; LF = litterfall; SS40, SS80 = soil solution at 40 and 80 cm; SF = streamflow; FF = forest floor; 0–60 = soil pools from 0 to 60 cm depth. Standing stock of potassium in soil is the exchangeable pool only. Reprinted from McDowell (2001).

Data from a montane site in Puerto Rico clearly support this generalization. At the site, rainfall is 3–4 m annually, soils are Ultisols, with an age of approximately 30 million years, and only potassium shows a large standing stock in biomass relative to soil (Silver et al., 1994; Figure 8.1). Nitrogen standing stock is almost 10 times higher in soils than in biomass (Figure 8.1), as are KCl-extractable Ca and Mg (Silver et al., 1994). Fluxes of N in litterfall are only 1% of total soil pools of N, and losses of N in stream water are less than precipitation inputs (Figure 8.1).

Nutrient Losses Under Background Conditions

Controls on Nutrient Losses

Over three decades ago, Gibbs (1970) proposed that variability in stream and river chemistry could be described as the result of three major factors, rock weathering, precipitation inputs, and evaporation/concentration (Figure 8.2). Evaporation/concentra-ncentration drives the chemistry of runoff in arid regions, where concentrations of all constituents can become exceedingly high due to evaporative concentration of solutes. For tropical rain forests, with their high precipitation inputs and high runoff, evaporative concentration is not a major driver of stream chemistry (Figure 8.2). Precipitation inputs can have a significant influence on stream chemistry in coastal areas, due to the high concentrations of marine aerosols found in tropical maritime precipitation (e.g., McDowell et al., 1990). This marine influence is clearly shown in the position of several Caribbean streams (Quebarda Sonadora and Rio Icacos) on the Gibbs diagram; they are midway between dominance by precipitation and dominance by rock weathering (Figure 8.2). Although not explicitly recognizing the effects of disturbance on stream chemistry and

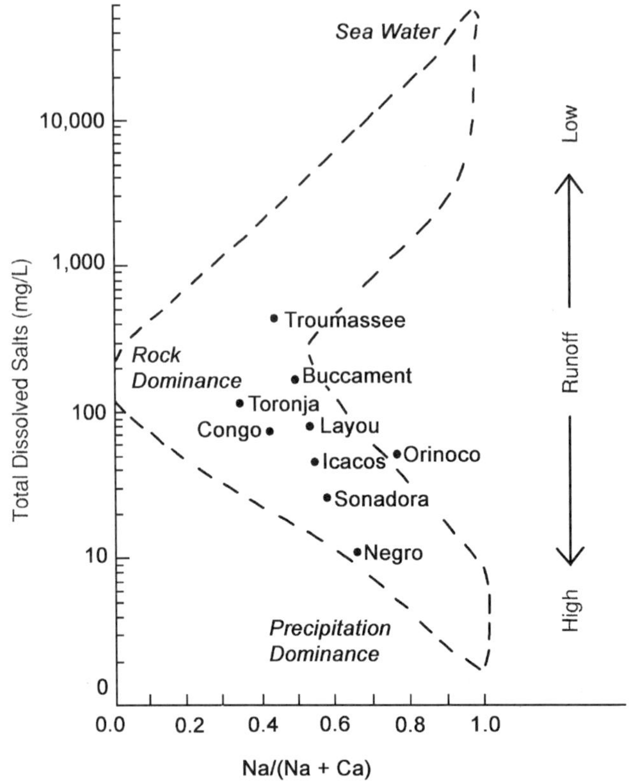

Figure 8.2. Sources of variation in streamwater chemistry as described by Gibbs (1970). River and streamwater chemistry is affected by evaporative concentration (upper arm, sea water), atmospheric deposition (lower arm, precipitation dominance) and rock weathering (intermediate, rock dominance). Modified from Gibbs (1970) and McDowell et al. (1995). Some major tropical rivers (Orinoco, Congo, Negro; Gibbs, 1970) as well as small Caribbean streams and rivers (Troumassee, Buccament, Toronja, Layou, Icacos, Sonadora; McDowell et al., 1995) are shown.

export (or any temporal variation in stream chemistry), the Gibbs formulation provides a useful template within which controls on stream chemistry can be characterized. It is particularly appropriate in this chapter, as it clearly articulates how processes occurring in a watershed (atmospheric deposition, weathering, and the hydrologic cycle) link terrestrial processes to stream chemistry.

Weathering is an important determinant of elemental losses from most tropical rain forest catchments. Data on rates of weathering in catchments around the globe show that the highest rates are found in areas with the high rainfall and temperature characteristic of tropical rain forests (White and Blum, 1995). In a detailed study of the Rio Icacos watershed in Puerto Rico, White et al. (1998) and Murphy et al. (1998) found that this rain forest watershed has the highest weathering rate of any granitic terrain in the world. Silica flux from a watershed is a good proxy for weathering rates, because it is derived almost entirely from weathering and is a dominant constituent of many rock types. Base cations are strongly related to silica concentrations in small Caribbean watersheds, suggesting that weathering is the primary driver of variability in nutrient flux among these tropical watersheds (Figure 8.3).

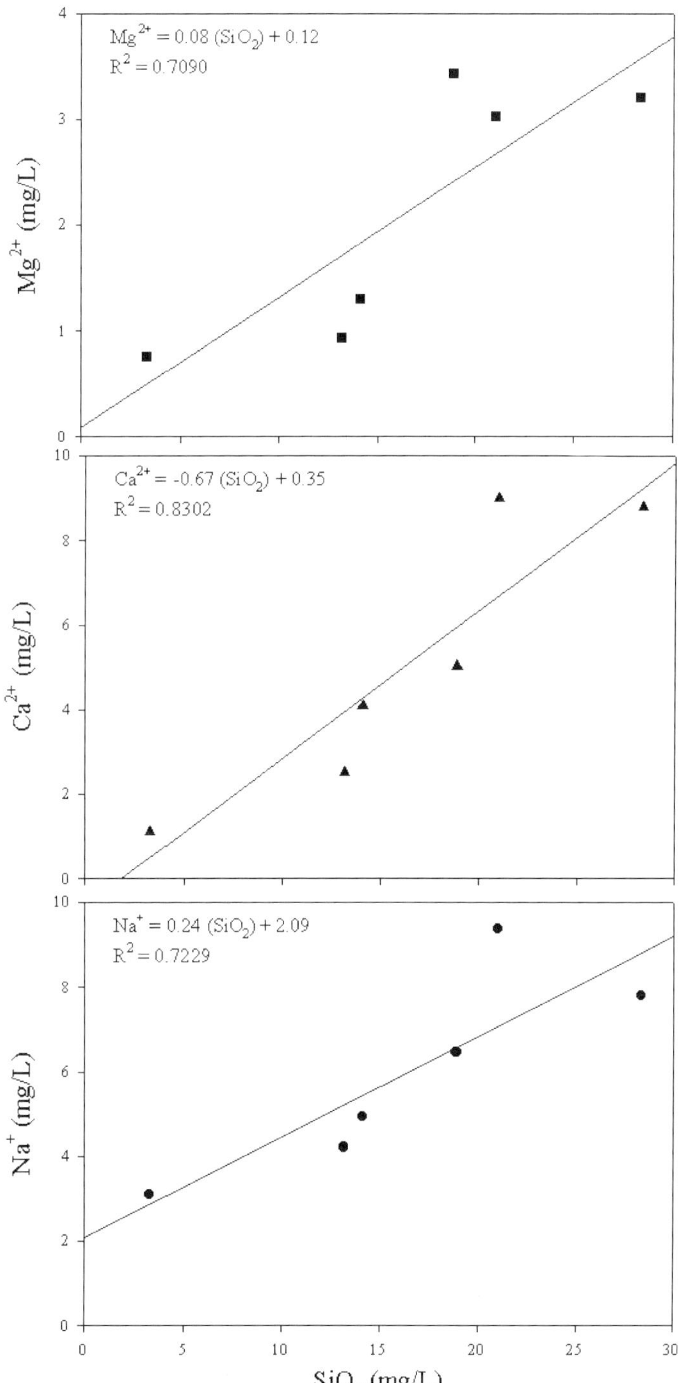

Figure 8.3. Sodium, calcium, and magnesium concentrations as a function of SiO_2 concentration in Caribbean streams and rivers. Data are flow-weighted mean annual concentrations from McDowell and Asbury (1994) and McDowell et al. (1995). Regression lines are included for relationships where $p < 0.05$. Potassium showed no statistically significant relationship with SiO_2 concentration.

A valuable case study of the effects of weathering on nutrient availability in tropical rain forests is provided by the natural experiment under way in the Hawaiian Islands. There, individual islands with similar parent material have been formed from volcanic eruptions along the edge of a tectonic plate. Studies of nutrient dynamics address the importance of weathering in providing elements to terrestrial ecosystems over geologic time (Chadwick et al., 1999). Based on their results, it appears that newly exposed land surfaces undergo a steady progression in nutrient availability over time. Soon after eruption, initial weathering of the volcanic substrate produces a material high in phosphorus, resulting in nitrogen limitation to plants. As soil organic matter builds up and soils develop, nitrogen accumulates from biological nitrogen fixation, and neither nitrogen nor phosphorus alone limits plant productivity. Ultimately, when rates of weathering are limited by the production of fresh rock surfaces and previously weathered phosphorus is lost in runoff and erosion from the site, phosphorus becomes limiting to plant growth. At this point in the weathering cycle, resupply of phosphorus is largely limited to airborne sources, in particular, dust transported across the Pacific from Asian deserts. Although calcium from weathering sources has been depleted, marine aerosols contribute enough calcium in atmospheric deposition to remove any calcium limitation. Although not studied directly, hydrologic nutrient losses from this Hawaiian geologic chronosequence should reflect this weathering regime, with decreasing elemental losses from the oldest and most highly weathered landscapes.

For some tropical regions of volcanic origin, geothermal sources can result in exceptionally high nutrient concentrations. The best-studied example of this is found in the Atlantic lowlands of Costa Rica, where a series of papers by Pringle and colleagues have documented the impacts of deep geothermally modified groundwaters on streams draining the flank of the Cordillera Central (Pringle, 1991; Pringle et al., 1993; Pringle and Triska, 1991). Their work shows that adjacent and seemingly similar streams can have differences in phosphate, calcium, and silica concentrations of up to several orders of magnitude due to this volcanic influence. These volcanically derived solutes often have significant biological effects on the receiving streams, increasing primary productivity or changing algal species composition (Pringle et al., 1993). The extent to which geothermal modification of stream nutrient export occurs in other tropical regions is largely unknown.

The Gibbs (1970) formulation of controls on river chemistry is most appropriate for elements whose flux is driven by geochemical processes, because it ignores the influence of biotic processes on stream and river chemistry. For nitrogen and phosphorus in particular, biotic processes in the watershed or stream channel are likely to have an important effect on river chemistry that often may override the influence of weathering, evaporative concentration, and precipitation inputs (e.g., Peterson et al., 2001). This has been clearly demonstrated by studies of nitrogen dynamics in Puerto Rican rain forest streams. Nitrate concentrations in Quebrada Sonadora, for example, are affected by leaf fall in the stream's watershed, with decreases in nitrogen concentration during peak leaf fall (McDowell and Asbury, 1994). In-stream processes can also have an effect on nitrogen export. Recent work with ^{15}N added at tracer levels shows that *in situ* nitrification is extremely rapid, converting essentially all of the ambient NH_4 to NO_3 in little more than 100 m of travel downstream (Merriam et al., 2001; Figure 8.4).

Microbial processes occurring at redox transition zones such as the aquatic–terrestrial interface can have a particularly important effect on nutrient losses from forested watersheds. These transition zones, located in the riparian zone (the stream edge or valley floor) or hyporheic zone (that portion of the stream bed and adjacent channel with significant

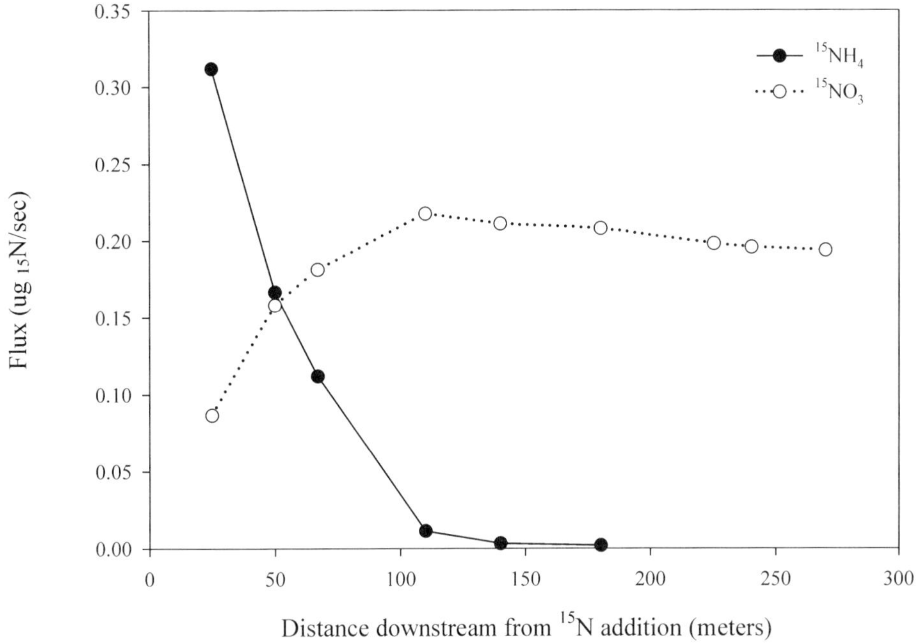

Figure 8.4. Retention of $^{15}NH_4$, and production of $^{15}NO_3$ due to nitrification in a small rain forest stream in Puerto Rico. Isotopically labeled NH_4 was added at tracer concentrations at a single point upstream and nitrogen fractions were measured at downstream stations. Loss of $^{15}NO_3$ is statistically significant and is due to denitrification, microbial immobilization, or uptake by autotrophs (modified from Merriam et al., 2001).

interchange between surface and subsurface flow; Findlay, 1995), are particularly important in regulating the form and quantity of nitrogen and dissolved organic carbon (McDowell et al., 1992, 1996; McClain et al., 1994; Chestnut and McDowell, 2000). In a sandy-bottom rain forest stream in Puerto Rico, they have been shown to dramatically reduce the export of nitrogen from the basin (Figure 8.5). Without the net nitrogen loss of nitrogen from solution occurring at this transition zone, basinwide fluxes of nitrogen in stream water would have increased by up to eight-fold (Chestnut and McDowell, 2000). This result suggests that nitrogen losses from tropical forests will be greatly affected if hydrologic flow paths or riparian vegetation are altered such that riparian-zone nitrogen removal no longer occurs.

Magnitude of Nutrient Flux from Tropical Rain Forests

For most elements, tropical forests have relatively high nutrient losses relative to their temperate counterparts (Table 8.1). The large nutrient losses from small tropical catchments listed in Table 8.1 are due to the combination of high weathering rates, high runoff, and relatively high nitrogen (Vitousek, 1984), typical of many tropical rain forests. They are also typical of losses from much larger tropical watersheds. The Orinoco, for example, has losses very similar to smaller watersheds throughout the American and Asian tropics (Table 8.1). Only calcium and magnesium losses from the Lake Calado watershed, Manaus, Brazil, are substantially lower than those found in other temperate and tropical forests.

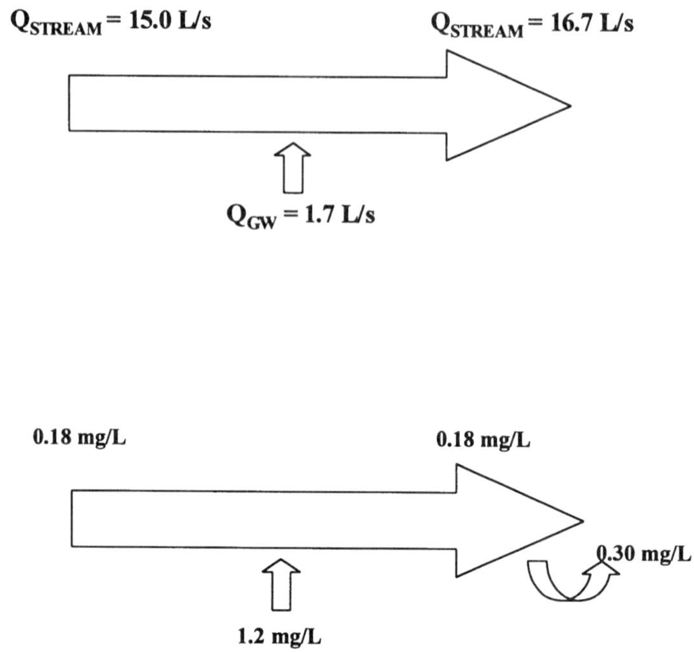

Figure 8.5. Magnitude of N retention/loss along near-stream flowpaths in the Rio Icacos basin, Puerto Rico. Upper panel shows stream discharge (Q; L/s) at the top and bottom of 100-m study reach, showing net groundwater inputs (1.7 L/s) along the reach. Lower panel shows stream total dissolved N concentration (0.18 mg/L) at the upstream and downstream ends of the study reach. Curved arrow shows N concentration (0.30 mg/L) that would result in stream if all N in incoming groundwater were delivered to the stream channel, unaffected by riparian/hyporheic immobilization or denitrification. Reprinted from McDowell (2001).

A remarkably consistent pattern has recently been documented for nitrogen losses from tropical forests throughout the Americas. By compiling data on nitrogen export in streams and rivers draining a wide variety of catchments, Lewis et al. (1999) showed that nitrogen losses are driven primarily by stream flow, with the wettest sites showing the highest nitrogen losses. Despite large differences in elevation, area, vegetation, and bedrock among the study sites, nitrogen concentrations were relatively constant, resulting in nitrogen fluxes directly related to stream discharge (Lewis et al., 1999). This suggests strong biotic control over nitrogen concentrations in drainage waters, which is particularly striking given the large nitrogen pools in soils relative to losses in streamwater (as illustrated by the Puerto Rican rain forest, Figure 8.1).

Although most work to date has focused on linking stream nitrogen fluxes to terrestrial vegetation dynamics (as pioneered by the whole-watershed approach of Bormann and Likens, 1967), microbial processes in the stream channel itself may be particularly important in regulating nitrogen flux from tropical watersheds. Work by Brandes et al. (1996) provides isotopic evidence that in-stream nitrification, rather than nitrogen delivery from the landscape, controls NO_3 concentrations in their Amazonian study site. This finding is corroborated by work in small Puerto Rican streams, which shows that nitrification is exceedingly rapid and complete (Merriam et al., 2002), that a large fraction of stream nitrogen flux is derived from the nitrification of NH_4 entering the stream in groundwater (Peterson et al., 2001), and that stream nitrate concentrations are

Table 8.1. Nutrient Flux from Tropical and Temperate Forests with Relatively Little Anthropogenic Disturbance

Name	Location	NO_3-N	NH_4-N	DON	TDP	Na	K	Ca	Mg	DOC	SiO_2	Runoff
W3	Malaysia[1]	0.38	0.29	1.93	0.19	30.7	12.4	19.7	17.3	86	104	196
W6	Malaysia[1]	0.10	0.22	4.15	0.07	7.1	3.2	3.1	3.2	194	13	195
Kali Mondo	Indonesia[2]				0.70	27.4	22.0	29	30.5		538	359
Pedra Preta	Amapa, BR[3]	0.71	0.24			15.1	4.7	22.1	7.5			135
Calado	Manaus, BR[4]	2.67	0.18	0.76	0.05	2.6	0.5	0.5	0.3			165
Tempisquito	Costa Rica[5]	6.10		3.00	0.57	114	26	192	70	37		290
Temp. Sur	Costa Rica[5]	4.90		2.10	0.33	339	64	442	137	27		430
Kathia	Costa Rica[5]	5.60		1.50	0.34	110	21	151	57	19		310
Marilin	Costa Rica[5]	4.00		1.90	0.46	123	41	174	67	24		160
El Jobo	Costa Rica[5]	4.30		2.00	0.34	104	35	133	53	26		140
Zompopa	Costa Rica[5]	6.00		3.40	0.43	164	31.0	199	73	43		280
Icacos	Puerto Rico[6]	2.54	0.68	4.79	0.08	161	17.3	96	35	94	486	368
Sonadora	Puerto Rico[6]	1.39	0.30	3.74	0.05	96	5.5	44.1	28.1	74	180	438
Toronja	Puerto Rico[6]	0.90	0.26	2.80	0.03	113	4.9	87.6	62.7	33	325	175
Orinoco	Venezuela[7]	0.95	0.41	1.92	0.24	17.5	7.8	30.9	7.9	52	70	119
Hubbard Brk.	N. Hampshire, USA[8]	0.60	0.10	1.30	0.02	7.2	1.9	13.7	3.1	29	38	83
Coweeta	N. Carolina, USA[9]	0.05	0.04		0.02	8.7	4.1	5.3	3.3	15	59	119
HJ Andrews	Oregon, USA[10]	0.81		0.69	0.70	52.8	9.5	123	8.6		114	237

Note: All values for flux are kg ha^{-1} yr^{-1}. Runoff is cm. DON = dissolved organic nitrogen; TDP = total dissolved phosphorus; DOC = dissolved organic carbon.
[1]Grip et al. (1994)
[2]Bruijnzeel (1983)
[3]Forti et al. (2000)
[4]Williams and Melack (1997)
[5]Newbold et al. (1995)
[6]McDowell and Asbury (1994)
[7]Lewis and Saunders (1989)
[8]Campbell et al. (2000); Likens et al. (1977)
[9]Swank and Waide (1987)
[10]Sollins et al. (1980)

higher than those in the adjacent groundwater or in stream sediments (Chestnut et al., 2000; McDowell et al., 1992). If these results are broadly applicable, they suggest that nitrate export from tropical watersheds is much more sensitive to stream nitrogen dynamics, and less sensitive to terrestrial nitrogen dynamics, than previously believed.

Lewis et al. (1999) also showed that inorganic nitrogen losses are typically equaled by particulate and dissolved organic losses, resulting in only about 35% of total nitrogen loss occurring as the most commonly measured forms, NO_3 and NH_4. Early workers assumed that these organic fractions were small, or of limited ecological significance. Recent studies of both tropical and temperate watersheds show that dissolved organic nitrogen is often a major fraction of nitrogen export (McDowell and Asbury, 1994; Hedin et al., 1995; Campbell et al., 2000), and that it is biologically available (Seitzinger and Sanders, 1997). Thus, there is a clear need for additional work on the export of organic nitrogen, especially following disturbance. Without measuring particulate and dissolved organic nitrogen, an accurate picture of the total nitrogen loss from a site following disturbance cannot be obtained.

Nutrient Flux Following Natural Disturbances

The literature on nutrient losses from tropical rain forests following natural disturbances is relatively sparse. Catastrophic fires (Sanford et al., 1985), landslides (Walker et al., 1996), and hurricanes (Lugo et al., 1983) are the natural disturbances with potentially significant impacts on nutrient losses. Of these major natural disturbances, hurricanes have been best studied. For Hurricane Hugo in Puerto Rico (1989), long-term studies of nutrient losses from eight watersheds show that there was a distinct increase in nutrient losses following the hurricane, but that losses returned to predisturbance levels for most elements in 1 or 2 years (Schaefer et al., 2000). A similar time course was observed for changes in groundwater nutrient concentrations following the hurricane (McDowell et al., 1996). The extent to which these results from Puerto Rico can be generalized to other tropical environments is unknown. I know of no other studies that document the response of wet tropical forests to large-scale natural disturbances.

Nutrient Flux Following Anthropogenic Disturbances

Anthropogenic disturbances are wide-ranging in nature, duration, and intensity. As one might anticipate, changes in nutrient flux in response to these disturbances are not well characterized. In fact, as is the case in temperate regions, characterizing the response to disturbance is often a *post hoc* process, with few controlled experimental studies. Although strong inferences can be made from such studies, they are not as compelling as the results of experimental work or before–after comparisons with paired reference watersheds.

The impacts of point-source inputs such as sewage, sewage treatment plant effluent, and industrial effluent have been described for several tropical rivers. Although data are typically not reported as fluxes, various authors have shown that urbanization and sewage inputs lead to the expected increases in concentrations of organic matter, nitrogen, and phosphorus (e.g., Maier and de Chiara, 1980; Khan, 1992; Martinelli et al., 1999). The effects of agriculture and other nonpoint sources on tropical water quality are not well known. Downing et al. (1999) hypothesize that altered land use will result in large changes in nutrient delivery to tropical waters, and corresponding changes in aquatic ecosystems. Neill et al. (2001) found that conversion of tropical forest to pasture typically causes a decrease in nitrate concentrations and an increase in phosphate concentrations. McDowell et al. (1995) showed that the concentration of nitrogen in small Caribbean watersheds is strongly related to population density, as Cole et al. (1994) have shown to be the case for large tropical basins. The best-described anthropogenic disturbance is logging; several studies have shown that concentrations of nitrogen, phosphorus, and potassium can increase 10-fold or more after forest harvesting (Malmer and Grip, 1994; Williams and Melack, 1997; Forti et al. 2000), but that the response is relatively short-lived (several years at most; Malmer and Grip, 1994).

A comparison of natural and anthropogenic disturbances in humid tropical watersheds suggests that the magnitude of their effects on nutrient losses is similar. As shown in Figure 8.6, similar increases in nitrogen losses are observed following forest harvesting at several sites and hurricane deforestation. Both logging and hurricanes are disturbances of high intensity and short duration, suggesting that duration of disturbance is one of the most important features in determining long-term effects on nutrient losses. Conversion of forest

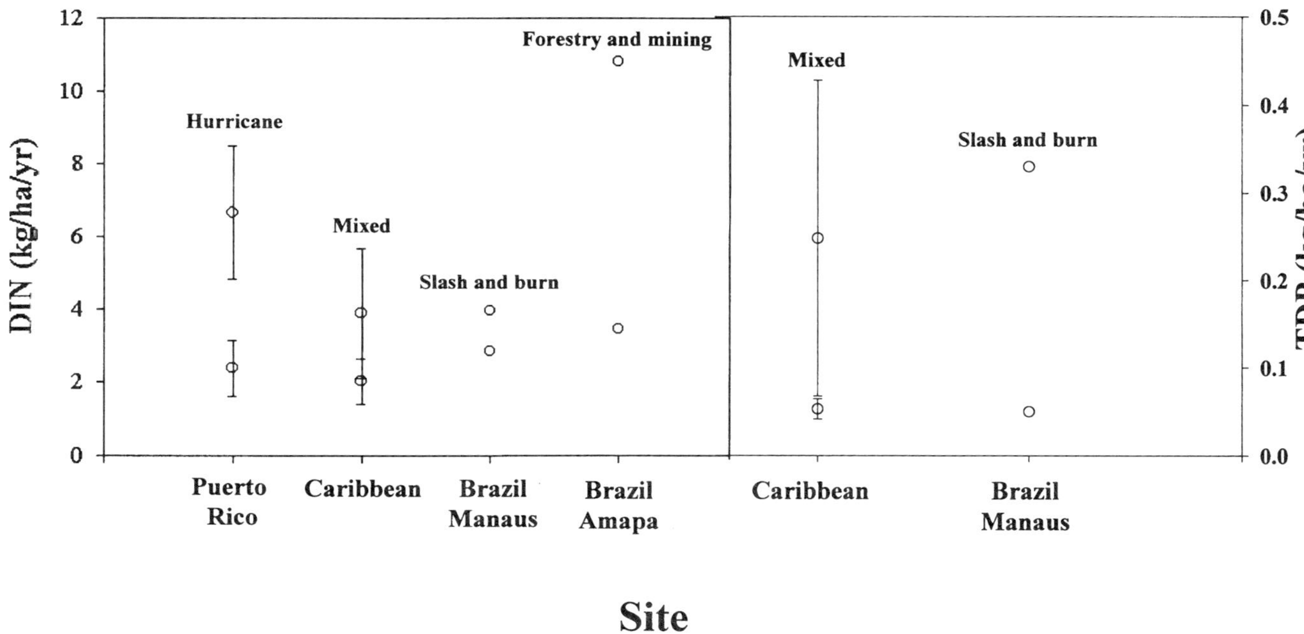

Figure 8.6. Magnitude of disturbance response (elemental export as kg ha^{-1} yr^{-1}) for various disturbance types in wet tropical forests. Circles represent results from single watersheds; error bars are the 95% confidence interval for results from multiple watersheds. In each case, lower data set is predisturbance; higher values are for postdisturbance conditions. Losses of dissolved inorganic N (DIN) and total dissolved P (TDP) are compared for hurricanes in Puerto Rico (Schaefer et al., 2000), mixed land use in the Caribbean with various population densities (McDowell et al., 1995; McDowell and Asbury, 1994), forest felling followed by burning in Manaus, Brazil (Williams and Melack, 1997), and forestry and mining in Amapa, Brazil (Forti et al., 2000).

to agricultural use is likely to have the most significant long-term effects on nutrient export, because the disturbance is both high intensity and long duration. The effects of long-term anthropogenic disturbance on nutrient export from humid tropical watersheds is not known, but it is likely to be large and qualitatively different from the temperate zone response (Matson et al., 1999; Downing et al., 1999, McDowell, 2001).

Landscape Management to Minimize Nutrient Fluxes

Management strategies to limit the effects of disturbance must rely on a sound understanding of the mechanisms governing water quality. Three primary elements drive the response of a landscape to disturbance. First, disruption of hydrologic flow paths in the watershed due to soil compaction, earth moving, and so on may alter the hydrologic regime (Mumeka, 1986) and cause increased elemental losses due to increased sediment transport and shortened hydrologic flow paths. Increased erosion is common in many disturbances, and is typically observed following land clearing and conversion to agriculture in the tropics (Dunne, 1979). Second, disruption of vegetative cover may result in a striking loss of elements under biotic control (especially nitrogen), and this may have secondary effects on other nutrients as well, due to the effects of acidification resulting from nitrification (e.g., McDowell et al., 1996; Williams and Melack, 1997). Third, disruption of the terrestrial–aquatic interface either through loss of vegetation or changes in hydrologic regime, may have disproportionately large impacts on water quality, because this interface is typically a control point that regulates nutrient losses from tropical landscapes (McDowell et al., 1992; McClain et al., 1994; Chestnut and McDowell, 2000).

One strategy that effectively limits nutrient losses is to replant disturbed areas with fast-growing trees. Plantation forests in Indonesia, for example, have been shown to sequester significant amounts of phosphorus, potassium, calcium, and magnesium in their biomass (Bruijnzeel, 1983). Calcium export in drainage water was 19 kg ha^{-1} yr^{-1}, for example, while accumulation in the biomass was 57 kg ha^{-1} yr^{-1}. Similar results were obtained for potassium and phosphorus; greater amounts of both were sequestered in biomass than lost in drainage waters. A second strategy is to enhance elemental retention at the land–water interface through the use of vegetated buffer strips or constructed wetlands. These can trap sediment and enhance denitrification, leading to reduced nutrient loading to surface waters (e.g., Dillaha et al., 1989). Unlike plant uptake, however, denitrification leads to a net loss of nitrogen from the biosphere to the atmosphere, and thus may contribute to declining site fertility.

References

Binkley, D., and Brown, T. C. (1993). Forest practices as nonpoint sources of pollution in North America. *Water Resources Bulletin* 29: 729–740.

Bormann, F. H., and Likens, G. E. (1967). Nutrient cycling. *Science* 155: 424–429.

Brandes, J. A.; McClain, M. E., and Pimentel, T. P. (1996). ^{15}N evidence for the origin and cycling of inorganic nitrogen in a small Amazonian catchment. *Biogeochemistry* 34: 45–56.

Brinkmann, W. L. F. (1983). Nutrient balance of a central Amazonian rainforest: comparison of natural and man-managed systems. In: R. Keller (Ed.), Hydrology of humid tropical regions. (pp. 153–163). IAHS Publ. No. 140; Paris: International Association of Hydrological Sciences.

Bruijnzeel, L. A. (1983). The chemical mass balance of a small basin in a wet monsoonal environment and the effect of fast-growing plantation forest. In: B. W. Webb (Ed.), *Dissolved loads of rivers and surface water quantity/quality relationships.* (pp. 229–239). IAHS Publ. No. 141. Paris: International Association of Hydrological Science.
Campbell, J. L., Hornbeck, J. W.; McDowell, W. H.; Buso, D. C.; Shanley, J. B., and Likens G. E. (2000). Dissolved organic nitrogen budgets for upland, forested ecosystems in New England. *Biogeochemistry* 49: 123–142.
Chadwick, O. A., Derry, L. A., Vitousek, P. M., Huebert, B. J., and Hedin, L. O. (1999). Changing sources of nutrients during four million years of ecosystem development. *Nature* 397: 491–497.
Chestnut, T. J., and McDowell, W. H. (2000). C and N dynamics in the riparian and hyporheic zones of a tropical stream, Luquillo Mountains, Puerto Rico. *Journal of the North American Benthological Society* 19: 199–214.
Cole, J. J., Peierls, B. L., Caraco, N. F., and Pace, M. L. (1994). Nitrogen loading of rivers as a human-driven process. In: M. J. McDonnell, and S. T. A. Picket (Eds.), *Humans as components of ecosystems: the ecology of subtle human effects and populated areas* (pp. 141–147). New York: Springer-Verlag.
Dillaha, T. A., Reneau, R. B., Mostaghimi, S., and Lee, D. (1989). Vegetative filter strips for agricultural nonpoint source pollution control. *Transactions of the American Society of Agricultural Engineering* 32: 513–519.
Downing, J. A., McClain, M., Twilley, R., Melack, J. M., Elser, J., Rabalais, N. N., Lewis, W. M. Jr., Turner, R. E., Corredor, J., Soto, D., Yanez-Arancibia, A., Kopaska, J. A., and Howarth, R. W. (1999). The impact of accelerating land-use change on the N-cycle of tropical aquatic ecosystems: current conditions and projected changes. *Biogeochemistry* 46: 109–148.
Dunne, T. (1979). Sediment yield and land use in tropical catchments. *Journal of Hydrology* 42: 281–300.
Findlay, S. 1995. Importance of surface–subsurface exchange in stream ecosystems: The hyporheic zone. *Limnology and Oceanography* 40: 159–164.
Forti, M. C., Boulet, R., Melfi, A. J., and Neal, C. (2000). Hydrogeochemsitry of a small catchment in northeastern Amazonia: A comparison between natural with deforested parts of the catchment (Serra Do Navio, Amapa State, Brazil). *Water, Air, and Soil Pollution* 118: 263–279.
Furch, K. (1984). Water chemistry of the Amazon basin: The distribution of chemical elements among freshwaters. In: *The Amazon* (pp. 167–199). Dordrecht: Dr. W. Junk.
Gibbs, R. J. (1970). Mechanisms controlling world water chemistry. *Science* 170: 1088–1090.
Grip, H., Malmer, A., and Wong, F. K. (1994). Converting tropical rain forest to forest plantation in Sabah, Malaysia: 1. Dynamics and net losses of nutrients in control catchment streams. *Hydrological Processes* 8: 179–194.
Hedin, L. O., Armesto, J. J., and Johnson, A. H. (1995). Patterns of nutrient loss from unpolluted, old-growth temperate forests: Evaluation of biogeochemical theory. *Ecology* 76: 493–509.
Hornbeck, J. W., Smith, C. T., Martin, C. W., Tritton, L. M., and Pierce, R. S. (1990). Effects of intensive harvesting on nutrient capitals of three forest types in New England. *Forest Ecology and Management* 30: 55–64.
Howarth, R. W., Billen, G., Swaney, D., Townsend, A., Jaworski, N., Lajtha, K., Downing, J. A., Elmgren, R., Caraco, N., Jordan, T., Berendse, F., Freney, J., Kudeyarov, V., Murdoch, P., and Zhao-Liang, Z. (1996). Regional nitrogen budgets and riverine N and P fluxes for the drainages to the North Atlantic Ocean: Natural and human influences. *Biogeochemistry* 35: 75–139.
Hynes, H. B. N. (1975). The stream and its valley. *Verhandlugen Internationale Vereinigung fur Theoretische und Angewandte Limnologie* 19: 1–15.
Khan, I. S. A. N. (1992). A study on the impact of urban and industrial development in the Linggi River basin, Malaysia: 2. Chemical environment. *Internationale Revue der Gesamten Hydrobiologie* 77: 391–419.
Lewis, W. M., Jr., Melack, J. M., McDowell, W. H., McClain, M., and Richey, J. E. (1999). Nitrogen yields from undisturbed watersheds in the Americas. *Biogeochemistry* 46: 149–162.
Lewis, W. M. Jr., and Saunders, J. F. I. (1989). Concentration and transport of dissolved and suspended substances in the Orinoco River. *Biogeochemistry* 7: 203–240.
Likens, G. E., Bormann, F. H., Pierce, R. S., Eaton, J. S., and Johnson, N. M. (1977). *Biogeochemistry of a forested ecosystem.* New York: Springer-Verlag, New York.
Likens, G. E., Driscoll, C. T., and Buso, D. C. (1996). Long-term effects of acid rain—response and recovery of a forest ecosystem. *Science* 272: 244–246.
Likens, G. E., Bormann, F. H., Pierce R. S., and Reiners, W. A. (1978). Recovery of a deforested ecosystem. *Science* 199: 492–496.
Lugo, A. E., Applefield, M., Pool, D. J., and McDonald, R. B. (1983). The impact of Hurricane David on the forests of Dominica. *Canadian Journal of Forest Research* 13: 201–211.
Lugo, A. E., Rogers, C. S., and Nixon, S. W. (2000). Hurricanes, coral reefs and rainforests: Resistance, ruin and recovery in the Caribbean. *Ambio* 29: 106–114.

Maier, M. H., and de Chiara, E. G. (1980). Estudo limnologico de um trecho do Rio Moji-Guacu: III. Acao de efluentes de usina acucareira (agua de lavagem de cana). *Boletin Instituto da Pesca* 7: 93–102.
Malmer, A., and Grip, H. (1994). Converting tropical rainforest to forest plantation in Sabah, Malaysia: II. Effects on nutrient dynamics and net losses in streamwater. *Hydrological Processes* 8: 195–209.
Martinelli, L. A., Krusche, A. V., Victoria, R. L., Camargo, P. B., de Bernardes, M., Ferraz, E. S., de Moraes, J. M., and Ballester, M. V. (1999). Effects of sewage on the chemical composition of Piracicaba River, Brazil. *Water, Soil and Air Pollution* 110: 67–79.
Matson, P. A., McDowell, W. H., Townsend, A. R., and Vitousek, P. M. (1999). The globalization of nitrogen deposition: ecosystem consequences in tropical environments. *Biogeochemistry* 46: 67–83.
McClain, M. E., Richey, J. E., and Pena Pimental, T. (1994). Groundwater nitrogen dynamics at the terrestrial-lotic interface of a small catchment in the central Amazon basin. *Biogeochemistry* 27: 113–127.
McDowell, W. H. (1998). Internal nutrient fluxes in a tropical rain forest. *Journal of Tropical Ecology* 14: 521–536.
McDowell, W. H. (2001). Hurricanes, people, and riparian zones: Controls on nutrient losses from forested Caribbean watersheds. *Forest Ecology and Management* 154: 443–451.
McDowell, W. H., and Asbury, C. E. (1994). Export of carbon, nitrogen, and major ions from three tropical montane watersheds. *Limnology and Oceanography* 39: 111–125.
McDowell, W. H., Bowden, W. B., and Asbury, C. E. (1992). Riparian nitrogen dynamics in two geomorphologically distinct tropical rain forest watersheds—subsurface solute patterns. *Biogeochemistry* 18: 53–75.
McDowell, W. H., Gines-Sanchez, C., Asbury, C. E., and Ramos-Perez, C. R. (1990). Influence of seasalt aerosols and long range transport on precipitation chemistry at El Verde, Puerto Rico. *Atmospheric Environment* 24A: 2813–2821.
McDowell, W. H., Lugo, A. E., and James, A. (1995). Export of nutrients and major ions from Caribbean catchments. *Journal of the North American Benthological Society* 14: 12–20.
McDowell, W. H., McSwiney, C. P., and Bowden, W. B. (1996). Effects of hurricane disturbance on groundwater chemistry and riparian function in a tropical rain forest. *Biotropica* 28: 577–584.
Medina, E., and Cuevas, E. (1989). Patterns of nutrient accumulation and release in Amazonian forests of the upper Rio Negro basin. In: J. Proctor (Ed.), *Mineral nutrients in tropical forest and savanna ecosystems* (pp. 217–240). Special Publications Series of the British Ecological Society No. 9. London: Blackwell Scientific.
Merriam, J. L., McDowell, W. H., Tank, J. L., Wolheim, W. M., Crenshaw, C. L., and Johnson, S. L. (2002). Characterizing nitrogen dynamics, transport, and retention in a tropical rainforest stream using an in situ ^{15}N addition. *Freshwater Biology* 47: 143–160.
Mumeka, A. (1986). Effect of deforestation and subsistence agriculture on runoff of the Kafue River headwaters, Zambia. *Hydrological Sciences Journal* 31: 543–544.
Murphy, S. F., Brantley, S. L., Blum, A. E., White, A. F., and Dong, H. (1998). Chemical weathering in a tropical watershed, Luquillo Mountains, Puerto Rico: II. Rate and mechanism of biotite weathering. *Geochimica et Cosmochimica Acta* 62: 227–243.
Neill, C., Deegan, L. A., Thomas, S. M., and Cerri, C. C. (2001). Deforestation for pasture alters nitrogen and phosphorus in small Amazonian streams. *Ecological Applications* 11: 1817–1828.
Nystrom, M., Folke, C., and Moberg, F. (2000). Coral reef disturbance and resilience in a human-dominated environment. *Tree* 15: 413–417.
Peterson, B. J., Wollheim, W., Mulholland, P. J., Webster, J. R., Meyer, J. L., Tank, J. L., Grimm, N. B., Bowden, W. B., Vallet, H. M., Hershey, A. E., McDowell, W. H., Dodds, W. K., Hamilton, S. K., Gregory, S. V., and D'Angelo, D. J. (2001). Stream processes alter the amount and form of nitrogen exported from small watersheds. *Science* 292: 86–90.
Pringle, C. M. (1991). Geothermally modified waters surface at La Selva Biological Station, Costa Rica: Volcanic processes introduce chemical discontinuities into lowland tropical streams. *Biotropica* 23: 523–529.
Pringle, C. M., Rowe, G. L., Triska, F. J., Fernandez, J. F., and West, J. (1993). Landscape linkages between geothermal activity and solute composition and ecological response in surface waters draining the Atlantic slope of Costa Rica. *Limnology and Oceanography* 38: 753–774.
Pringle, C. M., and Triska, F. J. (1991). Effects of geothermal groundwater on nutrient dynamics of a lowland Costa Rican stream. *Ecology* 72: 951–965.
Proctor, J. (1983). Mineral nutrients in tropical forests. *Progress in Physical Geography* 7: 422–431.
Proctor, J. (1987). Nutrient cycling in primary and old secondary rain forests. *Applied Geography* 7: 135–142.
Ranger, J., and Turpault, M.-P. (1999). Input–output nutrient budgets as a diagnostic tool for sustainable forest management. *Forest Ecology and Management* 122: 139–154.
Sanford, R. L. Jr., Saldarriaga, J., Clark, K. E., Uhl, C., and Herrera, R. (1985). Amazon rain-forest fires. *Science* 227: 53–55.

Schaefer, D. A., McDowell, W. H., Scatena, F. N., and Asbury, C. E. (2000). Effects of hurricane disturbance on stream water concentrations and fluxes in eight tropical forest watersheds of the Luquillo Experimental Forest. *Journal of Tropical Ecology* 16: 189–207.

Seitzinger, S., and Sanders, R. W. (1997). Contribution of dissolved organic nitrogen from rivers to estuarine eutrophication. *Marine Ecology Progress Series* 159: 1–12.

Silver, W. L., Scatena, F. N., Johnson, A. H., Siccama, T. G., and Sanchez, M. J. (1994). Nutrient availability in a montane wet tropical forest: Spatial patterns and methodological considerations. *Plant and Soil* 164: 129–145.

Sollins, P., Grier, C. C., McCorison, F. M., Cromack, K. Jr., Fogel, R., and Fredriksen, R. L. (1980). The internal element cycles of an old-growth Douglas-fir ecosystem in western Oregon. *Ecological Monographs* 50: 261–285.

Stark, N. M., and Jordan, C. F. (1978). Nutrient retention by the root mat of an Amazonian rain forest. *Ecology* 59: 434–437.

Swank, W. T., and Waide, J. B. (1988). Characterization of baseline precipitation and stream chemistry and nutrient budgets for control watersheds. In: Forest Hydrology and Ecology at Coweeta, W. T. Swank, and D. A. Crossley, Jr. (Eds.), (p. 57–79). New York: Springer-Verlag.

Vitousek, P. M. (1984). Litterfall, nutrient cycling, and nutrient limitation in tropical forests. *Ecology* 65: 285–298.

Walker, L. R., Zarin, D. J., Fetcher, N., Myster, R. W., and Johnson, A. H. (1996). Ecosystem development and plant succession on landslides in the Carribean. *Biotropica* 28: 566–576.

White, A. F., and Blum, A. E. (1995). Climatic effects on chemical weathering in watersheds: Application of mass balance approaches. In: S. T. Trudgill (Ed.), Solute modelling in catchment systems (pp. 101–131). New York: Wiley.

White, A. F., Blum, A. E., Schulz, M. S., Vivit, D. V., Sonestrom D. A., Larsen, M., Murphy, S. F., and Eberl, D. (1998). Chemical weathering in a tropical watershed, Luquillo Mountains, Puerto Rico: I. Long-term versus short-term weathering fluxes. *Geochimica Cosmochimica Acta* 62: 209–226.

Whitmore, T. C. (1989). Tropical forest nutrients, where do we stand? A *tour de horizon*. In: J. Proctor (Ed.), *Mineral nutrients in tropical forest and savanna ecosystems* (pp. 1–13). London: Blackwell Scientific.

Williams, M. R., and Melack, J. M. (1997). Solute export from forested and partially deforested catchments in the central Amazon. *Biogeochemistry* 38: 67–102.

9

Ozone Stress Impacts on Plant Life

Jürgen Bender and Hans-Joachim Weigel

Introduction

Ozone (O_3) is regarded as the main photochemical oxidant present in the troposphere that is predominantly produced by photochemical reactions involving volatile organic compounds (VOCs), nitrogen oxides, and sunlight. Tropospheric and ground O_3 levels in many regions of the northern hemisphere, and particularly in rural areas, have increased significantly during the past 100 years (Lefohn, 1992; Volz and Kley, 1988). For example, data from high-elevation sites in Europe show an annual increase of 1.6% (Marenco et al., 1994; Figure 9.1). From 1887 to 1910, annual mean O_3 concentrations in Europe were considered to be in the range of 5 to 12 ppb (Volz and Kley, 1988). Present-day tropospheric O_3 background concentrations are generally above 25 ppb (Stockwell et al., 1997), and current mean O_3 concentration values in nonurban areas are in the range of 40 to 75 ppb during the vegetation period (April–September), with 20–35 ppb annual means. Because of the dependency of O_3 formation on solar radiation energy, O_3 concentrations vary considerably in time and space, and show annual and diurnal patterns, with high concentrations during the growing season (April–September) and during the afternoon hours (see Figure 9.4). Quantification of O_3 effects on vegetation is thus extremely difficult as a result of this high variability.

Ozone, an all pervasive, secondary air pollutant, has long been recognized as phytotoxic at elevated concentrations. Numerous studies on O_3 effects on plants have been conducted since the phytotoxicity of the gas was first described in the 1950s (Richards et al., 1958; Heggestad and Middleton, 1959). O_3 has been shown to affect plant growth at all biological levels, ranging from subcellular effects to effects on whole ecosystems (Table 9.1). Evidence from experiments over the last 40 years with a wide range of plant species, as well as field observations, clearly show that current ambient O_3 levels are high enough to suppress crop yields of sensitive species in many industrialized areas (Heck et al., 1988; Fuhrer et al., 1997) and to affect growth and development of trees

Jürgen Bender and Hans-Joachim Weigel • Institute of Agroecology, Federal Research Centre for Agriculture, Bundesallee 50, 38116 Braunschweig, Germany; e-mail: Juergen.Bender@fal.de

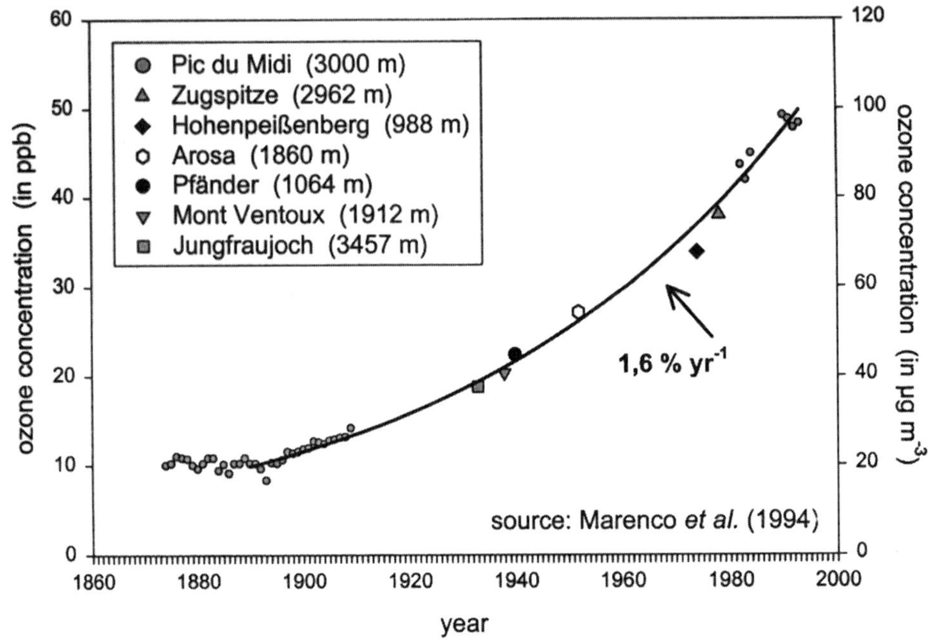

Figure 9.1. Estimate of O_3 evolution in the free atmosphere from measurements at various European stations at high altitude. Adapted from Marenco et al. (1994).

(Sandermann et al., 1997). More recently, evidence has indicated that O_3 may also have negative effects on plants species of nonwoody (semi) natural vegetation (Davison and Barnes, 1998).

This chapter provides a brief overview on the impact of O_3 on vegetation, with a focus on agricultural plants and natural (and seminatural) vegetation. In this review, we attempt to assess progress, and highlight problems and uncertainties in our understanding of the impact of O_3 stress on plant life.

Table 9.1. Overview of the Effects of O_3 on Plants

- ◆ Changes in membrane permeabilities
- ◆ Altered enzyme activities
- ◆ Ultrastructural changes in organelles
- ◆ Changes in photosynthesis and respiration
- ◆ Alterations in assimilate partitioning and allocation
- ◆ Visible foliar injury
- ◆ Alterations in growth and development
- ◆ Enhanced senescence
- ◆ Growth and yield reductions
- ◆ Changes in tissue and yield quality
- ◆ Alterations in the reproductive capacity (flowering, seed production)
- ◆ Changes in the susceptibility to biotic and abiotic stresses
- ◆ Altered competive abilities in plant communities
- ◆ Changes in species composition and diversity
- ◆ Changes in plant succession

Classification of Effects

It is important to note that the different effects of O_3 on the various organizational levels (Table 9.1) may be classified according to the the evaluation of the observed effect. Effects are usually classified as *injury* or *damage*. Injury encompasses all measurable plant reactions that impair the intended use of the plant (e.g., yield). Injury responses can include changes in metabolism, reduced photosynthesis, leaf necrosis, or reduced growth if yield is not impaired. In contrast, damage includes all effects that reduce intended human use (economic production, genetic resources, or cultural values) or value of the plant to the ecosystem (ecological structure and function) (Guderian et al., 1985). *Direct* O_3 effects on plants include visible leaf damage, premature aging, and reduced growth and biomass, as well as impairment of photosynthesis and carbon allocation within the plant (Krupa and Manning, 1988; Runeckles and Chevone, 1992). The pollutant is also considered to affect plant growth *indirectly* by causing alterations in the sensitivity of plants to other biotic and abiotic stresses.

Plant responses to O_3 exposures are further classified as acute or chronic. *Acute* responses, such as visible symptoms of foliar injury, occur within a short time period of hours and days, and are induced by relatively high O_3 concentrations. However, there is evidence that moderate, enhanced O_3 levels are more effective in eliciting injury symptoms (e.g., on tobacco leaves) than the highest pollutant concentrations (Tonneijck and Bugter, 1991; Krupa et al., 1995). This phenomenon is mainly attributed to atmospheric conditions (e.g., high vapor pressure deficit) or high soil moisture deficits that often coincide with high O_3 levels and result in lower O_3 flux and/or stomatal closure by the plant (see Chapter 3). *Chronic* exposures, in contrast, involve relatively low mean O_3 concentrations with periodic, highly variable episodes that usually last over longer time periods.

Uptake of O_3 and Mechanisms of Action at the Cellular Level

O_3 enters a plant primarily via the stomata. Cuticular conductance for O_3 is a few orders of magnitude smaller than stomatal conductance (Kerstiens and Lendzian, 1989) and is generally neglected in the analysis of O_3 uptake, although some evidence indicates that cuticular waxes can be partially degraded by O_3 (Runeckles, 1992). Thus, all environmental factors that affect stomatal conductance greatly determine the rate of O_3 uptake. At the canopy level, the vertical flux of O_3 into the canopy is determined by both the biological activity and the sink properties of the canopy itself (i.e., the plant's physiological activity and the canopy architecture) as well as the transfer properties of the atmosphere (e.g., turbulence) (Grünhage and Jäger, 1994).

Once inside the plant, O_3 is largely destroyed in the apoplast, so the plant internal O_3 concentration is very close to zero (Laisk et al., 1989). O_3 also dissolves in extracellular fluid and induces a diverse range of extra- and intracellular effects, including generation of reactive oxygen species and hydrogen peroxide, resulting in oxidative stress. Increased production of reactive oxygen intermediates, and oxygen toxicity is considered a common feature of O_3 stress (Elstner, 1990; Foyer et al., 1994). Plants have evolved protective scavenging systems in response to reactive oxygen species and hydrogen peroxide, respectively (Halliwell and Gutteridge, 1985). Antioxidants such as ascorbate and glutathione, and enzymes such as ascorbate peroxidase, glutathione reductase, and superoxide dismutase, that function in this capacity (Foyer et al., 1994; Figure 9.2) have repeatedly been shown to be affected by O_3 (Kangasjärvi et al., 1994; Pell et al., 1997). Changes in antioxidant levels due to O_3 have been suggested to be cellular repair processes that alleviate oxidative damage (Mehlhorn and Wellburn, 1994).

However, contradictory results have been obtained with respect to the direction and extent of these effects (Mehlhorn and Wellburn, 1994; Polle, 1998). On the one hand, responses of cellular antioxidants to O_3 seem to depend on the species investigated, the stage of plant development, and the degree of the pollutant stress (Price et al., 1990; Bender et al., 1994; Polle, 1998). On the other hand, positive correlations between the level of antioxidant molecules/enzymes and plant resistance to O_3 have been reported (e.g., Guri, 1983; Tanaka et al., 1985). The antioxidative defense system is considered a promising target for genetic engineering to aid our understanding of the mechanisms of O_3 action, as well as to generate plants with greater resistance to O_3 stress (Barnes et al., 1999). Most transformation studies examining oxidative stress tolerance in transgenic plants have focused on the overexpression of isoforms of superoxide dismutase and glutathione reductase. However, transgenic plants overproducing these particular antioxidative enzymes rarely displayed enhanced O_3 resistance, indicating that interactions with other components of the entire antioxidative defense system may be involved in scavenging the reactive oxygen intermediates effectively (Foyer et al., 1994; Harris and Bailey-Serres, 1994). The picture is further complicated by the fact that compartmentalization may play an important role in scavenging reactive oxygen species. Much research has focused on the intracellular antioxidant defense system because of its potential importance in the protection of chloroplasts, cytosol, and other cellular compartments. Recent evidence, however, indicated that the primary site of O_3 action may already be in the cell wall region (apoplast), that is, O_3 or its reactive products are mainly scavenged before reaching the plasmalemma. The leaf apoplast is known to contain several well-documented antioxidants, including polyamines, ascorbate peroxidase, Cu/Zn superoxide dismutase, and ascorbate (Dietz, 1997). Among these extracellular antioxidants, the apoplastic

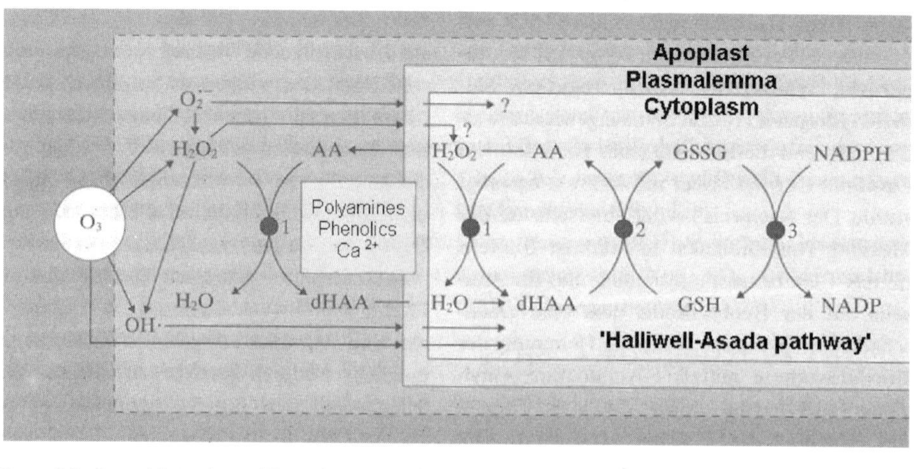

Figure 9.2. Potential reactions of O_3 and oxyradicals within the apoplast and symplast, involving ascorbate (AA), dehydroascorbate (dHAA), reduced (GSH), and oxidized (GSSG) glutathione as part of the Halliwell–Asada pathway. Polyamines, phenolics and Ca^{2+} ions are peroxidase activators. Enzyme 1 = ascorbate peroxidase; enzyme 2 = dehydroascorbate reductase; enzyme 3 = glutathione reductase. Adapted from Fuhrer (1996), based on Runeckles and Chevone (1992).

ascorbate pool seems to be a particularly important factor mediating O_3 tolerance (Barnes et al., 1999). Figure 9.2 summarizes the fate and the possible reactions of O_3 after its enty into the plant.

Effects of O_3 on Plant Physiology

The oxidative stress imposed by O_3 at the biochemical level is reflected at higher levels of organization by decline in the photosynthetic capacity, increased respiration, changed patterns of carbon distributions, and accelerated leaf senescence and foliar injury. Cell injury or death of plant tissue occurs when the O_3 uptake exceeds the detoxification and repair system (Tingey and Andersen, 1991). Because the initial site of injury caused by O_3 or its reactive products may be the plasmalemma, membrane changes in permeability, breakdown in compartmentation, and ionic and metabolic disturbances within the cell can occur (Heath, 1988). There is evidence that ethylene and salicyclic acid act as second messenger molecules induced by O_3 itself or O_3-derived active oxygen species; that is, signal chains (resembling those induced by pathogens) may be involved in elucidating intracellular O_3 effects (Sandermann, 1996). The resulting perturbation in metabolic function can then lead to higher susceptibility to other abiotic and biotic stressors (Sandermann, 1998). The sequence of events in the toxicity of O_3 is not yet fully clear, however. Heath and Taylor (1997) proposed the three most important effects caused by damage to the plasma membrane: (1) loss of membrane integrity, (2) loss of transport abilities and ionic balances, and (3) production of a signal (messenger), which leads to a wounding response (Mudd, 1998).

Intracellular effects include inhibition of chloroplast functions and pigment loss. Reductions in net photosynthetic rate due to O_3 exposure have been related to decreased levels and activity of Rubisco, and impaired electron transport (Pell et al., 1997). O_3-induced reduction in net photosynthesis (e.g., as measured by gas-exchange) has been documented in many studies, although the relationship between O_3 exposure and photosynthesis is not well known (U.S. Environmental Protection Agency, 1996). Besides a decline in the photosynthetic capacity of individual leaves, a decrease in stomatal conductance and increased rates of maintenance respiration may also contribute to this reduction in net photosynthesis (Darrall, 1989). O_3, in addition to decreasing the total amount of CO_2 assimilated, can reduce leaf duration (i.e., accelerated leaf senescence) and alter the pattern by which the reduced amount of assimilate is partitioned throughout the plant. There is good evidence that O_3 reduces resource allocation to the roots and reproductive organs in favor of shoot growth (Cooley and Manning, 1987; Miller, 1988), which has serious consequences in plant productivity. The key effects of O_3 on carbon assimilation and allocation are illustrated in Figure 9.3.

Interactions with Other Factors

Under field conditions, plant reactions to O_3 exposure are determined by the particular growth conditions that comprise a wide variety of additional abiotic (growth temperature, solar radiation intensity, water vapor pressure deficit [VPD], soil moisture and soil nutrient supply, and additional soil and air pollutants) and biotic (stage of plant development, pests, and diseases) variables. These all may interact in a complex way and

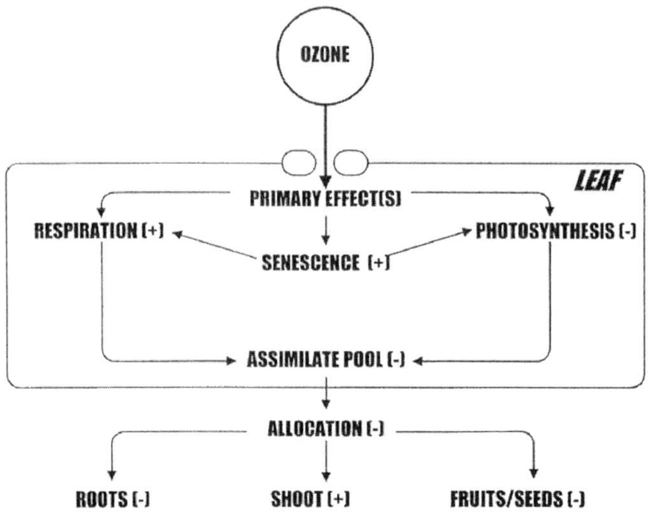

Figure 9.3. General scheme of the key effects of O_3 on the carbon assimilation and allocation in plants. Adapted from Fuhrer (1996).

considerably modify any O_3 effect. Here, only selected examples of these complex interaction are given.

Water Supply

Soil moisture and VPD are key determinators of a plant's water status via effects on water uptake, transport, and transpiration. Water supply and VPD directly affect stomatal conductance and hence the uptake and effects of O_3. Under certain weather conditions, high O_3 concentrations often coincide with low soil water status and high VPD, respectively. On the other hand, O_3 has been shown to modify stomatal aperture, affecting transpiration. Thus, a complex interaction between these two factors may result. For example, under restricted water supply, a reduced stomatal aperture may protect plants from O_3 effects by reducing O_3 uptake (Tingey and Hogsett, 1985), whereas an O_3 effect on stomatal aperture may alleviate the water stress conditions (Darrall, 1989). Moreover, although both O_3 and water stress favor the formation of activated oxygen species and affect the status of antioxidants (Smirnoff, 1993), water stress favors the allocation of assimilated carbon into roots. Conversely, O_3 stress is known to favor shoot growth at the expense of root growth. Under high solar radiation levels, both stresses may reduce CO_2 assimilation and result in photoinhibitory processes. Overall, experimental evidence points to a protective effect of low soil water supply and high VDP against adverse O_3 effects, although synergistic effects of water and O_3 stress cannot completely be ruled out.

Pests and Diseases

Biological components of the environment of plants include insects, fungi, bacteria, and viruses. O_3 can indirectly affect plant systems by modifying plant-parasitic and virus disease activity. Interactions between biotic factors and O_3 sensitivity have been studied in several crop species (Heagle, 1973; Manning and von Tiedemann, 1995); however, the available evidence is far from representative of the wide spectrum of plant diseases, so predictions of O_3 effects on particular plant-pathogen systems are not yet possible (Violini, 1995; Manning and von Tiedemann, 1995; U.S. Environmental Protection Agency, 1996). It is generally accepted that reduced vitality due to O_3 stress can make plants more susceptible to plant pathogens. Increased susceptibility after O_3 exposure has been reported for necrotrophic pathogens, whereas obligate biotrophic infections tend to be diminished by O_3 (Manning and von Tiedemann, 1995; Dowding 1988). With regard to insect pathogens, there is a general observational trend that some pests may prefer and grow better feeding on O_3-stressed plants, but in other observations, insect growth was not changed (U.S. Environmental Protection Agency, 1996). Viral infection often provides some protection from O_3 injury; however, the type and degree of protection depend on the specific host and virus (Violini, 1995; Manning and von Tiedemann, 1995). Although the available information is clearly insufficient to understand the importance of interactions between O_3 and biotic factors, it is suggested that these indirect effects may be more important under certain circumstances than the direct effects of the pollutant on plants.

Other Pollutants

Plants in their natural environment are seldom exposed to O_3 alone, but almost always to a number of additional airborne compounds (SO_2, NO/NO_2, HF, NH_3, heavy metals, acid rain, Persistent Organic Pollutant (POPs). The pattern of these mixtures of O_3 and other pollutants varies spatially and temporally (e.g., co-occurrences can be simultaneous, sequential, or any combination of the two). Currently, a simultaneous occurrence of O_3 and other gaseous air pollutants such as SO_2, NO, NO_2, or NH_3 at phytotoxic levels is rather unusual. Co-occurrences of O_3, SO_2, and NO_2 are of only short duration and are by far less frequent than sequential or combined sequential/concurrent exposures (Figure 9.4).

Exposure of plants to mixtures of air pollutants has been shown to modify the magnitude and nature of the response to individual pollutants. Interactions of two air pollutants in a mixture (pollutants A and B) can be described (U.S. Environmental Protection Agency, 1984) as *no joint action*, when one of the pollutants induces no plant responses, and as *joint action*, when both pollutants result in some plant responses. Joint action includes the subcategories (1) *additive* effects, when $effect_{AB}$ is equal to $effect_A + effect_B$ (which is not an interaction); and (2) *interactive* effects, when $effect_{AB}$ is not equal to $effect_A + effect_B$. If the sum of the two $effect_{AB}$ is less than $effect_A + effect_B$, the interaction is termed *antagonistic* ("less than additive"). Alternatively, if the sum of the two $effects_{AB}$ is greater than $effect_A + effect_B$, the interaction is called *synergistic* ("more than additive").

Research on plant responses to O_3 in combinations with other pollutants has primarily focused on interactions with only one other pollutant (Wolfenden et al., 1992; Bender and Weigel, 1993, 1994; Barnes and Wellburn, 1998), and very few studies have considered multiple pollutant interactions. Only examples of studies with crop species including both types of combinations are discussed here.

Both O_3 and NO_2 are constituents of "photochemical smog," and this combination must be regarded as one of the most common under field conditions. In many regions the concentrations of these gases vary during the day in a sequential pattern (Figure 9.4). Previous studies with mixtures of O_3 and NO_2 were carried out under simultaneous rather than sequential exposure regimens of both gases. Such pollutant mixtures caused

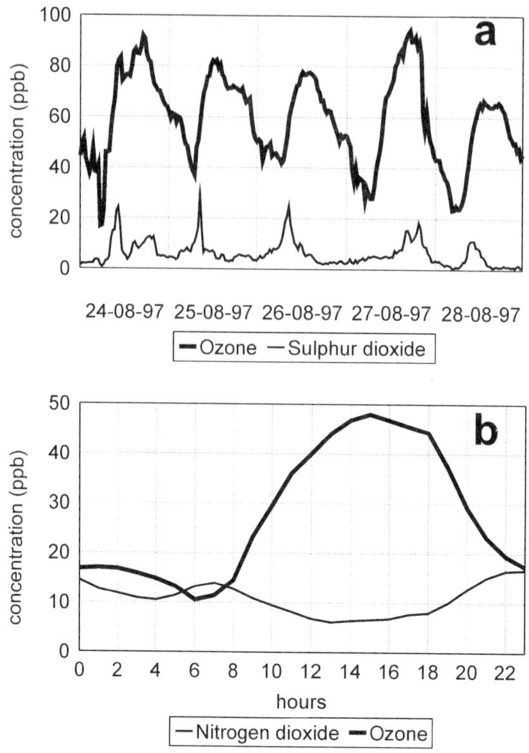

Figure 9.4. (a) Example of a period of co-occurrences of O_3 and SO_2 at Hasenholz (Brandenburg, Germany) in 1997 (U. Dämmgen, unpublished results), and (b) mean diurnal profile of O_3 and NO_2 ambient concentrations at the Federal Agricultural Research Center, Braunschweig (Bender et al., 1991).

reductions in growth and yield in a number of species (Kress and Skelly, 1982; Reinert, 1984; U.S. Environmental Protection Agency, 1984). However, the majority of these studies employed pollutant concentrations considerably higher than those observed even under worst-case pollutants scenarios. Short-term exposure to high concentrations of both gases may result in synergistic (more than additive) effects on plants (Guderian and Tingey, 1987).

More realistic studies using sequential exposures and near-ambient O_3 concentrations were carried out with the crop species wheat, barley, bush beans, tomato, and oil-seed rape (Runeckles and Palmer, 1987; Goodyear and Ormrod, 1988; Adaros et al., 1991a, 1991b; Bender et al., 1991). In general, these experiments provided only little evidence of statistically significant interactions (i.e., additive effects were predominate). When interactions were observed, the mode of interaction was mostly antagonistic, that is, the presence of NO_2 counteracted the effects of O_3 (Bender and Weigel, 1994).

Studies dealing with O_3 and two other gaseous pollutants usually have involved SO_2 and NO_2, because these gases were regarded as the most important pollutants affecting terrestrial ecosystems during the past decades. Adaros et al. (1991a, 1991b) investigated growth and yield responses of some crop species to O_3, SO_2 and NO_2, singly and in all combinations in open-top chambers under near-ambient pollutant exposure conditions. For spring rape (*Brassica napus*), the authors reported positive effects on yield with SO_2 or NO_2 alone, but antagonistic effects with the combined exposure to SO_2 plus NO_2. When O_3 exposure was combined with either SO_2 or NO_2, again, antagonistic effects were observed. In spring barley and spring wheat, Adaros et al. (1991b) found mainly antagonistic effects for any of the combinations of the three gaseous pollutants, SO_2, NO_2 or O_3. Only in one of the spring wheat cultivars tested (*Triticum aestivum* cv. Turbo) were synergistic effects on yield caused by the combination of O_3 with SO_2 and/or NO_2.

Effects of O_3 on Crop Yields

For agriculture, chronic O_3 effects are of particular concern, because they are due to O_3 exposures for weeks, months, or over the entire lifecycle of the plant. This exposures can lead to foliar chlorosis, enhanced senescence, or reductions in growth and yield.

Growth and yield effects of O_3 are the end product of the aforementioned biochemical and physiological responses related to the uptake of O_3 and its effects on assimilation, biosynthesis, and translocation. Studies with crops emerged from the speculation of possible adverse effects of O_3 on crop production and now provide the largest source of information on the impacts of O_3 on plants. A widely used experimental approach in the last 25 years is the use of open-top chambers (Heagle et al., 1973; Mandl et al., 1973) as an O_3 exposure system for field-related studies. Two major O_3 crop loss assessment programs were established. The U.S. National Crop Loss Assessment Network (NCLAN) (Heck et al., 1988) and the European Open-Top Chamber Program (EOTCP) (Jäger et al., 1993). A major aim of both NCLAN and EOTCP was to investigate the relationships between O_3 concentrations and yield by applying a range of artificially generated O_3 concentrations to either charcoal-filtered (CF) or nonfiltered (NF) ambient air. Both programes have provided evidence of significant yield losses due to current ambient O_3 exposures in several major U.S. and European crops, including soybean (*Glycine max*), bean (*Phaseolus vulgaris*), and wheat (*Triticum aestivum*); however, there were considerable differences between species, and even between cultivars within one species (see Figure 9.5).

Data from these experiments were subsequently used to develop statistical models to define cause–effect relationships between crop loss and O_3 exposure. Such relationships were the basis of O_3 exposure indices that can be used for air quality regulatory purposes to protect plants and vegetation against adverse O_3 effects. Lee et al. (1988), who analyzed the NCLAN data for five different crops from different sites and years, concluded that indices that accumulate exposure over time and emphasize higher concentrations (e.g., sum of all concentrations >60 ppb; SUM60), exhibit a better fit to the data than mean O_3 concentrations. In Europe, where wheat experiments provided the most comprehensive data sets on exposure–response relationships analysis, researchers confirmed that yield reduction was highly correlated with the cumulative exposure above a threshold of 40 ppb (Figure 9.6; Fuhrer, 1994). The exposure index (critical level) is referred to as the AOT40 (accumulated exposure over a threshold of 40 ppb calculated for daylight hours). Whereas SUM60 represents a sum of all hourly concentrations >60 ppb, AOT40 represents the sum of the difference between 40 ppb and hourly concentrations >40 ppb. The current critical level for crops in Europe was defined as an AOT40 value of 3000 ppb.h calculated over 3 flexible months, relating to the most important growth period of the respective crops (Kärenlampi and Skärby, 1996).

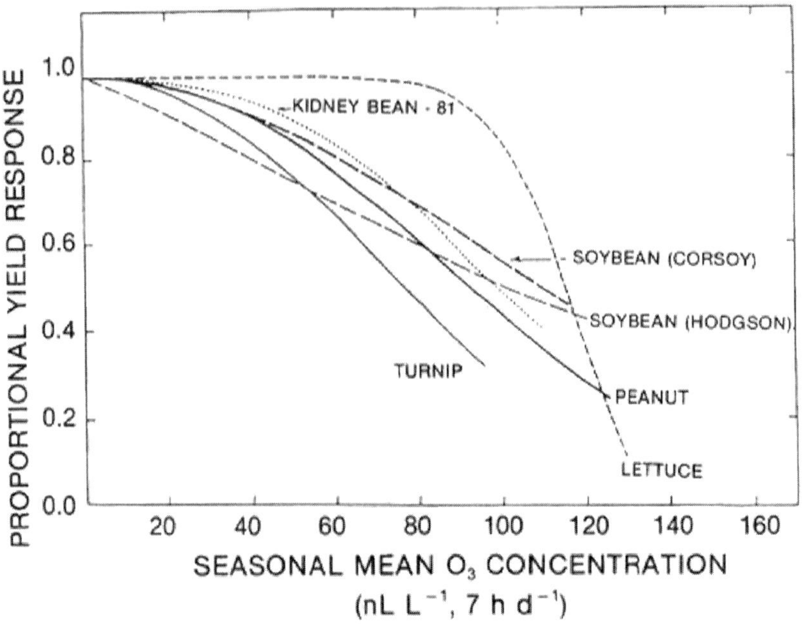

Figure 9.5. Proportional yield response curves for dicotyledonous crops in the NCLAN experiments. The curves were fitted using the Weibull model. Adapted from Heagle et al. (1988).

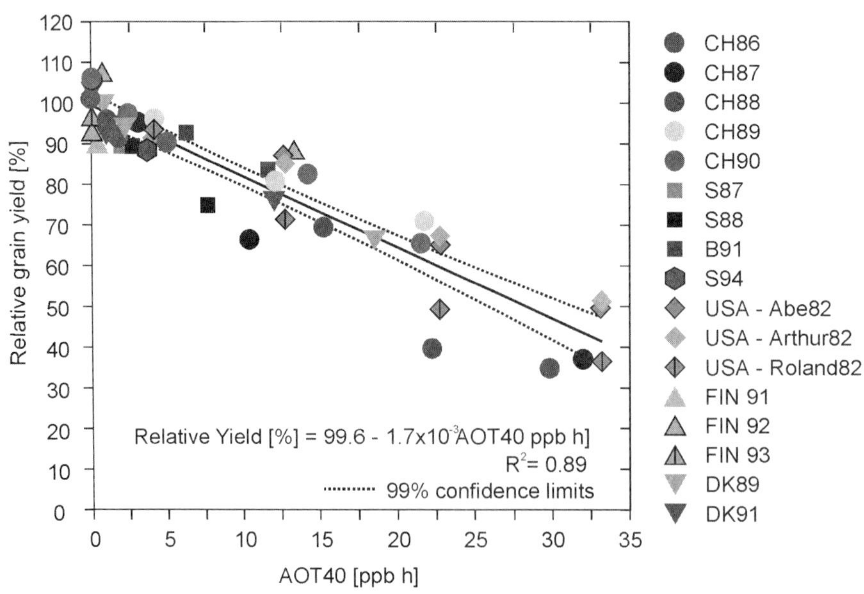

Figure 9.6. The relationship between relative grain yield of spring wheat and O_3 exposure as AOT40 over 3 months, based on data from several European and US open-top chamber experiments. Adapted from Fuhrer (1994).

Critical levels can then be used to assess the geographical area at risk by comparison with measured or modeled AOT40 values (i.e., by estimating the area of the critical level exceedance). However, there are a number of uncertainties associated with the a use of a single value (e.g., for yield reductions in crops). First, the level is deduced only from fumigation experiments with wheat in open-top chambers, so species-specific factors are ignored. Second, there are considerable differences in microclimate between chamber-grown plants and plants growing outside, because of differences in aerodynamic and boundary layer resistances, and thus O_3 uptake, which make it difficult to extrapolate results from chamber experiments to ambient air conditions. Third, it is well established that plant responses are more closely related to the internal O_3 dose, or the instantaneous flux through the stomata, than to the ambient O_3 exposure, which further restricts the application of the chamber-based relationships to ambient air conditions. Also, there is strong evidence of sensitivity of crops to O_3 changes with the developmental stage (i.e., plants may be sensitive at some growth stages and tolerant at others), which implies that the timing of O_3 stress episodes determines the impact of O_3 on growth and productivity. Several studies, for example, have shown that crops are more sensitive in terms of yield responses during reproductive than during vegetative development (Bender, 1993; Vandermeiren et al., 1995; Soja, 1996). It is thus clear that local factors such as crop species and cultivars, phenological stage, soil moisture deficit, co-occurence with other pollutants, and climatic factors (vapor pressure deficit and wind speed) must be taken into account in calculating site-specific critical levels of ozone (Fuhrer et al., 1997).

Effects of O_3 on Seminatural Vegetation

In contrast to the amount of existent information about effects of O_3 on agricultural plants, much less is known about effects on natural vegetation (or more accurately, seminatural, i.e., noncultivated vegetation). For example, a literature survey of more than 1,500 articles on ozone effects on plants revealed that, until 1998, only about 50 publications reported on O_3 effects on native, wild plant species. Currently, the O_3 sensitivity of about 250–300 species has been tested (Glombitza et al., 1998). Many of these studies involved short-term fumigation of isolated taxa in closed environments or laboratory chambers with the evaluation of the relative sensitivity to O_3 based mostly on the incidence of visible symptoms after acute exposure to high O_3 concentrations (Treshow and Stewart, 1973; Cornelius et al., 1985; Ashmore et al., 1987; Nebel and Fuhrer, 1994). Although these and other studies demonstrated wide variation in the O_3 sensitivity between species of seminatural communities (Davison and Barnes, 1998), they contributed little to our understanding how species respond in real communities. This is not only because of the unrealistic O_3 exposure conditions employed in these experiments, but also because a range of additional factors in the field (e.g., competition, mineral nutrition, management practices, soil water deficits, pests, and diseases) may influence plant response to O_3 (Davison et al., 1999). Results from some of the few longer term studies suggest that many wild plant species are as sensitive as the most sensitive crops (see Figure 9.7; Table 9.2). Of the 96 species listed in Table 9.2, 40 species responded with leaf injury and 42 species with a more than 10% reduction of leaf biomass, whereas 9 species (mainly grasses) responded with a higher leaf biomass (>10%) under elevated O_3 (Franzaring et al., 1998).

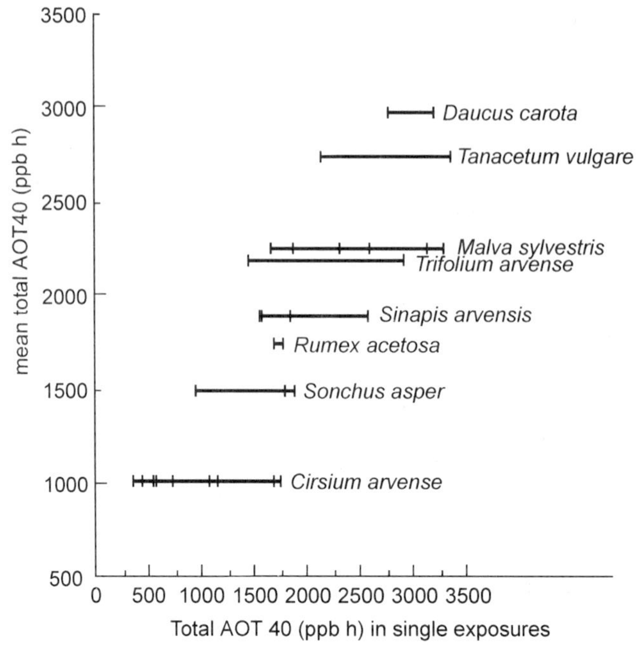

Figure 9.7. Total AOT40 values in single O_3 exposures required for the incidence of visible symptoms in wild plant species (each solid line represents an individual exposure) against the mean total AOT40s (average of AOT40s of single exposures (Bergmann et al., 1999).

A fundamental question in assessing O_3 effects on this type of vegetation concerns consideration of aspects of performance or key responses to O_3: What could be a suitable end point? Visible injury provides no direct evidence for an ecological impact of O_3, particularly because many studies show that symptoms of injury are not necessarily associated with growth reductions (Bergmann et al., 1996; Davison and Barnes, 1998). However, foliar injury is the first clear and unequivocal sign of the presence of phytotoxic levels of O_3, and injured tissues may offer a conduit for an enhanced pathogenic attack.

Moreover, there are significant differences between species with respect to the sensitivity of different end points toward O_3 exposure. As shown in Figure 9.8, the relative sensitivities of different species in terms of leaf biomass were different from those inferred from total, aboveground biomass and seed or flower production, indicating that species can differ widely in their relative allocation of resources to vegetative growth and reproductive organs under the impact of O_3 (Bergmann et al., 1998). Seed output, for example, is definitely an important parameter of ecological significance for annuals, which depend on regular seed production in order to survive, but for perennials, other changes in plant performance due to O_3 (e.g., root responses) may be more relevant (Davison and Barnes, 1998).

The most important impact of O_3 on plant communities may be through effects on growth and productivity of individual species, resulting in shifts in species composition and changes in genetic composition (i.e., losses of biodiversity). To date, there are only a few O_3 exposure studies on herbaceous plant communities (e.g., using open-top chambers and relatively simple species mixtures). Species of grassland communities of temperate

Table 9.2. Summary of Results from Long-Term Fumigation Experiments Using Species from the European Flora According to the Response Parameters Visible Injury and Growth

Reaction parameter: tested plant species:	Visible injury	Ref.	Leaf biomass (deviation from control)	Ref.	Reaction parameter: tested plant species:	Visible injury	Ref.	Leaf biomass (deviation from control)	Ref.
Achillea millefolium			–	3	*Hypericum pulchrum*			–	6
Agrostis capillares			±	6	*Knautia arvensis*	x	3	–	3
Agrostemma githago			±	2	*Koeleria macrantha*			+	6
Agrostis stolonifera			±, –	2,6	*Leontodon hispidus*			±, –	2,6
Alopecurus pratensis			±,±	2,6	*Lolium perenne*			+ ,±	3,6
Anthoxanthum odoratum			±	2	*Lotus corniculatus*	x	3	+ , –	3,6
Anthyllis vulneraria			±	2	*Lychnis flos-cuculi*			+	3
Arrhenatherum elatius	x	3	±, – , –	3,5,6	*Malva sylvestris*	x	1	–	1
Betula pubescens	x	4	±	4	*Malva moschata*			±	1
Brachypodium pinnatum			±	6	*Matricaria discoidea*	x	1	–	1
Briza media			±, –	2,6	*Matricaria perforata*			+	1
Bromus arvensis			±	2	*Onobrychis sativa*	x	3	–	3
Bromus erectus			+	3	*Papaver argemone*			±	1
Campanula rotundifolia			–	6	*Papaver dubium*			–	1
Capsella bursa-pastoris			±	1	*Papaver rhoeas*			±,±	1,2
Carum carvi	x	3	±	3	*Phalaris arundinacea*			±	2
Centaurea cyanus			±	2	*Phleum alpinum*	x	2	±	2
Centaurea jacea	x	3	–	3	*Phleum pratense*	x	4	–	4
Cerastium fontanum			– ,±	5,6	*Plantago lanceolata*	x	3,4	±,±, – , – , –	2–6
Crepis biennis			±	3	*Plantago major*	x	1	+ , –	1,5
Chamomilla recutita	x	1	–	1	*Plantago media*			±	2
Chenopodium album	x	5	+ ,±	1,2	*Poa annua*			±,±	2,5
Chrysanth. leucanthemum	x	3,4	–	3	*Poa palustris*			±	2
Chrysanthemum segetum			±	2	*Poa pratensis*			– , – ,±	3,4,6
Cirsium arvense	x	1	±	1	*Polygonum viviparum*			±	4
Cynosurus cristatus			±	6	*Potentilla erecta*	x	4	±	4
Dactylis aschersonii	x	2	±	2	*Rumex acetosa*	x	1,5	– ,±,±	1,2,6
Dactylis glomerata	x	2	±, – , – , –	2,3,4,6	*Rumex acetosella*	x	5	–	5
Danthonia decumbens			+	6	*Rumex crispus*	x	1	±	1
Daucus carota	x	1	±	1	*Rumex obtusifolius*	x	1	+ , + ,±	1,3,5
Deschampsia flexuosa			+	6	*Salvia pratensis*			±	3
Dianthus deltoides			±	2	*Sanguisorba minor*			±	6
Epilobium hirsutum	x	5	±	5	*Senecio vulgaris*			– ,±	1,2
Euphorbia peplus			–	1	*Silene acaulis*			–	4
Festuca ovina			+ , –	2,6	*Silene dioica*			+	3
Festuca pratensis			±, –	2,4	*Silene vulgaris*			±	2
Festuca rubra			– , – ,±	3,4,6	*Sinapis arvensis*	x	1	±	1
Galinsoga parviflora	x	1	–	1	*Solanum nigrum*			–	1
Galium saxatile			–	6	*Solidago virgaurea*	x	4	±	4
Hieracium pilosella			±, –	2,6	*Sonchus asper*	x	1	–	1
Holcus lanatus			–	6	*Stellaria media*			±	1
Hordeum murinum	x	5	±	5	*Tanacetum vulgare*	x	1	±	1
Hypochoeris radicata			±	2	*Tragopogon orientalis*	x	3	–	3
Hypericum perforatum	x	4	±	4	*Taraxacum officinalis*	x	3	–	3
					Teucrium scorondonia			–	5
References:	Chamber:		Concentration:	Control:	*Trifolium arvense*	x	1	–	1
1) Bergman et al. (1998)	OTC		0.6*AA + 30 ppb	30 ppb	*Trifolium dubium*	x	6	–	6
2) Pleijel et al. (1997)	OTC		1.5*AA	CF	*Trifolium pratense*	x	3,6	– , –	3,6
3) Grub et al. (1997)	OTC		1.5*AA	CF	*Trifolium repens*	x	3,6	– , –	3,6
4) Mortensen (1992)	Closed		15–80 ppb	CF	*Trisetum flavescens*			±	3
5) Reiling et al. (1992a)	Closed		70 ppb	CF	*Urtica dioica*			– , –	1,5
6) Ashmore et al. (1996)	Closed		80 ppb	CF	*Vicia sativa*	x	6	–	6

Modified from Franzing et al. (1998).
x = incidence of foliar injury; + = growth stimulation of leaf biomass by more than 10%; ± = difference in leaf biomass <10%; – = reduction of leaf biomass by more than 10%.
OTC: open top chambers; CF: charcoal-filtered air.

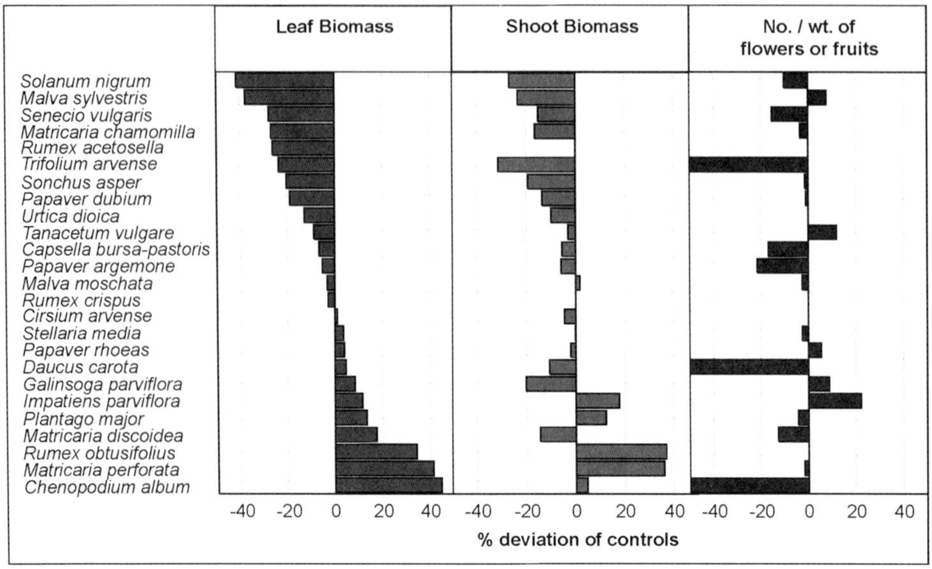

Figure 9.8. Relative effects (% of controls) of O_3 treatment on leaf biomass, shoot biomass and number or weight of flowers and fruits in native herbaceous species (Bergmann et al., 1998).

regions have been investigated most frequently. Investigations on simple grass–clover mixtures indicate that the effects of O_3 depend on the relative sensitivity of the competing species, and that grass species tend to be less sensitive than forbs (Davison and Barnes, 1998). Studies on artificial grassland communities provide further evidence that species composition may be affected by O_3, because the biomass of forbs such as clover decreases with increasing O_3 concentrations, whereas the total biomass of the plant community remain unaffected (Mortensen, 1992; Ashmore and Ainsworth, 1995; Ashmore et al., 1996; Nussbaum et al., 1995; Warwick and Taylor, 1995).

There is also evidence that some plant populations may evolve resistance to O_3 (Davison and Barnes, 1998; Barnes et al., 1999), implying that the pollutant may have the potential to reduce the frequency of sensitive genotypes in the population. A few studies have addressed this issue by exposing natural vegetation experimentally to elevated O_3 levels. Duchelle et al. (1983), for example, investigated the effects of O_3 on the productivity of natural vegetation in a high meadow in the Shenandoah National Park (USA), and Barbo et al. (1998) exposed an early successional forest community to different levels of O_3. Their findings suggest that O_3 may induce changes in species performance, canopy structure, species richness, and diversity.

Despite this lack of consistent information of relevance to field conditions, provisional critical levels for ozone have also been established for natural vegetation (Fuhrer and Achermann, 1999). For annuals, the critical level, expressed as AOT40, is the same as for crops (3,000 ppb.h over 3 months), and the proposed response parameter with highest ecological significance to plant species with a short lifecycle is seed output. For perennials, a provisional AOT40 value of 7,000 ppb.h over a time window of 6 months was proposed, and the important parameter for long-lived species was suggested to be root and shoot biomass. However, the uncertainties associated with the use of single values in assessing sensitivity are similar to those for crops, and the implications of effects of O_3 pollution on communities, community structure, and functioning remain to be determined.

References

Adaros, G., Weigel, H. J., and Jäger, H. J. (1991a). Single and interactive effects of low levels of O_3, SO_2 and NO_2 on the growth and yield of spring rape. *Environmental Pollution* 72: 269–286.

Adaros, G., Weigel, H. J., and Jäger, H. J. (1991b). Concurrent exposure to SO_2 and/or NO_2 alters growth and yield responses of wheat and barley to low concentrations of O_3. *New Phytologist* 118: 581–591.

Ashmore, M. R., and Ainsworth, N. (1995). The effects of ozone and cutting on the species composition of artificial grassland communities. *Functional Ecology* 9: 708–712.

Ashmore, M. R., Dalpra, C., and Tickle, A. K. (1987). Effects of ozone and calcium nutrition on native plant species. In: P. Mathy (Ed.), *Air pollution and ecosystems* (pp. 647–652). Dordrecht: Reidel.

Ashmore, M. R., Power, S. A., Cousins, D. A., and Ainsworth, N. (1996). Effects of ozone on native grass and forb species: A comparison of responses of individual plants and artificial communities. In: *Critical levels for ozone in Europe: Testing and finalizing the concepts* (pp. 193–197). Kuopio, Finland: UNECE Workshop Report, University of Kuopio.

Barbo, D. N., Chappelka, A. H., Somers, G. L., Miller-Goodman, M. S., and Stolte, K. W. (1998). Diversity of an early successional plant community as influenced by ozone. *New Phytologist* 138: 653–662.

Barnes, J., Bender, J., Lyons, T., and Borland, A. (1999). Natural and man-made selection for air pollution resistance. *Journal of Experimental Botany* 50: 1423–1435.

Barnes, J., and Wellburn, A. R. (1998). Air pollutant combinations. In: L. J. De Kok, and I. Stulen (Eds.), *Responses of plants to air pollution and global change* (pp. 147–164). Leiden: Buckhuys.

Bender, J. (1993). Physiological and biochemical studies within the CEC open-top chamber programme. In: H. J. Jäger, M. Unsworth, L. De Temmerman, and P. Mathy (Eds.), *Effects of air pollution on agriculatural crops in Europe* (pp. 339–358). Air Pollution Research Report No. 46. Brussels: Commission of the European Communities.

Bender, J., and Weigel, H. J. (1993). Crop responses to mixtures of air pollutants. In: H. J. Jäger, M. Unsworth, L. De Temmerman, and P. Mathy (Eds.), *Effects of air pollution on agricultural crops in Europe*. (pp. 445–453). Air Pollution Research Report No. 46. Brussels: Commission of the European Communities.

Bender, J., and Weigel, H. J. (1994). The role of other pollutants in modifying plant responses to ozone. In: J. Fuhrer, and B. Achermann (Eds.), *Critical levels for ozone* (pp. 240–247). Liebefeld: Schriftenreihe der FAC.

Bender, J., Weigel, H. J., and Jäger, H. J. (1991). Response of nitrogen metabolism in beans (*Phaseolus vulgaris L.*) after exposure to ozone and nitrogen dioxide, alone and in sequence. *New Phytologist* 119: 261–267.

Bender, J., Weigel, H. J., Wegner, U., and Jäger, H. J. (1994). Response of cellular antioxidants to ozone in wheat flag leaves at different stages of plant development. *Environmental Pollution* 84: 15–21.

Bergmann, E., Bender, J., and Weigel, H. J. (1996). Effects of chronic ozone stress on growth and reproduction capacity of native herbaceous plants. In: M. Knoflacher, J. Schneider, and G. Soja (Eds.), *Exceedance of critical loads and levels* (pp. 177–1850). Wien: Tagungsberichte Umweltbundesamt.

Bergmann, E., Bender, J., and Weigel, H. J. (1998). In: W. Nentwig, and H. M. Poehling (Eds.), *Zur Ozonempfindlichkeit von Wildpflanzen* (p. 88). Bern: Verlag Agrarökologie.

Bergmann, E., Bender, J., and Weigel, H. J. (1999). Ozone threshold doses and exposure-response relationships for the development of ozone injury symptoms in wild plant species. *New Phytologist* 144: 423–435.

Cooley, D. R., and Manning, W. J. (1987). The impact of ozone on assimilate partitioning in plants: A review. *Environmental Pollution* 47: 95–113.

Cornelius, R., Faensen-Thiebes, A., and Meyer, G. (1985). Einsatz von *Nicotiana tabacum* L. BelW3. *Staub-Reinhaltung Luft* 45: 59–61.

Darrall, N. M. (1989). The effect of air pollutants on physiological processes in plants. *Plant Cell and Environment* 12: 1–30.

Davison, A. W., Ashmore, M. R., Bender, J., Chappelka, A. H., and Weigel, H. J. (1999). Critical levels for semi-natural vegetation. In: J. Fuhrer, and B. Achermann (Eds.), *Critical levels for ozone: Level II* (pp. 73–76). Environmental Documentation No. 15. Bes, Swiss Agency for Environmental Forest Landscape.

Davison, A. W., and Barnes, J. D. (1998). Effects of ozone on wild plants. *New Phytologist* 139: 135–151.

Dietz, K. J. (1997). Functions and responses of the leaf apoplast under stress. *Progress in Botany* 68: 227–254.

Dowding, P. (1988). Air pollutant effects on plant pathogens. In: S. Schulte-Hostede, N. M. Darrall, L. W. Blankl, and A. R. Wellburn (Eds.), *Air pollution and plant metabolism* (pp. 329–380). London: Elsevier Applied Science.

Duchelle, S. F., Skelly, J. M., Sharick, T. L., Chevone, B., Yang, Y. S., and Nielessen, J. E. (1983). Effects of ozone on the productivity of natural vegetation in a high meadow of the Shenandoah National Park of Virginia. *Journal of Environmental Management* 17: 299–308.

Elstner, E. F. (1990). *Oxygen—biochemistry, biology, medicine*. Mannheim: BI-Wissenschaftsverlag.
Foyer, C. H., Descourvieres, P., and Kunert, K. J. (1994). Protection against oxygen radicals: An important defense mechanism studied in transgenic plants. *Plant, Cell, and Environment* 17: 507–523.
Franzaring, J., Bergmann, E., Bender, J., and Weigel, H. J. (1998). The impact of ozone on natural vegetation—an ecological discussion of European studies. *Verhandwingen Gesellschaft für Ökologie* 28: 355–362.
Fuhrer, J. (1994). The critical level for ozone to protect agricultural crops—an assessment of data from European open-top chamber experiments. In: J. Fuhrer and B. Achermann (Eds.), *Critical levels for ozone: A UN-ECE workshop report. Schriftenreihe der FAC Liebefeld* 16: 42–57.
Fuhrer, J. (1996). Gasförmige Luftschadstoffe. In: C. Brunold, A. Regsegger, and R. Brändle (Eds.), *Stress bei Pflanzen*. (pp. 221–246). Bern: Verlag Paul Haupt.
Fuhrer, J. (1997). Ozone sensitivity of managed pastures. In: P. N. Chereminisoff (Ed.), *Ecological advances and environmental impact assessment* (pp. 681–706). Houston: Gulf.
Fuhrer, J., and Achermann, B. (Eds.). (1999). Critical levels for ozone-level II. Environmental Documentation 115. Bern: Swiss Agency for Environment, Forest and Landscape.
Fuhrer, J., Skärby, L., and Ashmore, M. R. (1997). Critical levels for ozone effects on vegetation in Europe. *Environmental Pollution* 97: 91–106.
Glombitza, S., Bender, J., and Weigel, H. J. (1998). *Bewertung von Ozonwirkungen auf mitteleuropäische Pflanzenarten unter Verwendung existierender Literaturdatenbanken*. Abschlußbericht zum Projekt "Angewandte Ökologie." Braunschweig: Bundesforschungsanstalt für Landwirtschaft, Institut für Agrarökologie, [in German].
Goodyear, S. N., and Ormrod, D. P. (1988). Tomato response to concurrent and sequential NO_2 and O_3 exposures. *Environmental Pollution* 51: 315–326.
Grub, A., Bungener, P., Contat, F., Nussbaum, S., Endtner, V., and Fuhrer, J. (1997). Pollution atmospherique et biodiversite floristique. *Revue Suisse Agricultural* 29: 165–171.
Grünhage, L., and Jäger, H. J. (1994). Influence of the atmospheric conductivity on the ozone exposure of plants under ambient conditions: Considerations for establishing ozone standards to protect vegetation. *Environmental Pollution* 85: 125–129.
Guderian, R., and Tingey, D. T. (1987). *Notwendigkeit und Ableitung von Grenzwerten fr Stickoxide*. Umweltbundesamt, Berichte. Berlin: Erich Schmidt Verlag.
Guderian, R., Tingey, D. T., and Rabe, R. (1995). Effects of photochemical oxidants on plants. In: R. Guderian (Ed.), *Air pollution by photochemical oxidants: Formation, transport, control, and effects on plants* (pp. 129–335). Ecological Studies No. 52, Berlin: Springer-Verlag.
Guri, A. (1983). Variation in glutathione and ascorbic acid content among selected cultivars of *Phaseolus vulgaris* prior to and after exposure to ozone. *Canadian Journal of Plant Science* 63: 733–737.
Halliwell, B., and Gutteridge, J. M. C. (1985). *Free radicals in biology and medicine*. Oxford, UK: Clarendon Press.
Harris, M. J., and Bailey-Serres, J. N. (1994). Ozone effects on gene expression and molecular approaches to breeding for air pollution resistance. In: A. Singh Basra (Ed.), *Stress-induced gene expression in plants* (pp. 185–207). Singapore: Harwood Academic.
Heagle, A. S. (1973). Interactions between air pollutants and plant parasites. *Annual Review of Phytopathology* 11: 365–488.
Heagle, A. S., Body, D. E., and Heck, W. W. (1973). An open-top field chamber to assess the impact of air pollution on plants. *Journal of Environmental Quality* 2: 365–368.
Heagle, A. S., Kress, L. W., Temple, P. J., Kohut, R. J., Miller, J. E., and Heggestad, H. E. (1988). Factors influencing ozone-dose yield response relationships in open-top field chamber studies. In: W. W. Heck, O. C. Taylor, and D. T. Tingey (Eds.), *Assessment of crop loss from air pollutants* (pp. 141–179). London: Elsevier Applied Science.
Heath, R. L. (1988). Biochemical mechanisms of pollutant stress. In: W. W. Heck, O. C. Taylor, and D. T. Tingey (Eds.), *Assessment of crop loss from air pollutants* (pp. 259–286). London, New York: Elsevier Applied Science.
Heath, R. L., and Taylor, G. E. (1997). Physiological processes and plant responses to ozone exposure. In: H. Sandermann, A. R. Wellburn, and R. L. Heath (Eds.), *Forest decline and ozone* (pp. 317–368). Ecological Studies No. 127. Berlin: Springer.
Heck, W. W., Taylor, O. C., and Tingey, D. T. (Eds.). (1988). *Assessment of crop loss from air pollutants*. London, New York: Elsevier Applied Science.
Heggestad, H. E., and Middleton, J. T. (1959). Ozone in high concentrations as cause of tobacco leaf injury. *Science* 129: 208–210.

Jäger, H. J., Unsworth, M., De Temmerman, L., and Mathy, P. (Eds.). (1993). *Effects of air pollution on agricultural crops in Europe*. Air Pollution Research Report No. 46. Brussels: Commission of the European Communities.
Kangasjärvi, J., Talvinen, J., Utriainen, M., and Karjalainen, R. (1994). Plant defence systems induced by ozone. *Plant, Cell and Environment* 17: 783–794.
Kärenlampi, L., and Skärby, L. (Eds.). (1996). *Critical levels for ozone in Europe: Testing and finalizing the concepts*. UN-ECE workshop report. Finland: University of Kuopio.
Kerstiens; G., and Lendzian, K. J. (1989). Interactions between ozone and plant cuticles. I. Ozone deposition and permeability. *New Phytologist* 112: 13–19.
Kress, L. W., and Skelly, J. M. (1982). Response of several eastern forest tree species to chronic doses of ozone and nitrogen dioxide. *Plant Disease* 66:1149–1152.
Krupa, S. V., Grünhage, L., Jäger, H. J., Nosal, M., Manning, W. J., Legge, A. H., and Hanewald, K. (1995). Ambient ozone (O_3) and adverse crop response: A unified view of cause and effect. *Environmental Pollution* 87: 119–126.
Krupa, S. V., and Manning, W. J. (1988). Atmospheric ozone: Formation and effects on vegetation. *Environmental Pollution* 50: 101–137.
Laisk, A., Kull, O., and Moldau, H. (1989). Ozone concentration in leaf intercellular air spaces is close to zero. *Plant Physiology* 90: 1163–1167.
Lee, E. H., and Bennett, J. H. (1982). Superoxide dismutase: a possible protective enzyme against ozone injury in snap beans (*Phaseolus vulgaris L.*). *Plant Physiology* 69: 1444–1449.
Lee, E. H., Tingey, D. T., and Hogsett, W. E. (1988). Evaluation of ozone exposure indices in exposure–response modelling. *Environmental Pollution* 53: 43–62.
Lefohn, A. S. (1992). The characterization of ambient ozone exposure. In: A. S. Lefohn (Ed.), *Surface level ozone exposures and their effects on vegetation* (pp. 31–92). Chelsea, UK: Lewis.
Mandl, R. H., Weinstein, L. H., McCune, D. C., and Keveny, M. (1973). A cylindrical open–top chamber for the exposure of plants to air pollutants in the field. *Journal of Environmental Quality* 2: 371–376.
Manning, W. J., and von Tiedemann, A. (1995). Climate change: Potential effects of increased atmospheric carbon dioxide (CO_2), ozone (O_3), and ultraviolet-B (UV-B) radiation on plant diseases. *Environmental Pollution* 88: 219–245.
Marenco, A., Gouget, H., Philippe, N., Pages, J. P., and Karcher, F. (1994). Evidence of a long-term increase in tropospheric ozone from Pic du Midi data series: Consequences: Positive radiative forcing. *Journal of Geophysical Research* 99: 16.617–16.632.
Matters, G. L., and Scandalios, J. G. (1987). Synthesis of isoenzymes of superoxide dismutase in maize leaves in response to O_3, SO_2, and elevated O_2. *Journal of Experimental Botany* 38: 842–852.
Mehlhorn, H., and Wellburn, A. R. (1994). Man-induced causes of free radical damage to plants: O_3 and other gaseous pollutants. In: C. H. Foyer and P. M. Mullineaux (Eds.), *Causes of photooxidative stress and amelioration of defense systems in plants* (pp. 155–175). Boca Raton, FL: CRC Press.
Miller, J. E. (1998) Effects of photosynthesis, carbon allocation and plant growth associated with air pollution stress. In: W. W. Heck, O. C. Taylor, and D. T. Tingey (Eds.), *Assesment of crop loss from air pollutants* (pp. 287–314). London: Elsevier Applied Science.
Mortensen, L. (1992). Effects of ozone on growth of seven grass and one clover species. *Acta Agronomica Scandivica (Section B): Soil and Plant Science* 42: 235–239.
Mudd, J. B. (1998). Foreword: On ozone. In: L. J. Kok and I. Stulen (Eds.), *Responses of plants to air pollution and global change* (pp. xiii–xix). Leiden: Backhuys.
Nebel, B., and Fuhrer, J. (1994). Intra- and interspecific differences in ozone sensitivity in semi-natural plant communities. *Angewandte Botanik* 68: 116–121.
Nussbaum, S., Geissmann, M., and Fuhrer, J. (1995). Ozone-exposure relationships for mixtures of perennial ryegrass and white clover depend on ozone exposure patterns. *Atmospheric Environment* 29: 989–995.
Pell, E. J., Schlagnhaufer, C. D., and Arteca, R. N. (1997). Ozone-induced oxidative stress: mechanism of action and reaction. *Physiologia Plantarum* 100: 264–273.
Pleijel, H., and Daniellson, H. (1997). Growth of 27 herbs and grasses in relation to ozone exposure and plant strategy. *New Phytologist* 135: 361–367.
Pleijel, H., Karlsson, G. P., Sild, E., Daniellson, H., Skärby, L., and Sellden, G. (1996). Exposure of a grass-clover mixture to ozone in open-top chambers—effects on yield, quality and botanical composition. *Agriculture, Ecosystems and Environment* 59: 55–62.
Polle, A. (1998). Photochemical oxidants: Uptake and detoxification mechanisms. In: L. J. De Kok and I. Stulen (Eds.), *Responses of plant metabolism to air pollution and global change* (pp. 95–116). Leiden: Backhuys.

Price, A., Lucas, P. W., and Peter, J. L. (1990). Age dependent damage and glutathione metabolism in ozone fumigated barley: A leaf section approach. *Journal of Experimental Botany* 41: 1309–1317.
Rao, M. V., and Dubey, P. S. (1990). Biochemical aspects (antioxidants) for development of tolerance in plants growing at different low levels of ambient air pollutants. *Environmental Pollution* 64: 55–59.
Reiling, K., and Davison, A. W. (1992). The response of native herbaceous species to ozone: Growth and fluorescence screening. *New Phytologist* 120: 29–37.
Reinert, R. A. (1984). Plant response to air pollutant mixtures. *Annual Reviews in Phytopathology* 22: 421–442.
Richards, B. L., Middleton, J. T., and Hewitt, W. B. (1958). Air pollution with relation to agronomic crops versus oxidant stiple of grape. *Agronomy Journal* 50: 559–561.
Runeckles, V. C. (1992). Uptake of ozone by vegetation. In: A. S. Lefohn (Ed.), *Surface level ozone exposures and their effects on vegetation* (pp. 157–188). Chelsea, UK: Lewis.
Runeckles, V. C., and Chevone, B. I. (1992). Crop responses to ozone. In: A. S. Lefohn (Ed.), *Surface level ozone exposures and their effects on vegetation* (pp. 189–270). Chelsea, UK: Lewis.
Runeckles, V. C., and Palmer, K. (1987). Pretreatment with nitrogen dioxide modifies plant response to ozone. *Atmospheric Environment* 21: 717–719.
Sandermann; H. (1996). Ozone and plant health. *Annual Reviews in Phytopathology* 34: 347–366.
Sandermann, H. (1998). Ozone: an air pollutant acting as a plant signalling molecule. *Naturwissenschaften* 85: 369–375.
Sandermann, H., Wellburn, A. R., and Heath, R. L. (Eds.). (1997). *Forest decline and ozone.* Ecological Studies No. 127. Berlin: Springer-Verlag.
Smirnoff, N. (1993). The role of active oxygen in the response of plants to water deficit and desiccation. *New Phytologist* 125: 27–58.
Soja, G. (1996). Growth stage as a modifier of ozone response in winter wheat. In: M. Knoflacher, J. Schneider, and G. Soja (Eds.), *Exceedance of critical loads and levels* (pp. 155–163). Conference Papers, Vol. 15. Vienna: Federal Environmental Agency.
Stockwell, W. R., Kramm, G., Scheel, H. E., Mohnen, V. A., and Seiler, W. (1997). Ozone formation, destruction and exposure in Europe and the United States. In: H. Sandermann, A. R. Wellburn, and R. L. Heath (Eds.), *Ozone and forest decliine* (pp. 1–38). Ecological Studies No. 127. Berlin: Springer-Verlag.
Tanaka, K., Suda, Y., Kondo, N., and Sugahara, K. (1985). O_3 tolerance and the ascorbate-dependent H_2O_2 decomposing system in chloroplasts. *Plant Cell Physiology* 26: 1425–1431.
Tingey, D. T., and Andersen, C. P. (1991). The physiological basis of differential plant sensitivity to changes in atmospheric quality. In: G. E. Taylor, L. F. Pitelka, and M. T. Clegg (Eds.), *Ecological genetics and air pollution* (pp. 209–235). New York: Springer-Verlag.
Tingey, D. T., and Hogsett, W. E. (1985). Water stress reduces ozone injury via stomatal mechanism. *Plant Physiology* 77: 944–947.
Tonneijck, A. E. G., and Bugter, R. J. F. (1991). Biological monitoring of ozone effects on indicator plants in The Netherlands: Initial research on exposure–response functions. *VDI-Berichte* 901: 613–624.
Treshow, M., and Stewart, D. (1973). Ozone sensitivity of plants in natural communities. *Biological Conservation* 5: 209–214.
U. S. Environmental Protection Agency. (1996). *Air quality criteria for ozone and related photochemical oxidants* (Vol. II EPA/600/P–93/004bF). Washington, DC: Author.
U. S. Environmental Protection Agency. (1984). *A review and assessment of the effects of air pollutant mixtures on vegetation—research recommendations.* EPA-600/3-84. Corvallis, OR: Author.
Vandermeiren, K., De Temmerman, L., and Hookham, N. (1995). Ozone sensitivity of *Phaseolus vulgaris* in relation to cultivar differences, growth stage and growing conditions. *Water, Air and Soil Pollution* 85: 1455–1460.
Violini, G. (1995). Ozone and plant pathogens: An overview. Rivista di Patologia Vegetale. S.V. 5: 113–130.
Volz , A., and Kley, D. (1988). Evaluation of the Montsouris series of ozone measurements made in the nineteenth century. *Nature* 332: 240–242.
Warwick, K. R., and Taylor, G. (1995). Contrasting effects of tropospheric ozone on five native herbs which coexist in calcareous grassland. *Global Change Biology* 1: 143–151.
Wolfenden, J., Wookey, P. A., Lucas, P. W., and Mansfield, T. A. (1992). Action of pollutants individually and in combination. In: J. R. Barker and D. T. Tingey (Eds.), *Air pollution effects on biodiversity* (pp. 72–92). New York: Van Nostrand Reinhold.

10

The Effects of Ultraviolet B Radiation on Crop Plants

Alyson K. Tobin

Introduction

Plants require light as both a source of energy for photosynthesis and as a signal to regulate their development. The quality (wavelength) of light is an important influence on whether the plant responds photosynthetically or photomorphogenically. Higher plants use light within the wavelength range of 400–700 nm for photosynthesis (defined as *photosynthetically active radiation*; PAR). Photomorphogenic responses, such as altered growth and/or development, tend to result from absorption of light within a much more specific range, usually as a result of absorption by photoreceptors, including phytochrome, which absorbs red (660 nm) and far-red (730 nm) light; a blue-light photoreceptor (cryptochrome); and an as yet unidentified ultraviolet B (UV-B) photoreceptor. In this chapter, the specific responses of crop plants toward UV-B (280–320 nm) are described. In addition to photomorphogenic responses, plants may be affected as a result of the direct absorption of UV-B by a range of important molecules, including proteins, nucleic acids, and auxins. Consequently, UV-B has the potential to cause significant damage to a plant cell. Given the absolute requirement for sunlight for photosynthesis, plants have developed a range of protective and repair processes to minimize the damaging effects of UV-B. These adaptive and acclimative processes are described, along with evidence of how crop plants are likely to be affected if UV-B increases.

Global Ultraviolet Radiation

The UV radiation region of the solar spectrum is classified into three categories: UV-C (200–280 nm), UV-B (280–320 nm), and UV-A (320–400 nm). Of these, UV-C is

Alyson K. Tobin • Plant Science Laboratory, Sir Harold Mitchell Building, School of Biology, University of St Andrews, St Andrews, KY16 9TH, Scotland, UK; e-mail at6@st-andrews.ac.uk entirely absorbed by ozone in the lower mesosphere and upper stratosphere, and only UV-A and UV-B reach the Earth's surface. The stratospheric ozone layer absorbs a significant proportion of the UV-B from the sun, and damage to the ozone layer has resulted in increased UV-B at ground level.

Natural variation in UV-B occurs across the globe, due to a number of factors. Seasonal and latitudinal variations of UV-B occur as a result of differences in the Earth–sun distance, changes in the solar zenith angle, and differences in ozone column depth. The sun–Earth distance is at a minimum in early January (the perihelion) and at a maximum in early July (the aphelion), differing by about 3.4% between these two extremes (Madronich, 1993). The solar zenith angle changes over the course of 24 hours and also varies with the time of year and with latitude. The smaller the zenith angle (i.e., the lower the sun is in the sky) the greater the depth of atmospheric gases (e.g., stratospheric ozone) the radiation has to pass through before it reaches the Earth. This means that UV-B levels at the Earth's surface will be lowest when the sun is lowest in the sky. An added factor is the natural variation in ozone column depth, being at its lowest in the tropics throughout the year and at a maximum in the Arctic in the northern spring period of March and April. This difference in ozone column depth, together with the solar angle, explains why UV-B levels are highest at the Equator. Damage to the ozone layer, as a result of pollution (mainly by the release of chlorofluorocarbons, CFCs), has resulted in the lowest ozone column depths in the world occurring at the southern polar region, particularly during the southern spring period of September–October in the Antarctic (Farman et al., 1985). In addition to these geographic and seasonal differences, local factors such as cloud cover, pollution levels, altitude, and albedo all contribute to variations in UV-B levels. Uniform cloud cover results in reduced levels of UV-B, whereas broken clouds have a more complex effect, either increasing or decreasing UV-B, depending on their structure (Madronich, 1993). Frederick and Lubin (1988) estimated that clouds reduce UV-B levels by 10 to 30% at ground level, depending on latitude. Aerosols and pollution, including tropospheric ozone, reduce UV-B levels, so poor air quality in urban regions may well mask the effects of changing stratospheric ozone concentrations. Albedo, defined as the ratio of reflected to incident radiation, can have a significant impact on UV-B levels, depending on the local terrain. Albedo can be high in rocky regions, where snow is particularly reflective of UV-B, whereas vegetation has a low albedo (Madronich, 1993). Increased solar radiation with altitude is termed the *altitude effect* and is primarily a consequence of the smaller mass of air at higher altitudes. The altitude effect depends on solar elevation, due to the varying amount of irradiated air mass, and is higher at lower solar elevation. Other effects at high altitude, especially in the high mountains, include the high albedo of snow-covered surfaces and the purer air quality in these regions (Blumthaler, 1993).

Mode of Action of UV-B: Specific interaction with Proteins, Nucleic Acids, and Pigments

As already mentioned, UV-B radiation is absorbed by a range of components within the cell, including proteins, nucleic acids, and pigments. Depending on the amount of UV-B absorbed, this can result in damage—particularly to proteins and DNA—resulting in delayed growth and/or cell death. Pigments, in particular the flavonoids, provide protection by attenuation of UV-B reaching vulnerable tissue, such as the mesophyll, and by scavenging free-radicals (Bommann et al., 1997).

Damage to Photosynthetic Proteins

There have been several reports of damage to key photosynthetic proteins, for example, the Dl protein of PSII, in a number of different species. Marty of these studies used extremely high UV-B irradiances, frequently in association with low PAR levels. These are discussed in an earlier review by Fiscus and Booker (1995). For example, although Strid and Porra (1992) reported major damage to PSII in peas exposed to UV-B radiation, this occurred in low light (PAR of 150 μmoles m^{-2} s^{-1}) and with extremely high UV-B levels (equivalent to up to 9 times the UV-B exposure under a clear sky at midlatitudes (Fiscus and Booker, 1995). More recent studies have used lower levels of UV-B exposure, and some have reported effects on PSII under field conditions. Booij-James et al. (2000) reported damage to the Dl protein of PSII under relatively modest UV-B fluxes in *Arabidopsis thaliana*, but only in mutants deficient in protective phenolic pigments. Krause et al. (1999) carried out field studies in which they exposed shade- and sun-grown seedlings of tropical trees to direct sunlight in the presence or absence of filters to screen out either UV-B, UV-A, or both. They found that whereas shade leaves were vulnerable to UV-B-induced damage to PSII (as measured by chlorophyll fluorescence) sun leaves, although still affected, were less so. This was thought to be due to a greater accumulation of protective pigments in the leaves at the top of the tree canopy. Thus, exposure to naturally occurring levels of UV-B radiation in the tropics appears to result in damage to PSII, particularly in shade leaves exposed to full sunlight.

In a field study of the two species of higher plant found in Antarctica, *Colobanthus quitensis* and *Deschampsia antarctica*, Xiong and Day (2001) found reductions in the quantum yield of PSII in plants exposed to full sunlight compared to controls under UV-B-absorbing filters. The authors concluded, however, that this was not due to PSII damage but was more likely due to effects of UV-B on the “dark reactions” of photosynthesis. This could arise as a result of damage to Rubisco, for example, as also reported in the earlier literature by Jordan et al. (1992), although these experiments were carried out under UV-B levels far in excess of those likely to occur in the natural environment. In contrast to these reports of UV-B damage, Pinto et al. (2000) found that an initial exposure to UV-B actually protected wheat leaves from the photoinhibitory damage of exposure to high intensities of white light that caused PSII damage. This is just one example of the conflicting reports of UV-B effects on plant processes, illustrating the difficulty in making generalizations about plant responses. One of the complications, as discussed later, is the range of experimental conditions and species used. In a relatively recent review, Allen et al. (1998) conclude that field-based studies using modulated UV-B treatments tend not to show any significant disruption to photosynthesis in crops or natural vegetation. In field-grown peas, for example, where a modulated increase in UV-B of 30% was provided by supplementary lighting, Allen et al. (1999) found no effect on photosynthetic processes, such as PSII or Rubisco activity. The ability to produce protective flavonoid pigments in response to enhanced UV-B exposure (as discussed later) probably protects most plants from damage to photosynthetic proteins, provided they have had sufficient time to respond. The study of shade leaves in the tropics, already discussed, does illustrate that a sudden increase in UV-B exposure might result in photosynthetic impairment in the field (Krause et al., 1999).

Absorbance by Nucleic Acids

DNA and RNA absorb UV-B radiation, which can cause damage. The most abundant form of damage resulting from UV-B irradiation is the pyrimidine dimer, which accounts for up to 98% of the damage, with single strand breaks and DNA-protein crosslinks forming the remaining 2%. Two forms of dimer occur, the cyclobutane pyrimidine dimer (CPD) and the pyrimidine (6-4) pyrimidone dimer (or 6-4 photoproduct), with the former contributing 75% of the total dimers formed under UV-B (Mitchell and Nairn, 1989). A single pyrimidine dimer, if left unrepaired, is sufficient to prevent transcription (Britt, 1996). Plants, like other organisms, have efficient repair pathways in order to minimize the accumulation of damage within their genomes. These include light-dependent repair by photolyase, which absorbs photons within the wavelength range of 350–450 nm and directly reverses the damage by splitting one dimer for each photon absorbed. Two distinct photolyases have been discovered in plants, one specific for CPDs (Taylor et al., 1996; Ahmad et al., 1997) and the other for 6-4 photoproducts (Sakamoto et al., 1999). In contrast to the light-dependent, photoreactivation repair by photolyases, dark repair of DNA damage is a multienzyme process that cleaves out the lesions and replaces them. This is termed *excision repair* and includes *base excision repair* and *nucleotide excision repair* pathways. DNA repair in plants has been the subject of a number of recent reviews to which the reader is referred for further details (Britt, 1996, 1999; Taylor et al., 1997; Jansen et al., 1998; Vonarx et al., 1998; Tuteja et al., 2001).

That the capacity to repair DNA is important in surviving exposure to UV-B is apparent from studies of mutant plants. Landry et al. (1997) have found that an *Arabidopsis* mutant, defective in type II CPD photolyase, is particularly prone to UV-B-induced damage when placed under white light following UV-B exposure. Similarly, a CPD photolyase-deficient mutant of rice (*Oryza sativa L.*), Norin-1, also shows enhanced sensitivity to UV-B (Hidema et al., 1997), as discussed later.

The extent to which DNA damage occurs in higher plants growing in their natural environment has, so far, been the subject of only a small number of studies. Stapleton et al. (1997) detected CPDs in maize (*Zea mays L.*) plants growing in natural solar radiation. Accumulation was greatest toward the end of the day but there was no accumulation of damage over the growing season. They found that the plants had the capacity for photoreactivation repair of DNA damage throughout their tissue and concluded that there was sufficient repair activity to prevent accumulation of damage in normal sunlight. Rousseaux et al. (1999) found that in the spring of 1997, despite frequent cloud cover, the passage of the ozone hole over Tierra del Fuego resulted in significant increases in DNA damage in the native plant *Gunnera magellanica*. Ries et al. (2000) found, using sun simulators, that elevated solar UV-B increased the frequency of recombination and resulted in a marked induction of photolyase and *rad51* genes (involved in dark repair) in *Arabidopsis* and tobacco plants. The authors concluded that increases in UV-B radiation might affect genome stability in plants. Hidema et al. (1999) compared rice plants growing indoors and outdoors, with and without supplementary UV-B radiation. Plants growing outdoors with supplementary UV-B (1 W m^{-2}) accumulated significantly higher levels of CPDs than those grown without supplementary UV-B. Plants grown indoors accumulated lower levels of CPDs that increased with increasing UV-B. The same group found that the UV-sensitive rice cultivar, Norin-1, deficient in CPD photolyase, treated with supplementary UV-B (nonmodulated) in the field, lost a significantly greater yield than the more resistant variety, Sasanishiki, although there was no indication that DNA damage was the primary reason for this difference (Kumagai et al., 2001).

It would appear from these studies that plants have the capacity to repair quite substantial amounts of DNA damage, yet it remains to be seen whether, and under what conditions, repair capacity might be exceeded in the field.

Protective Pigments

Flavonoid pigments, which absorb strongly within the UV-B waveband, have been detected in virtually all higher plants and, in many cases, have been found to accumulate in response to UV-B radiation (reviewed by Bornmann et al., 1997). There is abundant evidence of their protective role in attenuating UV-B penetration into vulnerable tissue. Particularly convincing evidence of this function has come from analysis of mutants. As already discussed, *A. thaliana* mutants deficient in flavonoid production are significantly more vulnerable to PSII damage under UV-B (Booij-James et al., 2000). The same mutant (tt5, lacking chalcone isomerase and hence having reduced flavonoid content) showed reduced growth in response to UV-B under field conditions compared to the wild type (Fiscus et al., 1999). An *A. thaliana* mutant with elevated flavonoid content shows considerable tolerance to UV-B damage at levels that would kill wild type plants (Bieza and Lois, 2001). Further evidence of flavonoid protection is provided by a recent field study comparing different cultivars of soybean (*Glycine max* L.). Significant differences were found in UV penetration of cultivars with different leaf phenolic content (Mazza et al., 2000). Those with the lowest levels of protective pigments also showed increased UV-B transmittance into their tissue, together with increased levels of CPD formation.

As well as the protective function of flavonoids in attenuating UV-B, particularly as a result of their location within the epidermal tissue, there is increasing evidence that flavonoids also protect against oxidative damage resulting from UV-B irradiation. For example, in both *Brassica napus* and *Hordeum vulgare*, UV-B has been found to increase the concentration of flavonoids (quercetin and lutonarin, respectively) with additional hydroxyl groups, and thus increase antioxidant properties (Liu et al., 1995; Reuber et al., 1996; Bornmann et al., 1997).

UV-B Effects on Crop Plants

Experimental Approaches

The responses of plants to UV-B are many and varied, with differences having been reported both within and between species. One of the difficulties in reviewing the literature in this field is the fact that experimental conditions differ considerably between studies, making it almost impossible to make direct comparisons even when the same species and cultivar have been examined. The different experimental approaches have been discussed in detail by Corlett et al. (1997) and Fiscus and Booker (1995), and are briefly summarized here. Two main approaches have been taken: treatment within controlled environment cabinets (with artificial lighting and temperature regulation), and field-based treatments. Of the latter, various methods have been used to modulate the UV-B reaching the plant. Either additional UV-B has been provided, by supplementary lighting above the crops, delivered at a constant level (nonmodulated) or else regulated to respond to the incident radiation to provide, for example, 15% increase above ambient UV-B. The latter (modulated) is by far the best approach, because it ensures a proportionate increase in UV-B delivered to the plant. This does, however, require the use of relatively sophisticated equipment both to monitor the incident light and to regulate the output of the lamps. The danger of using nonmodulated UV-B treatments is that on cloudy days, when solar input is reduced, supplementary UV-B will be far more damaging because of the high UV-B/PAR in these conditions. Experiments carried out in controlled environment cabinets suffer from the fact that artificial lighting rarely provides PAR levels approaching those of natural sunlight and UV-B damage is more likely under low light. Nevertheless, they do allow for greater control over environmental conditions such as temperature and day length, and provide a useful means of testing mechanistic responses, without the confounding effects of a fluctuating environment, as occurs in the field. So all of these experimental approaches have their uses, but it is important to bear in mind these differences when making comparisons between studies.

Experimental Evidence of the Impact of UV-B on Crops

Recent reviews of this topic include those of Corlett et al. (1997), Caldwell et al. (1998), Jansen et al. (1998), Gaberscik (1999), and Paul (2001); hence, this chapter concentrates on significant develops in the past 3–5 years and focuses on two aspects: cultivar differences and interactions with other stresses.

Differences between Cultivars

Despite the difficulty of comparing studies that use different light treatments, sufficient evidence of genotypic variation of UV-B sensitivity, in most cases, has come from comparative studies using identical growth treatments. Torabinejad and Caldwell (2000) found differences in the UV tolerance among seven *Arabidopsis* genotypes taken from different European sites, on the basis of morphological traits. Pinto et al. (2000) compared seven Chilean cultivars of wheat (*Triticum aestivum* L.) in controlled environment cabinets with supplementary UV-B. Three varieties were more tolerant than the others, although all showed reduced growth in response to UV-B. Yuan et al. (2000) compared 20 different wheat varieties in the field in China and found that 19 of the 20 varieties showed a negative response to elevated UV-B (provided at 5.00 kJ m^{-2}). Effects included reduced growth and grain yield, although these varied according to the stage of development, with younger plants showing a more consistent response. One cultivar showed a positive response to enhanced UV-B, whereas the seven most tolerant cultivars were from lower latitude regions of China. The most sensitive cultivars originated in the high latitude regions, suggesting that selective breeding may already have resulted in the generation of relatively UV-B tolerant wheat cuitivars, without this having been part of the breeding program.

Several recent studies have identified differences in the UV-B response among rice (*Oryza sativa* L.) cultivars. For example, Dai et al. (1994) tested 188 different cultivars in controlled environment cabinets and found that six of these showed positive responses, with enhanced growth under elevated UV-B (13.0 kJ m^{-2} day^{-1}). These more tolerant cultivars originated from high-elevation regions of Bangladesh, again indicating a relationship between tolerance and selective breeding in regions with high ambient UV-B. These studies confirm earlier reports, albeit with different rice cultivars grown under different conditions, of differences in the susceptibility of rice cultivars to UV-B damage. Teramura et al. (1991) compared 16 rice cultivars from seven different geographical regions and found that approximately one-third of those tested showed a statistically significant decrease in total biomass with UV-B radiation. Dai et al. (1992) also reported differences in the sensitivity of different cultivars of rice toward UV-B. This diversity exhibited by rice in response to increased levels of UV-B suggests that selective breeding might be successfully used to develop UV-B-tolerant cultivars.

A possible early screening method to select for tolerant maize (*Zea mays*) cultivars has been tested by Cartwright et al. (2001). Four cultivars were examined for UV-induced changes in leaf rolling, biomass, fluctuating leaf asymmetry, and DNA damage. Short term dose–response curves were constructed for each response. The four varieties each exhibited a different pattern of tolerance; not only would this be useful in selective breeding programs, but it also provides a means of identifying the underlying mechanisms behind the UV-B response. Differences between maize cultivars have also been reported by Correia et al. (1999). A comparison of eight cultivars found significant differences, under quite extreme UV-B treatments, but nevertheless indicate genotypic variation in UV-B sensitivity of maize.

These cultivar differences are not exclusive to cereals. SaileMark and Tevini (1997) compared *Phaseolus vulgaris* cultivars from central and southern Europe, grown in greenhouses with UV-B radiation equivalent to that found at ambient levels in the southern European location, and controls with 10% lower UV-B. They found that the southern European cultivars were slightly less sensitive than the others, although there was reduced growth and yield in all cultivars.

The variety of experimental treatments, as already discussed, makes it difficult to provide mechanistic explanations for these cultivar differences. Some suggestions have arisen from previous studies. Dai et al. (1997), for example, suggested that UV-B affected susceptible rice cultivars in particular by reducing the activities of enzymes involved in antioxidant production, such as catalase and ascorbic acid peroxidase. Supporting evidence for the operation of this free-radical-scavenging pathway in response to UV-B exposure has come from field studies of barley mutants deficient in catalase. (Mazza, 1999) found that, although both wild type and mutant lines showed reduced growth under enhanced UV-B, the catalase-deficient mutants showed a greater reduction (32% compared with 20% reduction in the wild type). Solar UV-B caused measurable DNA damage in leaf tissue, even though both lines increased UV-absorbing pigments in response to UV-B. Whereas the wild type increased catalase and ascorbate peroxidase activities under UV-B, the mutant that had higher ascorbate peroxidase activity did not. Thus, antioxidants would appear to be important in protecting plants from UV-B damage, but it remains to be seen whether this is the fundamental reason for cultivar differences. Many of the known features of UV-B-tolerant plants, such as the presence of protective flavonoid pigments, and surface structures such as leaf waxes and hairs, as well as repair processes such as DNA repair and protein turnover, will also be important.

Interactions with Other Stresses

An early indication of the significant impact of other environmental stresses on the response of plants to UV-B came from the field studies of soybean crops carried out by Teramura et al. (1990). Field trials were carried out over five growing seasons and indicated not only cultivar differences but also effects of drought on the response to UV-B. The relatively sensitive soybean variety, Essex, showed reduced yields in four out of the five seasons, compared to the low UV-B controls. In the season in which Essex showed increased yield in response to UV-B, there had been a drought during the early season of growth. The opposite happened to the tolerant cultivar, Williams, which responded positively to UV-B in all but the dry year. No explanation for this interaction is given, but it does indicate that other environmental factors can exert a marked effect on the UV-B response. Allen et al. (1999) carried out a similar study on pea (*Pisum sativum* L.) plants grown in the field and subjected to a modulated 30% increase in ambient U.K. summer UV-B radiation, together with a mild drought stress. They found no significant effect of UV-B on CO_2 assimilation or on a range of photosynthetic components (e.g., Rubisco, P5II); neither did UV-B have any significant impact on plant biomass, leaf area, or partitioning. No UV-B and drought interactions were observed on photosynthesis or plant biomass. So, in contrast to the observed effects on soybean by Teramura's group, a 30% increase in UV-B had no effect on productivity in either droughted or well-watered pea plants in the field.

The response of weed species toward elevated UV-B may contribute to reduced crop yields. Andreasen et al. (1999) tested the UV-B response of a variety of weeds commonly found among crops, under greenhouse conditions. These included annual bluegrass (*Poa annua* L.), common groundsel (*Senecio vulgaris* L.), shepherd's purse (*Capsella bursa-pastoris* L.), and small nettle (*Urtica urens* L.). These differed in their sensitivity to UV-B, with annual bluegrass being the least sensitive. Comparison with two crops grown under the same conditions (*Brassica napus* L. and *Pisum sativum* L.) revealed that the former was about sevenfold more sensitive than the latter. This differential response of the weeds and crops may have important implications for crops growing in high UV-B environments. Barnes et al. (1995) tested mixtures and monocultures of wheat and wild oat (*Avena fatua*), a common weedy competitor of wheat, by exposing them to enhanced UV-B radiation (simulating a 20% reduction in stratospheric ozone). Results from 2 years of field studies revealed significant effects of UV-B on canopy structure, with wheat gaining an advantage over the weed in terms of relative leaf cover and light interception. In this case, increased UV-B would appear to improve the chances of the crop outcompeting the weed.

Biotic interactions are also likely to be affected by changes in UV-B. Paul (2000) reviewed the effect of UV-B on plant-pathogen interactions and concluded that although UV-B could be seen to affect the early pathogen infection—usually by affecting the pathogen directly—the response to projected ozone depletion is small compared to the variation in UV-B encountered during seasonal effects and changes in cloud cover. Despite the initial inhibition in infection, the overall trend was increased susceptibility for infection. Paul suggested that this might be due to UV-B-induced changes in leaf surface properties, making them more prone to fungal infection. Paul et al. (1997) also reviewed the possible impact of UV-B on herbivore–plant interactions.

Conclusions

The natural variation in UV-B reaching the Earth's surface as a result of seasonal, latitudinal, and local factors means that plants are exposed to wide-ranging doses of UV-B, depending on where they grow. Increased UV-B, as a consequence of stratospheric ozone depletion, may have an impact on plant growth, but there is no clear consensus as to the extent to which this is likely to occur. Responses of crops to increased UV-B exposure indicate large variations between different cultivars, with some indication of improved UV-B resistance in those cultivars selectively bred in regions with high ambient UV-B. Although this was not a parameter used in the selection of these cultivars, nevertheless, it may have contributed to their success during field trials. UV-B effects on crop growth must also be considered in relation to other environmental stresses, both biotic and abiotic, because these have been found to influence the response. Despite the increasing amount of information on UV-B and its effects on plants. there is still a dearth of long-term studies carried out in the field under realistic levels of UV-B, and without this information, it is not possible to draw firm conclusions about the likely effects of stratospheric ozone depletion on crops.

References

Ahmad, M., Jarillo, J. A., Klimczak, L. J., Landry, L. G., Peng, T., Last, R. L., and Cashmere, A. R. (1997). An enzyme similar to animal type II photolyases mediates photoreactivation in *Arabidopsis*. *Plant Cell* 9: 199–207.

Allen, D. J., Nogues, S., and Baker, N. R. (1998). Ozone depletion and increased UV-B radiation: Is there a real threat to photosynthesis? *Journal of Experimental Botany* 49: 1775–1788.

Allen, D. J., Nogues, S., Morison, J. I. L., Greenslade, P. D., McLeod, A. R., and Baker, N. R. (1999). A thirty percent increase in UV-B has no impact on photosynthesis in well-watered and droughted pea plants in the field. *Global Change Biology* 5: 235–244.

Andreasen, C., Hansen, L., and Streibig, J. C. (1999). The effect of ultraviolet radiation on the fresh weight of some weeds and crops. *Weed Technology* 13: 554–560.

Barnes, P. W., Flint, S. D. and Caldwell, M. M. (1995). Early-season effects of supplemented solar UV-B radiation on seedling emergence, canopy structure, simulated stand photosynthesis and competition for light. *Global Change Biology* 1: 43–53.

Bieza, K., and Lois, R. (2001). An *Arabidopsis* mutant tolerant to lethal ultraviolet-B levels shows constitutively elevated accumulation of flavonoids and other phenolics. *Plant Physiology* 126: 1105–1115.

Blumthaler, M. (1993). Solar UV measurements. In: M. Tevini (Ed.), *UV-B radiation and ozone depletion: Effects on humans, animals, plants, microorganisms and materials* (pp. 71–94). Boca Raton, FL: Lewis.

Booij-James, I. S., Dube, S. K., Jansen, M. A. K., Edelman, M. and Mattoo, A. K. (2000). Ultraviolet-B radiation impacts light-mediated turnover of the photosystem II reaction center heterodimer in *Arabidopsis* mutants altered in phenolic metabolism. *Plant Physiology* 124: 1275–1283.

Bornmann, J. F., Reuber, S., Cen, Y.-P., and Weissenböck, G. (1997). Ultraviolet radiation as a stress factor and the role of protective pigments. In: P. J. Lumsdon (Ed.), *Plants and UV-B: Responses to environmental change* (pp. 157–168). Cambridge, UK: Cambridge University Press.

Britt. A. B. (1996). DNA damage and repair in plants. *Annual Review of Plant Physiology and Plant Molecular Biology* 47: 75–100.

Britt, A. B. (1999). Molecular genetics of DNA repair in higher plants. *Trends in Plant Science* 4: 20–25.

Caldwell, M. M., Bjorn, L. O., Bornman, J. F., Flint, S., Kulandaivelu, G., Teramura, A. H. and Tevini, M. (1998). Effects of increased solar ultraviolet radiation on terrestrial ecosystems. *Journal of Photochemistry and Photobiology B–Biology* 46: 40–52.

Cartwright, H. N., Baucom, C., Singh, P., Smith, K. L., and Stapleton, A. E. (2001). Intraspecific comparisons reveal differences in the pattern of ultraviolet radiation responses in four maize (*Zea mays* L.) varieties. *Journal of Photochemistry and Photobiology B–Biology* 62: 88–96.

Corlett, J. E., Stephen, J., Jones, H. G., Woodfin, R., Mepsted, R., and Paul, N. D. (1997). Assessing the impact of UV-B radiation on the growth and yield of field crops. In: P. J. Lumsden (Ed.), *Plants and UV-B: Responses to environmental change* (pp. 195–211). Cambridge, UK: Cambridge University Press.

Correia, C. M., Areal, E. L. V, Torres-Pereira, M. S., and Torres-Pereira, J. M. G. (1999). Intraspecific variation in sensitivity to ultraviolet-B radiation in maize grown under field conditions; II. Physiological and biochemical aspects. *Field Crops Research* 62: 97–105.

Dai, Q. J., Coronel, V. P., Vergara, B. S., Barnes, P. W. and Quintos, A. T. (1992). Ultraviolet-B radiation effects on growth and physiology of 4 rice cultivars. *Crop Science* 32: 1269–1274.

Dai, Q. J., Peng, S. B., Chavez, A. Q., and Vergara, B. S.. (1994). Intraspecific responses of 188 rice cultivars to enhanced UV-B radiation. *Environmental and Experimental Botany* 34: 433–442.

Dai, Q. J., Yan, B., Huang, S. B., Liu, X. Z., Peng, S. B., Miranda, M. L. L., Chavez, A. Q., Vergara, B. S., and Olszyk, D. M. (1997). Response of oxidative stress defense systems in rice (*Oryza sativa*) leaves with supplemental UV-B radiation. *Physiologia Plantarum* 101: 301–308.

Farman, J. C., Gardiner, B. G., and Shanklin, J. D. (1985). Large losses of total ozone in Antarctica reveal seasonal ClO_x/NO_3 interactions. *Nature* 315: 207–210.

Fiscus, E. L., and Booker, F. L. (1995). Is increased UV-B a threat to crop photosynthesis and productivity? *Photosynthesis Research* 43: 81–92.

Fiscus, E. L., Philbeck, R., Britt, A. B. and Booker, F. L. (1999). Growth of *Arabidopsis* flavonoid mutants under solar radiation and UV filters. *Environmental and Experimental Botany* 41: 231–245.

Frederick, J. E., and Lubin, D. (1988). The budget of biologically active ultraviolet radiation in the earth-atmosphere system. *Journal of Geophysical Research* 93: 3825–3832.

Gaberscik, A. (1999). UV-B radiation—how burning is it for plants? *Phyton-Annales Rei Botanicae* 39: 35–39.

Hidema, J., Kang, H. S., and Kumagai, T. (1999). Changes in cyclobutyl pyrimidine dimer levels in rice (*Oryza sativa* L.) growing indoors and outdoors with or without supplemental UV-B radiation. *Journal of Photochemistry and Photobiology B–Biology* 52: 7–13.

Hidema, J., Kumagai, T., Sutherland, J. C., and Sutherland, B. M. (1997). Ultraviolet B-sensitive rice cultivar deficient in cyclobutyl pyrimidine dimer repair. *Plant Physiology* 113: 39–44.

Jansen, M. A. K., Gaba, V., and Greenberg, B. M. (1998). Higher plants and UV-B radiation: Balancing damage, repair and acclimation. *Trends in Plant Science* 3: 131–135.

Jordan, B. R., He, J., Chow, W. S. and Anderson, J. M. (1992). Changes in mRNA levels and polypeptide subunits of ribulose 1,5 bisphosphate carboxylase in response to supplementary ultraviolet-B radiation. *Plant Cell and Environment* 15: 91–98.

Krause, G. H., Schmude, C., Garden, H., Koroleva, O. Y., and Winter, K. (1999). Effects of solar ultraviolet radiation on the potential efficiency of photosystem II in leaves of tropical plants. *Plant Physiology* 121: 1349–1358.

Kumagai, T., Hiclema, J., Kang, H. S., and Sato, T. (2001). Effects of supplemental UV-B radiation on the growth and yield of two cultivars of Japanese lowland rice (*Oryza sativa* L.) under the field in a cool rice-growing region of Japan. *Agriculture Ecosystems and Environment* 83: 201–208.

Landry, L. G., Stapleton, A. E., Lim, J., Hoffman, P., Hays, J. B., Walbot, V., and Last, R. L. (1997). An *Arabidopsis* photolyase mutant is hypersensitive to ultraviolet-B radiation. *Proceedings of the National Academy of Sciences USA* 94: 328–332.

Liu, L., Gitz, D. C., III, and McClure, J. W. (1995). Effects of UV-B on flavonoids, ferulic acid, growth and photosynthesis in barley primary leaves. *Physiologia Plantarum* 93: 725–733.

Madlronich, S. (1993). UV radiation in the natural and perturbed atmosphere. In: M. Tevini (Ed.), *UV-B radiation and ozone depletion: Effects on humans, animals, plants, microorganisms, and materials* (pp. 17–69). Boca Raton, FL: Lewis.

Mazza, C. A., Battista, D., Zima, A. M., Szwarcberg-Bracchitta, M., Giordano, C. V., Aevedo, A., Scopel, A. L. and Ballaré, C. L. (1999). The effects of solar ultraviolet-B radiation on the growth and yield of barley are accompanied by increased DNA damage and antioxidant responses. *Plant Cell and Environment* 22: 61–70.

Mazza, C. A., Boccalandro, H. E., Giordano, C. V., Battista, D., Scopel, A. L., and Ballare, C. L. (2000). Functional significance and induction by solar radiation of ultraviolet-absorbing sunscreens in field-grown soybean crops. *Plant Physiology* 122: 117–125.

Mitchell, D. L., and Nairn, R. S. (1989). The biology of the (6-4) photoproduct. *Photochemistry and Photobiology* 49: 805–819.

Paul, N. D. (2000). Stratospheric ozone depletion, UV-B radiation and crop disease. *Environmental Pollution* 108: 343–355.

Paul, N. D. (2001). Plant responses to UV-B: Time to look beyond stratospheric ozone depletion? *New Phytologist* 150: 5–8.

Paul, N. D., Rasanayagam, S., Moody, S. A., Hatcher, P. E. and Ayres, P. G. (1997). The role of interactions between trophic levels in determining the effects of UV-B on terrestrial ecosystems. *Plant Ecology* 128: 296–308.

Pinto, M., Lizana, C., Pastenes, C., Riquelme, A., and Berti, M. (2000). Effect of the ultraviolet-B radiation on growth and photosynthesis of seven varieties of wheat (*Triticum aestivum* L.). *Revista Chilena De Historia Natural* 73: 55–66.

Reuber, S., Bornman, J. F., and Weissenbock, G. (1996). A flavonoid mutant of barley (*Hordeum vulgare* L.) exhibits increased sensitivity to UV radiation in the primary leaf. *Plant Cell and Environment* 19: 593–601.

Ries, G., Heller, W., Puchta, H., Sandermann, H., Seidlitz, H. K. and Hohn, B. (2000). Elevated UV-B radiation reduces genome stability in plants. *Nature* 406: 98–101.

Rousseaux, M. C., Ballaré, C. L., Giordano, C. V., Scopel, A. L., Zima, A. M., Szwarcberg-Bracchitta, M., Searles, P. S., Caldwell, M. M., and Diaz, S. B. (1999). Ozone depletion and UV-B radiation: Impact on plant DNA damage in southern South America. *Proceedings of the National Academy of Sciences USA* 96: 15310–15315.

SaileMark, M., and Tevini, M. (1997). Effects of solar UV-B radiation on growth, flowering and yield of central and southern European bush bean cultivars (*Phaseolus vulgaris* L.). *Plant Ecology* 128: 114–125.

Sakamoto, A., Tanaka, A., Tano, S., Nakajima, S., Yamamoto, K., and Watanabe, H. (1999). The genomic organization of the *Arabidopsis* 6-4 photolyase gene (accession no. AB017331) (vol. 118, p. 1101, 1998). *Plant Physiology* 119: 807.

Stapleton, A. E., Thornber, C. S., and Walbot, V. (1997). UV-B component of sunlight causes measurable damage in field-grown maize (*Zea mays* L.): Developmental and cellular heterogeneity of damage and repair. *Plant Cell and Environment* 20: 279–290.

Strid, Å., and Porra, R. J. (1992). Alterations in pigment content in leaves of *Pisum sativum* after exposure to enhance UV-B radiation. *Plant and Cell Physiology* 33: 1015–1023.

Taylor, R. M., Tobin, A. K., and Bray, C. M. (1996). Nucleotide sequence of an *Arabidopsis* cDNA at-phII (accession no. X99301) encoding a protein with high homology to the Class II CPD photolyases present in higher eukaryotes. *Plant Physiology* 112: 862 (Accession no. PGR96-083).

Taylor, R. M., Tobin, A. K. and Bray, C. M. (1997). DNA damage and repair in plants. In: P. J. Lumsden (Ed.), *Plants and UV-B: Responses to environmental change* (pp. 53–76). Cambridge, UK: Cambridge University Press.

Teramura, A. H., Sullivan, J. H., and Lyndon, J. (1990). Effects of UV-B radiation on soybean yield and seed quality: A 6-year field study. *Physiologia Plantarum* 80: 5–11.

Teramura, A. H., Ziska, L. H. and Sztein, A. E. (1991). Changes in growth and photosynthetic capacity of rice with increased UV-B radiation. *Physiologia Plantarum* 83: 373–380.

Torabinejad, J., and Caldwell, M. M. (2000). Inheritance of UV-B tolerance in seven ecotypes of *Arabidopsis thaliana* L. Heynh: And their F-1 hybrids. *Journal of Heredity* 91: 228–233.

Tuteja, N., Singli, M. B., Misra, M. K., Bhalla, P. L., and Tuteja, R. (2001). Molecular mechanisms of DNA damage and repair: Progress in plants. *Critical Reviews in Biochemistry and Molecular Biology* 36: 337–397.

Vonarx, E. J., Mitchell, H. L., Karthikeyan, R., Chatterjee, I., and Kunz, B. A. (1998). DNA repair in higher plants. *Mutation Research—Fundamental and Molecular Mechanisms of Mutagenesis* 400: 187–200.

Xiong, F. S., and Day, D. A. (2001). Effect of solar ultraviolet-B radiation during springtime ozone depletion on photosynthesis and biomass production of Antarctic vascular plants. *Plant Physiology* 125: 738–751.

Yuan, L., Yanqun, Z., Haiyan, C., Jianjun, C., Jilong, Y., and Zhide, H. (2000). Intraspecific responses in crop growth and yield of 20 wheat cultivars to enhanced ultraviolet-B radiation under field conditions. *Field Crops Research* 67: 25–33.

11

Vegetation Ecology and Regeneration of Tropical Mountain Forests

Rainer W. Bussmann

Introduction

Tropical mountain forests, found mainly at altitudes between 1,200 and 3,500 m (LaBastille and Pool, 1978; Stadtmüller, 1987; Food and Agriculture Organization, 1992) are among the species-richest ecosystems worldwide. Particularly the Eastern Andean region is one of the major "biodiversity hotspots" (Myers et al., 2000). In striking contrast to tropical lowland rain forests, these tropical montane forests have received only marginal attention in science and society until recently, despite their ecological and economic importance as water catchments and erosion barriers. The neotropical montane region is considered one of the 12 major crop–gene centers of the world (Churchill et al., 1995). At the same time, mountain forests are especially sensitive ecosystems due to their steep relief, which allows extreme erosion under a high rainfall regime. Due to increased population pressure and resource use (firewood, mineral resources, pastures, agriculture), montane forests are dwindling more and more rapidly (Figure 11.1). The Food and Agricultural Organization (FAO; 1993) estimates that tropical mountain forests comprise about 11% of the worlds tropical forest resources and suffer an annual deforestation of about 1.1%. Mountain forests generally occur on more humid mountains, and most frequently in Latin America and Southeast Asia, where mountain areas are most widespread, with outposts in tropical Africa (Stadtmüller, 1987). Most studies carried out in tropical ecosystems have focused on the lowland rain forests, where most research stations are located (Leigh, 1999). Even there, very little is known about the regeneration processes

Rainer W. Bussmann • Department of Plant Physiology, University of Bayreuth, 95440 Bayreuth, Germany; e-mail: ceja.andina@t-online.de

Figure 11.1. Fragmented montane cloud forest in southern Ecuador.

in the ecosystem (Finegan, 1996), and fairly nothing about its functioning. In tropical mountain ecosystems, studies concentrated mainly on the alpine zone, whereas the often-inaccessible forest belt, with its extreme species richness, has hardly been studied. The African and Andean zones are no exception. Although, in particular, Andean forests host extreme alpha-, beta- and gamma-diversity (Barthlott et al., 1996; Ibisch, 1996) that is often comparable or higher than species counts for Amazonian areas (Balslev et al., 1998), no comprehensive studies of a tropical mountain forest ecosystem have been undertaken. Even broad-focused projects such as "Ecoandes" in Colombia (van der Hammen et al., 1983; van der Hammen and Ruiz, 1984; van der Hammen, Diaz-Piedrahita, et al., 1989; van der Hammen and Dos Santos, 1995) included the forest regions only marginally. The few publications concerning the mountain forest vegetation of Ecuador mainly contain species lists or mention the montane region in short comparisons to the forests of the Amazon basin (Grubb et al., 1963; Grubb and Whitmore, 1966; Ek, 1997). First attempts to bring the frugal existing knowledge together were made only recently (Hamilton et al., 1994; Churchill et al., 1995). The deficit of scientific knowledge particularly includes all aspects of regeneration of montane forests, as well as succession processes after any kind of natural or man-made impact. The habitat requirements and the potential for rejuvenation of important tree species are almost unknown.

This present chapter focuses on a comparison of the montane forest ecosystems of southern Ecuador and East Africa as examples of tropical mountain forests in general. These systems are compared floristically, with respect to their vegetation ecology, and finally, their regeneration processes are assessed.

Study Areas

The border region of Ecuador and Peru is one of the most biologically diverse areas worldwide, and thus a "biodiversity hotspot" *par excellence*. Low passes in the Andean chain allow an easy exchange between the floras and faunas of the Amazon Basin and the Pacific lowlands. Additionally, the region reveals a very fast transition between the humid mountain forests of the northern Andes and the dry, deciduous forests of the northern Peruvian lowlands. Until the recent past, the Podocarpus National Park and the study area have been almost unknown scientifically. Few studies deal with the flora of Loja province (Espinosa 1948a, 1948b; Emperaire and Friedberg, 1990; Øllgaard and Madsen, 1993; Ulloa and Moeller-Jørgensen, 1993; Jørgensen and Ulloa, 1994; Madsen and Øllgaard, 1994; Bussmann and Lange, 1998a; 1998b; Jørgensen and León-Yánez, 1999) or attempt short descriptions of the area and its vegetation (Espinosa, 1989/1992; Madsen, 1989; 1991; Bøgh 1992). The most recent approach for the classification of the vegetation of Ecuador (Sierra, 1999) lists all montane forests between 1,800 and 3,000 m altitude as *bosque de neblina montano*, without further distinction.

Studies of composition and regeneration of the forest vegetation of Reserva Biológica San Francisco have been carried out since 1997 as part of the German Research Foundation Project, "Functionality in a Tropical Mountain Forest: Diversity, Dynamic Processes and Use–Potential under Ecosystem Aspects." The intention of the study presented here is to characterize the different forest types and their regeneration stages along ecological gradients.

Located between the provincial capitals Loja and Zamora, the private Reserva Biológica San Francisco covers 1,000 ha of the northern slopes of Cordillera de Consuelo in Zamora–Chinchipe Province, Ecuador, adjacent to Podocarpus National Park (Figure 11.2), the only protected area in southern Ecuador. Ranging from 1,800 to 3,150 m, it contains a complete, pristine transect of the montane forest of the region, as well as comparable anthropogenically disturbed areas in the close vicinity. The topography is extremely steep, with slope inclinations that reach 90°, and mainly range between 40 and 60°.

In the southern part of Ecuador and northern Peru, there are a number of lower ridges with deep and dry valleys. The highest elevation is 4,600 m and in most parts an elevation of 4,000 m is not exceeded. The area is the lowest part of the Andes near the equator. Whereas the substrate of the northern Andes are of Quaternary volcanic origin, the southern part is composed of pre-Cretaceous to Tertiary material (Hall, 1977). The geological substrate consists mainly of sandstones and phyllites, the soils of Dystrudepts, Humaquepts and Petraquepts (Schrumpf, 1999; Schubert, 1999). Emck (personal communication, 2001) reports an average annual precipitation of about 2,500 mm in the lower regions (Figure 11.3), and more than 5,000 mm in the uppermost regions of the reserve, with mean annual temperatures between 15 and 17°C, and 9 and 11°C, respectively. Particularly the peak regions are covered by dense clouds or mist most of the year. The main rainfall occurs February to March and particularly June to September, with both periods followed by a dry season, which can lead to a negative water balance, especially from October to January.

The natural forests of East Africa have been subjected to heavy tree felling for decades. Moreover, due to selective logging, they also have undergone significant changes in their species composition. Although private tree planting of exotic species has reduced

Figure 11.2. Study area in Latin America.

the fuelwood deficit considerably, the destruction of the natural forests has increased drastically in the last decades. Currently, the annual deforestation rate is 1.5%, partly driven by a human population growth rate of up to 3.8%. If current rates of forest clearing continue, no natural forest will remain by the year 2040 (Barnes, 1990).

The climate of East Africa shows the chraracteristic pattern of inner tropical regions, with large diurnal temperature oscillations known as *Tageszeitenklima* (Troll, 1959), but a small amplitude of the monthly means in the course of the year. The only seasonality originates from the distribution of precipitation; the region is characterized by two rainy seasons following the position of the innertropical convergence zone (ITC). The "long rains" last from March to June, whereas the "short rains" occur from October to

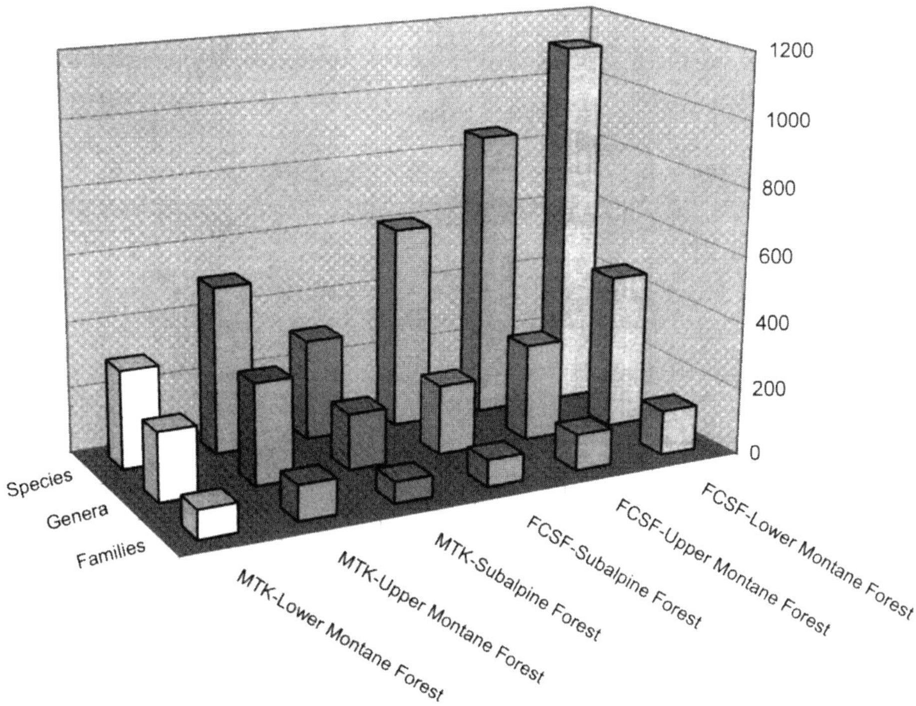

Figure 11.3. Number of plant families, genera, and species in the different forest zones of Mount Kenya (MTK) and Reserva Biológica San Francisco (FCSF).

November. Due to the equatorial position, the daylight period is nearly constant year-round, which is an important factor for vegetation growth.

Today, approximately 2% of East Africa is still covered with indigenous forests (Doute et al., 1981). In addition to providing timber and firewood, the forests represent the largest water catchments systems.

The results of the phytosociological fieldwork conducted in Kenya and Ethiopia from 1992 to 1996 have been presented in a wide array of publications (Bussmann, 1994, 1997; Bussmann and Beck, 1995a; Bytebier and Bussmann, 2000/2002). Ecological observations on growth and germination of east African forest species are currently in progress (Bussmann 1999/2001a; Bussmann and Lange 1999/2000). Descriptions of the regeneration processes in various East African forest types were supplied by Bussmann and Beck (1995b, 1995c) and Lange et al. (1997), based on field data, particularly from Mt. Kenya.

Materials and Methods

For the phytosociological analysis of the East African forests from 1992 to 1996, 933 relevés were established and analyzed. In Ecuador, starting in 1997, after a detailed floristic inventory based on random samples, 14 plots were established in intervals of 200 m altitude, following two transects, to investigate the stand structure of the area. In each altitudinal zone, two plots of 20×50 m (1,000 m^2) were established in a homogenous forest area. In each area, height and diameter at breast height (dbh) of all living and dead tree species with a dbh >10 cm were measured. In each plot, five additional subplots of 2×2 m were established, to get data on tree regeneration.

Additionally, 307 plant–sociological relevés in physiognomically homogenous forest areas, and 76 plots on natural, as well as 40 plots on antropogenically induced landslides were established. All plots were selected and sampled according to the methods of Braun-Blanquet (1964) and Mueller-Dombois and Ellenberg (1974), as slightly modified by van der Hammen, Muller-Dombois, et al. (1989). Because all relevés were sampled at least twice and in different seasons, the species inventory should be more or less complete in most cases.

During the extensive phytosociological fieldwork, special attention was given to regeneration patterns and succession processes in the forests; these observations have led to the regeneration cycles presented here.

To portray the structure of the different forest associations and some of their transitions, stand profile diagrams were established. Transects of 60 m length and 20 m width were photographed and mapped, whereby stem and crown diameter, and height of all trees were measured and plotted schematically.

For further floristic and stand structure analysis, a 1 ha plot was recently established in cooperation with the National Herbarium of Ecuador (Neill) to note height and dbh of all living and dead tree species with a dbh >10 cm.

Mount Kenya and ECSF: Two Tropical Montane Forest Floras in Latin America and Africa, and Their Ecological Significance

Forest Vegetation Units in South Ecuador and East Africa

Extremely steep slopes and deeply incised riverine valleys, providing a mosaic of different microclimates, as well as frequent natural landslides, lead to a very variable mosaic of vegetation units in the southern Ecuadorian region. Based on the dominance-oriented floristic data from the 0.1 ha plots and small relevés, the montane forests of the area have been grouped into three main formations, each covering a specific altitudinal range (Bussmann, 2001 b).

The prevalent forest formation in southern Ecuador at altitudes from 1,850 to 2,100 m, with an extremely diverse, 2-storied tree stratum reaching 25 m, and emergent trees up to 35 m tall in riverine valleys, and mean tree cover of 100%, is the lower montane broad-leaved forest (*Ocotea–Nectandra* forest), extending up to 2,300 m at the bottom of wind-protected riverine valleys. Species belonging to families found mainly at lower altitudes (e.g., Cyclanthaceae, Lauraceae, Hymenophyllaceae) are very common in this formation, whereas representatives of the flora of higher regions are very rare. Undisturbed tracts of the *Ocotea–Nectandra* forests are mainly encountered on very steep slopes, with an inclination of 30–50° or more, as well as in almost inaccessible valleys. In easier to reach areas, this formation has been almost entirely destroyed by human activities and replaced by secondary forest. Particularly interesting is the frequent occurrence of *Alzathea verticillata* (Alzatheaceae). Members of this family so far have only been encountered in montane forests of Peru. Additionally, the understory is characterized by large tree ferns (Cyatheaceae). The epiphyte- and ground flora contains mostly ferns (particularly Lomariopsidaceae) and Orchidaceae. In contrast to higher altitudes, Bromeliaceae are less frequent. The natural regeneration of gaps in these forests leads to a typical mosaic climax. Where the climax forest has been destroyed by natural fires, the tree stratum is dominated by large palms (Arecaceae, *Dictyocaryum lamarckianum*). Characteristically, the herb stratum is dominated by different species of bamboo (*Chusquea* spp.), as well as large Gleicheniaceae forming almost impenetrable thickets. The regeneration of the climax species seems to be inhibited in these areas. In areas with formerly strong human influence (slash and burn, clearcutting), a completely different, monotypic secondary forest develops. These forests have one single tree stratum completely dominated by the 10–12 m tall stems of *Axinea quitensis* (Melastomataceae). Species diversity is much lower. Almost no epiphyte species are encountered, and the ground flora is extremely impoverished. No data are available so far to clarify whether this secondary community is stable as such, or might develop to the original climax stage after time.

At altitudes above 2,100 m, up to about 2,750 m, the upper montane forest (*Purdiaea nutans, Myrica pubescens, Myrsine andina* forest), a monotypic formation, with only one tree stratum, with stems between 5 and 10 m, sometimes up to 15 m tall, replaces the *Ocotea–Nectandra* forest. Lowland species are completely disappearing. The canopy is completely dominated by the twisted stems of *Purdiaea nutans* (Cyrillaceae), which has its main distribution in northern Peru. A very diverse stratum of small treelets and shrubs occurs, and the undergrowth is often completely dominated by large Bromeliads. Above 2,450 m, particularly on wind-exposed ridges, a transition to the *subalpine elfin* forest can be observed. Frequent landslides foster the dynamics in the upper montane forest and are inevitable for the maintenance of its high floristic diversity. In overaged forests, species diversity declines drastically. In these landslides, succession leads to the monotypic climax, but no final data on the length of the successional process are available at present. On natural clearings, in flat ridgetop areas, which are most probably the effect of fires, the tree stratum almost disappears, with only few old specimens of *Purdiaea* remaining, making way for a dense layer of the grasses, suppressing almost all other species in the ground layer. In extremely dry years, fires occur also on steeper, wetter slopes. However, the intensity of these fires seem to be much lower than on ridgetops; thus, only the shrub and herb layers are destroyed.

The uppermost forest belt of the study area is formed by the subalpine elfin forest. This formation—more like an impenetrable bushland than a forest—is closely dovetailed with the adjacent Páramo region. The "timberline" in the area is mainly induced by strong winds. Wind-protected areas are densely covered with Yalca vegetation, whereas more open, wind-exposed regions at the same altitude are covered with grassy Páramo vegetation. A real timberline thus does not exist in the study area. From an altitude of 2,450 m, patches of Yalca occur already in the upper montane forest. The only 1–2 m wide crowns of the dominant—particularly Cunoniaceae, Clusiaceae, Clethraceae, and many small Melastomataceae—form a very dense canopy, allowing only little light to reach the ground. The stems of these "trees" thus protrude from a literally meter-deep carpet of mosses, whereas in contrast to the rich woody flora, almost no herbal species are found in the ground layer.

In East Africa, the montane forest vegetation has been studied intensively in the last decade (Bussmann, 1994, 1997; Bussmann and Beck 1995a, 1995b, Bussmann and Lange, 1999). The evergreen submontane forest, with its two-storied tree layer, is found mainly on the very wet southeastern and southern slopes of Mount Kenya at altitudes between 1,550 and 2,550 m, growing on humic Niti- and Acrisols, and receiving an annual rainfall of 1,500–2,500 mm.

In contrast to Latin America, a wider array of montane forest communities can be found. The evergreen and evergreen xeromorphic montane forests growing on humic acrisols (Speck, 1986) with rainfall between 700 and 1,500 mm, represent the typical vegetation of the altitudinal range between 2,500 and 2,950 m. In these forests, the pencil cedar (*Juniperus procera*) grows about 50-m tall under favorable conditions. Other important tree species, mainly of the lower canopy, are the African olive (*Olea europaea* ssp. *africana*) and *Podocarpus latifolius*. Evergreen montane bamboo forests, dominated by the African bamboo, *Sinarundinaria alpina*, are found on various East African mountains. Bamboo forests are restricted to a relatively narrow ecological range, depending mainly on a soil temperature of 10–15°C and the presence of very deep volcanic soils, namely, humic Andosols. Cyclic changes of the bamboo vegetation lead to regular changes of the floristic composition. One single growth cycle of *Sinarundinaria alpina* is supposed to take 15 (Clayton, 1970) to about 40 years [Wimbush (1945), Agnew (1985), and my own observations indicate even longer periods].

Finally, the timberline is formed by the evergreen subalpine forests, extending from 2,900 to 3,300 m above sea level, where low clouds and mist are frequent. The dominant tree is the "Kosso", *Hagenia abyssinica* (Bruce) J.G. Gmel., of the rose family.

The Montane Forest Floras of Souther Ecuador and Central Kenya as Examples of Tropical Mountain Forests: A Comparison

Due to the fact that more or less complete inventories of the floras of specific tropical areas are extremely rare, attempts to compare the floras of ecologically similar but biogeographically independent areas do not exist. Belonging to completely distinct geographical and phytogeographical regions, the mountain areas of Southern Ecuador and Central Kenya are evolutionarily independent.

The total number of plant species, genera, and families recorded in the mountain forests of the southern Ecuadorian Andes is much higher than that in central Kenya (Table 11.1, Figure 11.3). This effect is mainly caused by the extremely high epiphyte, fern, and tree species numbers found in all altitudinal zones in Ecuador, which are up to 10 times higher than in East Africa. Parasites and lianas show no significant differences in either mountain ecosystem, whereas herbal species are much more common on Mt. Kenya. (Figure 11.4).

Table 11.1. Forest Floras and Life Forms of Mount Kenya (MTK) and Reserva Biológica San Francisco (FCSF)

	Epiphytes	Ferns (without epiphytic and tree-like species)	Herbs	Shrubs	Trees	Lianas	Parasites
FCSF—lower montane forest	339	108	186	134	318	26	4
FCSF—upper montane forest	283	63	159	131	192	24	13
FCSF—subalpine forest	120	37	132	131	172	15	2
MTK—lower montane forest	34	32	98	37	70	22	6
MTK—upper montane forest	7	25	298	105	47	17	6
MTK—subalpine forest	5	17	224	43	10	6	0

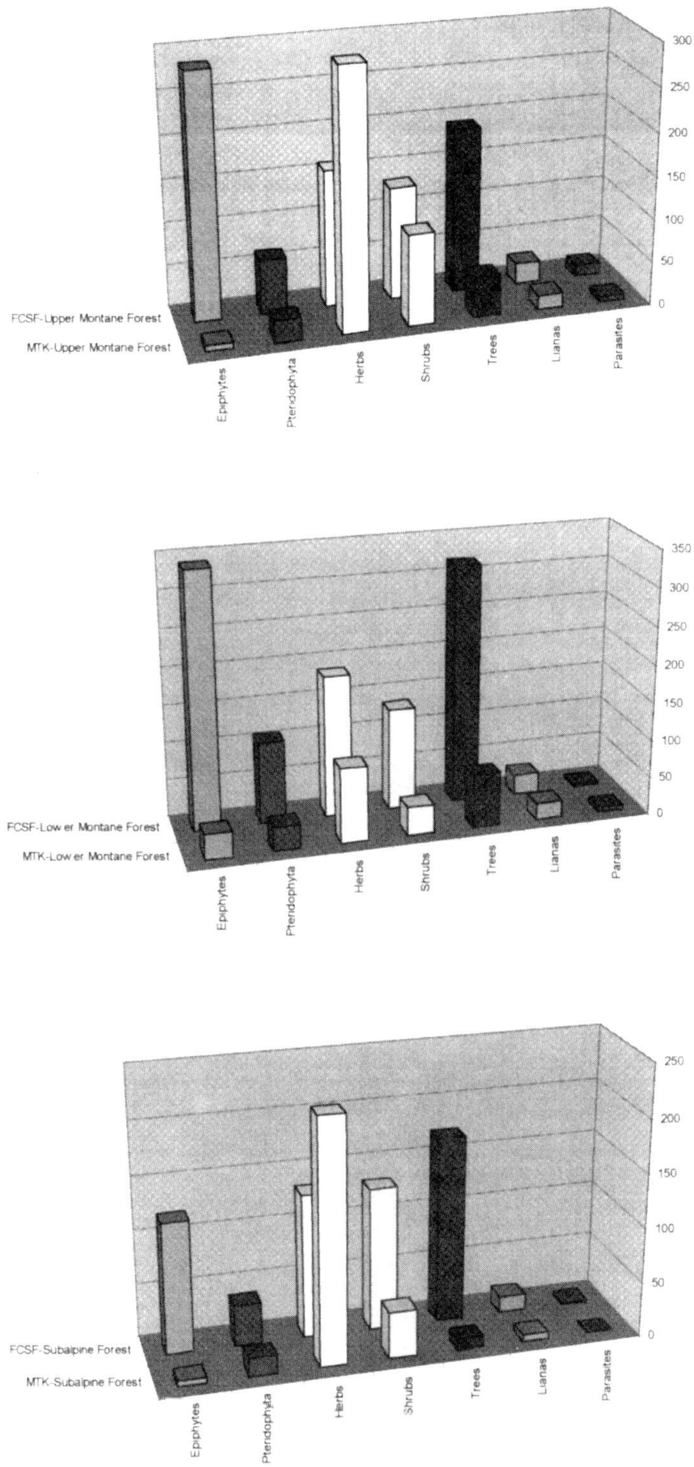

Figure 11.4. Number of species with different life forms in the different forest zones of Mount Kenya (MTK) and Reserva Biológica San Francisco (FCSF).

The number of species recorded in different elevational zones show very similar trends in both ecosystems. Species richness is always highest in the lower montane forest region and declines considerably toward the subalpine forests. In southern Ecuador, this trend can be observed especially for epiphytes, ferns, and tree species, whereas herbs, shrubs, and lianas show almost constant species numbers, with a slight decrease toward higher altitudes. Interestingly, the number of parasites is highest in the upper montane forest. On Mt. Kenya, a similar trend occurs. However, herbs and shrub species show an extreme rise in species richness in midelevations on Mt. Kenya. This might be explained by a high amount of forest gaps, mainly created by an extremely abundant megafauna, allowing a lush regeneration of a wide range of species.

The majority of plant families—105 out of 195—occurs in both mountain systems, whereas 51 families are found only in Ecuador, and 39, only in East Africa. These are, however, mainly species poor groups, apart from Alstroemeriaceae, Bromeliaceae, and Cunoniaceae, which show an extreme adaptive radiation in the New World. There is a high overlap in genera, but only few species are found in both regions. Many families show a very similar number of species. Nevertheless, a number of exceptions are found. Among the flowering plants, the families Aquifoliaceae, Araceae, Araliaceae, Arecaceae, Berberidaceae, Bromeliaceae, Campanulaceae, Clusiaceae, Cunoniaceae, Ericaceae, Gentianaceae, Gesneriaceae, Lauraceae, Melastomataceae, Monimiaceae, Moraceae, Orchidaceae, Piperaceae, and Polygalaceae are found with a multitude of species in the Andes, whereas they are often monotypic, poorly represented, or absent in the forests of Mt. Kenya. The most striking families are surely laurels and orchids, with 20–25 times more species growing in Ecuador. Fabaceae and Lamiaceae are the only large families with more representatives in African mountain forests due to a wider availability of open habitats. A very similar distribution can be observed among the pteridophytes, in which all major groups are much more species-rich in South America (Table 11.2).

Regeneration Ecology of Tropical Montane Forests

Typical Mosaic Climax Regeneration in Montane Broad-leaved Forests in Africa and Latin America

Only few natural remnants of lower montane forests remain after centuries of logging. Under natural conditions, these forests show a typical mosaic regeneration cycle, in which gaps resulting from natural treefall or larger natural clearings are randomly closed by the species arriving first. In most cases, however, a typical set of fast-growing pioneer species acts as shade trees, under which the typical, shade-tolerant climax species develop.

The east African lower montane forests are characterized by huge camphor trees (*Ocotea usambarensis*). Nearly everywhere only very old *Ocotea* trees, approximately 300–600 years old (stem diameter, 1.2–2.5 m), are found. Very few scattered young trees (less than 100 years old) and seedlings are encountered. The main flowering of the dioecious *Ocotea usambarensis* occurs between early March and late April, at the beginning of the long rains. A second flowering can be observed after the start of the short rains in November. In most cases, however, only a few trees flower during these periods, whereas every 5–7 years, there is mass flowering. Most of the produced fruits are attacked by gall insects when still unripe or are eaten by birds. The few remaining seeds are only viable for a few days, even under the most favorable conditions, such as high

Table 11.2. Comparison of the Plant Families of Mount Kenya (MTK) and Reserva Biológica San Francisco (FCSF)

	Reserva Biológica San Francisco (FCSF)								Mount Kenya (MTK)							
	E	F	H	S	T	L	P	Total	E	F	H	S	T	L	P	Total
Acanthaceae			4	7				11				12			1	13
Actinidiaceae					7			7								0
Alstroemeriacea			1			14		14								0
Alangiaceae								0						1		1
Alismataeeae								0				1				1
Alzateaceae					1			1								0
Amaranthaceae			5					5				8				8
Amaryllidaceae								0				1				1
Anacardiaceae					6			6					1			1
Annonaceae					4			4						2		2
Anthercicaceae								0				1				1
Apiaceae			21					21				21				21
Apocynaceae					1	1		2					3	2	2	7
Aquifoliaceae					19			19						1		1
Araceae	30		8					38				2				2
Araliaceae					20			20						6		6
Arecaceae					14			14								0
Asclepiadaceae						9		9				1			7	8
Asparagaceae								0				4				4
Asphodelaceae								0				1				1
Aspleniaceae	21	12						33		6	12					18
Asteraceae			83	40	1	5		129				74	32	3	1	110
Balanophoraceae							2	2								0
Balsamiaceae								0				7				7
Basellaceae						2		2				1				1
Begoniaceae			6					6				2				2
Berberidaceae				9				9					1			1
Bignoniaceae					4			4						3		3
Blechnaceae	3	10						13			3					3
Bombacaceae					1			1								0
Boraginaceae			2		4			6				8		2		10
Brassicaceae			1					1				12				12
Bromeliaceae	77		14					91								0
Brunelliaceae	1				3			4								0
Buddleiaceae					2			2								0
Burmanniaceae			1					1								0
Cactaceae	3							3								0
Caesalpiniaceae								0					1			1
Callitrichaceae								0				1				1
Campanulaceae			8	21				29				9	4		1	14
Cannaceae			1					1								0
Cannellaceae								0						1		1
Capparidaceae								0					1			1
Caprifoliaceae					7			7				1				1
Caryophyllaceae			4					4				11				11
Cecropiaceae					9			9								0
Celastraceae					5			5					4	1		5
Chenopodiaceae								0				1				1
Chloranthaceae					11			11								0
Chrysobalanaceae					2			2								0

Table 11.2. (*continued*)

	Reserva Biológica San Francisco (FCSF)								Mount Kenya (MTK)							
	E	F	H	S	T	L	P	Total	E	F	H	S	T	L	P	Total
Clethraceae					7			7								0
Clusiaceae					27			27				2		4		6
Commelinaceae			7					7			7					7
Connaraceae								0						2		2
Convolvulaceae								0			3			1	1	5
Coriariaceae				1				1								0
Cornaceae			1					1					1			1
Costaceae			1					1								0
Crassulaceae								0			9					9
Cucurbitaceae			6					6						7		7
Cunoniaceae					19			19								0
Cupressaceae								0					2			2
Cuscutaceae							1	1								0
Cyatheaceae					23			23					2			2
Cyclanthaceae	1		4					5								0
Cyperaceae			36					36			32					32
Cyrillaceae					1			1								0
Davalliaceae	3							3	1							1
Dennstaedtiaceae		16						16		5						5
Dichapetalaceae								0					1			1
Dicksoniaceae		1						1								0
Dioscorcaceae							5	5								0
Dipsacaceae								0			2					2
Dracaenaceae								0					2			2
Dryopteridaceae	19	54						73	9	12						21
Ebenaceae					1			1				1	2			3
Elaeocarpaceae					4			4								0
Equisetaceae		1						1								0
Eremolepidaceae							2	2								0
Ericaceae	11			53				62				4	1			5
Eriocaulaceae			5					5			2					2
Euphorbiaceae			2	5	16			23			2	11	9			22
Fabaceae			5	3	3			11			17	18	1			36
Flacourtiaceae				3	4			7					2			2
Fumariaceae								0			2					2
Gentianaceae			15	9				24			7					7
Geraniaceae			4					4			6					6
Gesneriaceae	11		4	10				25			2					2
Gleicheniaceae		13						13		2						2
Grossulariaceae				5	4			9								0
Haloragaceae			4					4								0
Heliconiaceae				1				1								0
Hamamelidaceae								0				1				1
Hymenophyllaceae	25							25	5							5
Hypoxidaceae								0			1					1
Icacinaceae								0					1			1
Iridaceae			2					2			7					7
Isoetaceae		1						1								0
Juncaceae			10					10			6					6
Lacistemataceae					1			1								0
Lamiaceae				5				5			13	23				36

Table 11.2. (*continued*)

	Reserva Biológica San Francisco (FCSF)								Mount Kenya (MTK)							
	E	F	H	S	T	L	P	Total	E	F	H	S	T	L	P	Total
Lauraceae					51			51					2			2
Lecytidaceae					2			2								0
Lemnaceae								0			1					1
Lentibulariaceae			5					5			1					1
Liliaceae								0			2					2
Linaceae								0			1					1
Loasaceae			3					3								0
Loganiaceae				1				1					5			5
Lophosoriaceae		1						1								0
Loranthaceae					1		3	4							6	6
Loxomataceae		1						1								0
Lycopodiaceae	11	27						38	1	3						4
Lythraceae								0			1					1
Magnoliaceae					1			1								0
Malpighiaceae					1	5		6								0
Malvaceae			3					3			8					8
Marattiaceae		5						5					1			1
Marcgraviaceae				2				2								0
Melanthiaceae			2					2								0
Melastomataceae				49	67	1		117				1				1
Meliaceae					17			17					5			5
Melianthaceae								0					1			1
Menispermaceae						4		4						4		4
Mimosaceae					9			9					1			1
Monimiaceae					10			10					1			1
Moraceae					10			10				1	1			2
Myricaceae					2			2				2	1			3
Myristicaceae					3			3								0
Myrsinaceae					15			15				2	1			3
Myrtaceae					17			17					2			2
Nyctaginaceae			1					1								0
Ochnaceae								0					1			1
Olacaceae					1			1					1			1
Oleaceae								0				1	2			3
Oliniaceae								0					1			1
Onagraceae			2	5				7			1					1
Ophioglossaceae	1	1						2								0
Opiliaceae			1					1								0
Orchidaceae	318		52					370	6		7					13
Orobanchaceae								0							1	1
Osmundaceae		1						1								0
Oxalidaceae			8					8			2					2
Passifloraceae				1		1		2						3		3
Phytolaccaceae			2					2			1					1
Pinaceae					2			2					2			2
Piperaceae	28		8	17	3			56	1		2					3
Plantaginaceae			2					2			1					1
Poaceae			76					76			56		1			57
Podocarpaceae					3			3					2			2
Polygalaceae			1	10				11			1					1
Polygonaceae			6					6			7					7

(*continued*)

Table 11.2. (*continued*)

	Reserva Biológica San Francisco (FCSF)								Mount Kenya (MTK)							
	E	F	H	S	T	L	P	Total	E	F	H	S	T	L	P	Total
Polypodiaceae	61	8						69	8							8
Portulacaceae			1					1								0
Primulaceae			1					1			3					3
Proteaceae				3	6			9				1	1			2
Pteridaceae		27						27		13						13
Ranunculaceae			5			1		6			8			2		10
Resedaceae								0			1					1
Rhamnaceae				2				2				3				3
Rhizophoraceae								0					1			1
Rosaceae			5	14	2			21			7	11	1			20
Rubiaceae			4	29	23			56			12	16	7			35
Rutaceae					4			4					7	1		8
Sabiaceae					4			4								0
Sapindaceae				1	2	1		4								0
Santalaceae								0				1				1
Sapindaceae								0				1	2			3
Sapotaceae					5			5					2			2
Schizeaceae		1						1								0
Scrophulariaceae			14	2				16			15	1	1			17
Selaginaceae								0				1				1
Selaginellaceae		7						7		2						2
Simarubaceae					2			2					1			1
Smilacaceae						4		4						2		2
Solanaceae			4	3				7			6	6				12
Sterculiaceae					2			2					3			3
Styracaceae					3			3								0
Symplocaceae					12			12								0
Theaceae					7			7								0
Thelypteridaceae		10						10		6						6
Thymeleaceae								0				1	3			4
Tiliaceae					1			1				4				4
Typhaceae								0			1					1
Ulmaceae					3			3					2			2
Urticaceae			3	6				9			11					11
Valerianaceae			11					11			3					3
Verbenaceae			1					1			1	2	4			7
Violaceae			2		1			3			2					2
Viscaceae							8	8							3	3
Vitaceae							2	2				1		3		4
Vittariaceae	3							3	1							1
Vochysiaceae					1			1								0
Winteraceae					1			1								0
Xyridaceae			1					1								0
Zingiberaceae			3					3			1					1
Number of species	**627**	**197**	**492**	**317**	**526**	**53**	**18**	**2230**	**38**	**58**	**462**	**161**	**116**	**37**	**11**	**884**

E: epiphytes, F: pteridophytes (without epiphytic taxa), H: herbs (without epiphytic taxa), S: shrubs, T: trees, L: lianas and winders, P: parasites.

humidity and radiation. Therefore, regeneration of *Ocotea* by seedlings is very rare (Bussmann and Beck, 1995b; Bussmann and Lange, 1999). Notwithstanding the uniform age structure, numerous *Ocotea* saplings originating from old roots can be observed. Seedlings and young specimens of the other tree species of the camphor forests are found everywhere, indicating the exceptive character of *Ocotea* itself, whereas other species follow normal regeneration patterns and therefore form a typical mosaic structure (Figure 11.5). Equally old populations of *Ocotea usambarensis* trees die over a largely extended period of time (1), creating a mosaic of gaps in which the Euphorbiacean *Macaranga kilimandscharica* starts growing as secondary species (2) on small scale, and does not change the structure of the forest to a significant extent. Regeneration of *Ocotea* from seedlings is negligible, but numerous saplings from the roots of a fallen tree, no longer suppressed by the mother tree, start growing (3). *Macaranga kilimandscharica* hereby plays an important role as shade tree, as *Ocotea* saplings apparently do not tolerate full sun. After the breakdown of the relatively short-living *Macaranga* trees, the well-established young *Ocotea* trees close the gaps and by shading prevent further germination or establishment of *Macaranga kilimandscharica* (4). Thus, *Ocotea* stands of a more or less uniform age structure result. In Ethiopia, *Ocotea usambarensis* is replaced by *Ocotea kenyensis*.

Under conditions of heavy logging and over-exploitation, a different successional cycle was encountered in wide parts of the camphor forests, leading to a drastical change of the vegetation structure. The removal of *Ocotea usambarensis* on a large scale (5) leads at once to numerous, large gaps, with many additional trees killed during felling, removal, and transport of *Ocotea*. The common logging practice of uprooting the *Ocotea* trees prevents the regeneration by root saplings. The big gaps are fast closed by *Macaranga kilimandscharica* and, at higher altitudes, by *Neoboutonia macrocalyx*. Because camphor does not regenerate in these man-made "gaps" after breakdown of the secondary trees, these forests regenerate in a cycle of their own (6). The possibilities for a regeneration to primary forest are unknown, and would at the least require very long time spans, and a complete logging stop around the gaps, to allow colonization.

The high population density of big herbivores, particularly elephants, was claimed to be responsible for the failure of *Ocotea* regeneration. The influence of megaherbivores on vegetation has, however, mainly been studied in savanna ecosystems (Buechner and Dawkins, 1961; Kortland, 1984), and only few data exist on forest habitats (Holloway, 1965; Schmitt, 1991; Schmitt and Beck, 1992). The argument that especially elephants are responsible for the failure of camphor regeneration could not be corroborated by the data. In areas with an estimated equally high elephant population (as counted by dung), regions without heavy logging showed considerable regeneration, whereas in logging areas, especially when the *Ocotea* trees had been felled by uprooting, nearly no regeneration was observed. Therefore, the main reasons for the failing regeneration of *Ocotea usambarensis* are overexploitation and unsustainable logging practices.

In southern Ecuadorian lower montane forests, gaps are rapidly colonized by Cecropiaceae (particularly *Cecropia montana*, with its shiny silvery leaves), *Piptocoma discolor* (Asteraceae), *Isertia laevis* (Rubiaceae), and *Heliocarpus americanus* (Tiliaceae). Rapidly, the canopy is being closed again, and climax species are developing. In the east African montane broad-leaved forests, regeneration also follows a dynamic pattern. Trees of all age classes are found in the forests, suggesting a typical mosaic climax. Due to the environmental conditions, especially the high rainfall, fires are rare and, whenever they occur, are limited to small areas. *Juniperus procera* occurs only sporadically here, because

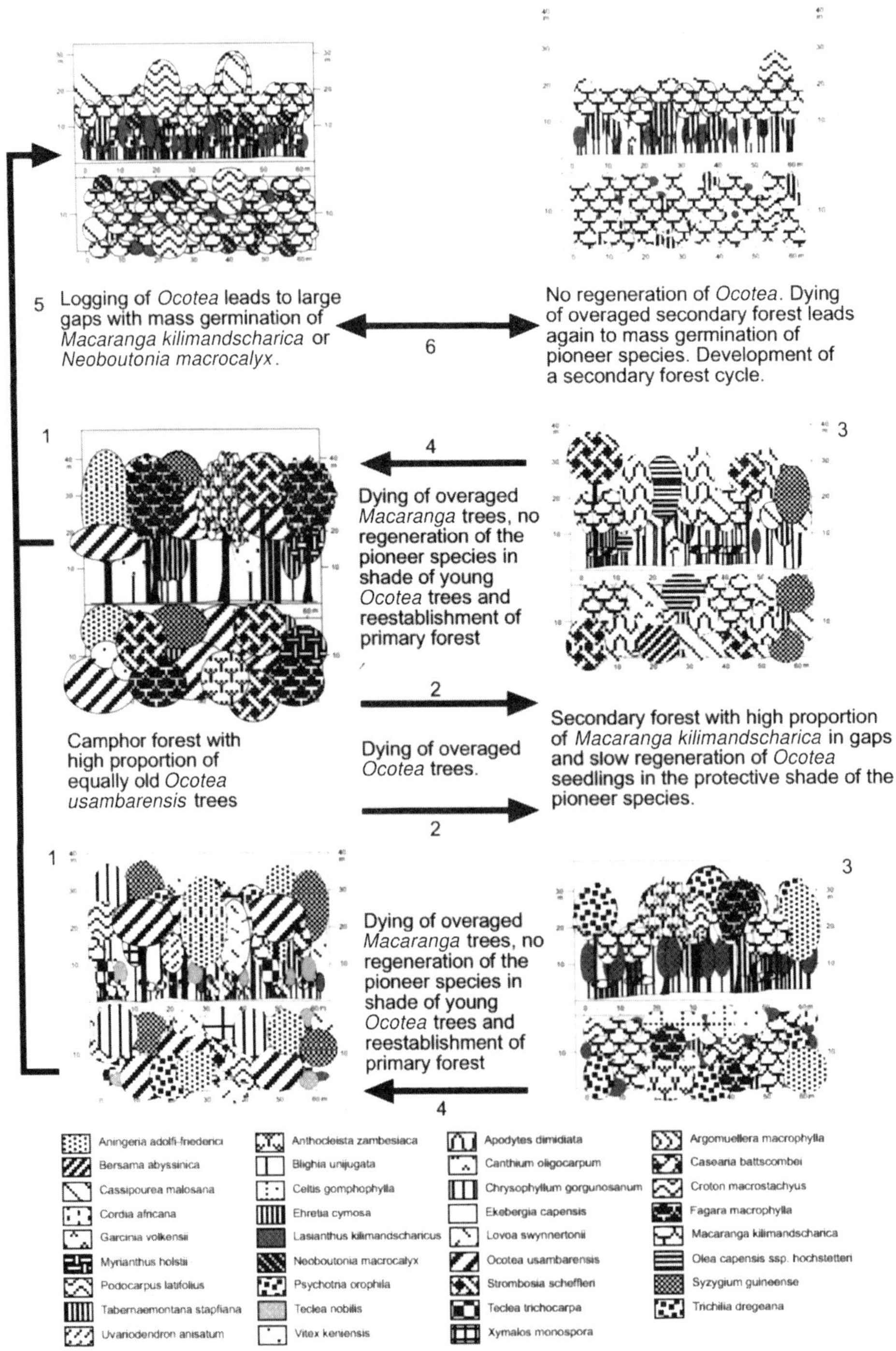

Figure 11.5. Mosaic climax regeneration cycle of the lower montane forests in East Africa.

its germination and regeneration is suppressed by heavy shade and the dense vegetation covering the forest floor. Regenerating cedar was only found in gaps where bright-light requiring seedlings compete with those of the broad-leaved species.

Forest Regeneration on Natural Landslides

Naturally occurring, often earthquake-triggered landslides (Figure 11.6) are a crucial factor for the dynamics and stability of tropical montane forest systems, and for the maintenance of high species diversity. In overaged forest communities, particularly in the upper montane forest, species diversity declines drastically. Research on Ecuadorian landslides has been carried out mainly by Stern (1995). Research on landslides in Bolivia was done by Kessler (1999), and in the central and southern Andes, by Ericksen et al. (1989). Other research on landslides in tropical mountain areas includes that of Garwood et al. (1979) in Panama, Guariguata (1990) in Puerto Rico, and Bartarya and Valdiya (1989) in the Lesser Himalaya in India. Keefer (1984) studied landslide areas all over the world. Landslides are very common in some tropical areas, where long and heavy rainfall and very steep slopes favor slides, as well as a high number of more or less intense earthquakes. In addition, the building of roads, and logging, and burning weaken the substrate. The material and degree of weathering of the geological substrate is also important (Brabb and Harrold, 1989).

In East Africa, most montane forests grow on relatively gentle slopes, and landslides are restricted to the steep flanks of river valleys, and are comparatively rare. After a slide event, the bare ground is rapidly colonized by the same pioneer species as in the mosaic climax regeneration cycle already described. In contrast, extremely steep slopes are the normal feature in the Andean region and many other tropical mountains. Here, natural

Figure 11.6. Natural landslides in tropical montane forests of southern Ecuador.

landslides are very common, and the main source of forest regeneration. Lower montane broad-leaved, as well as upper montane forests in southern Ecuador regenerate in a landslide-induced succession cycle (Figure 11.7).

In the first stage of regeneration after a slide event, bryophytes such as *Polytrichum* and a diverse lichen stratum patchily cover the bare ground. Some individual higher plants manage to settle (species composition by chance). The duration of this stage seems highly variable and depends on the intensity of activity going on at the slide. In most landslides, this stage usually does not take very long, and the cover of cryptogams is not well developed, because the settlement of higher plants is rather fast. After about 5 years, the first stage is replaced by the second.

Locations with bare ground, frequently combined with rocky and steep relief, are often ever-eroding. Here, the first stage of succession takes a much longer time due to water being collected in the zone above, rushing down over the ground, destroying fragile vegetation, and moving seeds and spores. Only some robust and runner shoot–building species can cope with this situation. In particular, *Baccharis genistelloides* and Lycopodaceae are found. A cover of *Polytrichum* and single individuals of Gleicheniaceae and other typical plants of the slides are present, together covering less than 5%. The advanced age of some of the higher plants shows that those sites were not created recently, but some time ago. A rocky relief hinders fast settlement. Strongly weathered but still compact silt to clay stone, with mainly capillary pores, results in high values of pF (field capacity) and causes a low availability of water for the plants, because the water is held strongly in the capillaries. Additionally, the ground has high swelling and contracting characteristics, making higher plant growth difficult, because the capillary roots are destroyed by those processes.

The accumulating planes show characteristics very similar to the eroding slopes. The second stage develops through the increase of vegetation density. A typical phenomenon is the accumulation of ever-growing patches, which occurs when individual plants that settled in the first stage spread via vegetative propagation. Single individuals of Gleicheniaceae or Lycopodiaceae spread via runner shoots and build patches. *Lycopodiella glaucescens* and *Lycopodium clavatum* spread more quickly than the Gleicheniaceae, with long, looping runner shoots, but they build stands of less density. On the slides at the eastern ridge, *Viola stipularis* builds dense carpets via runner shoots. The second stage is characterized primarily by competition between the pioneers of the first stage. The main distributing factor for the settlers of slides is the wind. Ferns, Asteraceae, Orchidaceae, and the genus *Pitcairnia* are distributed in this way. Melastomataceae, Ericaceae, Cyperaceae, Poaceae, and Violaceae are distributed by animals (birds, bats, or ants) (Fukarek, 1991–1994). The duration of the first and second stage is hard to estimate, as a wave-like sequence of vegetation change might occur. The abundance of single woody species points to an advanced age of the site, as well as the occasional occurrence of fairly thick layers of organic matter in second-stage positions. Seedlings of bushes such as *Tibouchina lepidota*, *Graffenrieda harlingii* (Melastomataceae), or *Bejaria aestuans* (Ericaceae) are frequently found under dense layers of *Sticherus*. These pioneer species are not found in the climax communities.

In the third stage, shrubs such as *Ageratina dendroides* (Asteraceae) take over. The ferns are not able to hinder the regeneration of forest for a very long period of time. The most important changes are found in species composition, and as well in a dramatic increase of epiphytic plants, again leading to the establishment of a mosaic climax. The Gleicheniaceae disappear slowly in the undergrowth and are replaced by terrestrial

Figure 11.7. Forest regeneration on natural landslides. Clockwise from left to right: Early first regeneration stage with high bryophyte and lichen cover; late first stage with invading ferns; second stage with dense cover of Gleicheniaceae; third stage with invading shrubs and tree species.

Bromeliaceae. Pioneer species of Melastomataceae such as *Brachyotum campanulare*, *Tibouchina lepidota*, and *Graffenrieda harlingii* disappear. Some Ericaceae survive in the undergrowth. *Sphyrospermum cordifolium* becomes more common. *Bejaria aestuans* disappears early in that stage and is replaced by species of *Cavendishia*, *Macleania*, and others. Piperaceae occur frequently in the undergrowth. Araceae become more common. Araliaceae (*Schefflera*) and Clusiaceae settle on spots with humid microclimate.

Regeneration of Monotypic Climax Forests

The Fire Cycles in the Monotypic East African Xeromorphic Montane Forests and Ecuadorean Upper Montane Forests

Fire apparently plays an important role in the regeneration cycle of the east African montane xeromorphic forests and the less steep parts of the upper montane forests in southern Ecuador.

Juniperus procera, the African pencil cedar, typically grows in the drier montane, submontane, and subalpine areas of the east African mountains between 1,000 and 3,000 m (Kerfoot, 1961). Temperature as well as rainfall apparently limit its spreading.

In his study of the South Aberdare Range, Wimbush (1937) suggested that without fire, the *Juniperus* forest is only a seral stage in the succession to a broad-leaved forest, and proposed a fire-induced regeneration cycle (Figure 11.8). After fire, herbaceous pioneer species and *Juniperus procera* seedlings invade the gap (1), leading to a regenerating forest (2), where broad-leaved species such as *Olea capensis* ssp. *hochstetteri*, *Olea europaea ssp. cuspidata*, and *Olinia rochetiana* become established too. After about 20 years, the *Juniperus* trees have grown about 9 m high. When a closed canopy is established, the cedar trees start to outgrow the other species, reaching about 30 m after 80–90 years. If the number of *Juniperus* trees established after a fire was high, a pure cedar forest (3) results by suppression of other trees. After about 200 years, the maximum height of about 40–46 m is reached. Although frequently infected by the fungus *Fomes juniperinus*, thinning the crowns, the *Juniperus* trees may live for another 300–700 years. In case a burnt area is mainly invaded by broad-leaved species, these create the canopy and only scattered *Juniperus* specimens emerge. As cedar cannot regenerate without fire, natural gaps are filled with broad-leaved species. Therefore, pure broad-leaved forests built the climax community (4) after some time. These broad-leaved forests represent a stable climax, if not destroyed by fire (Bussmann and Beck, 1995c; Bussmann and Lange, 1999).

At higher altitudes, particularly close to the timberline, even the climax vegetation remains dominated by *Juniperus*. After fire, a pioneer stage with a very high initial cover of *Juniperus* and *Myrsine africana* develops (1), where broad-leaved trees soon start growing, producing various independent climax subassociations (2), due to different ecological conditions and different seed intake. However, *Juniperus procera* remains always the dominant canopy species. These climax stages are open forests with only scattered shrubby undergrowth, allowing regeneration of all tree species. Thus, seedlings and young trees of *Juniperus* were often encountered in these forests, even when fire had been absent for a longer period (3). This indicates, that at higher altitudes, broad-leaved forests in the manner described by Wimbush (1937) cannot replace the *Juniperus* forests, even if burning has not occurred for a long time. If a fire occurs, a new regeneration cycle takes place.

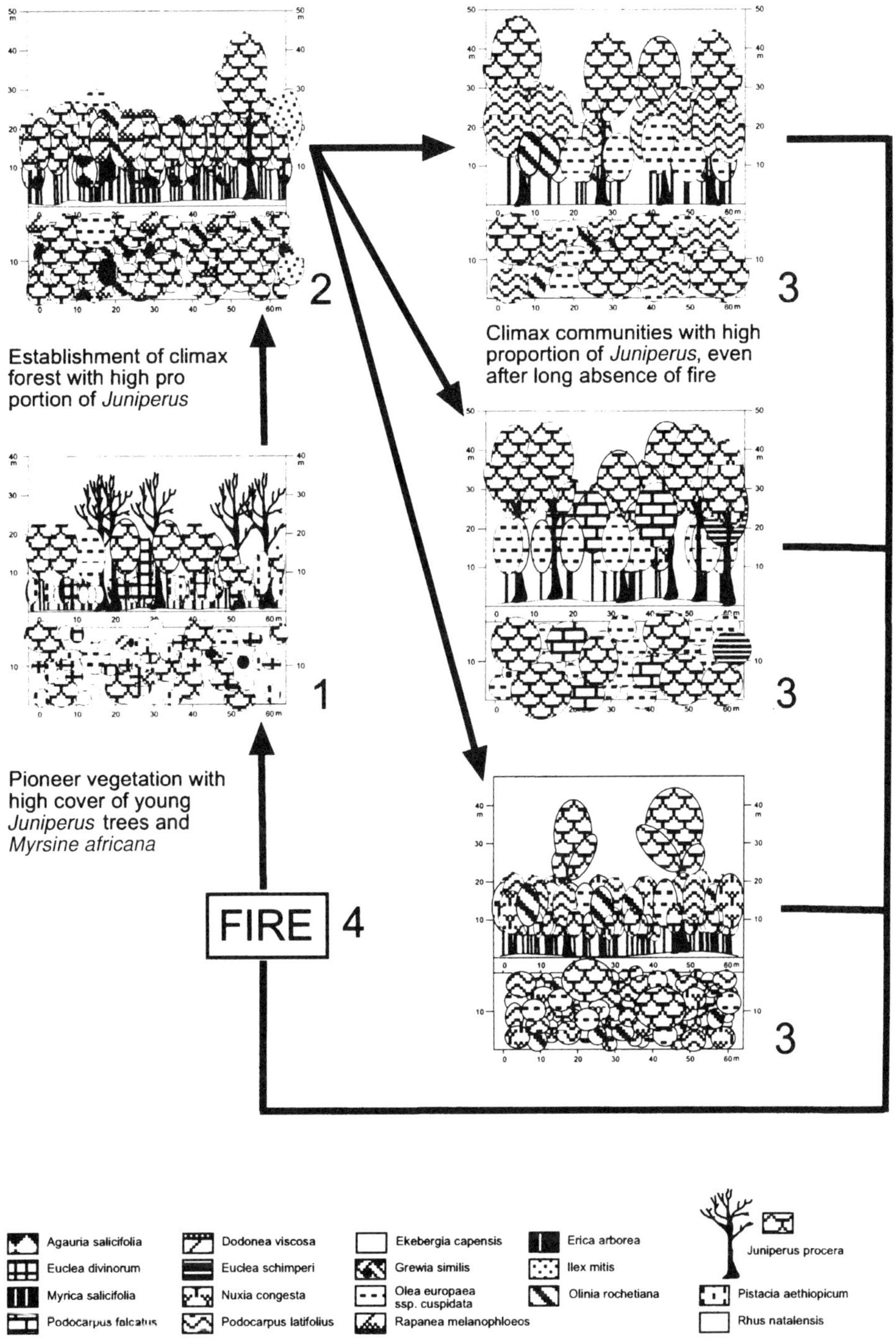

Figure 11.8. Monoclimax regeneration of upper montane forests in East Africa.

In overaged climax communities of the Ecuadorian upper montane forests, the ground layer is almost entirely dominated by large grasses, particularly *Neurolepis elata* (Bambusoideae), the canopy gets very depleted and open, with canopy cover as low as 5%. After mass flowering events of *Neurolepis*, as well as during extremely dry conditions, such as occur during El Niño/La Niña years, a large amount of dry organic matter accumulate, and provide ideal conditions for fires. After burning, the resulting open spaces are rapidly invaded by fern species, without a pronounced cryptogamic stage. Within a short time, woody species belonging to the surrounding primary forests start growing. After a few years, the characteristically dominating bromeliads occur. Due to the very low annual growth of most tree species in this region, a very dense shrub stratum develops and is maintained for many years. Slowly, *Purdiaea nutans* starts outgrowing all other species to form the typical monotypic climax. After about 400–500 years, this dense tree layer starts dying back, and the dense shrub and bromeliad layer is again replaced by the more light-tolerant grass species.

Cyclic Processes in the Bamboo Forests

The African Bamboo, *Sinarundinaria alpina*, inhabits the highland and mountain areas of East and Central Africa at altitudes between 1,800 and 3,400 m (Hedberg 1951; Dale and Greenway, 1961). Bamboo forests prefer deep, volcanic soils, especially andosols (Speck, 1986; and my own observations).

According to Wimbush (1945), bamboo shoots can reach a height of up to 18 m and a stem diameter of up to 10 cm. New vegetative shoots sprout from old rhizomes every 2–3 years. The main growing period starts some weeks after the long rains and lasts about 1–2 months. New shoots are encountered after each long rainy season. However, maximum rates of biomass production with most of the 4,500–7,000 culms produced per hectare require longer periods of favorable climate. When old culms die and fall over, young sprouts from the dense network of old rhizomes replace them, thus giving rise to a homogenous bamboo forests. In undisturbed areas, where the life span of a culm is estimated to be about 14 years, the numbers of dying culms roughly equal new ones. Flowering, after which the culms die, occurs in areas of about one to several hundred acres in more or less regular intervals of at least 40 years (Wimbush, 1945). On Mount Elgon, Clayton (1970) observed a flowering cycle of only 15 years.

Agnew (1985), referring to a long-term knowledge of the vegetation of the Aberdare Range, and after the analysis of ten 1 × 1 m quadrates, distinguished a bamboo cycle with five stages (Figure 11.9):

1. Flowering and Sambucus Stage. Within 2–3 years after mass flowering the culms die and fall, and light-demanding species invade, above all *Sambucus africana*. The soil pH increases again during this stage. The occurrence of trees such as *Podocarpus latifolius*, *Nuxia congesta*, and *Dombeya goetzii* in the bamboo therefore could be indicating the sites of frequent flowering.
2. Pioneer Stage. Regeneration from sections of old rhizomes and rhizome fragments, and occasionally from fallen culms, last about 3–5 years, during which dense thickets of 1–3 m tall *Sinarundinaria* stems intersperse the still present *Sambucus africana* bushes of the previous cycle. No seedlings or even viable seeds have ever been reported from stands of *Sinarundinaria*.

Figure 11.9. Bamboo regeneration cycle. Clockwise from top: Young bamboo shoots growing from old rhizomes; *Sambuscus africana* with regenerating bamboo; building stage with fast-growing culms; dense mature stage.

3. Building Stage. During this period, the growing bamboo exceeds the average plant height in the stand. The vegetation becomes floristically depleted as the bright light–requiring species of the pioneer stage disappear and more shade-tolerant species of the following stage, which might be survivors of a previous cycle, gain momentum. The pH of the soil starts decreasing.
4. Mature Stage. No further increase in culm height takes place during this stage. The culms show even distribution and the soil pH decreases further.

In general, the sequences described by Agnew (1985) are also found on Mt. Kenya (Bussmann, 1994; Bussmann and Beck, 1995a; Bussmann and Lange, 1999), Mt. Nyiro in northern Kenya (Bytebier and Bussmann, 2000/2002), and the Harenna Escarpment in the Ethiopian Bale Mountains (Bussmann, 1997). In Ethiopia, no distinct *Sambucus*-stage could be observed. Apparently *Sambucus africana* does not occur naturally in Ethiopia; however, after dieback of the flowered bamboo, other light-demanding species occur, until the pioneering stage takes over again.

On Mt. Kenya and the neighboring Aberdare Range, bamboo stands are mainly found on the wetter western, southern, and southeastern slopes and are absent in the northern areas. From this distribution, it is concluded that *Sinarundinaria alpina* requires precipitation of about 800–1,000 mm per year, although arid periods are tolerated. In clear correspondence to the observations of Wimbush (1945), new culms were mainly produced after prolonged rainy areas. During the period 1992–1996, the main culm production occurred in large areas in May and June 1995, after 9 months of unusually heavy rainfall. This, and the fact that the newly produced culms were mostly much larger and taller than the old ones, clearly depicts the dependency of bamboo growth on rainfall duration and intensity.

Only small areas of flowering or postflowering bamboo could be found, and no synchronization of flowering among different populations (Mt. Kenya, Aberdares, Mt. Nyiro, Harenna Escarpment (Bale Mountains, Ethiopia) could be observed. In clear contrast to the occurrence of mass-flowering effects suggested by Wimbush (1945), this phenomenon did not occur anywhere in the study area. Instead, bamboo flowered and later died back only in little patches, after periods of favorable climatic conditions. Therefore, it has to be concluded, that in contrast to many Asian and Latin American bamboo species, *Sinarundinaria alpina* shows no synchronized flowering at all. Despite the flowering and subsequent die back, seeds or seedlings of the species were never found, and exclusively vegetative regeneration of bamboo was observed in the whole growing area.

Of special interest is the occurrence of large areas of bamboo forests in Ethiopia. Although covering wide areas (Harenna; Bussmann, 1997), these forests either were not recognized as specific forest types or were not mentioned at all by earlier authors. The large bamboo populations of south and central Kenya (Mt. Kenya, Aberdares, Mau, about 1,000 km south of Harenna) are linked to Harenna only by a small population occurring on top of Mt. Nyiro at the southern end of Lake Turkana, at least 500 km away from the Bale Mountains. As mentioned earlier, viable seeds of *Sinarundinaria alpina* have never been found, and the species is known to regenerate only vegetatively; thus propagation by animals seems very unlikely. Therefore, the large gaps between the main population and even these outposts pose some interesting questions regarding the ecology and former distribution of *Sinarundinaria*, as influenced by the palaeoclimate of the region.

Regeneration in the Subalpine *Hagenia abyssinica* Forests

Already Schmitt (1991) and Bussmann (1994) have supposed a cyclic regeneration of ***Hagenia abyssinica*** depending on fire.

At first sight, it seems fairly incomprehensible that fire should be the decisive factor for the regeneration of the *Hagenia* forests. They are predominately located in the misty regions of the East African mountains, where high precipitation and frequent cloud formations might prevent the vegetation and dead plant material from becoming sufficiently dry to burn. Martin (1982) however, mentioned that fires tend to be most frequent in areas with a combination of comparatively high rates of primary production, slow decomposition rates, and occasional drying of the fuels. Even Houston (1995) stated that the alternation of wet and dry seasons produces ideal conditions for fires in wet regions. On Mt. Kenya, there are rainfall anomalies in form of severe droughts, for example, the one occurring in 1890. In recent times, the rainfall in 1990 was likewise extremely low, leading to the assumption that extraordinary dry periods favoring the outbreak of fires occur in approximately 100-year time interval. In sufficiently wet years, a relatively high plant productivity, together with the large litter production of an adult *Hagenia*, which is shedding the older leaves continually (Miehe and Miehe, 1994) and the low decomposition rates in the cold climate of the subalpine forests, result in the accumulation of fuel. This encourages the spreading of fires in extremely dry years. The existence of charcoal layers in the soil profiles dug within the *Hagenia* forests indicate that fires are really a characteristic feature of these misty forests. Field observations indicated that regeneration stages of *Hagenia abyssinica* always grow together with either young seedlings of *Juniperus procera* and/or fire-indicating vegetation. Furthermore, the equal ages of the *Hagenia* forest stands suggest that the Kosso is not regenerating successively, but all of a sudden, after a disturbing event like fire. Germination tests indicated that the *Hagenia* seeds were not germinating unless the dense undergrowth was removed. Fire is the only natural factor that is able to destroy the undergrowth of larger areas, and thus provide an environment with reduced competition, suitable for germination of the seeds. Germination therefore depends on a fire event, that removes the competing undergrowth and further stimulates the germination of the Kosso with a sudden heat surplus.

The regeneration cycle of the *Hagenia* forests (Lange et al., 1997; Bussmann and Lange, 1999) and other areas is shown in Figure 11.10). Stage 1 of the cycle describes the climax forests with adult Kosso trees of uniform size and age, and *Juniperus procera* as codominating tree species. The canopy of these forests is often nearly 100% closed, and the undergrowth consists mainly of a dense herb and grass layer, and higher bushes are rarely found. The conditions of high competition and intensive shading of the ground are not suitable for the germination of *Hagenia abyssinica*. A fire event disturbs the climax community by destroying herbs and leaf litter in the undergrowth, as well as all roots in the upper soil layers, whereas the adult Kosso trees remain more or less severely damaged but still alive in most cases. Miehe and Miehe (1994) assumed adult *Hagenia* trees to be almost fire resistant with their bark peeling in large pieces, preventing fire from reaching essential vascular strands of the tree. Supporting this assumption, many Kosso trees in the forests show often heavy signs of former burning, but they remain green and even flower. After a fire, the heated ground, now bare of vegetation and only covered with black ash, provides ideal conditions for the germination of the Kosso seeds (stage 2). Whether stage 3 or 4 of the regeneration cycle develops in a certain area depends on the frequency of fires.

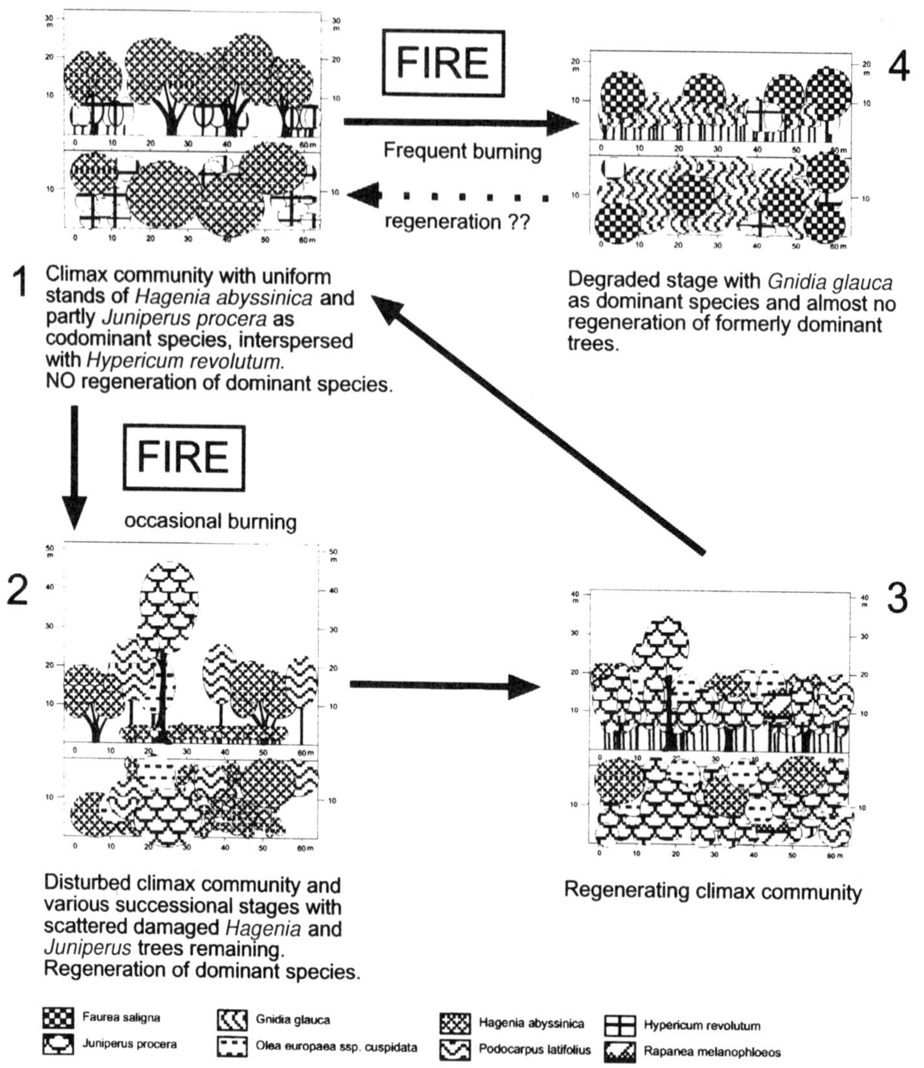

Figure 11.10. Regeneration cycle of subalpine forests in east Africa.

In case the time intervals between two fire events are sufficiently high (approximately 100 years), the young *Hagenia* trees are found scattered between a dense pioneer vegetation of about 5–6 m height. *Clutia abyssinica* (Euphorbiaceae), *Dombeya torrida* (Sterculiaceae), *Helichrysum schimperi* (Asteraceae) and *Leonotis mollissima* (Lamiaceae) are characteristic species in the impenetrable shrub thicket of this succession stage after fire. In contrast, more frequent burning results in a shift to a less high and dense vegetation dominated by *Gnidia glauca* (Thymeleaceae) (stage 4).

Acknowledgments

I gratefully acknowledge the continuing support by Deutsche Forschungsgemeinschaft (DFG) and Nature and Culture International (NCI). Special thanks are due to all my Ecuadorian, Kenyan, Ethiopian and German colleagues.

References

Agnew A. D. Q. (1985). Cyclic sequences in plant communities of the Aberdare mountains, Kenya. Journal of East African Natural History Society and National Museum 75: 1–12.

Balslev, H., Valencia, R., Paz y Miño, G., Christensen, H., and Nielsen, I. (1998). Species count of vascular plants in one hectare of humid lowland forest in Amazonian Ecuador. In: F. Dallmeier, and J. A. Comiseky (Eds.), *Forest biodiversity in North, Central and South America, and the Caribbean*, Paris: Pearl River.

Barnes, R. F. W. (1990). Deforestation trends in tropical Africa. *African Journal of Ecology*, 28: 161–173.

Bartarya, S. K., and Valdiya, K. S. (1989). Landslides and erosion in the catchment of the Gaula River, Kumaun Lesser Himalaya, India. *Mountain Research and Development* 9(4): 405–419.

Barthlott, W., Lauer, W., and Placke, A. (1996). Global distribution of species diversity in vascuar plants: Towards a world map of phytodiversity. *Erdkunde* 50: 317–327.

Bøgh, A. (1992). Composition and distribution of the vascular epiphyte flora of an Ecuadorian montane rainforest. *Selbyana* 13: 25–34.

Brabb, E. E., and Harrold, B. L. (1989). *Landslides: Extent and economic significance*. Rotterdam: Brookfield.

Braun-Blanquet, J. (1964). *Pflanzensoziologie* (3rd ed.). Wien, New York: Springer.

Buechner H. K., and Dawkins, H. C. (1961). Vegetation changes induced by elephants and fire in Murchinson Falls National Park, Uganda. *Ecology* 42: 752–766.

Bussmann, R. W. (1994). *The forests of Mount Kenya (Kenya): Vegetation, ecology, destruction and management of a tropical mountain forest ecosystem*. Doctoral dissertation, Universität Bayreuth, Bayreuth, Germany.

Bussmann, R. W. (1997). The forest vegetation of the Harenna escarpment (Bale Province, Ethiopia) syntaxonomy and phytogeographical affinities. *Phytocoenologia* 27(1): 1–23.

Bussmann, R. W. (1999) (printed 2001a). Growth rates of important East African montane forest trees, with particular reference to those of Mount Kenya. 89: 69–78.

Bussmann, R. W. (2001b). The montane forests of Reserva Biológica San Francisco (Zamora-Chinchipe, Ecuador)—vegetation zonation and natural regeneration. *Die ERDE* 132: 11–24.

Bussmann, R. W., and Beck, E. (1995a). The forests of Mount Kenya: A phytosociological synopsis. *Phytocoenologia* 25(4): 467–560.

Bussmann, R. W., and Beck, E. (1995b). Regeneration and succession processes in the cedar-forests (*Juniperion procerae*) of Mount Kenya. *Ecotropica* 1: 79–84.

Bussmann, R. W., and Beck, E. (1995c). Regeneration and succession processes in the *Ocotea*-forests (*Ocotetea usambarensis*) of Mount Kenya. *Verhandlungen der Gesellschaft für Ökologie* 24: 35–39.

Bussmann, R. W., and Lange, S. (1998a). *Expedición Rapido Asesso Cordillera Sabanilla—Inventario floristico*. Loja/Quito: Estaqción Centrifica San Francisco/Instituto Nacionál de Ecologia Faunay Arous Naturales.

Bussmann, R. W., and Lange, S. (1998b). *Inventario de las plantas del herbario de la Estación Cientĺfica San Francisco, Zamora-Chinchipe*. Loja: Estación Centrifica San Francisco.

Bussmann, R. W., and Lange, S. (1999). Succession and regeneration patterns of East African mountain forests. *Bayreuther Forum Ökologie* 64: 1–16.

Bussmann, R. W., and Lange, S. H. (1999) (printed 2002). Germination of important East African mountain forest trees. *Journal of East African Natural History* 89(1–2): 101–111.

Bytebier, B., and Bussmann, R. W. 2000 (printed 2002). Vegetation of Mount Nyiru (Samburu District, Kenya): A checklist and syntaxonomical survey. *Journal of East African Natural History* 89: 45–71.

Churchill, S. P., Balslev, H., Forero, E., and Luteyn, J. L. (Eds.). (1995). *Biodiversity and conservation of neotropical montane forests*. Bronx, NY: New York Botanical Garden.

Clayton, W. (1970). Gramineae (part 1). In: Turill W. B., Milne–Redhead E., Hubbard C. E., and Polhill E. (Eds.), *Flora of tropical East Africa*. Rotterdam: Balkema.

Dale, I. R., and Greenway P. J. (1961). *Kenya trees and shrubs*. Nairobi: Buchanan.

Doute, R., Ochanda, N., and Epp H. (1981). *Forest cover mapping in Kenya using remote sensing techniques*. Nairobi: Kenya Remote Sensing and Mapping Unit.

Ek, R. C. (1997). Botanical diversity in the tropical rain forest of Guayana. *Tropenbos-Guayana series* 4: 1–237.

Emperaire, L., and Friedberg, C. (1990). *Relevés floristiques des regions Piura (Perou) et de Loja (Ecuateur)*. Paris: Orstrom-Puc.

Ericksen, G. E., Ramirez, C. F., Concha, J. F., Tisnado M. G., and Urquidi, B. F. (1989). Landslide hazards in the central and southern Andes. In: Barabb and Harrod (Eds.), *Landslides: Extent and economic significance*. Rotterdam: Balkema.

Espinosa, B. (1948a). *Estudios botanicos en el sur del Ecuador: I. Loja-Catamayo-Malacatos-Vilcabamba*. Loja: Universidad Nacional.

Espinosa, B. (1948b). *Estudios botanicos en el sur del Ecuador: II. Herbarium Universitatis Loxoensis (Primer Inventario)*. Loja: Universidad Nacional.

Espinosa, G. D. A. (Ed.). (1989/1992). Parque Nacional Podocarpus. *Boletin informativo sobre biologia, conservacion y vida silvestre* (vols. 1–3).

Food and Agriculture Organization. (1992). Aménagement et conservation des forêts denses en Amérique tropicale. In: *Etude Forêts* 101. Rome: Author.

Food and Agriculture Organization. (1993). *Summary of the final report of forest resource assessment 1990 for the tropical world*. Rome: Author.

Finegan, B. (1996). Pattern and process in neotropical secondary rain forest: The first 100 years of succession. *Trends in Ecology and Evolution* 11: 119–124.

Fukarek, F. (1991–1994). *Urania Pflanzenreich*. Leipzig: Urania.

Garwood, N. C., Janos, D. P., and Brokaw, N. (1979). Earthquake-caused landslides: A major disturbance in tropical forests. *Science* 205: 997–999.

Grubb, P. J., Lloyd, J. R., Pennington, T. D., and Whitmore, T. C. (1963). A comparison of montane and lowland rain forest in Ecuador: I. The forest structure, physiognomy, and floristics. *Journal of Ecology* 51: 567–601.

Grubb, P. J., and Whitmore, T. C. (1966). A comparison of montane and lowland rain forest in Ecuador: II. The climate and its effects on the distribution and physiognomy of the forests. *Journal of Ecology* 54: 303–333.

Guariguata, M. R. (1990). Landslide disturbance and forest regeneration in the Upper Luquillo Mountains of Puerto Rico. *Journal of Ecology* 78: 814–832.

Hall, M. (1977). *El volcanismo en el Ecuador*. Quito: Abya Yala.

Hamilton, L. S., Juvik, J. O., and Scatena, F. N. (Eds.). (1994). *Tropical montane cloud forests: Ecological studies* 110. New York: Springer.

Hedberg O. (1951). Vegetation belts of the East African mountains. *Svensic Botanisk Tidskrjt* 45: 140–199.

Holloway, C. W. (1965). The impact of big game on the forest policy in the Mount Kenya and north-east Aberdare forest reserves. *East African Agriculture and Forestry Journal* 30(4): 370–389.

Houston, M. A. (1995). Biological diversity: The coexistance of species on changing landscapes. Cambridge, UK: Cambridge University Press.

Ibisch, P. L. (1996). *Neotropische Epiphytendiversität—das Beispiel Bolivien*. Wiehl: Martina Galunder Verlag.

Jørgensen, P. M., and León–Yánez, S. (1999). *Catalogue of vascular plants of Ecuador. Monographs in Systematic Botany from the Missouri Botanical Garden* (Vol. 75).

Jørgensen, P. M., and Ulloa C. (1994). *Seed plants of the high Andes*. AAU Report No. 34. Aarhus: Aarhus University.

Keefer, D. K. (1984). Landslides caused by earthquakes. *Geological Society of America* 95: 406–421.

Kerfoot, O. (1961). *Juniperus procera* Endl. (the African pencil cedar) in Africa and Arabia: I. Taxonomic affinities and geographical distribution. *East African Agricultural and Forestry Journal* 26: 170–177.

Kessler, M. (1999). Plant species richness and endemism during natural landslide succession in a perhumid montane forest in the Bolivian Andes. *Ecotropica* 5(2): 123–136.

Kortland, A. (1984). Vegetation research and the "bulldozer" herbivores of tropical Africa. In: A. C. Chadwick, and S. L. Sutton, (Eds.), *Tropical Rain Forest: The Leeds Symposium*, Leeds: Philosophical Society.

LaBastille, A., and Poole, D. J. (1978). On the need for a system of cloud-forest parks in Middle America and the Caribbean. *Environmental Conservation* 5(3): 183–190.

Lange S. H., Bussmann, R. W., and Beck, E. (1997). Ecology and regeneration of the subalpine *Hagenia abyssinica* forests of Mt. Kenya. *Botanica Acta* 110(6): 473–480.

Leigh, E. G. (1999). *Tropical forest ecology—a view from Barro Colordo Island*. Oxford, UK: Oxford University Press.

Madsen, J. E. (1989). Aspectos generales de la flora y vegetacion del Parque Nacional Podocarpus. *Boletin Informativo Sobre Biologia, Conservacion y Vida Silvestre*, 1: 59–74.

Madsen, J. E. (1991). *Floristic composition, structure and dynamics of an upper montane rain forest in Southern Ecuador*. Unpublished doctoral thesis, Aarhus: Botanical Institute of Aarhus University.

Madsen, J. E., and Øllgaard, B. (1994). Floristic composition, structure and dynamics of an upper montane rain forest in southern Ecuador. *Nordic Journal of Botany* 14(4): 403–423.

Martin, R. E. (1982). Fire history and its role in succession. In: J. E. Means (Ed.), *Forest succession and stand developement in the Northwest*, Corvalis: Forest Reserve Laboratory, Oregon State University.

Miehe, S., and Miehe G. (1994). *Ericaceous forests and heathlands in the Bale Mountains of South Ethiopia: Ecology and man's impact.* Hamburg: Stiftung Walderhaltung in Afrika und Bundesforschungsanstalt für Forst- und Holzwirtschaft.

Mueller-Dombois, D., and Ellenberg, H. (1974). *Aims and methods of vegetation ecology.* New York: Wiley.

Myers, N., Mittermeier, R. A., Mittermaier, C. G., da Fonseca, G. A. B., and Kent, J. (2000). Biodiversity hotspots for conservation priorities. *Nature* 403(25): 853–858.

Ohl, C., and Bussmann, R. W. (2000). Vegetation on natural landslides in the tropical mountain forest of southern Ecuador, with special consideration of the altitudinal gradient. *Verhandlungen der Gesellschaft für Ökologie* 30: 208.

Øllgaard, B., and Madsen, J. E. (1993). Inventario preliminar de las especies vegetales en el Parque Nacionál Podocarpus. *Revista de Difusión Técnica y Cientifica de la Facultad de Ciencias Agricolas, Universidad Nacional de Loja* 22/23 (1/2): 66–87.

Schmitt, K. (1991). The vegetation of the Aberdare National Park, Kenya. *Hochgebirgsforschung* 7: 1–250.

Schmitt K., and Beck, E. (1992). Die Zerstörung eines Bergwalds in Kenia: Beispiele des Zusammenwirkens von Mensch und Großwild. *Jahrbuch Verein z. Schutze der Bergwelt* 57: 207–226.

Schrumpf, M. (1999). *Genese, Klassifikation und Nährstoffgehalte tropischer Böden entlang eines Höhentransekts an der Andenostabdachung Südecuadors.* Diplomarbeit Universität Bayreuth, unpublished.

Schubert, C. (1999). *Klassifikation, Genese und Standorteigenschaften tropischer Gebirgsböden im lee der Cordillera Real, Süd-Ecuador.* Diplomarbeit Universität Bayreuth, unpublished.

Sierra, R. (Ed.). (1999). *Propuesta preliminar de un sistema de clasificación de vegetación para el Ecuador Continental.* Quito: Instituto Nacional de Ecología, Fauna y Vida Silvestre/Global Environment Facility.

Speck, H. (1986). Soils of the Mount Kenya Area, their formation, ecological, and agricultural significance. *African Studies Series* A1: 47–72.

Stadtmüller, T. (1987). *Cloud forests in the humid tropics: A bibliographic review.* Tokyo and Catie, Turrialba: United Nations University.

Stern, M. J. (1995). Vegetation recovery on earthquake-triggered landslide sites in the Ecuadorian Andes. In: S. P. Churchill, H. Balslev, E. Forero, and J. L. Luteyn, (Eds.), *Biodiversity and conservation of neotropical montane forests*, New York: Wiley.

Troll, C. (1959). Die tropischen Gebirge. Ihre dreidimensionale klimatische und pflanzengeographische Zonierung. *Bonner Geographisch Abhandlungen*. 25.

van der Hammen, T., Perez Precario, A., and Pinto E. P. (Eds.). (1983). *Studies on tropical Andean ecosystems* (Vol. 1). Vaduz: Cramer.

van der Hammen, T., and Ruiz, P. M. (Eds.). (1984). *Studies on tropical Andean ecosystems* (Vol. 2). Vaduz: Cramer.

van der Hammen, T., Diaz-Piedrahita, S., and Alvarez, V. J. (Eds.). (1989). *Studies on tropical Andean ecosystems* (Vol. 3). Vaduz: Cramer.

van der Hammen, T., and Dos Santos, A. G. (Eds.). (1995). *Studies on tropical Andean ecosystems* (Vol. 4). Vaduz: Cramer.

van der Hammen, T., Mueller-Dombois, D., and Little, M. A. (1989). *Manual of methods for mountain transect studies.* Paris: International Union for Bilogical Sciences.

Ulloa, C. U., and Moeller-Jørgenson, P. (1993). Arboles y arbustos de los Andes del Ecuador. *AAU Reports* 30: 1–263.

Wimbush S. H. (1937). Natural succession in the pencil cedar forests of Kenya Colony. *Empire Forestry Journal* 16: 49–53.

Wimbush, S. H. (1945). The African alpine bamboo. Kenya Colony. *Empire Forestry Journal* 24: 33–39.

12

Influences of Elevated Levels of CO_2 on Plants

Madhoolika Agrawal

Introduction

Humans have been exploiting the natural resources from the dawn of civilization, but the exploitation has accelerated at a very high rate since the Industrial Revolution. Atmosphere, hydrosphere, and lithosphere are utilized as resource as well as medium for discharge of waste products, which results in physicochemical changes in the quality of air, water, and soil that have harmful biological consequences. The chemical composition of the atmosphere is being modified by the injection of gases, particulates, and volatile substances that may upset the normal functioning of environment at varying degrees and run the risk of inducing irreversible ecological changes.

It has been recognized worldwide that changes in the chemical composition of the atmosphere are likely to alter the Earth's climate, which has severe implications on vegetation. Increases in the concentrations of carbon dioxide (CO_2) and other radiatively important gases in the atmosphere have become the most important scientific and political issues throughout the world. Gases that represent only a small percentage of the atmosphere play a major role in determining the temperature at the Earth's surface. These are called "greenhouse gases," because their role in retaining heat is similar to that of glass or a greenhouse.

The possibility of global climate change due to anthropogenic CO_2 has become a major scientific issue during the past decades (Schneider, 1989). Since the beginning of the Industrial Revolution, the burning of fossil fuels to release energy has increased the CO_2 concentration from about 270 to 365 ppm (i.e., a rise of more than 30%). The Mauna Loa, Hawaii, records showed a 12% increase in mean annual concentrations in 32 years, from 315 ppm to 353 ppm from 1959 to 1990 (Keeling and Whorf, 1994). Conway et al. (1994) reported an annual increase of 1.43 ppmv CO_2 at Cape Grim, Tasmania, Australia from

Madhoolika Agrawal • Department of Botany, Banaras Hindu University, Varanasi 221 005 India.

1985 to 1992. Marland et al. (1994) have estimated CO_2 emission from fossil fuels in India to be 32.1 million tons in 1960 and 192 million tons in 1991. With some uncertainty about the future projections of CO_2 concentration, it is expected that the concentration will reach 550 ppm by the middle of the 21st century (i.e., there will be a doubling of the preindustrial level).

Carbon dioxide is the third most abundant gaseous component of the Earth's atmosphere after nitrogen and oxygen, with an atmospheric background of 0.03% in clean air. CO_2 is released into the atmosphere as a result of both natural and anthropogenic activities. The main anthropogenic source of CO_2 emission is the combustion of carbon-based fuels. In 1993, approximately 5 gigatons of carbon were emitted globally from fossil fuel (Rotty and Masters, 1984). Other net CO_2 emissions from terrestrial biota (through processes such as respiration), including the clearing and burning of forests, have contributed significantly to the present atmospheric CO_2 concentration. Much of the CO_2 input to the atmosphere since 1800 may have been derived from forest and soil carbon based on the deconvolution of ^{13}C records of tree rings (Peng et al., 1983). Box models of oceans suggest that 60–70% of the CO_2 released remains in the atmosphere, with the remainder incorporated into the oceans (Baes and Mulholland, 1985).

Plant Responses to Elevated CO_2

The changes in CO_2 levels are expected to have far-reaching effects on the environment in terms of plant response. Increase in atmospheric CO_2 is cause of concern in terms of two interrelated reasons. First, the increase in availability of CO_2 will directly influence the growth and physiology of plants. Second, CO_2, a potent radiation absorbing gas, contributes to the increase in mean global surface temperature. Temperature is one of the important determinants of plant growth, function, and distribution. This chapter briefly reviews the impact of rising CO_2 on plants and its interaction with various environmental factors and rising air pollutants.

Rising CO_2 can influence the world ecosystem by direct effects on plant growth and development. The productivity of plants can be enhanced if more CO_2 is available. CO_2 fertilization of cultures in greenhouses has been known to occur for over a century (Kimball, 1983). Kimball's published literature review (1983) on mainly agricultural and horticultural plants showed that overall yield will be increased on average by about 36% if the present CO_2 concentration is doubled. This clearly shows that the actual CO_2 concentration of the troposphere seems to be suboptimal for plant growth. Keeling et al. (1989) suggested that the northern biosphere has increased its carbon fixation by 0.5% yr^{-1} over the past 20 years. Excellent reviews on the likely effects of elevated CO_2 on plants have been prepared by Kimball (1983), Allen (1990), Acock and Allen (1985), Idso (1989), Krupa and Kickert (1989), Ceulemans and Mousseau (1994), Atkinson (1996, 2000), Badiani et al. (2000), and Eamus (2000).

Physiological Responses

Photosynthesis

Photosynthesis (Ps) is not only the major physiological process by which plants sense changes in the atmospheric CO_2 concentration (Mott, 1990), but it is also the process *a priori* by which plant production is affected, and the process sets the upper limit on potential carbon uptake into vegetation. The direct effect of CO_2 is on Ps, which increases in C_3 plants on average by 52%, with a doubling of CO_2 (Cure and Acock, 1986), while on average 44% stimulation is apparent for trees. The increase in Ps may fall on average to a value around 29% of the original increment (Eamus and Jarvis, 1989). Eamus (2000) has shown increase of 20–50% in the assimilation of trees at a doubling of current atmospheric CO_2 concentration.

Experimental evidence has shown that Ps in C_3 plants responds more to the CO_2 enrichment than C_4 plants. The cause of increase in photosynthesis is CO_2 availability to the ribulose biphosphate carboxylase-oxygenase (Rubisco) enzyme, which is the key molecule in the fixation of CO_2 to sugars. This enzyme has a unique position, because it is the primary rate-limiting step in C_3 Ps. Its rate is enhanced by an increase in the concentration of rate-limiting substrate (i.e., CO_2) (Pearcy and Bjorkman, 1983). Rubisco acts as either carboxylase or oxygenase, depending on the partial pressures of CO_2 and O_2 at the active site. Increased CO_2 will therefore result in increased net CO_2 uptake by inhibiting oxygenation reaction. Competition between the two gaseous substrates, CO_2 and O_2, for the same enzyme site determines the flow of carbon between the reductive and oxidative modes of the Ps carbon cycle. The C_3 cycle is largely dependent on Rubisco activity.

The Ps in C_3 plants is strongly limited by the current makeup of the atmosphere, which contains a high concentration of O_2 relative to CO_2. At this level, the O_2 competes with the CO_2 for the enzyme site of Rubisco; consequently, photorespiration occurs simultaneously with Ps. In a CO_2-enriched environment, the possibility of the interference of O_2 is minimal; hence, photorespiration is reduced, resulting in an increase in net Ps (Pearcy and Bjorkman, 1983). It is suggested that a 25% increase in CO_2 concentration over the past century has increased net Ps by decreasing the wasteful process of photorespiration (Polley et al., 1993; Lawlor and Mitchell, 1991).

Bowes (1991) demonstrated that enhanced atmospheric CO_2 substantially modifies the amount or activation state of Rubisco in some species, whereas no change in either parameter was observed in others. CO_2 activates Rubisco through binding at a site discrete from that involved in the enzymatic carboxylation (Edwards and Walker, 1983). Species fixing of CO_2 by C_4 metabolism show very little response of Ps to an increase in CO_2 concentration (Pearcy and Bjorkman, 1983). This response can be explained by presence of CO_2 concentration close to its saturation level for phosphoenolpyruvate (PEP) carboxylase at the carboxylation site. This occurs because of its ability to raise local CO_2 internally by means of CO_2 pumping and Kranz anatomy. The carboxylase /oxygenase reaction is also absent with PEP carboxylase. Mauney et al. (1978) studied four species grown in greenhouses, comparing 660 ppm CO_2 and 330 ppm CO_2. Ps increases by 41% for soybean, 15% for cotton, 7% for sunflower, and only 2% for sorghum. Although most experiments on crop species reported greater Ps at high CO_2 compared at ambient CO_2, reports are also available that indicate enhancement of Ps being transient in some species (Smith et al., 1987; Long and Drake, 1992).

Short-term enhancement of Ps by elevated CO_2 has been reported by Pearcy and Bjorkman (1983), Strain et al., (1984), and Cure and Acock (1986). There are concerns that Ps may downregulate at long-term exposure to CO_2 enrichment (Eamus and Jarvis, 1989). In the studies with trees in which rooting volume was not limited, downregulation of Ps was not evident (Curtis et al., 1994). In other words, they became acclimatized; however, some species showed little or no adjustment (Carlson and Bazzaz, 1980). The degree of the adjustment of species under the enriched CO_2 environment is also influenced by other variables. Various reports showed that, initially, CO_2 enrichment brought about steep increase in Ps, which later decreased with time; however, the increase was always greater than that measured in plants growing at ambient CO_2 concentration (Campbell et al., 1988; Cure et al., 1989). There are also reports that Ps measured at high CO_2 concentration was similar to or less than that of plants grown at ambient CO_2 concentration (Havelka, Ackerson, et al., 1984a; Havelka, Wittenbach et al., 1984). In many C_3 plants, a nonacclimation of photosynthetic efficiency to increased CO_2 concentration has been observed (Campbell et al., 1988; Huber et al., 1984). Radin et al. (1987) found a wide range of response to CO_2 enrichment in cotton: assimilation rate not saturated at all by internal CO_2 concentrations, no decrease in photosynthetic capacity for most of the season, and maintenance of very high internal CO_2 concentration. Short-term increase in CO_2 has been shown to stimulate Ps in trees (Long, 1991), but CO_2 for longer periods has been found to affect the quantum yield by an acclimation process downregulating the photosynthetic activity. Spring wheat at optimal nutrition for 6 weeks at 700 ppm CO_2 showed an increase in Ps due to increased capacity for C fixation, total electron transport and Rubisco activity, and a concomitant partial reduction of photorespiration (Habash et al., 1995).

Ps response to enhanced CO_2 levels varied in different experimental conditions. Data from greenhouse studies are very different from those obtained from the field, even for the same crop. Regardless of the species involved, the gains in Ps generally led to a greater demand for nutrients, such as nitrogen and phosphorus. If these nutrients are in short supply, plant growth will be limited (Sage, 1994). The response to CO_2 is more pronounced under high levels of other resources, especially water, nutrients, and light (Bazzaz, 1990). In a meta-analysis of leaf gas exchange of trees in response to CO_2, Curtis (1996) showed that downward acclimation of Ps is usually found in studies using small pots in the growth chamber. Downward acclimation of assimilation rate may be correlated with reduced availability of water and nutrients (Will and Teskey, 1997), decrease in carboxylation efficiency due to decrease in Rubisco amount and activity, suppression of sucrose synthesis by accumulation of starch (Huber et al., 1984), reduced sink strength (Arp, 1991) and hormonal imbalances (Sage, 1994).

The percentage increase in Ps is highest at tropical latitudes. Differential sensitivity of plants to CO_2 is often explained by differences in phloem loading (Korner et al., 1995). Under CO_2-enriched conditions, herbaceous species representative of apoplastic phloem loading have larger accumulation of total nonstructural carbohydrate than woody species following symplastic loading (Eamus, 2000). Herbaceous species show higher magnitude of increase in Ps than woody species in CO_2 enriched environments.

Respiration

Changes in atmospheric CO_2 will have a profound impact on respiration. An increase in CO_2 through enhanced assimilation was found to increase the mass of nonstructural carbohydrate fraction in the leaf (Ryle et al., 1992), which itself can lead to a rise in dark respiration rate. Although this has occasionally been observed, the much more common response is that respiration rate declines or remains the same at elevated CO_2 concentrations (Bunce, 1994). Bunce (1990) found that the rate of respiration at 350 ppm CO_2 concentration ranged from 1.01 to 1.54 times that measured at 700 ppm in a study with soybean, tomato, and amaranthus. Oxygen uptake is also reduced by high CO_2 concentrations (Gale and Berry, 1992).

Recently a direct, instantaneous, and reversible effect of CO_2 on dark respiration rate has been demonstrated (Amthor et al., 1992). Amthor (1991) suggested that decline in respiration may be related to a direct CO_2 effect at any point of the respiratory enzymatic chain. In cotton, dark respiration has been significantly increased by elevated CO_2 and correlated with leaf expansion rate (Thomas et al., 1993). In *Castanea sativa*, dark respiration was significantly lower at elevated CO_2 during the first half of the season on both a whole-plant or leaf-area basis, but this effect changed at the end of the season (Mousseau, 1993). It has been shown that the leaf construction cost (growth respiration component) of *Pinus taeda* needles is affected by elevated CO_2, particularly when nitrogen availability is low (Griffin et al., 1993). When time-dependent changes in the CO_2 stimulation of growth are viewed relative to those of respiratory capacity, dark respiration rates correlate with size of nonstructural carbohydrate fraction in source leaves (Farrar and Williams, 1991). The response of plants with respect to respiration in CO_2-enriched environments is often contradictory. The increases in biomass of C_3 species grown under CO_2-enriched environments have been ascribed to reduced respiration (Mott, 1990). Lower CO_2 efflux rate of plants grown under 660 μmol CO_2 mol^{-1} compared to ambient (360 μmol CO_2 mol^{-1}) at night was reported by Ziska and Teramura (1992b).

Stomatal Response and Water Use Efficiency

Several experiments have shown that the enriched CO_2 environment causes a change in stomatal density of some species. Woodward (1987) determined stomatal densities on herbarium-stored specimens (from the year 1750) of various deciduous trees, such as beech, oak, poplar. lime, and sycamore, from the Midlands of England. Leaves of reproductive shoots were studied with the assumption that they had developed in full irradiance. The stomatal density was found to have changed by as much as 40% over 200 years. The magnitude of effects was variable with respect to different species. Woodward and Bazzaz (1988) found a slight decrease in stomatal density of several species. The high CO_2 concentration reduced the stomatal aperture by reducing the stomatal conduction (Cs) (Bazzaz, 1990). Allen (1990) reviewed stomatal response to elevated levels of CO_2 and found significant decline in Cs and transpiration (E) in most of the plant species. Mansfield et al. (1990) found that increases in CO_2 concentration cause partial stomatal closure and consequent increase of water use efficiency (WUE). Mansfield (1991) further suggested that changes in CO_2 concentration since pre industrial times have increased the WUE of trees to the extent that changes in stomatal density have occurred as a secondary response. Woodward and Kelley (1995) reported an average reduction in stomatal density of 14.3% with CO_2 enrichment, on the basis of a survey of 100 species. Reduction in stomatal density was observed in 74% of the studied cases. In observations made in controlled environments under CO_2 enrichment of 350–700 m mol mol^{-1}, average stomatal density declined by 9%, and 60% of the cases showed reductions in stomatal density. Amphistomatous samples had greater changes in stomatal density than the hypostomatous samples.

The ratio of CO_2 uptake to water vapor defines the efficiency with which a plant uses water. Due to reduced E and enhanced Ps, increase in ambient CO_2 results in higher WUE in plants (Acock and Allen, 1985; Heath and Mansfield, 2000). Strong evidence suggests that stomata respond more to internal than to external CO_2 concentration (Mott, 1990).

The reduction in E will eventually lead to favorable, instantaneous WUE, improved plant water status, higher carbon gain and biomass accumulation, and a lower seasonal water consumption rate (Reekie and Bazzaz, 1989; Tuba et al., 1994). Both C_3 and C_4 species experience increase in WUE via reduced E, and hence increased growth under conditions of limited water (Morison, 1985).

It has been suggested that the main factor reducing Cs at elevated atmospheric CO_2 is not the reduction in stomatal density or pore area, but probably the reduction in the stomatal opening (Paoletti and Gelline, 1993). Kimball (1983) estimated that doubling of the present CO_2 concentration would result in a reduction of the E by 34%. Stomatal conductance was estimated to decline between 30 and 40% in C_3 species at a doubling in atmospheric CO_2 levels (Eamus and Jarvis, 1989; Cure and Acock, 1986). A number of coniferous species from northern temperate forests, however, showed stomatal insensitivity to rising CO_2 (reviewed by Eamus and Jarvis, 1989).

The mechanism involved with stomatal response to CO_2 is not well understood (Mansfield et al., 1990). It has been proposed that CO_2 may inhibit an electrogenic pump in the plasmalemma, which extrudes H^+ ions from the guard cells, or CO_2-induced malate formation may reduce the K^+ uptake, hence leading to stomatal closure. Internal CO_2 concentration (Ci) has been viewed as important in terms of stomatal response and changes in assimilation rate at increasing CO_2. It has been suggested that stomatal conductance and Ps respond to ambient CO_2 depending on the ratio of internal to ambient CO_2. Response of stomata to increasing CO_2 prevails till a subsaturation level of internal CO_2 is present, but above the saturation point, stomata close, limiting transpiration but not Ps. This is particularly true for C_4 species. whereas in C_3 species, partial stomatal closure decreases assimilation (Atkinson, 1996). In C_3 species, WUE doubled due to both increased assimilation rate and decreased Cs, whereas in C_4 species, a greater decline in Cs was responsible for increased WUE rather than a small increase in assimilation rate. As a general rule, all plants growing at higher levels of atmospheric CO_2 would be better able to cope with drought stress by ameliorating osmoregulation and maintenance of higher turgor pressure (Woodward, 1987). However, lower E will lead to higher leaf temperature under high irradiance and low wind speed conditions (Idso et al., 1987).

Primary Metabolites

The enhanced CO_2 not only increases the rate of CO_2 uptake, but also alters the partitioning of newly fixed carbon into end products. Ho (1977) reported 40–45% increase in rate of carbon transport per unit fresh leaf weight in tomato plants exposed to 1,000 ppm CO_2, even though there was increased carbon accumulation. In *Lolium*, Wardlaw (1982) reported an extra photosynthate accumulation rather than translocation. Long-term exposure to elevated levels of CO_2 results in an increase in starch content of the leaves. However, some workers observed strong negative correlation between leaf starch concentration and Ps (Harley et al., 1992). The excessive accumulation of starch may disrupt the chloroplast structure; consequently, Ps may decline. However Vu et al. (1989) reported conflicting results in soybean plants, where increase in chlorophyll content of the plants exposed to high CO_2 concentration after 39 days of exposure was observed. The starch synthesis increased tremendously, although the starch did not impose adverse effect on chloroplast because of increased numbers of palisade cells, which provided more storage sites for starch. This may be the reason behind the nonacclimation of Ps in some plants. The extent of accumulation of starch on CO_2 enrichment largely depends on the developmental stage. The higher the sink demands, the lesser will be the accumulation of starch and vice versa. A negative correlation between the activity of sucrose phosphate synthetase (SPS), a key enzyme in rate control of sucrose synthesis, and leaf starch accumulation has been reported (Huber et al., 1984). Partitioning of assimilate is generally higher between storage pools of chloroplast starch than transport pools of sucrose.

Starch accumulation and its dark mobilization are not only linked to carbon exchange rate but are also responsible for eventual gain of dry matter in various organs and, therefore for increased growth and biomass. Several workers have reported that plants grown at elevated levels of CO_2 may have more carbohydrates available for the support of mycorrhizal or nitrogen-fixing bacteria, thereby improving soil nutrient status (Thomas et al., 1993). Carbon dioxide enrichment has increased mycorrhizal colonization of roots in short leaf pine (Norby et al., 1986). Studies with nitrogen-fixing plants have reported significant increases in N_2 fixation due to CO_2 enrichment (Vogel and Curtis, 1995). This phenomenon would compensate for nitrogen deficiencies as a result of CO_2-induced increases in legume N_2 fixation.

CO_2 enrichment has been found to increase concentrations of sucrose, glucose, and fructose in tomato (Behboudian and Tod, 1995). A similar response was observed for wheat leaves at elevated levels of CO_2 (Smart et al., 1994). In contrast Dijkstra et al. (1999) found no change in soluble sugars of wheat grain at double ambient CO_2 concentration. Other metabolites such as lipids in wheat (Karbati et al., 1990; Blumenthal et al., 1996) have been shown to experience no effect as a result of air enriched with CO_2. No effect on oil or fatty acid composition of maize or soybean seed with a CO_2-enriched environment has been reported (Rogers et al., 1986). The activity of SPS is negatively correlated with starch accumulation (Huber et al., 1984). The timing of export also influences the distribution of carbohydrate under increasing CO_2 levels (Cure et al., 1991). During light, assimilate goes more to root growth than to leaves, which may induce more water and nutrient uptake under CO_2-enriched environments. In fact the information on influence of CO_2 on assimilate transport is scarce and often contradictory.

One of the important consequences of atmospheric CO_2 enrichment is decreased foliar nitrogen concentration (Luo et al., 1994; Biswas et al., 1996). Cortrufo et al. (1998) reported that 82% of the published reports on nitrogen showed reduction in plant nitrogen under rising CO_2 levels, a reduction largely attributed to increase in total nonstructural carbohydrate levels due to enhanced growth in CO_2-enriched environments (Huber et al., 1984). It is also suggested that the biomass dilution effect is not the only phenomenon responsible for nitrogen reduction (Conroy et al., 1990). Conroy and Hocking (1993) have ascribed the reduction of foliar nitrogen to reduced translocation of nitrate due to reduced transpiration.

Foliar protein was also reduced under conditions of enriched CO_2 (Arnone et al.,1995; Sage et al., 1989). In the case of actively growing plant organ studies conducted in field, reductions in protein and nitrogen concentrations were reported less (Zanetli et al., 1996; reviewed by Idso and Idso, 2001). This response is correlated with good root growth, which can obtain needed nutrients from soil. In a long-term field study on sour orange trees subjected to 700 ppm CO_2 nitrogen concentration was significantly less than that in plants at 400 ppm CO_2 after 3 years, but after 8 years, the nitrogen content became identical at both the CO_2 levels. Grain protein concentration in tropical rice did not change in response to a 200 ppm increase in ambient CO_2, but 300 ppm caused 8% reduction in the concentration (Ziska et al., 1997). For wheat and barley, however, CO_2 enrichment reduced nitrogen as well as protein concentration (Monje and Bugbee, 1998). Reduction in protein content is found to reduce the baking quality of flour. The grain protein concentration was found to be affected by soil nitrogen availability also under a CO_2-enriched environment (Pleijel et al., 1999). Manderscheid et al. (1995) found that under a CO_2-enriched condition, more protein was used for essential amino acids in wheat grains compared to those growing in ambient air.

Antioxidants

Antioxidants protect plants against oxidative stresses caused naturally and by numerous environmental stresses (Foyer et al., 1994). Antioxidant levels have mostly been found to be reduced under CO_2-enriched air. Activities of catalase and superoxide dismutase, responsible for detoxifying H_2O_2 generated during photorespiration, and superoxide radicals, respectively were found to decrease in enriched CO_2 environments (Niewiandamska et al., 1999; Schwanz and Polle, 1998). These findings clearly indicate lower levels of oxidative stress in CO_2-enriched environments, and hence, lower levels of resources are needed for anitoxidative defense in plants. Vitamin C, an antioxidant, was found to show an increase (Schwanz et al., 1996) or no effect (Kimball and Mitchell, 1981). Idso and Idso (2001) have shown an increase in vitamin C of sour orange at 75% increase in the CO_2 content. It was further reported that in the years of high fruit production, the increase in vitamin C was of greater magnitude than that at low fruit production. Because plants produce more nicotinamide adenine dinucleotide phosphate (NADPH) under elevated levels of CO_2, it helps in maintenance of higher activities of glutathiane reductase, responsible for high redox states of both glutathione and ascorbate (Foyer et al., 1994; Rao et al., 1995). CO_2 enrichment has been found to increase numerous compounds having chemotherapeutic properties (Idso and Idso, 2001).

Secondary Metabolites

Given that atmospheric CO_2 enrichment usually promotes Ps rates, it may lead to manufacture of more carbon-based secondary compounds or phenolics (Penuelas et. al., 1996). Many studies have confirmed it, whereas others found either no change or a decrease in phenolics upon CO_2 enrichnient (reviewed by Idso and Idso, 2001). Woody perennials release considerable quantities of hydrocarbon into the atmosphere. The isoprene production in *Quercus rubra* doubled in response to a 26% increase in CO_2, whereas in *Eucalyptus globules, Quercus alba*, and *Mucuna pruriens*, the rate decreased (Sharkey et al., 1991). Penuelas and Estiarte (1998) found an increase in phenolics, particularly tannins and terpens, at elevated CO_2. Many of the secondary metabolites help plants survive attacks from herbivores and plant pathogens, and may act as allelopathic substance. So the changes in phenolics have an impact on palatability and digestibility for herbivores, decomposition of litter by microbes (Idso and Idso, 2001) and plant defenses against diseases (Malmstrom and Field, 1997).

Growth and Yield

The enriched CO_2 environment plays an important role in different stages of growth, as well as the economic yield, of agricultural plants. The growth responses include increase in leaf size (Riechers and Strain, 1988), leaf area index, leaf weight per unit area, and thickness; number of mesophyll cell layers (Acock and Allen, 1985), and branches and nodes (Allen, 1990); greater stem elongation; increased root: shoot ratios; altered time of flowering, and so on. The responses to enriched CO_2 may vary in different plant parts. Cure et al. (1989) found that in soybean plants, the emergence and expansion rates of leaves on main stem were increased, but leaf area was unaffected or showed very little response when exposed to high CO_2 concentrations. Leaf development on branches accounted for more than 40% of the total increase in leaf area.

The enhanced CO_2 level can also alter the flowering periods (Andersson, 1991); however, the response differs from species to species (Garbutt and Bazzaz, 1984). Corn, sunflower, and cotton plants showed slight delay in flowering on exposure to 1,000 ppm CO_2, whereas sorghum showed great delay in flower initiation (Hesketh and Hellmers, 1973). Paez et al. (1980) found hastening of flowering and bud initiation in Alaska pea. Hicklenton and Jolliffe (1980) observed early flowering in *Pharbitis* under noninductive, long-day condition at higher CO_2 concentration, whereas under short-day conditions delayed flowering was observed. Induction and formation of new secondary branches are reported due to extra carbon assimilation, leading to increase in biomass accumulation (Idso et al., 1991). At the end of the growing season, leaf abscission or senescence has been shown to be advanced due to elevated CO_2 levels in comparison to trees grown in ambient CO_2 (Evans and Lee, 1993). The leaf area phenology seems to be affected by modifying the nutritional status. The elevated CO_2 may influence different reproductive parameters, including seed number and size, and their nutrient content. Bazzaz (1990) reported an increase in fruit weight of *Datura stramonium* grown at elevated CO_2 concentrations. Plants grown in a CO_2-enriched environment produced thicker fruit walls, but seed size remained unaffected (Garbutt and Bazzaz, 1984). In the case of *Abutilon theophrasti*, flower, seed, and capsule numbers decreased, resulting in a decrease in total seed production. However, individual seed weight was higher in plants grown at elevated CO_2 concentrations.

Growth biomass, and yield for C_3 species are increased under CO_2-enriched environments (Krupa and Kickert, 1989; Mulchi et al., 1995; Deepak and Agrawal, 1999, 2001). It was estimated that a doubling of atmospheric CO_2 concentration from 340 to 680 ppm would increase the growth and yield of the major crops by 10–50% (Warrick et al., 1989). Kimball (1983), after analyzing 430 prior observations concluded that the mean yield response was 33% more with a doubling of atmospheric CO_2 concentration. An increase of about 14% in average yield with doubling of CO_2 in C_4 species was predicted (Kimball, 1986). Acock and Allen (1985) summarized their work and suggested an increase of yield up to 32%, on average, when plants were exposed to 660 ppm CO_2 in favorable conditions. Mauney et al. (1978) found an increase of 382,110 and 60% in dry weight, respectively, for soybean, cotton, and sunflower plants grown to maturity at 660 ppm CO_2, whereas sorghum did not show a statistically significant increase compared to plants grown at 330 ppm CO_2.

Idso et al. (1987) found an increase in biomass and yield by 82% and 87%, respectively, in cotton plants subjected to the doubling of ambient CO_2 concentration in open-top chambers. Rogers et al. (1986) observed increases in dry matter accumulation and seed yield by 22% and 7% respectively, in soybean plants at enhanced CO_2 levels. In the case of maize, at the doubled ambient CO_2 concentration, increases of 49% and 55%, respectively, in total dry matter and yield were reported (Rogers et al., 1983). Havelka, Wittenbach, et al. (1984) observed an increase of 20% in dry matter and 17% in grain yield of wheat plants on exposure to 1,200 ppm CO_2. Exposure of plants to 600 ppm CO_2 from germination to maturity led to yield increments of 30% and 32%, respectively, in HP1209 and M234 cultivars of wheat (Deepak and Agrawal, 1999) and 34% and 37%, respectively, in Bragg and PK472 cultivars of soybean (Deepak and Agrawal, 2001) compared to those grown at ambient CO_2 levels.

Wheat plants treated at 340, 485, 660 and 825 μmol CO_2 mol^{-1} at high and low water regimens showed 44% and 74% increases in grain yield, respectively, from lowest to highest CO_2 level (Chaudhuri et al., 1990). The significant increase in grain yield was mainly due to increased numbers of productive tillers bearing the ears (Hocking and Meyer, 1991; Deepak and Agrawal, 1999). Number of spikelets per head and weight of 1,000 seeds remained unchanged under increased CO_2 levels (Deepak and Agrawal, 1999). At 500 μmol CO_2 mol^{-1}, significant increases in grain yield, straw biomass, harvest index, and seed weight 1000^{-1} in wheat were reported compared to those grown at 350 μmol CO_2 mol^{-1} (Mulchi et al., 1995).

Increases of 9% and 29%, respectively, in dry matter and yield were reported for open-top-chamber-grown corn at double the CO_2 concentration. Cure and Acock (1986) calculated a 28% increase in biomass accumulation of C_3 crops on the basis of data on the responses of 10 important crops to doubling of CO_2. Biomass response to CO_2 doubling was low in C_4 species. In most cases, the increase in yield was not due to the increase in weight or size of the individual seed, but mainly due to the increase in number of structures. The increase in soybean yield at elevated CO_2 concentrations was also attributed to the production of more seeds containing pods rather than any increase in seed size itself (Acker et al., 1984).

Intraspecific variations in response to increased CO_2 were reported in 17 cultivars of rice with respect to growth and yield (Ziska et al., 1996). Krupa and Kickert (1989) reviewed the available literature on growth and yield responses of vegetation to elevated CO_2 and prepared the sensitivity ranking in relation to elevated CO_2 environments. The impact of CO_2 enrichment on yield is also dependent on the stage of growth at which it is applied. The preanthesis treatment of enhanced CO_2 increased grain number and weight, whereas postanthesis treatment increased percentage of filled grains (Cock and Yoshida, 1973). Similarly, Havelka, Wittenbach, et al. (1984) observed maximum yield response when the plants were exposed to 1,200 ppm CO_2 before anthesis. Ackerson et al. (1984) found that in soybean, CO_2 enrichment from early pod development until maturity, and from emergence to maturity, could increase yield by 27% and 36%, respectively.

The growth response of species also depends upon the growth form (i.e., indeterminate or determinate). Indeterminate species show larger increases in dry matter than do determinate species (Chaves and Pereira, 1992). Gardner et al. (1995) found that growth response of various poplar clones to elevated CO_2 was a reflection of their indeterminate growth habit. In a study with nine conifer species. Mortensen (1994) found that after one growth season with exposure at elevated CO_2, growth and total dry weight were enhanced for *Larix leptolepis, L. sibirica Picea glauca, P. sitchensis*, and *Psendotsuga menziesii*, whereas *Abies, Pinus sylvestris, Pinus contorta*, and *Pinus mugo* were not affected by elevated CO_2 levels.

Among various plant parts, roots often receive the largest enhancement when grown at elevated CO_2 (Batts et al., 1998; Rogers et al., 1994; reviewed by Pritchard and Rogers, 2000). Rattray et al. (1995) have reported that the proportion of ^{14}C translocation belowground almost doubled from 18% at ambient, to 34% under elevated CO_2. Increase in fine root biomass of free-air carbon dioxide enrichement (FACE)-grown wheat was reported (Wechsung et al., 1995). Increased root carbohydrates stimulate root growth by favoring lateral roots more than primary roots, leading to highly branched, shallower root systems. Root growth also depends upon the phenological stage of CO_2 exposure (Pritchard and Rogers, 2000). In woody perennials, also, elevated CO_2 increased the proportion of allocation to root and altered the root architecture (Mousseau and Sangier, 1992), including increase in fine root density (Fitter et al., 1997). This will have a beneficial impact on plant ability to acquire nutrients and water.

Relative growth rate (RGR) was found to be stimulated by an elevated CO_2 concentration, but the stimulation was transient (Poorter, 1993; Centritto et al., 1999 Marriott et al., 2001). The stimulation of RGR occurs primarily in the early stages of exposure. The decrease in RGR is linked with downregulation of Ps capacity (Sage, 1994), a reduction in net assimilation rate (NAR) (Poorter, 1993), and decreased leaf area ratio and specific leaf area (Poorter, 1993; Marriott et al., 2001).

Reproduction

Given that reproduction acts as a strong carbon sink, its response to elevated CO_2 may be an important determinant of species fitness. Increase in vegetative biomass is not always translated into reproductive output (Farnsworth and Bazzaz, 1995). Several studies have shown a direct relationship between increased dry matter production and seed yield (Reeves et al., 1994; Ziska et al., 1998; Deepak and Agrawal, 1999, 2001). CO_2 enrichment has been found to affect phenological development (Farnsworth and Bazzaz, 1995), total reproductive output (Reeves et al., 1994) seed number (Farnsworth and Bazzaz, 1995; Deepak and Agrawal, 1999, 2001), and seed quality (Bazzaz et al., 1992). Leishman et al. (1999) have conducted experiment with four annual C_3 plant species under ambient and ambient + 200 ppm CO_2 from seed germination to senescence. The plant species showed differential response to vegetative growth, phenology, and reproductive output. Elevated CO_2 had no effect on vegetative growth of *Cardamine hirsuta, Spergula arvensis*, or *Poa annua*. *Senecio vulgaris* produced larger leaves and greater biomass, but this increment was not translated into increased seed output, although seed mass was significantly increased. This clearly implies that reproductive response in not a simple translation of vegetative response. *S. vulgaris* and *P. annua* had faster maturation times.

Interaction of CO_2 and Other Environmental Factors

As plants are exposed not only to CO_2 but also to various environmental factors, it therefore becomes important to consider other factors in combination with enhanced levels of CO_2. The response of plants to elevated levels of CO_2 is dependent upon light levels (Pearcy and Bjorkman, 1983; Bazzaz, 1990) atmospheric temperature (Rawson, 1992; Rowland-Bamford, 2000; Ziska et al., 1996), and nutrient availability (Acock and Allen, 1985; Kimball, 1986; Conroy, 1992).

Interaction of CO_2 with temperature has also been found to show stimulation of biomass (Baker et al., 1992), decrease in yield (Wheeler et al., 1996), and no change (Ziska et al., 1997). It is suggested that up to a particular temperature, any increase can enhance the response to CO_2 enrichment, but any further temperature rise causes large negative effect (Rowland-Bamford, 2000).

Several researchers are of the opinion that the enhanced effect of elevated CO_2 disappears under nitrogen (N) and phosphorus (P) limitations (Conroy, 1992). The reduction in E at higher CO_2 concentration leads to the reduction of mass flow of nitrate, which ultimately reduces the N concentration in the foliage, roots, and grain of C_3 plants (Norby et al., 1986). The reduction in N content leads to reduction in protein content. An increase in N use efficiency was reported in *Quercus alba* in a nitrogen-limiting system (Cure et al., 1988). The decrease in leaf N can be linked to the redistribution of N, due to the reduced requirement for Rubisco protein. An accumulation of starch was found to be associated with reduction in Rubisco activity (Wong, 1990). The C : N ratio becomes higher in the case of elevated CO_2. The higher C : N ratio suggests that the leaves decompose more slowly, and that release of nutrients for uptake by plants may be delayed. Thus, elevated levels of CO_2 may affect nutrient cycling also. Changes in C : N ratio can affect the feeding of herbivores (Johnson and Lincoln, 1991). Evidence suggests that larger quantities of mineral nutrients are required for plants in order to take advantage of high CO_2 concentration (Kimball, 1986). Photosynthetic response of plants to increased CO_2 concentration also decreases under low P availability in soil (Conroy et al., 1988).

The elevated CO_2 may modify the effects of stress factors on plant growth. High CO_2 concentrations have been shown to alleviate water stress effects in several species (Allen, 1990). Higher CO_2 levels have been shown to ameliorate effects of high salinity (Bowman and Strain, 1987) by supplying extra energy for the maintenance of respiration, and by reducing the entry of salt into the plant due to reduced transpirational pull (Gale and Zeroni, 1985). Because C_3 plants responded more to CO_2 enrichment than do C_4 plants, C_4 crops will face more C_3 weed problems.

CO_2 enrichment can delay the onset of the effects of drought and the relative impact of CO_2 enrichment can be greater in drought than in well-watered conditions (Johnsen, 1992). CO_2 enrichment has the potential to increase the availability of carbon for solute accumulation induced by drought, and therefore ameliorate the impact (Morse et al., 1993). Tschaplinski et al. (1995) have, however, reported no amelioration of negative impact of drought on *Pinus taeda* growth.

Interactive Effects of Elevated CO_2 and Air Pollutants

Air pollutants are found in combination rather than alone. Mixtures of air pollutants and CO_2 compared to their individual effects may influence plants differently. The concentration of air pollutants, duration of exposure, species, and other factors may influence the interactive effects of different pollutants and CO_2 on plants. Carbon dioxide is one of the most important factors that can modify the sensitivity of plants to air pollutants. It plays a protective role in reducing the damage caused due to air pollutants in most of the cases (Allen, 1990; Mulchi et al., 1992; Deepak and Agrawal, 1999, 2001). Elevated CO_2 at 500 ppm concentration reduced leaf damage due to O_3 by 66% in tobacco plants and 14% in *Phaseolus vulgaris* L. (Heck and Dunning, 1967). Doubling of ambient CO_2 concentration reduced leaf injury due to SO_2 by 60% in alfalfa (Hou et al., 1977). It is generally believed that high CO_2-levels induced partial stomatal closure reduced the impact of air pollutants by regulating their fluxes (Allen, 1990).

Growth under a combination of high CO_2 and ultraviolet-B radiation (UV-B) showed that CO_2 not only counteracted the negative effects of UV-B, but the influence of high CO_2

in rice plants was also reduced due to UV-B (Ziska and Teramura, 1992a). Rao and Dekok (1994) and Sandhu et al. (1992) have reported the nullification of adverse SO_2 effects in plants grown under a combination of high CO_2 and SO_2. Increasing CO_2 concentration and concomitant stomatal closure of C_3 plants appear to reduce the sensitivity of these plants to SO_2 (Carlson and Bazzaz, 1982). However, the sensitivity of C_4 plants to SO_2 has been shown to increase at increasing levels of CO_2, but the reason behind this is not clear. Black (1982) reported that gradual increase in CO_2 concentration from 300 ppm to 430, 480, 580, and 600 ppm decreased the photosynthetic inhibition of plants exposed to 0.25 μmol mol^{-1} SO_2 in a progressive manner. High CO_2 levels might also limit SO_2 penetration into the chloroplast (Miszalski and Mydlarz, 1990). Deepak and Agrawal (2001) reported that elevated level of CO_2 ameliorated the toxic effect of SO_2 on two cultivars of soybean by increasing the photoassimilation availability and reducing the SO_2 flux. The greater photoassimilation availability also enabled the plants to better utilize the SO_2 as nutrient source. A similar response was also reported for two cultivars of wheat (Deepak and Agrawal, 1999), where combined treatment of CO_2 and SO_2 had maximum favorable effects on biomass and yield of plants.

McKee et al. (1997) found that stomatal exclusion is an important factor in reducing the detrimental effects of O_3 by CO_2 concentration. They found 25% reduction in effective O_3 dose due to CO_2 elevation, which gave 50–60% protection against the O_3 effect on Rubisco activity, and complete protection against O_3 effects on photoassimilation and biomass accumulation. It has also been suggested that elevated CO_2 may increase thickness of leaf and cell walls (Jin Xing and Yu-Xi, 1996), which might raise the threshold against damage due to air pollutants (McKee et al., 1997).

Recent studies on the impact of high CO_2 and air pollutant levels have suggested that the elimination of air pollutant toxicity at high CO_2 levels is largely due to the stimulation of internal detoxification mechanisms rather than reduced pollutant uptake (Mulchi et al., 1992). Elevated CO_2 has also been found to increase the ability of plants to tolerate O_3 toxicity (Rao et al., 1995). It has also been suggested that exposure to O_3 might reduce the positive influence of high CO_2. However, Rao et al. (1995) did not observe any adverse effect of O_3 on the CO_2-stimulated growth of wheat plants when exposed to 800 ppm CO_2 and 120 pbb O_3 concentrations in combination for prolonged period. Experiments with combined exposure of plants to elevated CO_2 and O_3 resulted in additive effects (Heagle et al., 1993). Tausz et al. (1996) observed no significant interactive effects when Norway spruce (*Pinus hybrid*, L.) plants were exposed to 800 ppm CO_2 and 0.06 ppm SO_2 in combination.

Elevated CO_2 has the potential to lessen both superoxide formation and photorespiration by competing with oxygen at the Rubisco active site, and increasing the overall use of reducing equivalents (Long, 1991). Elevated CO_2 may also increase the antioxidant enzyme activities inside cells and could thus provide a protective mechanism against oxidative stress (Rao et al., 1995). However, these mechanisms are still debatable. Rudorff et al. (1996a, 1996b) exposed wheat plants to 350 and 500 μmol CO_2 mol^{-1}, and 20 and 60 μmol O_3 mol^{-1}, singly and in combination in open-top chamber, and found that damaging effect of O_3 was counteracted; Ps, yield, and biomass accumulation were equivalent to controls. Barnes et al. (1995), however, found a very low level of protection against O_3 damage in five wheat cultivars at elevated CO_2 concentration. Higher levels of CO_2 increase the pools of carbohydrate for the repair of tissue damaged by pollutants and for antioxidants to detoxify the O_3 products in cell (Rudorff et al., 2000).

Conclusions

The response of plants to changes in physiological, metabolical, growth, and production parameters under elevated CO_2 is highly variable. Atmospheric CO_2 enrichment significantly enhances the growth and development of all plants. Indeterminate species show larger increases in dry matter than determinant species. Increase in vegetative biomass under CO_2-enriched environment is not always translated into reproductive output. Concentrations of plant constituents such as starch, sugar and phenolics increase, and protein, nitrogen, and many antioxidants often decrease at elevated CO_2 levels. Physiological processes such as Ps and WUE mostly increase, whereas respiration, photorespiration, stomatal conductance, and transpiration often decline in response to rising CO_2. Phenology is also affected by CO_2 enrichment. Most of this variability might be attributed to differences in experimental protocols, but more recent evidence based on FACE open-top chambers or experiments suggest that responses to high CO_2 levels are also species-specific. A number of environmental factors regulate the response of plants to CO_2 enrichment. A more comprehensive approach is required for a better understanding of these factors in modifying CO_2 response patterns. It is suggested that future studies of plant response should integrate root nutrient dynamics to predict long-term plant responses to nutrient acquisition.

Differential response patterns of plants to CO_2 also raise the possibility that elevated CO_2 might alter community composition by shifting the competitive interaction affecting competitive hierarchies and changes in abundance of species. Long-term ecological and genetic response of vegetation to rising CO_2 levels should be a major thrust of future research to predict responses of natural communities to global change.

Acknowledgment

The author thankfully acknowledges the financial assistance from Ministry of Environment and Forest, Government of India, New Delhi.

References

Ackerson, R. C., Havelka, U. D., and Boyle, M. G. (1984). CO_2 enrichment effects on soybean physiology. II. Effects of stage specific SO_2 exposure. *Crop Science* 24: 1150–1154.

Acock, B., and Allen, L. H. J. (1985). Crop responses to elevated carbon dioxide concentrations. In: B. R. Strain and J. D. Cure (Eds.), *Direct effects of increasing carbon dioxide on vegetation* (pp. 53–98). Washington, DC: U.S. Department of Energy.

Allen, L. H., Jr. (1990). Plant responses to rising carbon dioxide and potential interactions with air pollutants. *Journal of Environmental Quality* 19: 15–34.

Amthor, J. S. (1991). Respiration in a future, higher-CO_2 world. *Plant, Cell and Environment*, 14: 13–20.

Amthor, J. S., Koch, G. W., and Bloom, A. J. (1992). CO_2 inhibits respiration in leaves of *Rumex crispusi* L. *Plant Physiology* 71: 574–581.

Andersson, N. E. (1991). The influence of constant and diurnally changing CO_2 concentrations on plant growth and development. *Journal of Horticultural Science* 66: 569–574.

Arnone, J. A., III, Zaller, J. G., Ziegler, C., Zandt, H., and Korner, C. (1995). Leaf analytic and insect herbivore in model tropical plant communities after long term exposure to elevated CO_2. *Oecologia* 104: 72–78.

Arp, W. J. (1991). Effect of source-sink relations on photosynthetic acclimation to elevated CO_2. *Plant Cell and Environment* 14: 869–875.

Atkinson, C. J. (1996). Global changes in atmospheric carbon dioxide: The influence on terrestrial vegetation. In: M. Yunus and M. Iqbal (Eds.), *Plant response to air pollution* (pp. 99–134). Chichester, UK: Wiley.

Atkinson, C. J. (2000). Effects of climate change on the behavior of woody perennials. In: S. B. Agrawal and M. Agrawal (Eds.), *Environmental pollution and plant responses* (pp. 19–32). Boca Raton, FL: Lewis.

Badiani, M., Raschi, A., Paolacci, A. R., and Miglietta, F. (2000). Plant responses to elevated CO_2: A perspective from natural CO_2 springs. In: S. B. Agrawal and M. Agrawal (Eds.), *Environmental pollution and plant responses* (pp. 45–81). Boca Raton, FL: Lewis.

Baes, C. F., and Mulholland, P. J. (1985). Uptake of carbon dioxide by the ocean. In: J. R. Trabalka (Ed.), *Atmospheric carbon dioxide and the global cycle* DOE/ER 0239, Carbon Dioxide Research Division. Washington, DC: U.S. Department of Energy.

Baker, J. T., Allen, L. H., and Boote, K. J. (1992). Response of rice to carbon dioxide and temperature. *Agricultural Forest Meteorology* 60: 153–166.

Barnes, J. D., Ollerenshaw, J. N., and Whiffield, C. P. (1995). Effect of elevated CO_2 and/or O_3 on growth development and physiology of wheat (*Triticum aestivum* L.). *Global Change Biology* 1: 129–142.

Batts, G. R., Ellis, R. H., Morison, J. I. L., Nkema, P. N., Gregory, P. B., and Hadlex, P. (1998). Yield and partitioning of crops of contrasting cultivars of wheat in response to CO_2 temperature in field standees using temp graduate tunnels. *Journal of Agricultural Science* 130: 17–27.

Bazzaz, F. A. (1990). The response on natural ecosystems to the rising global CO_2 levels. *Annual Review of Ecology and Systematics* 21: 167–196.

Bazzaz, F. A., Ackerly, D. D., Woodward, F. I., and Rochefort, L. (1992). Carbon dioxide enrichment and dependence of reproduction on density in an annual plant and a simulation of its population dynamics. *Journal of Ecology* 80: 643–651.

Behboudian, M. H., and Tod, C. (1995). Postharvest attributes of virosa tomato fruit produced in an enriched carbon dioxide environment. *Horticulture Science* 30: 490–491.

Biswas, P. K., Hileman, D. R., Gnosh, P. P., Battacharya, N. C., and McCrimmon, J. N. (1996). Growth and yield responses of field grown sweet potato to elevated carbon dioxide. *Crop Science* 36: 1234–1239.

Black, V. J. (1982). Effects of sulfur dioxide on physiological processes in plants. In: M. H. Unsworth and D. P. Ormrod (Eds.), *Effects of gaseous pollutants in agriculture* (pp. 67–91). London: Butterworths.

Blumenthal, C., Rawson, H. M., Mckenzie, E., Gras, P. W., Barlow, E. W. R., and Wrigley, C. W. (1996). Changes in wheat grain quality due to doubling the level of atmospheric CO_2. *Cereal Chemistry* 73: 762–766.

Bowes, G. (1991). Growth elevated CO_2: Photosynthesis responses mediated through Rubisco. *Plant, Cell and Environment* 14: 795–806.

Bowman, W. D., and Strain, B. R. (1987). Interaction between CO_2 enrichment and salinity stress in the C_4 non-halophyte *Andropogon glometatus. Plant Cell and Environment* 10: 267–270.

Bunce, J. A. (1990). Short and long term inhibition of respiratory carbon dioxide efflux by elevated carbon dioxide. *Annals of Botany* 65: 637–642.

Bunce, J. A. (1994). Responses of respiration to increasing atmospheric carbon dioxide concentrations. *Physiologia Plantarum* 90: 427–430.

Campbell, W. J., Allen, L. H., Jr., and Bowes, G. (1988). Effects of CO_2 concentration on Rubisco activity, amount and photosynthesis in soybean leaves. *Plant Physiology* 88: 1310–1316.

Carlson, R. W., and Bazzaz, F. A. (1980). The effects of elevated CO_2 concentrations on growth, photosynthesis, transpiration and water use efficiency of plants. In: J. J. Singh and A. Deepak (Eds.), *Environmental and climatic impact of coal utilization* (pp. 609–623). New York: Academic Press.

Carlson, R. W., and Bazzaz, F. A. (1982). Photosynthetic and growth responses to fumigation with SO_2 at elevated CO_2 for C_3 and C_4 plants. *Oecologia* 54: 50–54.

Centritto, M., Lee, H. S. J., and Jarvis, P. G. (1999). Increased growth in elevated (CO_2): An early short term response? *Global Change Biology* 5: 623–633.

Ceulemans, R., and Mousseau, M. (1994). Effects of elevated atmospheric CO_2 on woody plants. *New Phytologist* 217: 425–446.

Chaudhuri, U. N., Kirkham, M. B., and Kanmasu, E. T. (1990). Root growth of winter wheat under elevated carbon dioxide and drought. *Crop Science* 30: 853–857.

Chaves, M. M., and Periera, J. S. (1992). Water stress, CO_2 and climate change. *Journal of Experimental Botany* 43: 1131–1139.

Cock, J. H., and Yoshida, S. (1973). Changing sink and source relations in rice (*Oryza sativa* L.) using carbon dioxide enrichment in the field. *Soil Science and Plant Nutrition* 19: 229–234.

Conroy, J. P. (1992). Influence of elevated atmospheric CO_2 concentrations on plant nutrition. *Australian Journal of Botany* 40: 445–456.

Conroy, J. P., and Hocking, P. (1993). Nitrogen nutrition of C_3 plants at elevated atmospheric CO_2 concentrations. *Physiologia Plantarum* 51: 171–174.

Conroy, J. P., Kuppers, M., Kuppers, B., Virgona, J., and Barlow, E. W. (1988). The influence of CO_2 enrichment, phosphorous deficiency and water use of *Pinus radiata*. *Plant, Cell and Environment* 11: 91–98.

Conroy, J. P., Melham, P. J., and Barlow, E. W. R. (1990). Increases in phosphorus requirement for CO_2-enriched pine species. *Plant Physiology* 92: 977–982.

Conway, T., J., Tans, P. P., and Waterman, L. S. (1994). Atmospheric CO_2 records from sites in the NOAA/CMDL air sampling network. In: T. A. Boden, D. P. Kaiser, R. J. Sepanski, and F. W. Stoss (Eds.), *Trends 93: A compendium of data on global change* (pp. 41–119). ORNL/CDIAC-65. Oak Ridge, TN: Carbon Dioxide Information Analysis Center, Oak Ridge National Laboratory.

Cortrufo, M. F., Ineson, P., and Scott, A. (1998). Elevated CO_2 reduces the nitrogen concentration of plant tissues. *Global Change Biology* 4: 43–54.

Cure, J. D., and Acock, B. (1986). Crop responses to carbon dioxide doubling: A literature survey. *Agriculture and Forest Meteorology* 30: 127–145.

Cure, J. D., Israel, D. W., and Rufty, T. W., Jr. (1988). Nitrogen stress effects on growth and seed yield of non nodulated soybean exposed to elevated carbon dioxide. *Crop Science* 28: 671–677.

Cure, J. D., Rufty, T. W., Jr., and Israel, D. W. (1989). Alteration in soybean leaf development and photosynthesis in a CO_2 enriched atmosphere. *Botanical Gazette* 150: 337–345.

Cure, J. D., Rufty, T. W., Jr., and Israel, D. W. (1991). Assimilate relations in source and sink leaves during acclimation to a CO_2 enriched atmosphere. *Physiologia Plantarum* 83: 687–692.

Curtis, P. S. (1996). A meta-analysis of leaf gas exchange and nitrogen in trees grown under elevated carbon dioxide. *Plant, Cell and Environment* 19: 127–137.

Curtis, P. S., O'Neill, E. G., Teeri, J. A., Jak, D. R., and Pregitzer, K. S. (1994). Below ground responses to rising atmospheric CO_2: Implications for plants, soil biota, and ecosystem processes. *Plant and Soil* 165: 1–6.

Deepak, S. S., and Agrawal, M. (1999). Growth and yield responses of wheat plants to elevated CO_2 and SO_2, singly and in combination. *Environmental Pollution* 104: 411–419.

Deepak, S. S., and Agrawal, M. (2001). Influence of elevated CO_2 on the sensitivity of two cultivars to sulphur dioxide. *Environmental Experimental Botany* 46: 81–91.

Dijkstra, P., Schapendonk, A. H. C. M., Groenwold, K. O., Jansen, M., and Vande Geijn, S. C. (1999). Seasonal changes in the response of winter wheat to elevated atmospheric CO_2 concentration grown in open-top chambers and field tracking chambers. *Global Change Biology* 5: 563–576.

Eamus, D. (2000). Some tree responses to CO_2 enrichment. In: S. N. Singh (Ed.), *Trace gas emissions and plants* (pp. 75–92). London: Kluwer/Academic.

Eamus, D., and Jarvis, P. G. (1989). The direct effects on increases in global atmospheric CO_2 concentration on natural and commercial temperate trees and forests. *Advances in Ecology Research* 19: 1–55.

Edwards, G. E., and Walker, D. A. (1983). *C_3-C_4: Mechanism, and cellular and environmental regulation of photosynthesis*. Oxford, UK: Blackwell Scientific.

Evans, L., and Lee, H. S. J. (1993). The effect of elevated CO_2 on the long term growth of birch. *Journal of Experimental Botany* 44: 12–18.

Farnsworth, E. J., and Bazzaz, F. A. (1995). Inter and inter-generic differences in growth, reproduction and fitness of nine herbaceous annual species grown in elevated CO_2 environment. *Oecologia* 104: 454–466.

Farrar, J. F., and Williams, M. L. (1991). The effects of increased atmospheric carbon dioxide and temperature on carbon partitioning, source–sink relations and respiration. *Plant, Cell and Environment* 14: 819–830.

Fitter, A. H., Graves, J. D., Wolfenden, J., Self, G. K., Brown, T. K., Bofie, D., and Mansfield, T. A. (1997). Root production and turnover and carbon budgets of two contrasting grasslands under ambient and elevated atmospheric carbon dioxide concentrations. *New Phytologist* 137: 247–255.

Foyer, C. H., Lelandaia, M., and Kunert, K. J. (1994). Photo oxidative stress in plants. *Physiologia Plantarum* 92: 696–717.

Gale, J., and Berry, J. (1992). The relation between net photosynthesis and dark respiration: A criterion for genetic and agrotechnical manipulation of basic plant productivity. *BARD Report* I: 1344–1387.

Gale, J., and Zeroni, M. (1985). Cultivation of plants in brackish water in controlled environment agriculture. In: R. C. Staples and G. Toenniessen (Eds.), *Salinity tolerance in plants: Strategies for crop improvement* (pp. 363–380). New York: Wiley Interscience.

Garbutt, K., and Bazzaz, F. A. (1984). The effects of elevated CO_2 on plants: III. Flower, fruit and seed production and abortion. *New Phytologist* 98: 433–446.

Gardner, S. D., Taylor, G., and Bosac, C. (1995). Leaf growth of hybrid poplar following exposure to elevated CO_2. *New Phytologist* 131: 81–89.

Griffin, K. L., Thomas, R. B., and Strain, B. R. (1993). Effects of nitrogen supply and elevated carbon dioxide on construction cost in leaves of *Pinus taeda* (L.) seedlings. *Oecologia* 95: 575–580.

Habash, D. J., Paul, M. J., Parry, M. A. J., Keys, A. J., and Lawlor, W. (1995). Increased capacity for photosynthesis in wheat grown at elevated CO_2: The relationship between electron transport and carbon metabolism. *Planta* 197: 482–489.

Harley, P. C., Thomas, R. B., Reynolds, J. F., and Strain, B. R. (1992). Modeling photosynthesis of cotton grown in elevated CO_2, *Plant, Cell and Environment* 15: 271–282.

Havelka, U. D., Ackerson, R. C., Boyle, M. G., and Wittenbach, V. A. (1984). CO_2 enrichment effects on soybean physiology: I. Effects on long-term CO_2 exposure. *Crop Science* 24: 1146–1150.

Havelka, U. D., Wittenbach, V. A., and Boyle, M. G. (1984). CO_2 enrichment effects on wheat yield and physiology. *Crop Science* 24: 1163–1168.

Heagle, A. S., Miller, J. E., Sherrill, D. E., and Rawling, J. O. (1993). Efffects of ozone and carbon dioxide mixtures on two clones of white clover. *New Phytologist* 123: 751–762.

Heath, J., and Mansfield, T. A. (2000). CO_2 enrichment of the atmosphere and the water economy of plants. In: S. B. Agrawal and M. Agrawal (Eds.), *Environmental pollution and plant responses* (pp. 33–43). Boca Raton, FL: Lewis.

Heck, W. W., and Dunning, J. A. (1967). The effects of ozone on tobacco and pinto bean as conditioned by several ecological factors. *Journal of Air Pollution Control Association* 17: 112–114.

Hesketh, J. D., and Hellmers, H. (1973). Floral initiation in four plant species growing in CO_2 enriched air. *Environmental Control Biology* 111: 51–53.

Hicklenton, P. R., and Jolliffe, P. A. (1980). Carbon dioxide and flowering in *Pharbitis nil* Choisy. *Plant Physiology* 66: 13–17.

Ho, L. C. (1977). Effects of CO_2 enrichment on the rates of photosynthesis and translocation of tomato leaves. *Annals of Applied Biology* 87: 191–200.

Hocking, P. J., and Meyer, C. P. (1991). Effects of CO_2 enrichment and nitrogen stress on growth, and partitioning of dry matter and nitrogen in wheat and maize. *Australian Journal of Plant Physiology* 18: 339–356.

Hou, L. Y., Hill, A. C., and Soleimani, A. (1977). Influence of CO_2 on the effects of CO_2 and NO_2 on alfalfa. *Environmental Pollution* 12: 7–16.

Huber, S. L., Rogers, H., and Israel, D. W. (1984). Effects on CO_2 enrichment on photosynthesis and photosynthate partitioning in soybean (*Glycine max*) leaves. *Physiologia Plantarum* 33: 95–101.

Idso, S. B. (1989). *Carbon dioxide and global change: Earth in transition*. AZ: IBP Press.

Idso, S. B., and Idso, K. E. (2001). Effects of atmospheric CO_2 enrichment on plant constituents related to animal and human health. *Environmental Experimental Botany* 45: 179–199.

Idso, S. B., Kimball, B. A., and Allen, S. G. (1991). CO_2 enrichment of sour orange trees: 2.5 years into a long-term experiment. *Plant, Cell and Environment* 14: 351–353.

Idso, S. B., Kimball, B. A., and Mauney, J. R. (1987). Atmospheric CO_2 enrichment effects on cotton midday foliage temperature: Implication for plant water use and crop yield. *Agronomy Journal* 79: 667–672.

Jin-Xing, L., and Yu-Xi, H. (1996). Structural response of soybean leaf to elevated CO_2 concentration. *Acta Botanica Sinica* 38: 31–34.

Johnsen, K. H. (1992). Growth and eco-physiological responses of black spruce seedlings to elevated CO_2 under varied water and nutrient additions. *Canadian Journal of Forestry Research* 23: 1033–1042.

Johnson, R. H., and Lincoln, D. R. (1991). Sagebush carbon allocation patterns and grasshopper nutrition: The influence CO_2 enrichment and soil mineral limitation. *Oecologia* 87: 127–134.

Karpati, E. M., Bekes, F., Lasztity, R., Orst, F., Sanied, I., and Mosonyl, I. (1990). Investigation of the relationship between wheat lipids and baking properties. *Acta Limentaria* 19: 237–260.

Keeling, C. D., Bacastow, R. B., Carter, A. F., Piper, S. C., Whorf, T. P., Heimann, M., Mook, W. G., and Roeloffzen, H. (1989). A three-dimensional model of atmospheric CO_2 transport bases on observed wind: I. Analysis of observational data. In: H. Peterson (Ed.), *Aspects of climate variability in the Pacific and the western Americas* (pp. 165–236). Geophysics Monograph 55. Washington, DC: American Geophysics Union.

Keeling, C. D., and Whorf, T. P. (1994). Atmospheric CO_2 records from sites in the SIO air sampling network. In: T. A. Boden, D. P. Kaiser, R. J., Sepanski, and F. W. Stoss (Eds.), *Trends '93: A compendium of data on global change* (pp. 16–26). Oak Ridge, TN: Carbon Dioxide Information Analysis Center, Oak Ridge National Laboratory.

Kimball, B. A. (1983). Carbon dioxide and agricultural yield: An assemblage and anlaysis of 430 prior observations. *Agronomy Journal* 75: 779–788.

Kimball, B. A. (1986). Influence of elevated CO_2 on crop yield. In: H. Z. Enoch and B. A. Kimball (Eds.), *Carbon dioxide enrichment of green house crops: Vol. 2. Physiology, yield and economics* (pp. 105–115). Boca Raton, FL: CRC Press.

Kimball, B. A., and Mitchell, S. T. (1981). Effect of CO_2 enrichment, ventilation and nutrient concentration on the flavor and vitamin C content of tomato. *Horticulture Science* 16: 665–666.
Korner, C. H., Pelaez-Riedl, S., and van Bel, A. J. E. (1995). CO_2 responsiveness of plants: A possible link to phloem loading. *Plant, Cell and Environment* 18: 595–600.
Krupa, S. V., and Kickert, R. N. (1989). The greenhouse effect: Impact of ultraviolet B (UV-B) radiation, carbon dioxide (CO_2) and Ozone (O_3) on vegetation. *Environmental Pollution* 61: 263–393.
Lawlor, D. W., and Mitchell, A. C. (1991). The effects of increasing CO_2 crop photosynthesis and productivity: A review of field studies. *Plant, Cell and Environment* 14: 807–818.
Leishman, M. R., Sanbrook, K. J., and Woody, R. M. (1999). The effects of elevated CO_2 and light environment on growth and reproductive performance of four annual species. *New Phytologist* 144: 455–462.
Long, S. P. (1991). Modification of the response of photosynthetic productivity to rising temperature by atmospheric CO_2 concentrations: Has its importance been underestimated? *Plant, Cell and Environment* 14: 729–739.
Long,, S. P., and Drake, B. G. (1992). Photosynthetic CO_2 assimilation and rising atmospheric CO_2 concentrations. In: N. R. Baker and H. Thomas (Eds.), *Crop photosynthesis and temporal determinants* (pp. 69–95). Amsterdam: Elsevier.
Luo, Y., Field, C. B., and Mooney, H. A. (1994). Predicting responses of photosynthesis and root fraction to elevated (CO_2): Interactions among carbon, nitrogen and growth. *Plant, Cell and Environment* 17: 1195–1204.
Malmstrom, C. M., and Field, C. B. (1997). Virus induced differences in the response of oat plants to elevated carbon dioxide. *Plant, Cell and Environment* 20: 178–188.
Manderscheid, R., Bender, J., Jaler, H. J., and Weigel, H. J. (1995). Effects of season long CO_2 enrichment on cereals: II. Nutrient concentrations and grain quality. *Agriculture Ecosystem Environment* 54: 175–185.
Mansfield, T. A. (1991). Global warming and the greenhouse effect. *Professional Horticulture* 5: 3–9.
Mansfield, T. A., Hetherington, A. M., and Atkinson, C. J. (1990). Some current aspects of stomatal physiology. *Annual Review of Plant Physiology and Plant Molecular Biology* 41: 55–75.
Marland, G., Andres, R. J., and Boden, T. A. (1994). Global, regional and national CO_2 emissions. In: T. A. Boden, D. P. Kaiser, R. J. Sepanski, and F. W. Stoss (Eds.), *Trends '93: A compendium of data on global change* (pp. 505–584). ORNAL/CKIAC-65. Oak Ridge, TN: Carbon Dioxide Information Analysis Center, Oak Ridge National Laboratory.
Marriott, D. J., Stirling, C. M., and Farrar, J. (2001). Constraints to growth of annual nettle (*Urtica urens*) in an elevated CO_2 environment: Decrease leaf area ratio and tissue N cannot be explained by ontogenetic drift or mineral N supply. *Physiologia Plantarum* 111: 23–32.
Mauney, J. R., Fry, K. E., and Guinn, G. (1978). Relationship of photosynthetic rate to growth and fruiting of cotton, soybean, sorghum and sunflower. *Crop Science* 18: 259–263.
McKee, I. F., Eiblmeier, M., and Polle, A. (1997). Enhanced ozone tolerance in wheat grown at an elevated CO_2 concentration. *New Phytologist* 137: 275–284.
Miszalski, Z., and Mydlarz, J. (1990). SO_2 influence on photosynthesis of tomato plants (*Lycopersicon esculantum* L.) at different CO_2 concentration. *Photosynthetica* 24: 2–8.
Monje, O., and Bugbee, B. (1998). Adaptation to night CO_2 concentration in an optimal environment: Radiation capture, canopy quantum yield and carbon use efficiency. *Plant, Cell and Environment* 21: 315–324.
Morison, J. I. L. (1985). Sensitivity of stomata and water use efficiency to high CO_2. *Plant, Cell and Environment* 8: 467–474.
Morse, S. R., Wayne, P., Migo, S. L., and Bazzaz, F. A. (1993). Elevated CO_2 and drought after tissue water relations of birch (*Belala populifolia* Marsh) seedling. *Oecologia* 95: 599–602.
Mortensen, L. M. (1994). Effect of carbon dioxide concentrations on biomass production and partitioning in *Betula pubesceus* seedlings at different ozone and temperature regions. *Environmental Pollution* 87: 337–343.
Mott, K. A. (1990). Sensing of atmospheric CO_2 by plants. *Plant, Cell and Environment* 13: 731–737.
Mousseau, M., and Sangier, B. (1992). The direct effect of increased CO_2 gas exchange and growth of forest tree species. *Journal of Experimental Botany* 43: 1121–1129.
Mousseau, M. (1993). Effects of elevated CO_2 on growth, photosynthesis and respiration of sweet chestnut (*Castanea sativa* Mill.). *Vegetatio* 104/105: 413–419.
Mulchi, C. L., Slaugher, L., Saleem, M., Lee, E. H., Pausch, R., and Rowland, R. (1992). Growth and physiological characteristics of soybean in open-top chambers in response to ozone and increased atmospheric CO_2. *Agriculture Ecosystem and Environment* 38: 107–118.

Mulchi, C. L., Rudorff, B., Lee, E. H., Rowland, R., and Pausch, R. (1995). Morphological responses among crop species to full-season exposures to enhanced concentrations of atmospheric CO_2 and O_3. *Water, Air and Soil Pollution* 85: 1379–1386.

Niewiadomska, E., Gaucher-Veilleux, C., Cheurier, N., Manffette, Y., and Dizengremel, P. (1999). Elevated CO_2 does not protect against ozone considering the activity of several antioxidant enzymes in the leaves of sugar maple. *Journal of Plant Physiology* 155: 70–77.

Norby, R. J., O'Neill, E. G., and Luxmore, R. J. (1986). Effects of atmospheric CO_2 enrichment on the growth and mineral nutrition of *Quercus alba* seedling in nutrient poor soil. *Plant Physiology* 82: 83–89.

Paez, A., Hellmers, H., and Strain, B. R. (1980). CO_2 effects on apical dominance in *Psium sativum, Physiologia Plantarum* 50: 43–46.

Paoletti, E., and Gelline, R. (1993). Stomatal density variation in beech and holm oak leaves collected over the last 200 years. *Acta Oecologia* 14: 173–178.

Pearcy, R. W., and Bjorkman, O. (1983). Physiological effects: In: E. R. Lemon (Ed.), CO_2 *and plant: The response to rising levels of atmospheric carbon dioxide* (Vol. 84, pp. 65–105). AAAS Selected Symposium.

Peng, T. H., Broecker, W. S., Freyer, H. D., and Trumbore, S. (1983). A deconvolution of the tree ring based and ^{18}C record. *Journal of Geophysical Research* 88: 3609–3620.

Penuelas, J., and Estiarte, M. (1998). Can elevated CO_2 affect secondary metabolism and ecosystem function? *Trends in Ecology Evolution* 13: 20–24.

Penuelas, J., Estiarte, M., Kimball, B. A., Idso, S. B., Pinter, P. J., Jr., Well, G. W., Garcia, R. L., Hansaker, D. J., Lamorte, R. L., and Hendrix, D. L. (1996). Variety of responses of plant phenolic concentration to CO_2 enrichment. *Journal of Experimental Botany* 47: 1463–1467.

Pleijel, H., Mortensen, L., Fuhrer, J., Ojanpera, K., and Danielsson, J. (1999). Grain protein accumulation in relation to grain yield of spring wheat (*Triticum aestivum* L.) grown in open-top chambers with different concentrations of ozone, carbon dioxide and water availability. *Agriculture Ecosystem Environment* 72: 265–270.

Polley, H. W., Johnson, H. B., Mayeux, H. S., and Maloue, S. R. (1993). Physiology and growth of wheat across a subambient carbon dioxide gradient. *Annals of Botany* 71: 347–356.

Poorter, H. (1993). Interspecific variation in the growth response on plants to an elevated ambient CO_2 concentration. *Vegetatio* 104/105: 77–97.

Pritchard, S. G., and Rogers, H. H. (2000). Spatial and temporal development of crop roots in CO_2 enriched environments. *New Phytologist* 147: 55–71.

Radin, J. W., Kimball, B. A., Hendrix, D. L., and Mauney, J. R. (1987). Photosynthesis of cotton plants exposed to elevated levels of carbon dioxide in the field. *Photosynthesis Research* 12: 191–203.

Rao, M. V., and DeKok, L. J. (1994). Interactive effecs of high CO_2 and SO_2 on growth and antioxidant levels in wheat. *Phyton* 36: 279–290.

Rao, M. V., Hale, B. A., and Ormord, D. P. (1995). Amelioration of ozone induced oxidative damage in wheat plants grown under high carbon dioxide. *Plant Physiology* 109: 421–432.

Rattray, E. A. S., Paterson, E., and Killham, K. (1995). Characterization of the dynamics of C partitioning within *Lolium perenne* and to the rhizosphere microbial biomass using ^{14}C pulse change. *Biology and Fertility of Soils* 19: 280–286.

Rawson, H. (1992). Plant responses to temperature under elevated CO_2. *Australian Journal of Botany* 40: 473–490.

Reekie, E. G., and Bazzaz, F. A. (1989). Competition and patterns of resource use among seedlings of five tropical trees grown at ambient and elevated CO_2. *Oecolgia* 79: 212–222.

Reeves, D. W., Rogers, H. H., Prior, S. A., Wood, C. W., and Runion, G. B. (1994). Elevated atmospheric carbon dioxide effects on surghum and soybean nutrient status. *Journal of Plant Nutrition* 17: 1939–1954.

Riechers, G. D., and Strain, B. R. (1988). Growth of blue grama (*Bouteloua bracilis*) in response to atmospheric carbon dioxide enrichment. *Canadian Journal of Botany* 66: 1570–1573.

Rogers, H. H., Cure, J. D., and Smith, J. N. (1986). Soybean growth and yield response to elevated carbon dioxide. *Agriculture, Ecosystem and Environment* 16: 113–128.

Rogers, H. H., Runion, G. B., and Krupa, S. V. (1994). Plant responses to antmospheric CO_2 enrichment with emphasis on roots and the rhizosphere. *Environmental Pollution* 83: 155–189.

Rogers, H. H., Bingham, C. E., Cure, J. D., Smith, J. N., and Surano, K. A. (1983). Response of selected plant species to elevated carbon dioxide in the field. *Journal of Environmental Quality* 12: 569–574.

Rotty, R. M., and Master, C. D. (1984). Carbon dioxide from fossil fuel CO_2 emission. TRO14, DOE/OR21400-2. Washington, DC: Carbon Dioxide Research Division, U.S. Department of Energy.

Rowland-Bamford, A. J. (2000). Plant responses to changing carbon dioxide and temperature. In: S. N. Singh (Ed.), *Trace gas emissions and plants* (pp. 63–71). London: Kluwer/Academic.

Rudorff, B. F. T., Mulchi, C. L., and Lee, E. H. (2000). Plant responses to elevated CO_2 and interactions with O_3. In: S. N. Singh (Ed.), *Trace gas emissions and plants* (pp. 155–173). London: Kluwer/Academic.
Rudorff, B. F. T., Mulchi, C. L., Lee, E. H., Rowland, R., and Pausch, R. (1996a). Effect of enchanted O_3 and CO_2 enrichment on plant characteristics in wheat and corn. *Environmental Pollution* 94: 53–60.
Rudorff, B. F. T., Mulchi, C. L., Lee, E. H., Rowland, R., and Pausch, R. (1996b). Photosynthetic characteristics in wheat exposed to elevated O_2 and CO_2. *Crop Science* 36: 1247–1251.
Ryle, G. J. A., Powell, C. E., Tewson, V. (1992). Effect of elevated CO_2 on the photosynthesis, respiration and growth of perennial ryegrass. *Journal of Experimental Botany* 43: 811–818.
Sage, R. F. (1994). Acclimation of photosynthesis to increasing atmospheric carbon dioxide. *Annual Review of Plant Physiology and Plant Molecular Biology* 44: 309–332.
Sage, R. F., Sharkey, T. D., and Seemann, J. R. (1989). Acclimation of photosynthesis to elevated CO_2 of five C_3 species. *Plant Physiology* 89: 590–596.
Sandhu, R., Li, Y., and Gupta, G. (1992). Sulphur dioxide and carbon dioxide induced changes in soybean physiology. *Plant Science* 83: 31–34.
Schneider, S. H. (1989). The changing climate. *Scientific American* 261: 38–47.
Schwanz, P., Kimball, B. A., Idso, S. B., Hendrix, D. L., and Polle, A. (1996). Antioxidants in sun and shade leaves of sour organge trees (*Citrus aurantium*) after long term acclimation to elevated CO_2. *Journal of Experimental Botany* 47: 1941–1950.
Schwanz, P., and Polle, A. (1998). Antioxidative systems, pigment and protein contents in leaves of adult Mediterranean oak species. (*Quercus pubescens* and *oilex*) with lifetime exposure to elevated CO_2. *New Phytologist* 140: 411–423.
Sharkey, T. D., Lorelo, F., and Delwiche, C. F. (1991). High carbon dioxide and sun shade effects on isoprene emissions from oak and aspen tree leaves. *Plant, Cell and Environment* 14: 333–341.
Smart, D. R., Chatterlon, N. J., and Bugbee, B. (1994). The influence of elevated CO_2 on non structural carbohydrate and friction distribution and friction accumulation in wheat canopies. *Plant, Cell and Environment* 17: 435–442.
Smith, S. D., Strain, B. R., and Sharkey, T. D. (1987). Effects of CO_2 enrichment on four Great Basin grasses. *Functional Ecology* 1: 139–143.
Strain, B. R., Sionit, N., and Cure, J. D. (1984). Direct effects of atmospheric CO_2 enrichment on plants and ecosystems: A bibliography with abstracts. ORNL/CKIC-13. Washington, DC: U.S. Department of Energy.
Tausz, M., DeKok, L., Stulen, I., and Ziska, L. H. (1996). Physiological responses or Norway spruce trees to elevated CO_2 and SO_2. *Journal of Plant Physiology* 94: 470–475.
Thomas, R. B., Reid, C. D., Ybema, R., and Strain, B. R. (1993). Growth and maintenance components of leaf respiration of cotton grown in elevated carbon dioxide partial pressure. *Plant, Cell and Environment* 16: 539–546.
Tschaplinski, T. J., Stewart, D. B., Hanson, P. J., and Norby, R. J. (1993). Interactions between drought and elevated CO_2 on growth and gas exchange of seedlings of three deciduous tree species. *New Phytologist* 129: 63–71.
Tuba, J., Szente, K., and Koch, J. (1994). Response of photosynthesis, stomatal conductance, water use efficiency and production to long-term elevated CO_2 in winter wheat. *Journal of Plant Physiology* 144: 661–668.
Vogel, C. S., and Curtis, P. S. (1995). Leaf gas exchange and nitrogen dynamics of N_2 fixing field grown *Alnus glutionsa* under elevated atmospheric CO_2. *Global Change Biology* 1:55–61.
Vu, J. C. V., Allen, L. H., and Bowes, G. (1989). Leaf ultrastructure, carbohydrates and protein of soybeans grown under CO_2 enrichment. *Environmental and Experimental Botany* 29: 141–147.
Wardlaw, I. F. (1982). Assimilate movement in *Lolium* and *Sorghum* leaves: III. Carbon dioxide concentration effect on the metabolism and translocation of photosynthates. *Australian Journal of Plant Physiology* 9: 705–715.
Warrick, R. A., Gifford, R. M. L., and Parry, M. L. (1989). CO_2 climatic change and agriculture. In: B. Bolin, B. R. Doos, J. Jager, R. A. Warrick (Eds.), *The greenhouse effect: Climatic change and ecosystems* (pp. 391–473). SCOPE-29. Chichester, UK: Wiley.
Wechsung, G., Wechsung, F., Wall, G. W., Adawsen, F. J., Kimball, B. A., Garcia, R. L., Pinter P. J., Jr., and Kartschall, T. (1995). Biomass and growth rate of spring wheat root system grown in free air CO_2 enrichment (FACE) and ample soil moisture. *Journal of Biogeography* 22: 623–631.
Wheeler, T. R., Batts, G. R., Ellis, R. H., Hakley, P., and Morison, J. I. L., (1966). Growth and yield of winter wheat (*Triticum aestivum*) crop in response to CO_2 and temperature. *Journal of Agriculture Science* 127: 37–48.
Will, R. E., and Teskey, R. D. (1997). Effect of elevated carbon dioxide concentration and root restrictions on net photosynthesis, water relations and four carbohydrate status of loblolly pine seedlings. *Tree Physiology* 17: 655–661.

Wong, S. C. (1990). Elevated atmospheric partial pressure of CO_2 and plant growth: II. Nonstructural carbohydrate content in cotton plants and its effect on growth parameters. *Photosynthesis Research* 23: 171–180.

Woodward, F. I. (1987). Stomatal numbers are sensitive to increase in CO_2 from preindustrial levels. *Nature* 327: 617–618.

Woodward, F. I., and Bazzaz, F. A. (1988). The responses of stomatal density to CO_2 partial pressure. *Journal of Experimental Botany* 39: 1771–1781.

Woodward, F. I., and Kelly, C. K. (1995). The influence of CO_2 concentration on stomatal density. *New Phytologist* 131: 311–327.

Zanetli, S., Hartwig, U. A., Luscher, A., Hebeisen, T., Frehner, M., Fischer, B. H., Hendrey, G. R., Blum, H., and Nosperger, J. (1996). Stimulation of symbiotic N fixation in *Trifolium repensus* L. under elevated atmospheric CO_2 in a grassland ecosystem. *Plant Physiology* 112: 575–583.

Ziska, L. H., Bonce, J. A., and Caulfield, F. (1998). Intraspecific variation in seed yield of soybean (*Glycine max*) in response to increased atmospheric carbon dioxide. *Australian Journal of Plant Physiology* 25: 801–807.

Ziska, L. H., Manalo, P. A., and Ordonez, R. A. (1996). Intraspecific variations in the response of rice (*Oryza sativa* L.) to increased CO_2 and temperature: Growth and yield response of 17 cultivars. *Journal of Experimental Botany* 47: 1353–1359.

Ziska, L. H., Namuco, O.,. Moya, T., and Quilang, J. (1997). Growth and yield response of field-grown tropical rice to increasing carbon dioxide and air temperature. *Agronomy Journal* 89: 45–53.

Ziska, L. H., and Teramura, A. H. (1992a). CO_2 enhancement of growth and photosynthesis in rice (*Oryza sativa*): Modification by increased ultraviolet-B radiation. *Plant Physiology* 99: 473–481.

Ziska, L. H., and Teramura, A. H. (1992b). Intraspecific variation in the response of rice (*Oryza sativa*) to increased CO_2: Photosynthetic, biomass and reproductive characteristics. *Physiologia Plantarum* 84: 269–274.

13

Plant Cover: Ecological Implications and Methodical Approaches

Hansjörg Dietz and Thomas Steinlein

Introduction

Plant or vegetation cover is usually regarded as the vertically projected area of all (living) aboveground plant parts as a percentage of the total ground area considered (i.e., intraspecific overlap is not taken into account). Plant cover also does not include belowgound plant parts. In this review, we use the term *plant cover* very broadly to include cover of mosses, lichens, algae, and cyanobacteria. In both ecological and economic contexts, plant or vegetation cover is widely used as a readily accessible parameter describing the dominance of certain plant species, plant life forms, or the degree of closure of the whole vegetation. In this respect, cover can be superior to density or productivity in different vegetation types, because density ignores size variability between individuals, and because it is difficult to determine individuals in clonal plant species. Furthermore, by using cover, it is possible to compare species of widely different growth forms (Daubenmire, 1968; Mueller-Dombois and Ellenberg, 1974). On the other hand, a major limitation of plant cover is that it represents a mostly three-dimensional structure in only two dimensions.

The factors determining plant cover at any place at a certain point in time are naturally very diverse (see e.g., Archibold, 1995). Generally, vegetation cover declines from the warm and humid parts of the world toward arid and/or cold regions, and the contribution of different taxonomic groups to vegetation cover often changes from dominance of higher plants to that of lower plants (i.e., mosses and lichens or soil crusts) (e.g., Ullmann and

Hansjörg Dietz • Geobotanisches Institut ETH, Eidgenössische Technische Hochschule Zürich, Zürichbergstr. 38, CH-8044 Zürich, Switzerland; e-mail: dietz@geobot.umnw.ethz.ch.
Thomas Steinlein • Lehrstuhl für Experimentelle Ökologie und Ökosystembiologie, Universität Bielefeld W4-107, Universitätsstr. 25, 33615 Bielefeld, Germany; e-mail: Tom.Steinlein@biologie.uni-bielefeld.de.

Büdel 2001). In addition to climatic factors, local conditions such as the type of bedrock and soil, and disturbances such as fire or anthropogenic land use, are of overriding importance. The range of substrates where vegetation cover in the widest sense can develop is almost limitless. Even on longer lived leaves, vegetation cover of cryptogamous species may develop as, for instance, successional communities of lichens on the leaves of *Capparis arborea* in the rainforest of Queensland, Australia (Rogers, 1995). Still other factors affecting vegetation cover and composition are herbivory and the developmental stage of the vegetation in primary or secondary succession.

Although the processes shaping vegetation development and structure are rather well documented, information on the ecological implications of plant cover is scattered and, to our knowledge, there does not yet exist an overview that compiles this information. Therefore, we do not focus on the development of plant cover in this chapter. Rather, it is our intention to present a broad and actual view of the ecological implications of variations in plant cover. The delicacy of this endeavor is that plant or vegetation cover is often used as a surrogate measure for other variables that might more directly influence the phenomena under study. For example, if nutrient availability in the soil is significantly related to plant cover, it is usually not plant cover *per se* that determines the nutrient status, but the related turnover of standing crop (biomass). An example of a more direct influence of plant cover on nutrient status would be nutrients leached from a leaf canopy that are absorbed by a moss layer (cf. e.g., Okland et al., 1999). In our review, we strive to present an overview of the more direct ecological effects of plant cover. However, in many cases, such as microhabitat choice in animals, ambiguities remain whether direct effects of plant cover (e.g., providing shade or shelter) or indirect effects (e.g., increased food supply underneath a plant canopy) play the most important role.

In addition, we present an overview of the diverse array of methods used to determine plant cover. We only briefly summarize and evaluate the variety of classic methods to determine plant or vegetation cover, because already extensive overviews deal with these methods (e.g., Mueller-Dombois and Ellenberg, 1974; Knapp, 1984a; Kent and Coker, 1992). Rather, we concentrate on more recent methodical refinements or developments, particularly those related to computerized analysis of images to determine plant or vegetation cover or methods of remote sensing techniques.

Even with these restrictions in coverage, the material presented in the remaining topics is highly selective in that not all ecological aspects of plant cover could be included in this review.

Characterization of Plant Cover in the Vegetation

Most communities of higher plants are structured vertically—not only forests (e.g., Smith, 1973; Bee et al., 1989) but also herbaceous vegetation such as chalk grassland (Mitchley, 1988) and even lawns (Roxburgh et al., 1993). Thus, the major limitation of plant cover as a measure of the composition and dominance of species, and the above-ground performance of the vegetation is that its significance decreases with increasing vertical structure unless the ecological question itself focuses on the canopy area of the vegetation. Even in low herbaceous vegetation, the relative vertical positions of the species may consistently differ, even at different phenological points in the mowing cycle (Roxburgh et al., 1993). Thus, with examination of canopy cover there is always the caveat to underestimate systematically the cover of species with low stature.

Vegetation density and its vertical structure are dependent on a wide variety of factors such as the climatic zone, soil conditions, herbivory, and disturbances (i.e., anthropogenic land use, fire frequency, and frequency of catastrophic events such as hurricanes or avalanches). They are also related to the successional status of the vegetation and the composition of life forms. Plant species cover may vary not only due to abundance, plant growth, and seasonal variation in aboveground performance or the vertical position of the species in the vegetation but also as a result of varying foliage inclination angles. Foliage inclination angles may vary to overshadow less tall neighbors (horizontal foliage), to let more radiation pass through to lower layers of the canopy (inclined foliage angles), or to keep the foliage perpendicular to the sun rays (solar tracking). Furthermore, some plant species turn their leaves vertically or position them in a north–south direction ("compass plants") to avoid high-irradiance stress or overheating in the middle of the day (see Caldwell, 1987; Koller, 1990).

Spatial patchiness in species composition can also be represented by varying plant cover. For example, in a California coastal grassland, different patch types (ranging in size from < 1 to $500\,m^2$) could be distinguished where the three most abundant grass species (*Anthoxanthum odoratum*, *Holcus lanatus*, and *Deschampsia holciformis*) each dominated their respective patches with approximately 90% cover (Peart, 1989a).

Environmental Effects of Vegetation Cover

Interrelationships with Climate

The feared consequences of the destruction of rain forests contribute to our conventional wisdom that plant cover considerably influences the climate by changing the meteorological conditions, not only on a global scale but also near to the land surface (e.g., Clark and Arritt, 1995; Pielke et al., 1998). In particular, vegetation cover affects boundary layer turbulence, evaporation rates, and absorption of solar radiation. These influences are exerted at different scales and with interrelated changes in vegetation cover and climate.

The observed increase in the atmospheric concentration of carbon dioxide and related climatic changes may have considerable impacts on the density of vegetation cover (e.g., Press et al., 1998; Robinson et al., 1998) which may in turn provide feedback on climate (Betts et al., 1997, 2000). For example, in a model scenario, warmer temperatures due to the glasshouse effect lead to increased LAI (leaf area index) at high latitudes. There, LAI increase may more than compensate for reduced transpiration by stomatal closure at the canopy scale. Under these conditions, LAI provides negative feedback on climatic warming through an increase in evaporation. Furthermore, increased LAI may cause higher temperatures in winter through a reduction in surface albedo (Betts et al., 2000). In a similar vein, from the results of a coupled biosphere–atmosphere model, Bounoua et al. (2000) suggest that increased vegetation density can cause both an increase in evapotranspiration and precipitation on a global scale. Evapotranspiration, however, may increase more than precipitation, resulting in a global soil water deficit. Furthermore, changes in the state of vegetation may also affect precipitation distribution.

Changes in climate may also affect the cover proportions of different vegetation layers. This effect was studied in subarctic alpine meadow and heath communities (i.e., in a zone where climate changes are predicted to be most pronounced) (Molau and Alatalo,

1998). Their results suggest that climate change leads to shifts in dominance of both ground and canopy layer species. In rich meadows, canopy closure of higher plants negatively affect the bottom (moss and lichen layer), whereas in poor heath communities with open cover, the ground layer is likely to respond positively or remain unchanged.

Effects on Microclimate

It is apparent that the most important direct ecological function of plant cover is the reduction of solar radiation below the canopy, resulting in aboveground competition (for light), particularly at productive sites with high standing crop (see Goldberg, 1990; Barton, 1994). Less than 5% of the photosynthetic active radiation (PAR) is transmitted by most leaves (Gates, 1980). Therefore, most of the solar radiation is absorbed in the topmost vegetation layer and, in dense vegetation, light availability decreases exponentially with increasing LAI from the top of the canopy (Caldwell, 1987; Werger and Hirose, 1988; Hirose and Werger, 1995). In mixed boreal forests, mean PPFD (photon flux density) varied from 2% at the forest floor under *Populus* forests to 15% above the understory vegetation cover under *Betula* forests. Furthermore, PPFD above the understory vegetation cover was significantly higher under shade-intolerant tree species such as *Populus*, *Betula*, and *Pinus* than under shade-tolerant conifers (Messier et al., 1998). Higher PPFD above the understory vegetation also led to increased PPFD at the forest floor. However, increasing cumulative total cover of the understory vegetation reduced PPFD at the forest floor. Differences in light availability in the soil depending on canopy cover may even determine algal density and biomass (Lukesova and Hofmann, 1996).

Plant cover and, in relation to it, the amount of solar radiation reaching the soil or litter layer below the canopy not only vary considerably between habitats but may also be highly variable within habitats. An example of considerable small-scale variations in plant cover and light transmittance through the canopy is reported for an abandoned pasture near Ontario, Canada (Reader and Buck, 1986). This pasture was characterized by many small depressions of 1–4 m diameter and 0.5–1.0 m depth. Measurements at the litter layer yielded a much higher proportion of the incident radiation at the top (ca. 70%) than at the bottom of the depressions (ca. 30%), and visible estimates of plant cover varied accordingly.

Absorption of light energy also results in substantial variations in soil heat flux between canopy gaps and underneath vegetation cover. This is apparent in many natural and agricultural landscapes with partial vegetation cover, particularly in arid and semiarid regions, where vegetation cover is low and clustered or "clumped" with large areas of exposed soil (Kustas et al., 2000). The lower temperatures and higher air humidity under plant cover can facilitate establishment or growth of desiccation-sensitive organisms or developmental stages (e.g., in "safe sites" for seedling recruitment and establishment of juvenile plants). For example, on the northern coast of Papua New Guinea the three wood-inhabiting fungi *Hexagonia tenuis*, *Polyporus philippinensis*, and *Microporus xanthopus* were found to be restricted to dense vegetation cover providing microclimatic conditions that prevent desiccation stress (Castillo and Demoulin, 1998).

Erosion

The two main causes for erosion are rain and wind energy. Erosion occurs when the soil lacks protective vegetative cover (Pimentel and Kounang, 1998). Therefore, vegetation cover can provide a low-cost, and pleasing solution to problems of erosion and slope instability, and is of greatest relevance in humid tropical areas (Collison et al., 1995). This function is not just exerted by higher plants. Even soil algae were found to be important in reducing soil erosion, especially on degraded land, where complete algal cover may develop and the considerable cover of higher plants may take years (Booth, 1941). Attempts to revegetate eroded land can be difficult, however. For example, in Iceland, barren lands were treated with aerial broadcasting of the grass *Festuca rubra* to stop further desertification by large-scale soil erosion (Greipsson and El Mayas, 1999). The resulting buildup of vegetation cover by *F. rubra* was only temporary, though, and did not initiate rapid natural succession.

Several studies document the fact that vegetated slopes have a lower risk of being affected by landslides. For example, in reserved forest at steep hillsides in the Kumaun Himalaya, landslide activity is negatively correlated with forest cover (Haigh et al., 1995). Likewise, landslides at coastal cliffs in Bangladesh were caused by not only the steepness of slopes and wave activity but also the lack of vegetation cover (Rahman and Khan, 1995).

Anthropogenic land use that disrupts surface vegetation cover results in increased soil erosion. In a field experiment using simulated rainfall at an intensity of 60 or 120 mm/h at hillslopes of buried-waste sites in the steppe region of the northwestern United States, soil loss rates were 100 to 1,000 times greater if the vegetation canopy and litter was removed than under naturally vegetated conditions. If only the canopy was removed, the soil loss rates were five times higher than under undisturbed conditions (Goff et al., 1993). Apparently, litter cover can be more efficient in preventing erosion than canopy cover. Yet, the amount of ground cover by litter itself depends on the canopy cover. On the other hand, in field experiments on erosion by overland-flow, grass stems clipped close to the ground reduced flow resistance and shear stress for significant sediment transport by up to an order of magnitude compared to the unclipped situation (complete grass cover, Prosser and Dietrich, 1995). In further field rainfall simulation trials in Spain, vegetation cover increased infiltration of water into the soil and reduced surface runoff and erosion as compared to bare soils. The variability of all these processes was also greatly reduced with increasing vegetation cover. There was an interaction of plant cover and soil type with plant cover being particularly effective in reducing high erosion rates of marl soil (Cerda, 1999).

Vegetation cover is also the key to controlling wind erosion by reducing or preventing soil loss, sandblasting effects, or contamination of water and food resources by wind-deposited material (Leys, 1991, and references therein). Wind, the second main erosion factor, is particularly important in semiarid regions. Sand dunes are among the most dynamic habitats greatly affected by the shaping forces of the wind constantly reshuffling sand in the system and reshaping the relief. Under these circumstances, vegetation cover is often particulary effective in subsiding substrate movement and stabilizing dunes. The dune crests and the upper slopes are particularly dependent on vegetation or ground cover by litter for reduced blowouts (Gares, 1992; Wiggs et al., 1995). Lancaster and Baas (1998) found that sediment flux by aerodynamic processes decreased negatively exponentially with increasing plant cover of salt grass on sand sheets at Owens Lake, California. Sand transport was effectively eliminated when the cover of salt grass was greater than 15%.

Water Budget, Nutrient Availabilty, and General Deposits

Changes in vegetation cover can have considerable influence on spatiotemporal patterns of water availablity. For example, in mountain ash forests in Victoria, stream flow from regrowth stands was found to be less than from the mature stands they replaced. This appeared to be related to the cover of regrowth stands that surpassed the cover of mature stands after 5–10 years of regeneration (Borg and Stoneman, 1991). In the arid Karoo shrubland in Africa, vegetation clearing increased the amplitude of soil moisture oscillations at 50 mm below the soil surface but reduced the rate of soil moisture depletion at 150 mm below the surface (Milton, 1995). In a rainfall simulation study in Australia, soil infiltration rates increased with developing plant cover over time. The highest infiltration rates were obtained at 100% cover (Loch, 2000). Generally, consideration of both vegetation cover and soil type is needed for an evaluation of the spatial pattern of soil infiltration properties (Brath and Montanari, 2000).

Although plant cover has no considerable direct influence on soil nutrient status in most cases, it can be indicative of soil nutrient status. For example, in a semiarid shortgrass steppe, patterns of discontinuous plant cover had a greater effect on ecosystem scale estimates of soil properties than the attributes of particular plant species, due to limitations in the decomposition by available water (Vinton and Burke, 1995). Likewise, due to the water-limited nature and discontinuous plant cover, spatial patterns in ecosystem properties are influenced more by the presence or absence of plants than by plant type in shortgrass steppe (Epstein et al., 1998). A more direct effect of plant cover is the modification of the amount of elements absorbed from throughfall precipitation of the vegetation canopy by the ground layer. In the moss *Hylocomium splendens*, leaching of canopy leaves caused considerable input of K but negative correlations of vascular plant cover with Pb, Cd, and Cu concentrations in the moss indicated that the amounts of elements intercepted by plant cover were higher than the amounts leached (Okland et al., 1999).

Likewise, the transfer of herbicides by runoff or leaching varies in dependence on plant cover (e.g., Schiavon et al., 1992). Deposition patterns of airborne pollutants that may affect the organisms also depend mainly on landscape configuration and plant cover (Rusek and Marshall, 2000). Dead vegetation cover can also influence deposits considerably. In the unburned tallgrass prairie, standing dead vegetation reduced not only the amount of water reaching the soil but also relative amounts of nitrogen in the throughfall, due to the scavenging action of microbes on the detritus (Seastedt, 1985). Biological control efforts can also be influenced by vegetation cover. Spores of an entomopathogenic fungus, *Metarhizium anisopliae*, are used as a biopesticide of the noxious migratory locust (*Locusta migratoria*). From field trials, researchers learned that application rates of *Metarhizium* must be higher in denser vegetation, because high cover reduces the likelihood of locusts receiving spores directly from the spraying operation or the vegetation (Scanlan et al., 2001).

Responses of the Organisms to Variations in Plant Cover

Species Diversity

According to the intermediate disturbance hypothesis (Connell, 1978; Huston, 1979), the diversity in a community is maximized at intermediate levels of disturbance because, under high disturbance intensity, only few species are present, whereas under very low disturbance intensity, species number is reduced by competitive displacement. Correspondingly, Grace and Pugesek (1997) found a lower species number in dense vegetation and hypothesized this to be the result of competitive displacement due to shading. Grytnes (2000) found a humpback relationship between the number of vascular plant species and plant cover within different alpine vegetation types in western Norway: Species number was highest at intermediate cover values in some cases. In other cases, species number increased monotonically with cover. In this study, species richness was slightly more strongly related to vegetation cover than to biomass. Thus, light availability appears to be an important factor influencing the species-richness pattern (Grytnes, 2000; see also Grace and Pugesek, 1997). The importance of the light climate is particularly apparent in the understory vegetation of forests. For example, in Canadian montane forests, plots with more open forest cover had greater lichen cover, and a higher lichen and vascular plant species diversity (Pharo and Vitt, 2000).

Another positive correlation of species occurrence with plant cover was observed for vesicular arbuscular mycorrhizae (VAM) and ectomycorrhizal fungi (ECM). Both groups were positively associated with vegetation cover and patches in highly disturbed areas (Alagely and Reeves, 1995; Bundrett et al., 1996).

In arid regions, species diversity was found to increase with plant cover, at least in open or sparse vegetation. For example, the numbers of some groups of invertebrates (planthoppers, weevils, and pseudoscorpions) were positively correlated with the diversity of perennial plant species and plant cover in old fields in the southern Karoo in Africa (Dean and Milton, 1995). The diversity of small mammals in the semiarid Karoo also increased with increasing plant cover, but only at low cover values, and decreased at cover values greater than 30%, a pattern that is similar in other desert systems (Kerley, 1992). Invertebrate diversity decreased sharply with decreasing plant cover under heavy grazing on a farm in Namaqualand, South Africa. Seymour and Dean (1999) argued that a lack of host plants and the loss of vertical structure in the overgrazed areas could be the reasons for their observation.

As a result of its temperature-moderating effects, plant cover has different effects on the community structure of ground-dwelling ants. In Mediterranean communities where potentially dominant species have a lower thermal tolerance, increased plant cover leads to lower evenness in the community because of a higher abundance of these species. Conversely, in other communities, a reduced abundance of dominant species was found at higher plant cover, and some of these had the highest diversity at intermediate plant cover (Retana and Cerda, 2000).

In all these examples, plant cover may influence species diversity directly through modification of the microhabitat conditions, but it may also just represent general resource availability (e.g., of food, nesting sites, or vertical habitat structure) that attracts species and permits higher species diversity.

The Role of Plant Cover in Competition and Facilitation

Competitive interactions between plant species are considered to be among the most important processes affecting the diversity and distribution of species and the structure of plant communities. There is still an ongoing debate about the relationship between competitive interactions and variations in biomass at the plant community level. Evidence was found for both an increase of competition with increasing biomass, that is, habitat productivity (e.g., Reader and Best, 1989; Campbell and Grime, 1992) and a lack of this relationship (e.g., Turkington and Klein, 1993; Reader et al., 1994). But it is obvious that the ability to overtop or shade competitors is crucial in dense and productive vegetation.

Dense cover in the upper vegetation layer (overstory species) usually reduces the cover of lower, understory vegetation layers. For example, in removal experiments in the transition zone between shrubs and herbaceous vegetation on a lakeshore, cover of the herbaceous species increased significantly in experimental openings in the shrub layer compared to control plots, provided that nutrient availability was not low (Keddy, 1989). To our knowledge, there is no competitive index explicitly using cover except the removal response coefficient used by Silander and Antonovics (1982). Most authors use biomass changes to describe competitive interactions, although they mostly state that cover, reproductive output, or survival may be used instead.

Increased competition for light by high plant cover typically results in taller plant growth (Caldwell, 1987). Using a game theoretic model, Givnish (1982) showed that the evolutionary stable strategy favors greater leaf heights (taller plants) in dense, productive vegetation but lower leaf heights in sparse vegetation. The reason for that is the higher proportion of biomass allocated to structural tissue in taller plants (Figure 13.1A), resulting in a lower net carbon gain from photosynthesis on a per-unit biomass basis. So in sparse vegetation, plants should be small. However, in dense vegetation, these structural costs trade off against reductions in photosynthesis by being overshadowed, favoring taller growth to hold the leaves higher than the neighbors (Figure 13.1B). Givnish (1982) found that maximum leaf height of forest herbs roughly doubled with each 7% increase in herbaceous cover along a forest gradient. However, in tall and dense vegetation, alternative growth strategies are also realized: In a study by Hirose and Werger (1995) the tallest species in a tall-herb community received a higher amount of solar radiation (PPFD) per leaf area. But per unit aboveground biomass, the absorbance was not higher than that of subordinate species. In plants that cannot cope with their taller neighbors in terms of vertical growth, increasing plant cover can induce changes in the leaf angles. For example, at productive sites, rosettes of the forb species *Bunias orientalis* (Brassicaceae) produce long leaves $\geqslant$ 40 cm that vary widely in the leaf angle (Dietz and Ullmann, 1997). Under

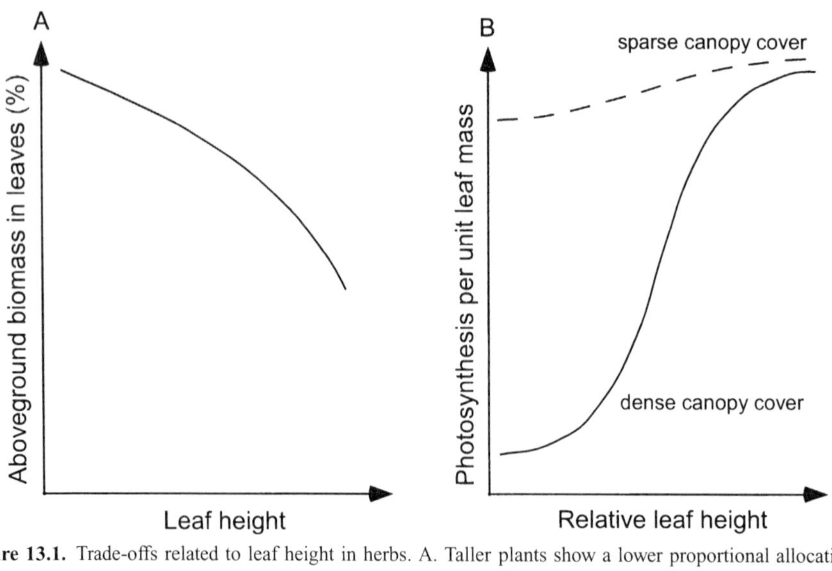

Figure 13.1. Trade-offs related to leaf height in herbs. A. Taller plants show a lower proportional allocation of biomass to leaves. B. Photosynthetic benefits per unit leaf biomass should be small under sparse herbaceous cover but large if the canopy is dense. Redrawn and modified from Givnish *American Naturalist* 120: 355. With permission from The University of Chicago Press.

dense cover, the leaves are almost erect, partly escaping overlap by the matrix vegetation. Competitive dominance is particularly evident where rapid growth of herbs or shrubs is combined with the capacity for clonal expansion. Under these conditions, floristic diversity is exceedingly low, and complete ground cover, and thus intensive shading, result in very low invasibility (Grime, 1987).

Plant cover may have not only competitive but also also facilitative effects (e.g., by promoting colonization of organisms by moderating harsh environmental conditions (see also next section)). For example, colonization of bare patches by both *Spartina patens* and *Distichlis spicata* facilitated the colonization of *Juncus gerardi*, the competitive dominant in dense vegetation. Bare patches in the high marsh are often hypersaline due to increased surface evaporation. Patch colonization by salt-tolerant competitive subordinates appears to reduce substrate salinity by passively shading the substrate which benefits colonization of the stronger competitor (Bertness, 1991).

The Role of Vegetation Cover in Plant Regeneration

Vegetation cover can inhibit or impair seed germination, seedling emergence, and seedling survival. Positive effects of reduced vegetation cover or larger gap size on seedling establishment may be mainly due to increased light availability (Goldberg and Werner, 1983). Thus, vegetation density appears to be a crucial factor for seedling establishment. For example, seedling emergence of *Prunella vulgaris* was inhibited by herbaceous cover in old-field habitats and by both litter and herbaceous cover in woodland habitats (Winn 1985). Seedling establishment of *Hieracium floribundum* growing in an abandoned Canadian pasture was higher in microsites with lower plant and litter cover (Reader and Buck, 1986). Likewise, seedling mortality of the invasive forb *Bunias orientalis* (Brassicaceae) was considerably higher in herbaceous areas with dense vegetation cover than in less dense or temporally open vegetation (Dietz and Steinlein, 1998). Survival of seedlings of four biennial forb species in abandoned fields in Michigan was also positively related to the amount of bare (uncovered) ground in the vegetation (Gross 1982). Whereas seedlings of the two large-seeded species *Daucus carota* and *Tragopogon dubius* survived under a wider range of plant cover, the two smaller seeded species *Verbascum thapsus* and *Oenothera biennis* survived only in bare ground.

The size of gaps in the vegetation canopy appears to be very important for seedling survival (e.g., Davis and Cantlon, 1969; Ehrenfeld, 1980; Goldberg and Werner, 1983). In a perennial grassland undergoing succession after the removal of domestic herbivores, the formation of openings in the vegetation cover by the death of individual bunchgrasses and gopher mounds resulted in a 6–2,500-fold colonization success of the best colonizers (*Holcus lanatus* and *Anthoxanthum odoratum*). These disturbances were necessary for changes in species dominance in the perennial vegetation (Peart, 1989b). Survival of the later plant stages seems to be less related to plant cover than to that of seedlings (Gross, 1982; Reader and Buck, 1986).

Vegetation cover can also positively affect seedling establishment (facilitation) if neighboring plants ameliorate harsh environmental conditions for the seedlings (Bertness and Callaway, 1994). For example, in a seasonally dry savanna ecosystem in Brazil, seedling establishment profited from higher moisture content of the soil, reduced soil temperatures, and higher nutrient availability under tree cover (Hoffmann, 1996). Similarly, seedlings of woody species were dependent on prior invasion of the clonal shrub *Rhus copallina* to be able to expand into climax tallgrass prairie, profiting from the improved soil moisture conditions and higher air humidity underneath the shrub canopy (Petranka and McPherson, 1979). Seedling survival of the endemic *Quercus douglasii* in California was much higher under shrubs than in the open grassland, apparently due to shade effects (Callaway, 1992). Seedling survival of the desert succulent *Agave desertii* in the Sonoran Desert of California was dependent on unusually wet years and on protection by nurse plants (*Hilaria rigida*). These were found to reduce leaf temperatures of *A. desertii* to below lethal levels (Jordan and Nobel, 1979).

Even the ground cover of the bryophyte layer may be important for species regeneration. For example, in chalk grasslands, the bryophyte layer was found to act as a seed trap. It can influence seedling emergence and mortality, and even reduce seed predation (Keizer et al., 1985; van Tooren, 1988).

Seasonal Patterns in Plant Cover

Development of plant species cover may be different between co-occurring species at some time during the growing season. For instance, cover proportions in the understory vegetation in a Canadian *Pinus contorta* stand changed very little between June and August but differed considerably from those in May and September (Kennedy and Addison, 1987). Such phenological displacements are often the result of niche differentiation, which, on a temporal scale, is one important mechanism to reduce competition between species with similar resource requirements and thus contributes to species coexistence and diversity (e.g., Harper, 1977; Burrows, 1990; Bazzaz, 1996). Differences in the seasonal pattern of shoot expansion and leaf development between plant species (e.g., Al Mufti et al., 1977; Grubb, 1977; Fowler and Antonovics, 1981; Negi et al., 1992) can be at least partly interpreted as niche differentiation in relation to plant cover as an important component of aboveground competition (see Harper, 1977, and references therein). One of the most prominent seasonal niche displacements in this respect is the early aboveground growth and flowering of geophytes in the temperate deciduous forest before the trees sprout, strongly reducing light availability for the understory vegetation (Archibold, 1995).

In addition, phenological displacements can promote the establishment of alien species in native plant communities. For example, in Californian pastures, introduced species showed delayed senescence of aboveground plant parts relative to native species (Crawley, 1987). Rosette growth of *Bunias orientalis* (Brassicaceae), a perennial forb that is invasive in western and northern Europe, was found to be two-phased under various habitat conditions (Figure 13.2; Dietz and Ullmann, 1997). The second flush of rosette growth occurred in autumn, in a period when cover of the matrix vegetation was reduced after warm and dry summer months. This phenological displacement appears to compensate partly for the weak competitive ability of *B. orientalis* for light in well-developed herbaceous vegetation.

Effects of Plant Cover on Animals

Not only plants but also animals interact in various ways with plant cover. For example, seed predators may be attracted by plant cover, but plant cover can also provide shelter from the eyes of predators, as well as herbivores, that may miss the juvenile growing underneath the canopy of its neighbors: In California, acorn predation was the primary cause for the mortality of *Quercus* seedlings growing underneath shrubs, whereas

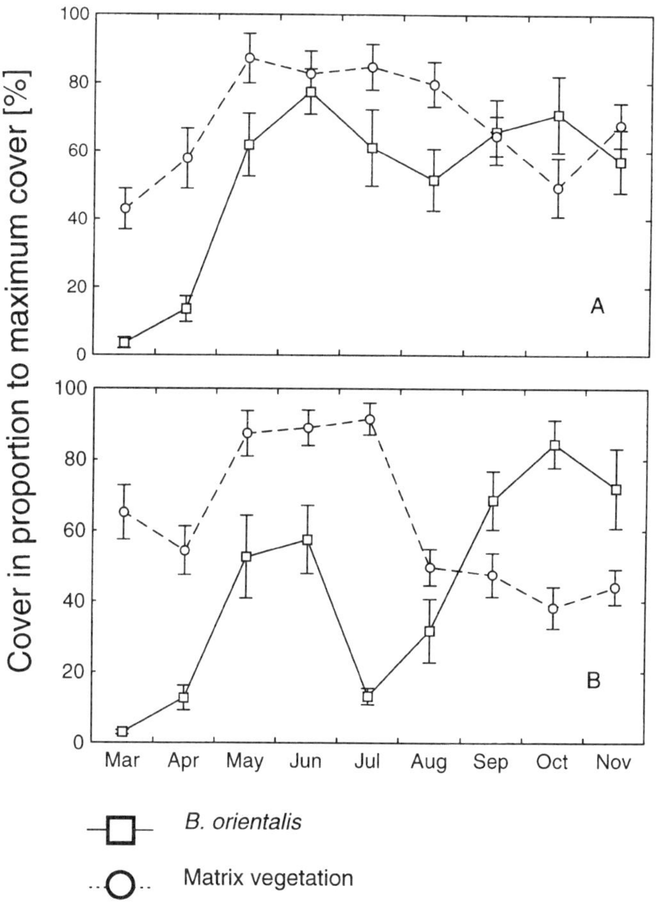

Figure 13.2. Cover development of the invasive forb *Bunias orientalis* (Brassicaceae) and of the matrix vegegation in plots in (A) nutrient-rich ruderal sites and (B) disused farmland (growth period, 1995). The values shown denote mean cover ± standard error in % of mean maximum cover in the plots. Modified from Dietz and Ullmann (1997) *Journal of Vegetation Science* 8: 842. With permission from Opulus Press AB.

shoot herbivory was much more frequent in the open grassland (Callaway, 1992). Thus, if the acorns underneath shrubs escaped seed predation, the seedlings profited from the shrub cover protecting them from shoot herbivores. In an experiment monitoring the removal of seeds from dishes by mice in manipulated old-fields, vegetation cover, time of day, and distance from the forest edge explained most of the variation in patterns of seed predation (Manson and Stiles, 1998). In differently managed wildflower strips, a positive correlation was observed between vegetation cover and seed predation by rodents. Likewise, increased seed predation of rodents under dense cover of brambles (*Rubus fruticosus* agg.) potentially contributes to reduced establishment of forest tree seedlings in montane forests in southwestern Germany. Similar relationships have been found in other ecosystems as a result of higher food availability and protection against predators for the rodents under dense canopies (Schreiner et al., 2000; Kollmann and Bassin, 2001, and references therein).

Besides these direct plant–animal interactions, animals may respond in a multitude of ways to changes in the vegetation structure. As one example, insect life cycle strategies are influenced by habitat permanence, habitat complexity, and resource diversity, factors that increase during succession, partly driven by changes in cover proportions of annual herbs, perennial herbs, grasses and woody plants (Brown, 1985). Frouz (1997) observed relationships between changes of two species of terrestrial chironomids (*Smittia* cf. *aterrima* and *Bryophaenocladius* cf. *virgo*) and changes of vegetation cover during succession, which indicated that successional changes of terrestrial chironomids can be explained by the effect of vegetation cover on oviposition habitat preference.

At least partly in relation to microclimatic and structural effects, plant cover influences the (micro)habitat choice (e.g., Borghi et al., 1991; Deplessis and Kerley, 1991) and the behavior of animals in general. For example, in shortgrass prairie beetle (*Eleodes* spp.) movements were strongly influenced by vegetation structure, with net displacements highest in bare ground and under grass cover, and lowest in cactus and shrub (Crist et al., 1992). In desertified grassland in Arizona, nests of the ant species *Myrmecocystus mexicanus* that were experimentally shaded either died or the entrances were moved away from the shade (Chew, 1995). Population levels and abundance of the common Mediterranean lacertid lizard, *Psammodromus algirus*, were most highly correlated with shrub cover, apparently related to thermoregulation, predator avoidance, and movement minimization (Diaz and Carrascal, 1991). Plant cover may also be important for the choice of nest sites in birds (e.g., Fasola and Canova, 1992; Bisson and Stutchbury, 2000).

Many animal species seek shelter under plant canopies to escape detection and/or attack by predators. In the Negev Desert highlands in Israel, large and small tenebrionid beetles (*Coleoptera*) showed different habitat preferences, with the large species being abundant only in areas with high plant cover, whereas areas with low plant cover were dominated by smaller species (Ayal and Merkl, 1994). The authors argued that the larger species are concealed under high plant cover and so can escape predation by birds. Low vegetation cover also contributed to egg loss by conspecific egg predation in herring gulls (*Larus argentatus*) on the Dutch Frisian Island of Terschelling (Brouwer and Spaans, 1994). If feeding stations for black-tailed jackrabbits (*Lepus californicus*) were positioned at different distances from perennial shrubs, jackrabbits consumed significantly more food at stations under shrub canopies than at stations 5 and 10 m from shrubs (Longland, 1991). The foraging activities of jackrabbits thus appear to be constrained by risk of predation. Martin and Lopez (1995) observed significantly longer approach and flight distances of lizards (*Psammodromus algirus*) in early spring (when most bushes were without leaves) than in summer. Correspondingly, within a season, approach and flight distances were longer in microhabitats with less vegetation cover. Even moonlight may be sufficient to influence animal behavior in patchy habitats compared to overcast nights, as in a study by Jensen and Honess (1995). In low vegetation with less cover, wood mice were only trapped in overcast nights.

Environmental Changes by Cryptogamous Cover

Whereas by far the most studies dealing with plant cover have examined vegetation of higher plants, there is also a range of studies that consider plant cover in cryptogamous communities (i.e., moss, lichen, and algal cover; e.g., Hurka and Winkler, 1973; Wirth and

Brinckmann, 1977; Keizer et al., 1985). Cryptogamous crusts can modify soil infiltration. In semiarid woodland in eastern Australia, cover of cryptogamous crusts had little effect on infiltration of soils with high physical condition, but infiltration rates on degraded surfaces were significantly improved by cryptogamous cover (Eldridge, 1993a). Furthermore, the presence of cryptogamous crusts increased infiltration significantly at some grazed rangelands but not at ungrazed sites (Eldridge, 1993b). Likewise, on sand dunes in Israel, water availability in the upper soil layer was higher under cyanobacteria crusts and, particularly, under moss cover than in uncovered areas (Kutiel, 1998).

Dark cryptogamous crusts can also reduce surface albedo and thus help to raise surface temperature sufficiently to support greater plant cover in cold climates. In the arctic polar desert in Canada, root growth and seedling survival of vascular plants probably benefited from such higher surface temperatures, whereas tissue temperatures of adult plants did not differ between cryptogamous and bare surfaces (Gold, 1998). However, cryptogamous cover may also have negative effects on vascular species. In glasshouse experiments, seedling emergence of *Arenaria serpyllifolia*, *Festuca ovina*, *Filipendula vulgaris*, and *Veronica spicata* was investigated comparatively in bare soil, and in bryophyte and lichen carpets (Zamfir, 2000). Under relatively dry conditions, seedling emergence was partly negatively affected even by thin moss cover but was not much different from bare soil in the moist treatment. The lichens inhibited only the emergence of the forbs. As seedling emergence in the presence of dead mosses and lichens did not differ from that on bare soil, the inhibitory effect of cryptogamous cover might have been due to some biotic factors.

Use of Plant Cover for Analyzing Vegetation Structure and Change

Plant species composition and abundance change during succession and as a result of different management regimens. In many cases, plant cover is a straightforward and suitable measure to monitor such changes and, therefore, has been used in a wide variety of studies. For example, over a period of 8 years, the cover of grasses in experimental heathland plots increased strongly upon addition of nutrient solution compared to both the cover of co-occurring dwarf shrubs and control plots (Berendse et al., 1994). Cover estimates were also used to analyze different management regimes (cutting, burning) on the dominance of the grass *Brachypodium pinnatum* that is increasing in nutrient-poor, calcareous grassland (Schlaepfer, 1997). Arnthórsdóttir (1994) used plant cover determined by the point–quadrat method (see below) to monitor species-specific recolonization of experimental patches in a mown grassland in Wales, depending on patch size and on the presence of the grass *Holcus lanatus*. In a field experiment on the establishment of the invasive plant *Bunias orientalis* L., cover analysis was used to contrast (aboveground) competition intensity by the matrix vegetation in an area with spontaneous vegetation with that in an area where green fodder and pasture plants were sown in to result in dense matrix vegetation (Dietz et al., 1999).

Species cover has also been used to examine species increase or decline along environmental gradients. In a study of epiphytic lichens, the cover of foliose forms on the trunk of avenue maple trees decreased with decreasing distance from the city of Freiburg (southwest Germany) due to increasing stress by SO_2 pollution at that time, whereas the cover of resistant crustose species increased in the same direction. Prach and Pysek (1994) used plant cover to analyze relative success of clonal plants in different stages of secondary succession. In the majority of the seres studied, clonal species dominated the vegetation cover, predominantly in moist sites, where the peak in their relative cover occurred earlier in succession. Whereas the clonal species reached dominance regardless of soil conditions, dominance of nonclonal species tended to be restricted to very wet and nutrient-poor sites. Furthermore, the dominance of clonal plant species was more permanent than that of nonclonal plants. Cover analyses were used to assess changes in the community composition (in functional groups) in a Mediterranean herbaceous community affected by cattle grazing managements (Sternberg et al., 2000). In this study, widely varying grazing regimens did not cause large changes in plant cover of functional groups, particularly in that of hemicryptophytic species. A reduction of cover in tall grasses, however, was correlated with an increase of some prostrate annuals and less palatable groups.

Vegetation cover and composition may be indicative of not only the actual site conditions, such as management practice and the dominant (overstory) species but may also reflect site history. For example, vegetation cover and composition in the herbaceous vegetation layer differed significantly between longleaf pine (*Pinus palustris*), loblolly (*Pinus taeda*), and slash pine (*Pinus elliottii*) forests in Georgia (Hedman et al., 2000). The lower ground cover and species richness of the afforested loblolly and slash pine forests compared to the longleaf pine forests, which had partly been naturally regenerated, were attributed more to differences in agricultural land-use history than to recent forest management or pine cover type.

Relation to Biomass and Density

In open vegetation (i.e., in the initial phase of succession or under limiting conditions that do not allow the development of dense vegetation), cover may be directly correlated to the biomass of the species involved. In this situation, the growth of the plants may be limited by the availability of water and belowground resources rather than by competition for light, especially in unproductive habitats, or the intrinsic growth rates of these species (early succession after disturbance in highly productive habitats). Plants growing under these conditions should ideally show a growth pattern of constant surface (cover) to biomass proportion (i.e., a linear relationship between these parameters). This would prevent changes in photosynthetic efficiency elicited by structural changes concurrent with growth and, from a methodical point of view, provide the researcher with a rapid means to estimate aboveground biomass (Röttgermann et al., 2000). To date, only a few studies have used this relationship to estimate biomass by cover analysis. Cover and density may be correlated in a similar way, as shown in many studies. However, because plant individuals show high phenotypic plasticity, these relationships are often poor. Furthermore, a higher overlap of plant structures (single leaves, branches, canopies) and the concurring increase in competition for light results in only weak correlations between biomass and cover.

The following examples from different vegetation zones demonstrate a close relationship between plant biomass and cover. All examples have in common that the study and/or growing site is not primarily limited by the availability of light. The first example (Figure 13.3A) is taken from the Nizzana dune system in Israel (data from Veste et al., 2000). An aerial photo was analyzed to determine cover of the vegetation (shrubs and herbs) in specific vegetation units (interdune, dune base, slope, mobile crest). Vegetation cover was compared to aboveground biomass of these units harvested within selected plots in the analyzed area. There was a significant correlation ($p < .01$) between cover and mean

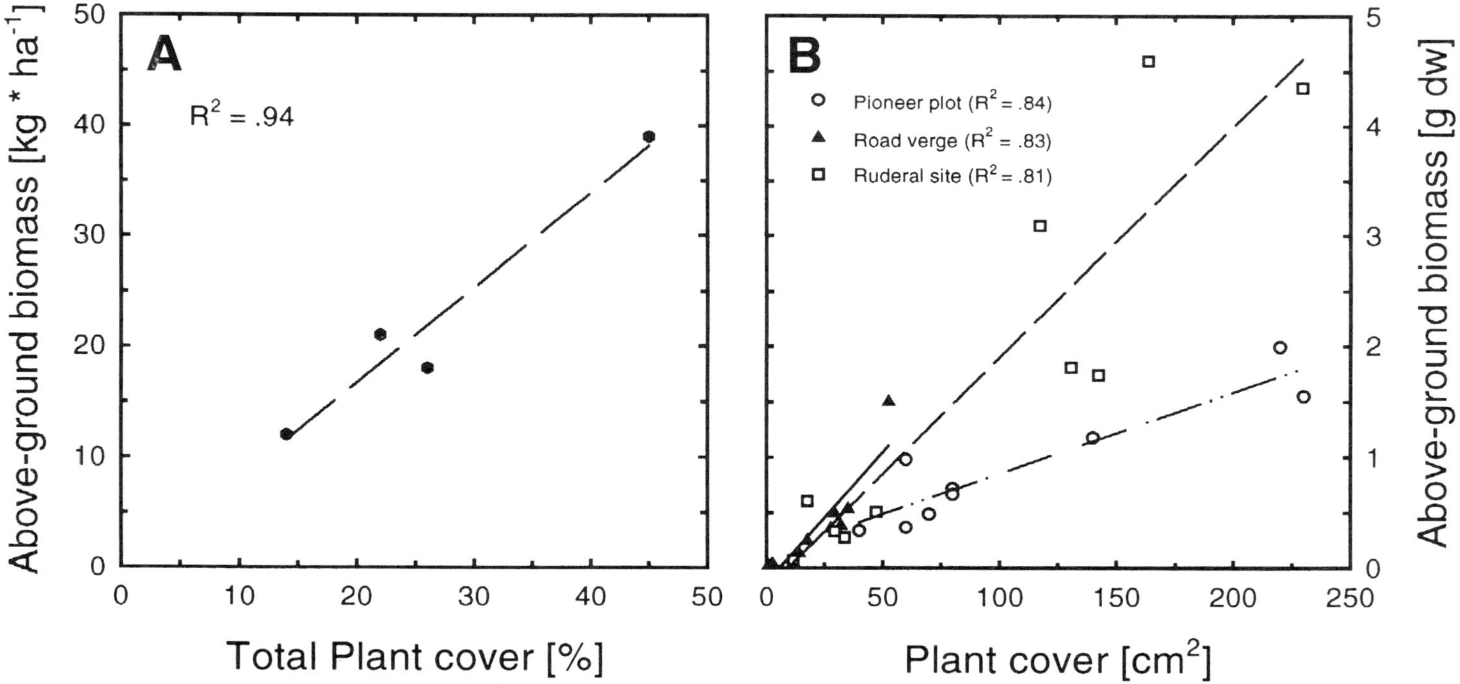

Figure 13.3. Linear regressions of plant/vegetation biomass on cover. A. Cover estimation via aerial photographs in selected plots (see text for details) in the Nizzana dune systems in Israel. Data from Veste et al. (2000) Gesellschaft für Ökologie, with permission from Blackwell Verlag, Berlin. B. Linear regression of biomass on cover of all plants of single species in different habitats. Data from Röttgermann et al. (2000). *Journal of Vegetation Science* 11(1): 147. With permission from Opulus Press AB, and Dettmer (2000).

Table 13.1. Coefficients (R^2) and Significance of Linear Regressions of Aboveground Biomass on Cover for Species with Different Growth Forms

Plant species	Growth form	Regression coefficient (R^2)	p value
Armeria maritima	Herb with grasslike, errect leaves	.75	$<.05$
Calluna vulgaris	Small, evergreen shrub	.96	$<.001$
Carex arenaria	Spacer sedge	.79	$<.001$
Epilobium parviflorum	Erect perennial shrub	.98	$<.01$
Hieracium pilosella	Dense rosette plant with sessile leaves	.97	$<.001$
Oxalis acetosella	Small herb	.96	$<.001$
Plantago major	Rosette plant	.95	$<.001$
Rumex acetosella	Semirosette plant with petiolate leaves	.95	$<.001$
Urtica dioica	Erect clonal plant with leafy stems	.77	$<.05$
Dryopteris carthusiana	Fern	.97	$<.001$
Polytrichum formosum	Moss	.93	$<.001$

Data from Dettmer (2000) and Röttgermann et al. (2000) with permission.

biomass and a linear regression coefficient (R^2) of .94. Similar results were attained in an early successional patch (1 m^2) on a ruderal site near Würzburg, Germany (Figure 13.3B; Röttgermann et al., 2000). The nine predominantly broad-leaved plant species growing in the patch showed a significant linear correlation between cover and aboveground biomass ($p < .001$) and a regression coefficient of .84. Analogous results were obtained at different sites near Bielefeld, Germany, by Dettmer (2000), who found correlations of biomass and plant species cover and linear regression coefficients of 0.7–0.9. In Figure 13.3B, the results of a road verge with a total coverage of 15%, and a ruderal site with a total cover of 60% confirm these results. In a further field experiment above- and belowground biomass of plant species with different growth forms and from different habitats was correlated with cover in different plots (Table 13.1; Dettmer, 2000). Slope coefficients varied, but for all growth forms (ferns and mosses included), a significant correlation between cover and aboveground biomass could be detected and linear regression coefficients ranged from .66 to .97. In all cases, plastic responses in aboveground biomass were paralleled by changes in plant cover. As expected, cover was a poor predictor variable for belowground biomass. This was, first and foremost, caused by the high variability of root–shoot ratios among conspecific plants.

It appears that the relationship of cover to (aboveground) biomass is fairly linear, especially in open herbaceous vegetation, irrespective of species and/or growth form. However, there a many situations that abolish these simple relationships, for example, changing biomass allocation patterns, such as increasing proportions of supporting tissue in taller plants, increased storage of resources, and shifts from vegetative to sexual reproduction.

Methodical Approaches to Plant Cover

At least the coarse methods of visual cover estimation that accept relatively low accuracy are rapid and straightforward, so they are extremely widely used to provide some quantitative measure of vegetation density and composition. The generally negative relationship between time and efficiency, and resolution and accuracy in the determination of plant cover is represented by a wide variety of methods based on the specific requirements in the trade-off gradient between time and efficiency and accuracy. Additionally, the choice of the method relies on the structure of the vegetation, because some methods are better suited for particular types of vegetation than the others. In very fine-grained or highly structured vegetation, the determination of plant species cover may not be appropriate at all. Conversely, the determination of percentage cover is particularly appropriate and useful for species-poor and less layered plant communities, for example, boreal plant communities (oligo- to mesotrophic forests with moss and lichen ground and open herb layers; Pakarinen, 1984) and other herbaceous vegetation with an open canopy (e.g., Röttgermann et al., 2000). Whichever method is used, one must consider that, due to variations in the phenological behavior of plant species, cover estimations at any one point of time in the growing season may yield an incomplete picture of relative species abundance or even presence and absence of some species, particularly for subordinate taxa and/or taxa that show up only during short periods of time (e.g., therophytes; Ramirez, 1984).

"Close" Sensing

Visual cover estimation, a rapid and very widespread method, uses a cover or combined cover abundance scale to classify the visual cover estimates. The most widely known scale, introduced by Braun-Blanquet (1928, 1964), uses two or three abundance scores for species with low dominance ($< 5\%$ cover) and four cover classes for species with higher dominance ($\geqslant 5\%$ cover). Another regularly used scale is the Domin scale, with 11 cover abundance scores (see Moore and Chapman, 1986). These two scales, along with the corresponding cover abundance values are tabulated in Table 13.2.

The use of cover classes is efficient and best suited for extensive surveys of different vegetation types (Pakarinen, 1984). The cover abundance scales have been widely adapted to meet specific needs, such as to improve resolution or to facilitate numerical analysis. For example, one of the refined cover scales, the decimal scale (e.g., Londo, 1976, 1984), uses intervals of 10% for values between 5% and 95% cover and may have a finer subdivision for values below 5%. Use of this scale related to actual cover values may overcome

Table 13.2. Two Commonly Used Cover/Abundance Scales

Cover category	Domin scale	Braun-Blanquet scale
+	1 individual	< 1% (1–5 individuals)
1	1–2 individuals	< 5% (6–50 individuals)
2	< 10%	5–25%
3	1–4%	26–50%
4	4–10%	51–75%
5	11–25%	76–100%
6	26–33%	—
7	34–50%	—
8	51–75%	—
9	76–90%	—
10	91–100%	—

At low cover values the number of individuals may be used rather than percent cover. Modified from Moore and Chapman (1986). *Methods in plant ecology* (p. 450). With permission from Blackwell Science Ltd., UK.

problems with numerical analysis in the analysis of permanent quadrats (Londo, 1976, 1984; for problems with rounding to cover classes, particularly at low cover values, see van der Maarel, 1979; Wildi, 1986). A compilation and description of cover and abundance scales used in vegetation surveys is provided in Küchler and Zonneveld (1988).

Visual cover estimation methods suffer from relatively high errors. The cover of species that flower conspicuously or are familiar to the observer may be overestimated compared to other co-occurring species (Kent and Coker, 1992). Visually estimated cover values also differ between observers (Clymo, 1980; Sykes et al., 1983; Kennedy and Addison, 1987). Interobserver variation in cover estimates was found to be higher than variations between cover estimates on the same quadrat obtained from single observers at different times. Interobserver variation was higher with fine-grained vegetation than with broad-leaved species. The average estimation error was much higher for species with low cover than for species with high cover values. Sykes et al. (1983) argued that higher accuracy in visual cover estimates could be gained by calibrating biases of single, consistent observers.

A more accurate method to assess plant cover that is particularly feasible for sparse vegetation or larger areas is the line intercept method. A measurement tape is laid out, and percentage cover of the species can be determined by summing up the sections on the tape that the species touch, and relating this length to the total length of the tape (Mueller-Dombois and Ellenberg, 1974; Knapp, 1984b; Kent and Coker, 1992).

The point–intercept method (Levy and Madden, 1933; Goodall, 1952, 1953) is a further widespread and relatively accurate approach to determine cover. In this case, a point (usually the tip of a needle) is lowered vertically through the vegetation, and all contacts of the point with plant parts are recorded. If this procedure is repeated sufficiently often (e.g., several hundred times), sampling the plot area in a regular or random fashion, the contact frequencies for each species correspond accurately to their cover values. However, there are considerable costs in attaining high accuracy with the point method, in that it is very time-consuming, particularly in vegetation with high structural complexity (Stampfli, 1991). Furthermore, the standardization of the method is complicated (e.g., regarding unpredictable factors such as wind; Stampfli, 1991), so that results are not comparable (Daubenmire, 1968). The statistical properties of the point sampling procedure have been quite extensively studied. If changes in the composition of the vegetation are to be detected over time, the precision of the method has to match the minimum changes that should be detectable (i.e., the confidence limits about the data must be smaller than the expected change in cover). Everson et al. (1990) discussed minimum sample sizes for point techniques and stated that nonrandom sampling and, particularly, monitoring less frequent species require larger sample sizes to achieve the same precision. In a similar vein, Knapp (1984b) stated that there would be only a relatively small number of points necessary for determination of cover values of more than 20%, but a high sample size for species covering less than 5%. An inverse point-sampling technique that fixes the number of hits required to achieve a specified boundary on the error of estimation, instead of fixing the sample size, was found to be comparable to or better than the classical point-sampling technique for determination of cover of rare species (Reich et al., 1994).

A further approach to record plant cover is to use a grid frame from which the outline of plants can be drawn to scale on a sheet of paper for subsequent cover analysis (quadrat-charting method; Daubenmire, 1968; Mueller-Dombois and Ellenberg, 1974). A variation of this method is particularly suited to determine the cover of epiphytic lichens, algae, or mosses on the trunks of trees. In this case, the outlines of lichen thalli traced onto transparent plastic foil wrapped around the trunk can be analyzed as mentioned earlier (Hurka and Winkler, 1973).

A rapid method to record vegetation cover in the field is to take photographs of the sampling plot (e.g., Molloy and Moran, 1991; Dietz and Steinlein, 1996; McCarthy and Zaniewski, 2001). This can be achieved by constructing simple camera stands, for example, tripods with telescopic legs or booms that provide flexibility in camera height and/or are easy to transport (Wimbush et al., 1967; Molloy and Moran, 1991). Alternatively, photographs can be taken directly by the observer by using household ladders and boards that support different vertical distances to the vegetation (Dietz and Steinlein, 1996). Taking photographs not only reduces fieldwork, which is particularly advantageous in periods of bad weather, but the pictures also preserve original raw data, rather than derived data sets, and can be analyzed at any time. The main problem in overhead cover photography is that the closer an object is to the camera, the larger it appears. Therefore, the actual canopy cover may be lower than it appears. If the average canopy height is known, however, the actual percent cover can easily be calculated using similar triangles (Thomas et al., 1988).

The plant outlines obtained from the photographs can be traced and cover analysis performed by superimposing a transparent grid sheet (e.g., Wimbush et al., 1967) or by planimetry. Berendse et al. (1994) projected the photographs on a grid and calculated plant cover by recording the species present at each intersection point of the gridlines. More modern methods to analyze cover photographs have used computerized image analysis (e.g., Stone et al., 1988; Dietz and Steinlein, 1996; Birdsall et al., 1997; Röttgermann et al., 2000; Vanha-Majamaa et al., 2000; McCarthy and Zaniewski, 2001). About 10 years ago, relatively high time consumption and costs of the digital image analysis procedure were the main drawbacks of the approach (e.g., Thomas et al., 1988). But since image analysis equipment has become far more sophisticated and widespread, prices have dropped steadily, so that digital image analysis is easily approached now.

Image analysis programs provide different approaches to assign units of area (pixels) to different plant species or types of cover (e.g., plant or litter cover). Vanha-Majamaa et al. (2000) used supervised classification and probability functions based on the pixels' spectral features (colors) to assign them to the different species. The proportion of the group of pixels in a class represents the cover of the corresponding species. This method does not discriminate sensitively between species unless the species show contrasting colors (Vanha-Majamaa et al., 2000). Dietz and Steinlein (1996) used semimanual techniques to determine plant cover. Software tools that delineate single areas based on contiguous zones of similar color, or that select areas throughout the image based on predefined color ranges, were combined with manual (mouse) techniques to trace plant outlines, where necessary. This approach allowed for appropriate discrimination between species and yielded fairly accurate results for open, mainly broad-leaved or simple layered vegetation. However, these techniques would become strenuous and inaccurate for fine-grained and multilayered vegetation (Dietz and Steinlein, 1996). For example, in a study on succession in heathland dependence on litter accumulation and fertilization, Berendse et al. (1994) combined the dwarf shrubs and grasses in two different groups for analysis, because the species within groups could not always be clearly distinguished in the photographs.

Image analysis of plant cover has often been used in applied plant science. For example, Birdsall et al. (1997) compared image analysis with visual estimation for determination of cover of the invasive perennial forb leafy spurge (*Euphorbia esula*) to find an efficient method to assess the effects of biological control agents on leafy spurge cover. They used a conventional reflex camera to take photographs and a frame-grabber board to digitize the color prints. By using image analysis software adjusted to discriminate between pixels belonging to leafy spurge cover and those associated with other cover types based on threshold grayscale values, the authors achieved average error levels similar to those of visual cover estimation. However, the resolution of the digitized images was relatively low (512×480 pixels) compared to current standard image resolutions processed. Still, the authors recommended image analysis as a measurement tool, because quantification was efficient (requiring smaller sample sizes to arrive at the same error values as in ocular estimation). Thomas et al. (1988) developed an image processing procedure to evaluate the percent cover of crops from overhead slide photographs and compared computer analysis with leaf area measurements and the visual estimation method. Compared to the leaf area measurements, digital image analysis produced acceptable cover estimates, whereas visual estimates overestimated cover. Molloy and Moran (1991) used manual counting and digital image analysis of vertical photographs to estimate crop residue cover to set up a manual for the estimation of sorghum and wheat/barley residue cover, with calibrated photographs at 10% cover intervals from 0 to 100%.

Cover photographs and image analysis were also utilized to obtain various measures of thalli in crustose lichen communities (McCarthy and Zaniewski, 2001). Images were taken of lichens growing on flat rock surfaces in two glacier forefields. The authors used Adobe Photoshop™ to digitally remove barren rock in the images and the (IDRISI) geographic information system (GIS) to measure thallus size, perimeter, compactness, and total lichenized area. GIS software such as IDRISI may for many users be less accessible for image analysis but it offers ways to automate different processes in image analysis that render the work particularly efficient (McCarthy and Zaniewski, 2001). Their technique yielded reproducible results and allowed them to distinguish very small differences in lichen cover, examine lichen colonization patterns, and explore new approaches in lichenometric dating.

Remote Sensing

With humans' increasing impact on global changes (global warming, increase in land use and levels of UV radiation), global changes in distribution of vegetation types and/or land cover become an issue of rising concern. Today, changes in regional or global land use and cover due to agriculture, forestry and human settlements are among the most evident impacts on the global environment. These processes have profound implications for the functioning of ecosystems, the biochemical fluxes in these systems, and the climate.

When considering changes in cover on such global or regional scales, remote sensing and mapping techniques are suitable methods. The use of satellite or airborne aerial images (towers, aircrafts, balloons) is an efficient and relatively inexpensive approach for either global or regional assessment of changes in cover or long-term monitoring, detection, and changes in habitats or surface areas. But remote sensing techniques or general circulation models also must sometimes lump functional ecological classes into very crude assemblages, thus sacrificing considerable biological detail (Levin, 1992). Remote sensing is most often used for economic reasons, mainly in agriculture (extent and degree of desertification, temporal and spatial patterns of inundation, estimation of biomass production or standing crop and final yield) but also for direct or indirect applications in ecological research.

In ecology remote sensing data are used (1) to interpret landscape patterns and changes, (2) to find correlations between abiotic (chemical, physical, geographic) and biotic parameters by using remote sensing in combination with georeferenced data, and (3) to extrapolate known relationships over a wider range of geographic areas or longer time periods (Kasischke et al., 1997). Remote sensing data are also used to understand processes linked to changes in landscape patterns, for example, detailed information about land use and cover changes necessary in order to implement environmentally sensitive policies and practices, and to monitor such policies (Flores and Martinez, 2000).

In general, there are two ways to access changes on Earth's surface on a regional or global scale: by means of (1) multispectral scanner data or (2) imaging radars or synthetic aperture radars (SARs). The latter system has received less attention than data from optical systems, because the interpretation of imaging radar data (recorded in multifrequency, polarimetric SAR data) is much more complex than "visible" and easily comprehensible multispectral image data. Furthermore, computer software for the analysis and processing of imaging radar data has been essentially rare, though, more recently, software systems have been developed. For more details on this method and current developments, see Waring et al. (1995) and Kasischke et al. (1997).

The most widespread system used to obtain multispectral scanner data is the Landsat Thematic Mapper™ satellite imaging system, which, launched in 1982, provides images with a spatial resolution of about 30 meters and a temporal revisit resolution of about 2 weeks. This system continuously collects images in six different wavelength bands: three visible bands (blue, green, and red), two mid-infrared bands, and one near-infrared band. Another satellite system in use, the Pathfinder AVHRR, also provides multispectral images. Many software packages and modeling systems have been developed to interpret these multispectral scanner data and to analyze the different images using brightness, color, and digital change images between bands to calculate cover-related coefficients. Other satellite systems in use have different spatial resolution (see Table 13.3).

When using multispectral imagery, changes in cover or the amount of vegetation present are expressed as vegetation indices that can be related to ecological variables such as biomass, LAI, vegetation cover, or absorbed PAR (see Table 13.4). The most widespread indices are (SR), Normalized Difference Vegetation Index (NDVI), and LAI (Lawrence and Ripple, 1997). SR and NDVI have been shown to be highly correlated to vegetation cover, LAI and/or aboveground biomass and other ecologically relevant variables (e.g., Anderson et al., 1993; Spanner et al., 1990; Lawrence and Ripple, 1997; Yoder and Waring, 1994).

One of the main problems for multispectral imaging and the calculation of vegetation parameters lies in removing background or scatter that may be partly caused by changes of climatic conditions, because short-term climatic changes may cause heterogeneous changes in surface brightness and thus imitate real cover changes. In addition, the presence of continuous cloud cover (e.g., in tropical and boreal regions) renders imaging with visible/near infrared systems difficult. In this case, only imaging yields reliable results (Kasischke et al., 1997).

At present, there is an urgent need for more complex tools to analyze heterogeneous vegetation scenes and to combine classical remote sensing, providing information about the biophysical properties of ecosystems analysis with data on energy, heat, and/or mass fluxes within these systems. Therefore, physically based, accurate models that can process

Table 13.3. Common Vegetation Indices Using Landsat^{TM} Bands and Correlations

Index	Name	Formula	Estimation by
SR	Simple ratio	$\frac{Band\ 4}{Band\ 3}$	
NDVI	Normalized difference. Vegetation index	$\frac{Band\ 4 - Band\ 3}{Band\ 4 + Band\ 3}$	
GVI	Greenness index or greenness vegetation index	−(0.2848*Band 1) −(0.2435*Band 2) −(0.5436*Band 3) +(0.7243*Band 4) +(0.0840*Band 5) −(0.1800*Band 7)	
LAI	Leaf area index = one-sided foliage area per ground area	$\frac{Leaf\ area\ [m^2]}{Ground\ area\ [m^2]}$	Correlations with NDVI
PAI	Plant area index = one-sided, plant area per ground area	$\frac{Plant\ area\ [m^2]}{Ground\ area\ [m^2]}$	Correlations with NDVI
FT	Total fractional cover	Areal proportion of the landscape occupied by green or nongreen vegetation	Correlations with NDVI

Band 1: 0.45–0.52 μm; Band 2: 0.52–0.60 μm; Band 3: 0.63–0.69 μm; Band 4: 0.76–0.90 μm; Band 5: 1.55–1.75 μm; Band 7: 2.08–2.35 μm).

Table 13.4. Spatial Resolution of Selected Remote Sensing Systems

	Landsat MSS	Landsat TM	SPOT XS	SPOT Pan	AVIRIS (airborne)	CASI (airborne)	Aerial photo
Spatial resolution (m)	80	30	20	10	20	10–0.5	User defined resolution

three-dimensional (3-D) structures of a regional ecosystem and link remote sensing to ecosystem process studies for a natural, complex scene are highly desirable (Qin and Gerstl, 2000). Qin and Gerstl developed such a model based on radiosity principles in combination with advanced computer graphic techniques.

References

Alagely, A. K., and Reeves, F. B. (1995). Inland sand dune mycorrhizae—effects of soil depth, moisture, and pH on colonization of *Oryzopsis hymenoides*. *Mycologia* 87: 54–60.

Al Mufti, M. M., Sydes, C. L., Furness, S. B., Grime, J. P., and Band, S. R. (1977). A quantitative analysis of shoot phenology and dominance in herbaceous vegetation. *Journal of Ecology* 65: 759–791.

Anderson, G. L., Hanson, J. D., and Haas, R. H. (1993). Evaluating Landsat Thematic Mapper derived vegetation indices for estimating above-ground biomass on semiarid rangelands. *Remote Sensing of Environment* 45: 165–175.

Archibold, O. W. (1995). *Ecology of world vegetation*. London: Chapman & Hall.

Arnthórsdóttir, S. (1994). Colonization of experimental patches in a mown grassland. *Oikos* 70: 73–79.

Ayal, Y., and Merkl, O. (1994). Spatial and temporal distribution of tenebrionid species (*Coleoptera*) in the Negev Highlands, Israel. *Journal of Arid Environments* 27: 347–361.

Barton, A. M. (1994). Gradient analysis of relationships among fire, environment, and vegetation in a southwestern USA mountain-range. *Bulletin of the Torrey Botanical Club* 121: 251–265.

Bazzaz, F. A. (1996). *Plants in changing environments*. Cambridge, UK: Cambridge University Press.

Bee, M. W., Wilson, J. B., and Mark, A. F. (1989). Stratification in a New Zealand rain forest. *Vegetatio* 79: 33–39.

Berendse, F., Schmitz, M., and de Visser, W. (1994). Experimental manipulation of succession in heathland ecosystems. *Oecologia* 100: 38–44.

Bertness, M. D. (1991). Interspecific interactions among high marsh perennials in a New-England salt-marsh. *Ecology* 72: 125–137.

Bertness, M. D., and Callaway, R. (1994). Positive interactions in communities. *Trends in Ecology and Evolution* 9: 191–193.

Betts, R. A., Cox, P. M., Lee, S. E., and Woodward, F. I. (1997). Contrasting physiological and structural vegetation feedbacks in climate change simulations. *Nature* 387: 796–799.

Betts, R. A., Cox, P. M., and Woodward, F. I. (2000). Simulated responses of potential vegetation to doubled-CO_2 climate change and feedbacks on near-surface temperature. *Global Ecology and Biogeography* 9: 171–180.

Birdsall, J. L., Quimby, P. C., Rees, N. E., Svejcar, T. J., and Sowell, B. F. (1997). Image analysis of leafy spurge (*Euphorbia esula*) cover. *Weed Technology* 11: 798–803.

Bisson, I. A., and Stutchbury, B. J. M. (2000). Nesting success and nest-site selection by a neotropical migrant in a fragmented landscape. *Canadian Journal of Zoology* 78: 858–863.

Booth, T. D. (1941). Algae as pioneers in plant succession and their importance in erosion control. *Ecology* 22: 38–46.

Borg, H., and Stoneman, G. L. (1991). Long-term implications for streamflow of changes in vegetation cover and stand height in regenerating karri stands in south-west western Australia. *Forest Ecology and Management* 40: 65–73.

Borghi, C. E., Giannoni, S. M., and Martinezrica, J. P. (1994). Habitat segregation of 3 sympatric fossorial rodents in the Spanish pyrenees. *International Journal of Mammalian Biology* 59: 52–57.

Bounoua, L., Collatz, G. J., Los, S. O., Sellers, P. J., Dazlich, D. A., Tucker, C. J., and Randall, D. A. (2000). Sensitivity of climate to changes in NDVI. *Journal of Climate* 13: 2277–2292.

Brath, A., and Montanari, A. (2000). The effects of the spatial variability of soil infiltration capacity in distributed flood modelling. *Hydrological Processes* 14: 2779–2794.

Braun-Blanquet, J. (1928). *Pflanzensoziologie—Grundzüge der Vegetationskunde* (1st ed.). Berlin: Springer.

Braun-Blanquet, J. (1964). *Pflanzensoziologie* (3rd ed.). Wien: Springer.

Brouwer, A., and Spaans, A. L. (1994). Egg predation in the herring gull *Larus argentatus*—Why does it vary so much between nests? *Ardea* 82: 223–231.

Brown, V. K. (1985). Insect herbivores and plant succession. *Oikos* 44: 17–22.

Bundrett, M. C., Ashwath, N., and Jasper, D. A. (1996). Mycorrhizas in the Kakadu region of tropical Australia: 2. Propagules of mycorrhizal fungi in disturbed habitats. *Plant and Soil* 184: 173–184.

Burrows, C. J. (1990). *Processes of vegetation change*. London: Unwin Hyman.

Caldwell, M. M. (1987). Plant architecture and resource competition. In: E. D. Schulze and H. Zwölfer (Eds.), *Potentials and limitations of ecosystem analysis* (pp. 164–170). Berlin: Springer.

Callaway, R. M. (1992). Effect of shrubs on recruitment of *Quercus douglasii* and *Quercus lobata* in California. *Ecology* 73: 2118–2128.

Campbell, B. D., and Grime, J. P. (1992). An experimental test of plant strategy theory. *Ecology* 73: 15–29.

Castillo, G., and Demoulin, V. (1998). Phenology of lignicolous basidiomycetes from Laing Island (Papua New Guinea). *Belgian Journal of Botany* 131: 237–243.

Cerda, A. (1999). Parent material and vegetation affect soil erosion in eastern Spain. *Soil Science Society of America Journal* 63: 362–368.

Chew, R. M. (1995). Aspects of the ecology of 3 species of ants (*Myrmecocystus* spp, *Aphaenogaster* sp) in desertified grassland in southeastern Arizona, 1958–1993. *American Midland Naturalist* 134: 75–83.

Clark, C. A., and Arritt, R. W. (1995). Numerical simulations of the effect of soil-moisture and vegetation cover on the development of deep convection. *Journal of Applied Meteorology* 34: 2029–2045.

Clymo, R. S. (1980). Preliminary survey of the peat-bog Hummell Knowe moss using various numerical methods. *Vegetatio* 42: 129–148.

Collison, A. J. C., Anderson, M. G., and Lloyd, D. M. (1995). Impact of vegetation on slope stability in a humid tropical environment—a modeling approach. *Proceedings of the Institution of Civil Engineers–Water Maritime and Energy* 112: 168–175.

Connell, J. H. (1978). Diversity in tropical rain forests and coral reefs. *Science* 199: 1302–1310.

Crawley, M. J. (1987). What makes a community invasible? In: A. J. Gray, M. J. Crawley, and P. J. Edwards (Eds.), *Colonization, succession and stability* (pp. 429–454). Oxford, UK: Blackwell Scientific Publications.

Crist, T. O., Guertin, D. S., Wiens, J. A., and Milne, B. T. (1992). Animal movement in heterogeneous landscapes—an experiment with *Eleodes* beetles in shortgrass prairie. *Functional Ecology* 6: 536–544.

Daubenmire, R. F. (1968). *Plant communities: A textbook of plant synecology.* New York: Harper and Row.

Davis, R. M., and Cantlon, J. E. (1969). Effect of size area open to colonization on species composition in early old-field succession. *Bulletin of the Torrey Botanical Club* 96: 660–673.

Dean, W. R. J., and Milton, S. J. (1995). Plant and invertebrate assemblages on old fields in the arid southern Karoo, South Africa. *African Journal of Ecology* 33: 1–13.

Deplessis, A., and Kerley, G. I. H. (1991). Refuge strategies and habitat segregation in 2 sympatric rodents *Otomys unisulcatus* and *Parotomys brantsii*. *Journal of Zoology* 224: 1–10.

Dettmer, L. (2000). *Biomassebestimmung mittels computergestützter Image-Analyse: Korrelationen an Standorten unterschiedlicher Ressourcenverfügbarkeit.* Magister thesis, University of Bielefeld, Bielefeld, Germany.

Diaz, J. A., and Carrascal, L. M. (1991). Regional distribution of a mediterranean lizard—influence of habitat cues and prey abundance. *Journal of Biogeography* 18: 291–297.

Dietz, H., and Steinlein, T. (1996). Determination of plant species cover by means of image analysis. *Journal of Vegetation Science* 7: 131–136.

Dietz, H., and Steinlein, T. (1998). The impact of anthropogenic disturbance on life stage transitions and stand regeneration of the invasive alien plant *Bunias orientalis* L. In: K. Edwards, I. Kowarik, U. Starfinger, and M. Williamson (Eds.), *Invasions: Ecological mechanisms and human responses* (pp. 169–184). Leiden: Backhuys.

Dietz, H., Steinlein, T., and Ullmann, I. (1999). Establishment of the invasive perennial herb *Bunias orientalis* L.: An experimental approach. *Acta Oecologica* 20: 621–632.

Dietz, H., and Ullmann, I. (1997). Phenological shifts of the alien colonizer *Bunias orientalis* L.: An image-based analysis of temporal niche separation. *Journal of Vegetation Science* 8: 839–846.

Ehrenfeld, J. G. (1980). Understory response to gaps of varying size in a mature oak forest. *Bulletin of the Torrey Botanical Club* 107: 29–41.

Eldridge, D. J. (1993a). Cryptogam cover and soil surface condition—effects on hydrology on a semiarid woodland soil. *Arid Soil Research and Rehabilitation* 7: 203–217.

Eldridge, D. J. (1993b). Cryptogams, vascular plants, and soil hydrological relations—some preliminary results from the semiarid woodlands of eastern Australia. *Great Basin Naturalist* 53: 48–58.

Epstein, H. E., Burke, I. C., and Mosier, A. R. (1998). Plant effects on spatial and temporal patterns of nitrogen cycling in shortgrass steppe. *Ecosystems* 1: 374–385.

Everson, T. M., Clarke, G. P. Y., and Everson, C. S. (1990). Precision in monitoring plant species composition in montane grasslands. *Vegetatio* 88: 135–141.

Fasola, M., and Canova, L. (1992). Nest habitat selection by 8 syntopic species of Mediterranean gulls and terns. *Colonial Waterbirds* 15: 169–178.

Flores, L. A., and Marinez, L. A. (2000). Land cover estimation in small areas using ground survey and remote sensing. *Remote Sensing of Environment* 74: 240–248

Fowler, N., and Antonovics, J. (1981). Competition and coexistence in a North Carolina grassland. *Journal of Ecology* 69: 825–841.

Frouz, J. (1997). The effect of vegetation patterns on oviposition habitat preference: A driving mechanism in terrestrial chironomid (*Diptera*: Chironomidae) succession? *Researches on Population Ecology* 39: 207–213.

Gares, P. A. (1992). Topographic changes associated with coastal dune blowouts at Island Beach state park, New Jersey. *Earth Surface Processes and Landforms* 17: 589–604.

Gates, D. M. (1980). *Biophysical ecology.* Berlin: Springer.

Givnish, T. J. (1982). On the adaptive significance of leaf height in forest herbs. *American Naturalist* 120: 353–381.

Goff, B. F., Bent, G. C., and Hart, G. E. (1993). Erosion response of a disturbed sagebrush steppe hillslope. *Journal of Environmental Quality* 22: 698–709.

Gold, W. G. (1998). The influence of cryptogamic crusts on the thermal environment and temperature relations of plants in a high Arctic polar desert, Devon Island, NWT, Canada. *Arctic and Alpine Research* 30: 108–120.

Goldberg, D. E. (1990). Components of resource competition in plant communities. In: J. B. Grace and D. Tilman (Eds.), *Perspectives in plant competition* (pp. 27–49). New York: Academic Press.

Goldberg, D. E., and Werner, P. A. (1983). The effects of size of opening in vegetation and litter cover on seedling establishment of goldenrods (*Solidago* spp.). *Oecologia* 60: 149–155.

Goodall, D. W. (1952). Some considerations in the use of point quadrats for the analysis of vegetation. *Australian Journal of Scientific Research Series B—Biological Research* 5: 1–41.

Goodall, D. W. (1953). Point quadrat methods for the analysis of vegetation. *Australian Journal of Botany* 1: 457–461.

Grace, J. B., and Pugesek, B. H. (1997). A structural equation model of plant species richness and its application to coastal wetland. *American Naturalist* 149: 436–460.
Greipsson, S., and El Mayas, H. (1999). Large-scale reclamation of barren lands in Iceland by aerial seeding. *Land Degradation and Development* 10: 185–193.
Grime, J. B. (1987). Dominant and subordinate components of plant communities: Implications for succession, stability and diversity. In: A. J. Gray; M. J. Crawley, and P. J. Edwards (Eds.), *Colonization, succession and stability* (pp. 413–428). Oxford, UK: Blackwell Scientific.
Gross, K. L. (1982). Colonizing abilities of biennial plant species in relation to ground cover: Implications for their distributions in a successional sere. *Ecology* 63: 921–931.
Grubb, P. J. (1977). The maintenance of species-richness in plant communities: The importance of the regeneration niche. *Biological Review* 52: 107–145.
Grytnes, J. A. (2000). Fine-scale vascular plant species richness in different alpine vegetation types: Relationships with biomass and cover. *Journal of Vegetation Science* 11: 87–92.
Haigh, M., Rawat, J. S., Rawat, M. S., Bartarya, S. K., and Rai, S. P. (1995). Interactions between forest and landslide activity along new highways in the Kumaun Himalaya. *Forest Ecology and Management* 78: 173–189.
Harper, J. L. (1977). *Population biology of plants*. London: Academic Press.
Hedman, C. W., Grace, S. L., and King, S. E. (2000). Vegetation composition and structure of southern coastal plain pine forests: An ecological comparison. *Forest Ecology and Management* 134: 233–247.
Hirose, T., and Werger, M. J. A. (1995). Canopy structure and photon flux partitioning among species in a herbaceous plant community. *Ecology* 76: 466–474.
Hoffmann, W. A. (1996). The effects of fire and cover on seedling establishment in a neotropical savanna. *Journal of Ecology* 84: 383–393.
Hurka, H., and Winkler, S. (1973). Statistische Analyse der Rindenbewohnenden Flechtenvegetation einer Allee Tübingens. *Flora* 162: 61–80.
Huston, M. (1979). A general hypothesis of species diversity. *American Naturalist* 113: 81–110.
Jensen, S. P., and Honess, P. (1995). The influence of moonlight on vegetation height preference and trappability of small mammals. *Mammalia* 59: 35–42.
Jordan, P. W., and Nobel, P. S. (1979). Infrequent establishment of seedlings of *Agave desertii* (Agavaceae) in the northwestern Sonoran Desert. *American Journal of Botany* 66: 1079–1084.
Kasischke, E. S., Melack, J. M., and Dobson, M. C. (1997). The use of imaging radars for ecological applications—a review. *Remote Sensing of Environment* 59: 141–156.
Keddy, P. A. (1989). Effects of competition from shrubs on herbaceous wetland plants: A 4-year field experiment. *Canadian Journal of Botany* 67: 708–716.
Keizer, P. J., van Tooren, B. F., and During, H. J. (1985). Effects of bryophytes on seedling emergence and establishment of short-lived forbs in chalk grassland. *Journal of Ecology* 73: 493–504.
Kennedy, K. A., and Addison, P. A. (1987). Some considerations for the use of visual estimates of plant cover in biomonitoring. *Journal of Ecology* 75: 152–157.
Kent, M., and Coker, P. (1992). *Vegetation description and analysis*. Boca Raton, FL: CRC Press.
Kerley, G. I. H. (1992). Ecological correlates of small mammal community structure in the semiarid Karoo, South Africa. *Journal of Zoology* 227: 17–27.
Knapp, R. (ed.) (1984a). *Handbook of vegetation science: IV. Sampling methods and taxon analysis in vegetation science*. The Hague: Dr. W. Junk Publishers.
Knapp, R. (1984b). Sample (relevé) areas (distribution, homogeneity, size, shape) and plotless sampling. In: R. Knapp (Ed.), *Handbook of vegetation science: IV. Sampling methods and taxon analysis in vegetation science* (pp. 35–44). The Hague: Dr. W. Junk Publishers.
Koller, D. (1990). Light-driven leaf movements. *Plant, Cell and Environment* 13: 615–632.
Kollmann, J., and Bassin, S. (2001). Effects of managment on seed predation in wildflower strips in northern Switzerland. *Agriculture, Ecosystems and Environment* 83: 285–296.
Küchler, A. W., and Zonneveld, I. S. (1988). Floristic analysis of vegetation. In: A. W. Küchler and I. S. Zonneveld (Eds.), *Vegetation mapping* (pp. 51–66). Dordrecht: Kluwer Academic.
Kustas, W. P., Prueger, J. H., Hatfield, J. L., Ramalingam, K., and Hipps, L. E. (2000). Variability in soil heat flux from a mesquite dune site. *Agricultural and Forest Meteorology* 103: 249–264.
Kutiel, P. (1998). Possible role of biogenic crusts in plant succession on the Sharon sand dunes, Israel. *Israel Journal of Plant Sciences* 46: 279–286.
Lancaster, N., and Baas, A. (1998). Influence of vegetation cover on sand transport by wind: Field studies at Owens Lake, California. *Earth Surface Processes and Landforms* 23: 69–82.

Lawrence, R. L., and Ripple, W. J. (1998). Comparison among vegetation indices and bandwise regression in a highly disturbed, heterogeneous landscape: Mount St. Helens, Washington. *Remote Sensing of Environment* 64: 91–102.

Levin, S. A. (1992). The problem of pattern and scale in ecology. *Ecology* 73: 1943–1967.

Levy, E. E., and Madden, E. A. (1933). The point method of pasture analysis. *New Zealand Journal of Agriculture* 46: 267–279.

Leys, J. F. (1991). Towards a better model of the effect of prostrate vegetation cover on wind erosion. *Vegetatio* 91: 49–58.

Loch, R. J. (2000). Using rainfall simulation to guide planning and management of rehabilitated areas: Part I. Experimental methods and results from a study at the Northparkes Mine, Australia. *Land Degradation and Development* 11: 221–240.

Londo, G. (1976). The decimal scale for relevés of permanent quadrats. *Vegetatio* 33: 61–64.

Londo, G. (1984). The decimal scale for relevés of permanent quadrats. In: R. Knapp (Ed.), *Sampling methods and taxon analysis in vegetation science* (pp. 35–44). The Hague: Dr W. Junk Publishers.

Longland, W. S. (1991). Risk of predation and food-consumption by black-tailed jackrabbits. *Journal of Range Management* 44: 447–450.

Lukesova, A., and Hoffmann, L. (1996). Soil algae from acid rain impacted forest areas of the Krusnehory Mts.: 1. Algal communities. *Vegetatio* 125: 123–136.

Manson, R. H., and Stiles, E. W. (1998). Links between microhabitat preferences and seed predation by small mammals in old fields. *Oikos* 82: 37–50.

Martin, J., and Lopez, P. (1995). Influence of habitat structure on the escape tactics of the lizard *Psammodromus algirus*. *Canadian Journal of Zoology* 73: 129–132.

McCarthy, D. P., and Zaniewski, K. (2001). Digital analysis of lichen cover: A technique for use in lichenometry and lichenology. *Arctic, Antarctic and Alpine Research* 33: 107–113.

Messier, C., Parent, S., and Bergeron, Y. (1998). Effects of overstory and understory vegetation on the understory light environment in mixed boreal forests. *Journal of Vegetation Science* 9: 511–520.

Milton, S. J. (1995). Spatial and temporal patterns in the emergence and survival of seedlings in arid Karoo shrubland. *Journal of Applied Ecology* 32: 145–156.

Mitchley, J. (1988). Control of relative abundance of perennials in chalk grassland in southern England: II. Vertical canopy structure. *Journal of Ecology* 76: 341–350.

Molau, U., and Alatalo, J. M. (1998). Responses of subarctic-alpine plant communities to simulated environmental change: Biodiversity of bryophytes, lichens, and vascular plants. *Ambio* 27: 322–329.

Molloy, J. M., and Moran, C. J. (1991). Compiling a field manual from overhead photographs for estimating crop residue cover. *Soil Use and Management* 7: 177–183.

Moore, P. D., and Chapman, S. B. (1986). *Methods in plant ecology.* Oxford, UK: Blackwell Scientific.

Mueller-Dombois, D., and Ellenberg, H. (1974). *Aims and methods of vegetation ecology.* New York: Wiley.

Negi, G. C. S., Rikhari, H. C., and Singh, S. P. (1992). Phenological features in relation to growth forms and biomass accumulation in an alpine meadow of the central Himalaya. *Vegetatio* 101: 161–170.

Okland, T., Okland, R. H., and Steinnes, E. (1999). Element concentrations in the boreal forest moss *Hylocomium splendens*: Variation related to gradients in vegetation and local environmental factors. *Plant and Soil* 209: 71–83.

Pakarinen, P. (1984). Cover estimation and sampling of boreal vegetation in northern Europe. In: R. Knapp (Ed.), *Handbook of vegetation science: IV. Sampling methods and taxon analysis in vegetation science* (pp. 35–44). The Hague: Dr W. Junk Publishers.

Peart, D. R. (1989a). Species interactions in a successional grassland: I. Seed rain and seedling recruitment. *Journal of Ecology* 77: 236–251.

Peart, D. R. (1989b). Species interactions in a successional grassland: III. Effects of canopy gaps, gopher mounds and grazing on colonization. *Journal of Ecology* 77: 267–289.

Petranka, J. W., and McPherson J. K. (1979). The role of *Rhus copallina* in the dynamics of the forest-prairie ecotone in north-central Oklahoma. *Ecology* 60: 956–965.

Pharo, E. J., and Vitt, D. H. (2000). Local variation in bryophyte and macro-lichen cover and diversity in montane forests of western Canada. *Bryologist* 103: 455–466.

Pielke, R. A., Avissar, R., Raupach, M., Dolman, A. J., Zeng, X., and Denning, A. S. (1998). Interactions between the atmosphere and terrestrial ecosystems: Influence on weather and climate. *Global Change Biology* 4: 461–475.

Pimentel, D., and Kounang, N. (1998). Ecology of soil erosion in ecosystems. *Ecosystems* 1: 416–426.

Prach, K., and Pysek, P. (1994). Clonal plants: What is their role in succession? *Folia Geobotanica and Phytotaxonomica* 29: 307–320.

Press, M. C., Potter, J. A., and Burke, M. J. W. (1998). Responses of a subarctic dwarf shrub heath community to simulated environmental change. *Journal of Ecology* 86: 315–327.

Prosser, I. P., and Dietrich, W. E. (1995). Field experiments on erosion by overland-flow and their implication for a digital terrain model of channel initiation. *Water Resources Research* 31: 2867–2876.

Qin, W., and Gerstl, S. A. W. (2000). 3-D scene modeling of semidesert vegetation cover and its radiation regime. *Remote Sensing of Environment* 74: 145–162.

Rahman, M. H., and Khan, Y. A. (1995). Landslides and stability of coastal cliffs of Coxs Bazar area, Bangladesh. *Natural Hazards* 12: 101–118.

Ramirez, C. (1984). Einfluss der Jahreszeit auf Vegetations-Aufnahmen von Rasengesellschaften mit Thero- und Geophyten. In: R. Knapp (Ed.), *Handbook of vegetation science: IV. Sampling methods and taxon analysis in vegetation science* (pp. 181–183). The Hague: Dr. W. Junk Publishers.

Reader, R. J., and Best, B. J. (1989). Variation in competition along an environmental gradient: *Hieracium floribundum* in an abandoned pasture. *Journal of Ecology* 77: 673–684.

Reader, R. J., and Buck J. (1986). Topographic variation in the abundance of *Hieracium floribundum*: Relative importance of differential seed dispersal, seedling establishment, plant survival and reproduction. *Journal of Ecology* 74: 815–822.

Reader, R. J., Wilson, S. D., Belcher, J. W., Wisheu, I., Keddy, P. A., and Tilman, D. (1994). Plant competition in relation to neighbor biomass: An intercontinental study with *Poa pratensis*. *Ecology* 75: 1753–1760.

Reich, R. M., Bonham, C. D., and Remingtojn, K. K. (1994). Comparison of classical vs inverse sampling for estimating plant cover using Monte-Carlo simulations. *Applied Mathematics and Computation* 64: 171–190.

Retana, J., and Cerda, X. (2000). Patterns of diversity and composition of Mediterranean ground ant communities tracking spatial and temporal variability in the thermal environment. *Oecologia* 123: 436–444.

Robinson, C. H., Wookey, P. A., Lee, J. A., Callaghan, T. V., and Press, M. C. (1998). Plant community responses to simulated environmental change at a high arctic polar semi-desert. *Ecology* 79: 856–866.

Rogers, R. W. (1995). Lichen succession on leaves of the rain-forest shrub, *Capparis arborea* (Capparaceae). *Australian Journal of Botany* 43: 387–396.

Röttgermann, M., Steinlein, T., Beyschlag, W., and Dietz, H. (2000). Linear relationships between aboveground biomass and plant cover in low open herbaceous vegetation. *Journal of Vegetation Science* 11: 145–148.

Roxburgh, S. H., Watkins, A. J., and Wilson, J. B. (1993). Lawns have vertical stratification. *Journal of Vegetation Science* 4: 699–704.

Rusek, J., and Marshall, V. G. (2000). Impacts of airborne pollutants on soil fauna. *Annual Review of Ecology and Systemation* 31: 395–423.

Scanlan, J. C., Grant, W. E., Hunter, D. M., and Milner, R. J. (2001). Habitat and environmental factors influencing the control of migratory locusts (*Locusta migratoria*) with an entomopathogenic fungus (*Metarhizium anisopliae*). *Ecological Modelling* 136: 223–236.

Schiavon, M., Portal, J. M., and Andreux, F. (1992). The movement of atrazine in the environment. *Agronomie* 12: 129–139.

Schlaepfer F. (1997). Influence of management on cover and seed production of *Brachypodium pinnatum* (L.) Beauv. in a calcareous grassland. *Bulletin of the Geobotanical Institute ETH* 63: 3–10.

Schreiner, M., Bauer, E.-M., and Kollmann, J. (2000). Reducing predation of conifer seeds by clear-cutting *Rubus fruticosus* agg. in two montane forest stands. *Forest Ecology and Management* 126: 281–290.

Seastedt, T. R. (1985). Canopy interception of nitrogen in bulk Porecipitation by annually burned and unburned tallgrass prairie. *Oecologia* 66: 88–92.

Seymour, C. L., and Dean, W. R. J. (1999). Effects of heavy grazing on invertebrate assemblages in the succulent Karoo, South Africa. *Journal of Arid Environments* 43: 267–286.

Silander, J. A., and Antonovics, J. (1982). Analysis of interspecific interactions in a coastal plant community—a perturbation approach. *Nature* 298: 557–560.

Smith, A. P. (1973). Stratification of temperate and tropical forests. *American Naturalist* 107: 671–683.

Spanner, M. A., Pierce, L. L., Peterson, D. L., and Running, S. W. (1990). Remote sensing temperate coniferous forest leaf area index: The influence of canopy closure, understory and background reflectance. *International Journal of Remote Sensing* 11: 95–111.

Stampfli, A. (1991). Accurate determination of vegetational change in meadows by successive point quadrat analysis. *Vegetatio* 39: 97–114.

Sternberg, M., Gutman, M., Perevolotsky, A., Ungar, E. D., and Kigel, J. (2000). Vegetation response to grazing management in a Mediterranean herbaceous community: A functional group approach. *Journal of Applied Ecology* 37: 224–237.

Stone, D. A., Lancashire, R. A., Sutherland, K. B., Niendorf, R. B., and Sampson, R. B. (1988). A low cost microcomputer-based image analysis system for the measurement of percent ground cover. *Research and Development in Agriculture* 5: 65–70.

Sykes, J. M., Horrill, A. D., and Mountford, M. D. (1983). Use of visual cover assessments as quantitative estimators of some British woodland taxa. *Journal of Ecology* 71: 437–450.

Thomas, D. L., da Silva, F. J., and Cromer, W. A. (1988). Image processing techniques for plant canopy cover evaluation. *Transactions of the ASAE* 31: 428–434.

Turkington, R., and Klein, E. (1993). Interactive effects of nutrients and disturbance: An experimental test of plant strategy theory. *Ecology* 74: 863–878.

Ullmann, I., and Büdel, B. (2001). Biological soil crusts of Africa. In: J. Belnap and O. L. Lange (Eds), *Biological soil crusts: Structure, function and management: Ecological studies* (Vol. 150, pp. 107–118). Berlin: Springer.

van der Maarel, E. (1979). Transformation of cover-abundance values in phytosociology and its effects on community similarity. *Vegetatio* 39: 97–114.

van Tooren, B. F. (1988). The fate of seeds after dispersal in chalk grassland: The role of the bryophyte layer. *Oikos* 53: 41–48.

Vanha-Majamaa, I., Salemaa, M., Tuominen, S., and Mikkola, K. (2000). Digitized photographs in vegetation analysis—a comparison of cover estimates. *Journal of Vegetation Science* 3: 89–94.

Veste, M., Littmann, T., Schultz, A., Eggert, K., Sommer, C., and Breckle, S.-W. (2000). Biomasseverteilung und deren räumliche Modellierung in Sanddünen der Negev-Wüste (Israel). *Verhandlungen der Gesellschaft für Ökologie* 30: 85.

Vinton, M. A., and Burke, I. C. (1995). Interactions between individual plant species and soil nutrient status in shortgrass steppe. *Ecology* 76: 1116–1133.

Waring, R. H., Way, J. B., and Hunt, E. R., Jr. (1995). Imaging radar for ecosystem studies. *Bioscience* 45: 715–723.

Werger, M. J. A., and Hirose, T. (1988). Effects of light climate and nitrogen partitioning on the canopy structure of stands of a dicotyledonous, herbaceous vegetation. In: M. J. A. Werger, P. J. M. van der Aart, H. J. During and J. T. A. Verhoeven (Eds.), *Plant form and vegetation structure* (pp. 171–181). The Hague: SPB Academic.

Wiggs, G. F. S., Thomas, D. S. G., and Bullard, J. E. (1995). Dune mobility and vegetation cover in the southwest Kalahari Desert. *Earth Surface Processes and Landforms* 20: 515–529.

Wildi, O. (1986). Analyse vegetationskundlicher Daten: Theorie und Einsatz statistischer Methoden. *Veröffentlichugen des Geobotanische Instituts Eidgenüssiche Technische Hochschule Stiftung Rübel, Zürich*. 90.

Wimbush, D. J., Barrow, M. D., and Costin A. B. (1967). Color stereography for the measurement of vegetation. *Ecology* 48: 150–152.

Winn, A. (1985). Effects of seed size and microsite on seedling emergence of *Prunella vulgaris* in four habitats. *Journal of Ecology* 73: 831–840.

Wirth, V., and Brinckmann, B. (1977). Statistical analysis of the lichen vegetation of an avenue in Freiburg (southwest Germany), with regard to injurious anthropogenous influences. *Oecologia* 28: 87–101.

Yoder, B. J., and Waring, R. H. (1994). The normalized difference vegetation index of small Douglas-fir canopies with varying chlorophyll concentrations. *Remote Sensing of Environment* 49: 81–91.

Zamfir, M. (2000). Effects of bryophytes and lichens on seedling emergence of alvar plants: Evidence from greenhouse experiments. *Oikos* 88: 603–611.

14

Ecology of Individuals

Individual-Based Models of Single Population Dynamics

Janusz Uchmański

Introduction

General Remarks

Life exists in the form of individuals. To express the same idea in a more orthodox way, nothing exists in nature except individuals. Individuals do not live in isolation. They interact with other individuals. These interactions influence their features and are the most important obstacle in understanding ecological processes. All terms we use in ecology, are results or by-products of individuals and interactions between them. Each ecologist performing experiments or observations deals with individuals or the effects of their activity. However, science means not only performing experiments and observations but also developing theories that explain them. In classical ecology, theories most often do not refer directly to individuals.

The individual-based approach in ecology is furthest developed in theoretical ecology, and primarily means special, individually focused ways of mathematically modeling the dynamics of ecological systems. The term *individual-based approach* was first used in ecology more than 10 years ago (Huston et al., 1988). In the same period the term *ecology of the individuals* was also introduced in ecological literature (Łomnicki, 1988). There are no great differences between these two terms. The first may suggest that there are other approaches. On the other hand, the second term suggests that there also exists some kind of mature theory. After more than 10 years of the history of the problem and many discussions (Koehl, 1989; DeAngelis et al., 1994; Judson, 1994; Uchmański and Grimm, 1996; Łomnicki, 1992, 1999), one can say that both suggestions are correct.

The history of the problem is of course much longer. Some papers that can now be assigned to the individual-based approach appeared much earlier, in 1950s or 1960s. Some

Janusz Uchmański • Department of Ecology and Environmental Protection, University of Podlasie, ul. Prusa 12, 08-110 Siedlce, Poland of them are still important for understanding ecology of individuals. So far, only a few meetings focused on the individual-based approach in ecology have been organized. I am aware of three of them: in 1990, in Knoxville, Tennessee (DeAngelis and Gross, 1992); in 1996, at MikoŁajki, Poland (Uchmański et al., 1999); and the symposium organized as a part of the VII International Congress of Ecology in Florence, in 1998 (Uchmański et al., 1999).

Classical Theoretical Ecology

Traditional ecology is centered on the dynamics of ecological systems. It can be described by different variables of which the most natural is the number of individuals in populations forming an ecological system. In classical theoretical ecology, the logistic equation is used to describe single-population dynamics, and the Lotka–Volterra models are applied to model community dynamics. These basic models describe features of an average individual. They also assume a very simple lifecycle of this average individual. Only elements of lifecycle such as birth, death, and the number of offspring are taken into account. In an extreme case, an average individual is born, immediately produces offspring, and dies shortly afterwards. Even in models with age, stage, or size structure, which are based on the assumption that reproduction and mortality depend on age, stage, or size of the individual, individual variability within each class is still neglected, and features of an average individual in each class are considered. Also, other classes of models that include more complex lifecycles (e.g., models by Kaiser, 1979; Matessi, 1982) and models describing so-called "physiologically structured populations" (Metz and Diekmann, 1986; Ebenman and Persson, 1988) are still based on an average individual.

The second general feature of classical ecological models is how resources are modeled. They can be modeled explicitly, but are modeled implicitly in most classical population models, using the concept of carrying capacity. For most types of biotic or abiotic resources, an implicit representation of resources is probably not sufficient (Kozlowski, 1980). Only explicit modeling of resource dynamics will clarify the relationship between individuals of population and resources.

In most classical population and community models, population density, which is a real number, is used as a state variable. This approach can be justified if the model describes large populations, because, for large numbers of individuals, the discreteness of individuals may be neglected. However, the validity of this argument is not absolute and depends on the character of interactions between individuals.

Introduction of explicit resource dynamics into classical ecological models is a relatively easy task (see, for instance, so-called resource–consumer models, or models describing energy and matter flows between trophic levels). There is also a group of classical models that use the number of individuals in the population (a natural number). They are mainly stochastic models used in extinction theory and population viability analysis (e.g., Wissel et al., 1994; Groom and Pascual, 1998) to describe the dynamics of small populations.

The assumption, which states that for a population model it is sufficient to describe features of an average individual, is much more deeply rooted in ecological literature. The most important consequence of the decision to use features of an average individual in population and community models is neglecting of differences among individuals. However, it contradicts even everyday observations. In nature, individuals differ not just in age and sex. They are different even if they are of the same age. There are two sources of differences in even-age populations. First, they may be casual differences. This can be illustrated, for instance, by the situation when all individuals have the same probability of producing an offspring, but it is impossible to predict which one will reproduce in a given, short-term interval. In this sense, individual variability is introduced in some stochastic population models (Stephan and Wissel, 1999), but these models usually use extremely simple lifecycles. Second, differences among individuals may be due to other than random, more important, factors. They may be an effect of competition between individuals. Many models, both for plants and animals, describe the development of differences among individuals due to competition. For example, analyses of changes in frequency distributions of body weights in even-age plant and animal populations in relation to the population density and resources are frequent (see later sections in this chapter). Models describing the development and change in individual variability within single generations can be iterated and thereby produce models of population dynamics incorporating not only more complex lifecycles of individuals but also differences among them.

Individual-Based Approach to Ecology: Different Definitions

The individual-based approach in ecology represents a way of thinking about ecological processes that primarily considers the fates of individuals in ecological systems. Applied to the dynamics of an ecological system, it analyzes at the beginning the properties of individuals and interactions among them, then uses this information as the basis to explain changes in sizes of populations in ecological systems.

As a young discipline with significant potential for the future development of theoretical ecology, the individual-based approach demands much discussion. For example, various authors define it in different ways. Some consider that each model with a little more information about individuals than that available in classical models is an individual-based model (Bolker et al., 1997). They believe that there is a continuous transition between a classical and individual-based way of modeling. For others, individual-based models must conform to very restrictive assumptions (Uchmański and Grimm, 1996). Each type can produce widely different assumptions and results.

A large part of the uncertainty about whether individual-based modeling constitutes a new approach results from imprecisely delimited boundaries between classical and individual-based modeling. The ofien-used differentiation of individual-based ecological models into so-called *i*-state and *i*-distribution models (Metz and Diekmann, 1986) reflects only a rather technical aspect of the model structure. The classification of ecological models can be much more meaningful when based on some biologically important attributes of the models. Four criteria have been proposed (Uchmański and Grimm, 1996): (1) the degree to which the complexity of the individual's lifecycle is reflected in the model; (2) whether dynamics of the resources (such as food, space, or habitat quality) are explicitly taken into account; (3) the use of real or natural numbers in representing the size of a population; and (4) the extent to which variability of individuals of the same age is considered. All models of theoretical ecology that describe dynamics of ecological systems may be classified according to these criteria.

As one can see from the description of classical models presented earlier, this type of model is characterized by (1) extremely simple lifecycles of individuals, (2) resource dynamics not taken explicitly into account, (3) the real number—the population density—representing the state variable of the model, and (4) individual variability (especially among individuals of the same age) not considered in the classical models. In contrast,

Uchmański and Grimm (1996) proposed to call the "narrowly defined" individual-based model one that fulfils the following conditions: It describes a population made up of individuals that may differ from one another (even when they are of the same age); it takes into account the complexity of the individual's lifecycle; it describes changes in numbers of individuals rather than in the population density; and it also takes resource dynamics explicitly into account. All other models that somehow deal with individuals but do not fulfil these four criteria differ only more or less slightly from classical models and can therefore be called "individual-oriented" models.

Content of this Chapter

A set of problems important to an individual-based approach in ecology has emerged over the last 10 or so years as a result of publications and discussions at conferences (cf. DeAngelis et al., 1994). The general problem—why apply an individual-based approach in ecology, and does it produce results that differ from results of classical ecology?—still needs more discussion. In addition, the set of basic problems involve growth of individuals, competition, and resource partitioning among competing individuals, especially in even-age populations; individual variability including weight distribution in even-age populations; and population dynamics generated by individual-based models. However, as we can see from the literature, general, theoretical questions are not the main aims of individual-based models. Most individual-based models are constructed to answer some applied questions (Bart, 1995; Dunning et al., 1995). They involve all the aforementioned problems, but in different ways and proportions. To clarify a presentation of individual-based modeling I present only one, very simple approach used to address questions concerning stability and regulation of population in this type of model. Based on energetic budgets of an individual, is weight-oriented—it assumes that the weight of the individual influences many important functions of the organism, and that conflicts between individuals can be decided on the basis of their weights.

The Individual: An Ecological Model

The equation presenting the energy balance of an organism can serve as a simple model of an individual. The energy is delivered to an organism in the process of assimilation. Respiration means the amount of energy needed for the cost of life. The difference between assimilation and respiration represents energy, which can be used for different life functions, for example, growth or reproduction. Let us denote the rate of assimilation by A and the rate of respiration by R. The difference between them is equal to the rate of production, P:

$$P = A - R \tag{1}$$

It can be divided into progeny production E and growth of the individual organism Δw:

$$P = E + \Delta w. \tag{2}$$

Theory of optimal life strategies excludes situations in which growth and reproduction take place simultaneously (Stearns, 1992). An individual starts its lifecycle from growth, then reproduces after reaching some weight. There can be many reproduction periods, each separated by the period of growth.

The rate of assimilation (A) and the rate of respiration (R) of an individual can be presented as power functions of its weight (w) (Sushchenya and Khmeleva, 1967; Duncan and Klekowski, 1975):

$$A = a_1 w^{b_1} \tag{3}$$

and

$$R = a_2 w^{b_2}, \tag{4}$$

where a_1, a_2, b_1, and b_2 are constants.

Both assimilation and respiration rates depend on environmental conditions. It can be assumed that in changing food conditions, A changes, but in such a way that only a_1 takes different values (Sushechnya, 1975). The Michaelis–Menten function (Murray, 1989) may be used to describe changes of assimilation rate with the amount (V) of food available in the environment:

$$a_1(V) = a_1^{\max} \frac{V}{G + V}, \tag{5}$$

where $a_1^{\max}$ is a maximal value of parameter a_1 for $V = \infty$ and G is the half-saturation constant. The other possibility is Ivlev function (Ivlev, 1961) of the form:

$$a_1(V) = a_1^{\max}(1 - e^{\xi V}), \tag{6}$$

where ξ is a constant and $a_1^{\max}$ is a maximal value of parameter a_1 when $V = \infty$.

Respiration rate depends strongly on temperature. Again, only the value of parameter a_2 changes with temperature. The Arrhenius relationship can be used in this case:

$$a_2(V) = a_2^{\max} e^{-\alpha/T}, \tag{7}$$

where $a_2^{\max}$ and α are constants. More simply, the Van't Hoff relationship can be also used in this case:

$$R(T_2) = Q_{T_2 - T_1} R(T_1), \tag{8}$$

where $Q_{T_2-T_1}$ specifies how many times respiration increases when temperature changes from T_1 to T_2.

According to equations (1) and (2), changes of environmental conditions, such as the amount of food or temperature, result in different weights of individuals growing in changed conditions.

All of these equations are expressions of the principle of energy conservation or the results of laboratory experiments conducted on single individuals, or on a group of individuals, but in the latter case, results were then averaged and recalculated per capita. This requires the introduction of the term *isolated individual* to denote an individual that is not interacting with other individuals in the population. Most ecophysiological experiments (which are also the basis for equations in this section) produce valid results for isolated individuals only.

Growth: Isolated Individuals

Two models of individual growth, unlimited and limited, are considered (Majkowski and Uchmański, 1980). Let us assume that individual growth is described by a limited curve. In the period when there is no reproduction, the rate of weight increase dw/dt is described by the following equation:

$$\frac{dw}{dt} = A - R, \tag{9}$$

or using equations (3) and (4) we obtain:

$$\frac{dw}{dt} = a_1 w^{b_1} - a_2 w^{b_2}. \tag{10}$$

Equation (10) describes an S-shaped growth curve with final weight w_∞ given by the equation:

$$w_\infty = \left(\frac{a_1}{a_2}\right)^{1/(b_2-b_1)}. \tag{11}$$

The final weight is reached asymptotically by the growing individual.

As an example of unlimited growth of an isolated individual, the exponential growth equation can be used:

$$\frac{dw}{dt} = rw, \tag{12}$$

where r is the relative growth rate. The solution to this equation is given by the following function:

$$w(t) = w_0 e^{r(t-t_0)}, \tag{13}$$

where w_0 is initial weight at initial time instant t_0.

Let us assume that the population consists of N individuals who grow in isolation, without interaction between them. Koyama and Kira (1956) considered the unlimited growth described by the exponential growth equation (12). After taking logarithms of the equation (13) they obtain:

$$\ln w(t) = \ln w_0 + r(t - t_0) \tag{14}$$

Let us assume that individuals in the population differ in r values. If r is randomly distributed according to normal distribution in the population, then we conclude that $\ln w(t)$ is also normally distributed. This means that $w(t)$ has a lognormal distribution of positive skewness. Koyama and Kira (1956) (and later Attchinson and Brown, 1957; Turner and Rabinowitz, 1983) concluded that skewed weight distribution should be expected for groups of individuals growing in isolation, provided that growth is exponential and relative growth rate has a normal distribution. Both skewness and variance of weight distribution increase with age of individuals. The same results can be obtained when initial weight w_0 has a normal distribution in the population, or when both parameters of the equation (13), relative growth rate r and initial weight w_0, are chosen from normal distribution.

A similar phenomenon is observed in the case of limited growth. If only initial weights are normally distributed, then weight distribution will have greater variance and skewness in the phase of rapid growth, but final weights for all individuals will be equal. Let us assume now that in addition values of parameter a_1 are normally distributed. In this situation, the distribution of final weights is also characterized by a positive value of coefficient of skewness, while distribution of initial weight is symmetric. Skewness and variance of weight distribution will change during individuals' growth, reaching a maximum in the phase of rapid growth. An increase in the standard deviation of a_1 distribution leads to an increase of the skewness coefficient of the final weight distribution (Uchmański and Dgebuadze, 1990).

Parameters a_1 and r can be interpreted as factors reflecting individual ability to consume and assimilate food. So for positive skewness of final weight distribution, it is enough to assume that a_1 or r are normally distributed in the population. This type of variation of a_1 and r can be of genetic origin only and need not be the result of interactions between individuals. The effect can be more pronounced when not only a_1 or r are taken from normal distribution but also other parameters of growth equation are random variables with normal distributions.

Competition: Unequal Resource Partitioning

Intraspecific competition is of special interest for the individual-based approach in ecology. Competition between members of a cohort is a basic situation. It is a very important source of individual variability.

Organisms are often limited in many dimensions. The available amount of resources when partitioned between present organisms is not sufficient for performing living functions by all individuals. Waiting for improved conditions is usually the wrong strategy, because the next limitation, time, starts to operate. Sufficient conditions of competition include, of course, the assumption that organisms are living together in space. Emigration is not always possible, and not always profitable. However, the term *competition* means more than just living together in space and time with limited resources. Two individuals compete with each other when one of them gains something also needed by the other, and in limited supply. The amount of resources taken by an individual as the result of competition is always smaller than the amount of resources available to individuals. But, more importantly, competition implies that resources, which are the subject of competition, are not evenly partitioned between competing individuals. Someone is getting more, another less. An uneven resource partitioning as the result of competition is one of the basic problems of the individual-based approach in ecology. The existence of uneven resource partitioning between competing individuals and its consequences for population processes were first stressed by Łomnicki (1978, 1980, 1988).

Only limited information based on experiments or observations is available about gains of individuals of the same species and the same age competing for resources. The assumption that light is unevenly partitioned among competing plants is intuitively plausible. For the same reason, we can expect the uneven resource partitioning in the case of sedentary animals. For other animals, our intuition fails.

Resource partitioning and differentiation among individuals of the same age are relatively well documented only in the case when there is social hierarchy or territoriality in the population. Jobling (1993), in the review of fish bioenergetics, argued that interactions among individuals of the same age lead to the development of social hierarchy. This, in turn, accounts for uneven resource partitioning and differences in body size among individuals if resources are in short supply. The most competitive individual, with the highest rank in the social hierarchy, is the heaviest.

Nakano (1995) examined local populations of the red-spotted masu salmon, *Oncorhynchus mason ishikawai*, in mountain streams in Japan. Direct observations of individually marked fishes revealed a permanent hierarchy. Individuals with the highest rank in this hierarchy were territorial and occupied the largest territories, located in places with most abundant food supply in the stream. They spent more time outside their shelters and consumed larger prey. This led to large differences in the amount of food consumed. There was a positive correlation between the rank of the fish in the social hierarchy and the amount of food consumed. The same relationship was observed between the rank of a fish and its body weight. Fishes with the lowest rank in social hierarchy were not territorial; they moved over much larger areas than territorial fishes and emigrated from local populations. Nakano concluded that due to differences among individuals and emigration of individuals with lowest rank in the social hierarchy, fluctuations in fish numbers in the local populations he studied were very low.

Yamagishi (1962) examined small, experimental populations of the rainbow trout, *Salmo gairdneri* fry. His results not only confirm the fact of the differentiation of body weights and food consumption in even-age fish populations but also show that the differentiation is greater when the habitat enables dominant fishes to monopolize the resources. Similarly, an increase in differences in growth rate among individuals and in consumption under deteriorating food conditions, as observed by Symons (1968), can also be due to the strengthening of the hierarchy (measured by the intensity of nipping) by food deprivation in the Atlantic salmon, *S. salar.* Magnuson (1962) reported similar effects in *Oryzias latipes*.

Competition between two individuals can be strong or weak. The strength of competition is determined by the complex interplay of environmental conditions and behavioral mechanisms, but in most competition models, it is characterized by a single parameter. Competition can be also symmetric or asymmetric. Intuitively, we expect that competition is symmetric if resources are equally divided between two similar individuals. But individuals can be different. Let us assume that two competing individuals have weights w_1 and w_2. Competition is symmetric if the ratio of the amount of resources taken by the first individual to the amount of resources taken by the second individual is proportional to w_1 divided by w_2. We call competition asymmetric if the greater individual takes unproportionally more resources than is determined by the ratio of their weights (Weiner, 1990; Thomas and Weiner, 1989; Schwinning and Weiner, 1998). And in the end, competition can be local or global. It is global if each individual competes with all other individuals in the population. Competition is local if each individual competes with only some individuals in the population. The degree of locality of the competition is measured by the ratio of the number of individuals with which an individual competes to the total number of individuals in the population.

Weight Differentiation in Even-Aged Populations

In nonreproductive periods, for instance, before maturity or between reproductive periods, the weight of an organism is a measure of its net energy gains [see equations (1) and (2)]. We should analyze the distribution of weights of organisms competing in even-age populations in order to obtain information about resource partitioning among them. Mathematical modeling may be useful in transforming the description of weight structure in even-age populations into the analysis of the process of resource partitioning between competing individuals.

Studies on the differentiation and frequency distributions of weights in even-age plant populations started in Japan in the 1950s. Different plant species were cultivated at variable densities in highly uniform habitats in which plants were evenly spaced. The results of this work were summarized by Koyama and Kira (1956). Similar results were obtained by Obeid et al. (1967), Ogden (1970), and White and Harper (1970). Weight distributions of plants under natural conditions were analyzed by Ford and Nowbould (1970, 1971), Ford (1975), and Mohler et al. (1978).

Investigations of the frequency distributions of individual weights in animals are very scarce. Yamagishi (1969) and Nakamura and Kasahara (1977a, 1977b, 1977c, 1977d) attempted to analyze systematically the structure of weights and its changes with time, density of population, and food concentration in even-aged fish populations. The results obtained by these authors are in agreement with results of experiments on plant population. It is also worth quoting the papers by Wilbur and Collins (1973) and Collins (1979), describing the frequency distributions of weights and their changes with time and density for tadpoles of two frog species.

A more complete list of literature concerning the frequency distributions of body weights for both plants and animals can be found in Uchmański (1985). The conclusions may be summarized as follows:

1. The frequency distributions of individual weights for even-age plant and animal population are most often positively skewed.
2. Symmetric distributions are characteristic of even-aged populations of organisms starting their individual development.
3. Skewness of weight distributions in even-aged populations varies with individual growth. Two kinds of changes in distribution shape can be distinguished. In the case of experimental cultures, the initially symmetric distributions develop a positive skewness, increasing in time. Under natural conditions, the same pattern is observed, but not exclusively. It also happens that the initially increasing positive skewness begins to decline after reaching a maximum, so that the final frequency distribution can be symmetric again. The decrease of weight-distribution skewness is correlated with increasing mortality.
4. The rate at which frequency distribution of individual weights develops increasing positive skewness, and the degree of skewness obtained largely depends on the population density and environmental conditions such as food quality, concentration and availability. High-density and poor-trophic conditions enhance the rate of skewness development and increase the value of skewness coefficient.
5. Weight variability in even-aged populations measured by the variability coefficient or the variance of weight distribution most often behaves like skewness. An increase in the density of an even-age population, or a deterioration of food conditions, often accounts for an increase in weight variability.

Growth of Competing Individuals

General Model

An isolated individual in optimal environmental conditions grows according to equation (10). Let us now assume that population consists of N competing individuals of the same age. If the amount of resources (V) in the environment is unlimited ($V = \infty$), individuals are not competing and each still grows according to equation (10). Individuals will compete when $V < \infty$. Competition means unequal resource partitioning. The equation describing growth of the i-th individual from the population of N competing individuals can be of the following form (Wyszomirski et al., 1999):

$$\frac{dw_i}{dt} = K_i a_1 w_i^{b_1} - a_2 w_i^{b_2}, \tag{15}$$

where competition coefficient K_i ($0 \leq K_i \leq 1$) describes the influence of competition on the amount of energy assimilated by i-th individual ($i = 1, \ldots, N$). There are no general rules for the form of coefficient K_i except two: $K = 1$ for all individuals when there is no competition, and K should have such values for competing individuals to make weight distributions produced by the model similar to weight distributions observed in even-aged natural or experimental populations (see the section on weight distribution in even-aged populations). In practical terms, K is defined by some "competition rules" specific for species and ecological situations considered in the model. In general, K can be a function of individuals age.

Global Competition

A population consists of N competing individuals, each characterized by weight (w). Let us assume that the outcome of competition depends on weights of competing individuals and focus our attention on the i-th individual, with weight w_i. Competition is global, so the i-th individual competes with all other individuals in the population. Competitive influence on the i-th individual is smaller if w_i, is greater, and it is greater if competitors are heavier. Then K_i can be described by the following function (Wyszomirski et al., 1999):

$$K_i = \frac{w_i^b}{w_i^b + S \sum_{j \neq i} w_j^b}, \tag{16}$$

where parameter S describes competition strength (the higher its value, the smaller the value of K) and parameter b characterizes the degree of competition asymmetry. If $b = 0$ then,

$$K_i = \frac{1}{1 + SN} \tag{17}$$

and each individual obtains the same part of resources (competition is absolutely symmetric). If $b = 1$, competition is relative-size symmetric. For higher values of b, competition is asymmetric. It becomes one-sided, when $b \to \infty$ (in a conflict between two individuals, the smaller individual loses everything, the bigger one practically nothing).

Values of parameter K can also be determined using so-called functions of resource partitioning (Uchmański, 1987). They present K as a function of two variables—weight of individual w, for which K value is calculated, and the amount (V) of resources available to all individuals in the environment. Positively skewed weight distributions and desired changes in their shapes with age, population density, and resource conditions are obtained if functions of resource partitioning fulfill a number of conditions: (1) Under constant resource conditions, K is a convex or linear nondecreasing function of weight; (2) an improvement in resource conditions is followed by a decrease in differences of K values between individuals with different weights; (3) a decrease in the amount of resources in the environment causes a greater decrease in K for individuals with smaller weights compared with those with greater weights.

Different types of competition can be described by these functions (Uchmański, 2000b). Scramble competition $-K$ values for all individuals are different and greater than zero, even for small values of V. Contest competition $-K$ values for individuals with weight greater then some threshold weight are positive and different, while those for the other individuals are equal to zero. Monopolization of resources $-K$ values for individuals with weights higher than some threshold weight are the same, not equal to zero, while for other individuals, they are equal to zero.

Local Competition

Competition in terrestrial plants and sedentary animals takes place between neighborhood individuals only. There are several mechanistic models describing resource partitioning between locally competing individuals.

Let us assume that the amount of resources available to a given individual is proportional to the amount of space ascribed to that individual. It can be defined as a so-called Voronoi (or Thyssen) polygon consisting of points in the plane that lies closer to the given individual than to other individuals in the population (Mithen et al., 1984; Czárán, 1996). The whole model space is then divided into individuals' polygons, and the amount of resources available exclusively to the given individual (or K value according to earlier notation) is related to the surface of its polygon.

The numerical procedure used for determining Voronoi polygons is usually the following: We draw sections connecting the position of an individual with the positions of all its neighbors. Then centers of these sections are determined, and lines perpendicular to these sections in their central points are drawn. Perpendicular lines surrounding the individual and the crossing points determine the Voronoi polygon of the individual. Such formulation makes competition very static. The amount of resources ascribed to an individual is constant during its growth. It will change, however, if we assume that the point at which the perpendicular line crosses the section connecting two neighbor individuals is situated not in the center but moved away from the individual with greater w (the section is divided proportionally to the weights of individuals situated at its ends).

Other models describing local competition use so-called zones of influence (ZOIs) (Hara, 1988), which, in the case of trees, can be interpreted as the horizontal extent of the crown or root system. An individual is represented by a circle with its center located at the point where the individual is situated. The area of the circle represents resources used by the individual, and its radius increases with its growth. Parameter K is a function of ZOI area. Each individual uses all resources from its own ZOI when individuals are small or population density is low. But sooner or later, ZOIs of neighbor individuals start to overlap. Then, an individual will use all resources from the nonoverlapping part of its ZOI, and only some part of the resources from an overlapping part. Partitioning of resources in overlapping parts of ZOIs depends on the degree of asymmetry of competition. Resources are partitioned there equally or in proportion to the weights of competing individuals, if competition is symmetric, or heavier individuals obtain disproportionately more resources, if competition is asymmetric.

Another way of calculating K values is to combine weight and distances to neighbor competitors directly in a simple formula (Benjamin and Hardwick, 1986). Let us focus our attention on i-th individual in the population of N locally competing individuals. Its weight is w_i. Weights of other individuals in the population are w_j, where $j = 1, \ldots, N$, and $j \neq i$. Distance between i-th and j, the individual's is d_{ij}. The K value for the j-th individual can be calculated from the slightly modified equation (18):

$$K_i = \frac{w_i^b}{w_i^b + S\sum_{j \neq i} \frac{w_j^b}{d_{ij}^2}}, \tag{18}$$

where, as earlier, S is a parameter characterizing the strength of competition and b, the degree of its asymmetry.

Size Distributions of Locally Competing Individuals

The simple rules describing skewness of weight distributions in even-age populations as a result of growth itself (see section on growth) and as monotonic, increasing function of population density and deteriorating resource condition (see section on weight differentiation) do not seem to hold generally. This pattern can be found most often in globally competing animals and regularly distributed in space plants, but is not generally typical for terrestrial plants and sedentary animals. In even-aged plant populations, skewness of size distribution is often a nonmonotonic function of age of individuals and population density (Mohler et al., 1978; Knox et al., 1989). Decrease of skewness (even the appearance of negative skewness) of stem diameter at breast height or mass distribution with age of individuals, followed by its increase, was found by Gates et al. (1983), Brand and Magnussen (1988), and Hara (1984a, 1984b, 1986, 1988). In other cases, increase in population density was not followed by higher size inequality (Turner and Rabinowitz, 1983; Schmitt et al., 1986; Geber, 1989), or bimodality of size distributions has been observed (Ford, 1975; Ford and Diggle, 1981; Rabinowitz, 1979; Nagashima et al., 1995), resulting from splitting the population into dominant and suppressed individuals.

A set of explanations under the common name "spatial effects" is very important if we try to understand results of competition of locally interacting individuals (Benjamin and Hardwick, 1986; Huston and DeAngelis, 1987; Hara, 1988, 1994; Weiner, 1990; Schwinning and Weiner, 1998; Wyszomirski et al., 1999). A simple example illustrates the problem. The positive skewness of size distribution is the result of asymmetric and global competition (see section on weight differentiation). But symmetric competition can also produce positive skewness of size distribution if competition is local and individuals are irregularly distributed in space (Wyszomirski, 1983, 1986, 1992; Miller and Weiner, 1989; Firbank and Watkinson, 1985; Bonan, 1988, 1991). In other words, spatial effects can mimic the results of asymmetric competition.

Spatial effects can also produce bimodality of weight distribution if competition is local, symmetric, and takes place between individuals randomly distributed in space (Huston, 1986). Wyszomirski et al. (1999) addressed a model to explain these effects, which are highly density-dependent. Competition between individuals randomly distributed in space was local and symmetric. At low overall population density, noncompeting individuals dominate (their ZOIs do not overlap). Competition only takes place between a small number of individuals forming locally dense clumps. Skewness of weight distribution is negative in this case. Increase of overall population density makes the group of competing individuals greater. Weight distribution becomes bimodal. Further increase of overall population density produces positively skewed weight distributions.

According to the hypothesis formulated by Hara and Wyszomirski (1994), the competitive asymmetry reduces spatial effects. Therefore, switching from symmetric to asymmetric competition in the model presented earlier makes the result of competition less sensitive to variability in local density of individuals. In initial stages of growth, asymmetric competition works on initial variability of individuals, enhances it, and if the degree of asymmetry is sufficiently high, it can later produce size variability only slightly affected by local clumping of individual (we can even expect monotonic changes of skewness with population density). The reduction in spatial effects is possible, additionally, when initial size variability is sufficiently high. If initial variability is strongly reduced, switching from symmetry to asymmetry affects the aforementioned nonmonotonic pattern of skewness changes only slightly (Wyszomirski et al., 1999, p. 266).

However, Wyszomirski et al. (1999) pointed out that

> "asymmetric competition implies two kinds of effects. Firstly, it promotes variation and skewness *because it is asymmetric*: large individuals have an advantage over smaller ones and their faster growth causes this advantage to magnify. If competition is very weak, the effect of asymmetry is only slightly pronounced.... Secondly, asymmetric competition hampers an increase in variation and skewness, simply *because it is competition*, i.e. a process slowing down the growth of individuals. If competition is very strong, the latter effect dominates. Only in the case of completely one-sided competition, monotonic increase in variation and skewness with increasing density might be found"

Both effects are observed in models describing results of global and local competition. Each of these effects expresses itself differently in different combinations of degree of asymmetry, degree of locality, and strength of competition. We can even expect nonmonotonic changes of size-distribution skewness with population density in the case of global competition, if asymmetry is not very high and the competition is strong.

When population density is high or competition is very asymmetric, small individuals obtain nothing or such a small amount of resources that they cannot live on it. It leads to mortality and self-thinning of the population. This process is typical for dense, even-age plant populations (Westoby, 1984; Harper, 1977) and was also observed in animal populations (Begon et al., 1986; Latto, 1994). Self-thinning strongly modified shapes of weight distributions by deleting the left side of weight distribution. It has to be taken into account during analysis of empirical data.

Growth Models: Prospects

Correlation between Parameters of the Growth Model

In even-age populations not only weight and assimilation rate of individuals are variable but the respiration rate also exhibits a considerable degree of variability. Klekowski et al. (1967) demonstrated it for oxygen consumption rates of adult individuals of *Trybolium castaneum*. Similar results were obtained by Cianciara (1980) for *Cloëon dipterum*.

There is a purely physiological positive relationship between the rate of growth and metabolism of growing individuals (Konarzewski, 1995). Additionally, competing with other individuals means not only uneven resource partitioning but also different energetic costs for individuals obtaining different amounts of resources during competition. Correlation between assimilation and respiration rates of individuals competing in an even-age population can be of different sign. Correlation can be positive—assimilation of greater amount of energy may be accompanied by greater energetic costs of obtaining it. Metcalfe et al. (1995) demonstrated for Atlantic salmon, *Salmo salar*, that the higher the standard metabolic rate, the more dominant the fish. On the other hand, individuals possessing a higher position in the social hierarchy also had higher growth rates in this case. However, for other species, we can expect negative correlation between assimilation and respiration rates.

The cost of being dominant and possessing greater amount of resources can be so high that the net energy gains of dominant individuals become smaller than gains of individuals occupying lower social positions. Yamagishi et al. (1974) described social relations in small laboratory populations of the freshwater eleotid goby, *Odontobutis obscurus*. The dominance order was recognized by observing aggressive and territorial behavior of fishes. The individual food intakes were also measured. The socially dominant fish defended the best territory with the highest food abundance and had the highest food intake. However, the second-ranking fish was heavier than the dominant fish, because the energy expenditure of the dominant fish was greater, and it was able to devote only a small part of its activity to resource competition, due to the necessity of maintaining its dominant social position. However, the dominant fish was characterized by a smaller variation in food intake during the experiment.

The correlation between assimilation and respiration rates changes the shape of weight distribution of competing individuals. Intuitively, a high positive correlation between assimilation and respiration rates should diminish variability and skewness of size distribution in an even-age population, and a high negative correlation should increase them. However, simulation shows that the results can be much more diverse. Some cases of symmetric competition were analyzed by Wyszomirski et al. (1999). It turns out that the size variability and skewness of size distribution depend on the type of growth (they are different in the case of limited and unlimited growth), the strength of competition (size variability can be equal to zero for some of its values), and, of course, the sign and strength of correlation between parameters a_1 and a_2 of the growth equation (10). Generally, different values of variability and skewness can be obtained for different sets of these parameters, including reversal in size order—individuals growing slowly in the initial stage of growth become greater in the final stage.

Energetic Reserves of an Organism

The balance equation (10), which can serve as the first approximation, is useful only when growth of individuals and their variability are not the main subject of the study, for instance, in modeling population dynamics. Its most important shortcoming is the assumption that the amount of energy assimilated by an individual in a unit of time is used immediately for sustaining its life functions. This is obviously not true. There is a time lag between assimilation of a given portion of energy and its utilization for growth, reproduction, or other functions; that is, the same energy portion assimilated in a given time instant is not simultaneously used for costs of life, for example, to cover the cost of finding and processing the assimilated energy.

This problem was investigated in many studies devoted to energetics of particular species [e.g., Fromm and Rasmussen's (1984) model for *Salmo gairdneri*)]. The general solution to the problem is shown by Kooijman (1993), who proposed the so-called dynamic energetic budgets (DEBs) to describe growth of individuals. DEBs consider the fact that organisms store resources and do not immediately transform the assimilated energy into their own bodies or offspring. Kooijman presented different DEB models for different types of organisms.

The question of how the application of this type of model influences dynamics of size distributions in even-age populations of competing individuals is difficult to answer. Nobody has addressed it yet. One answer, however, is intuitively clear. Let us assume that intensity of competition changes during the individual's life (population density or amount of resources in the environment are variable). If periods of high population density or low resource levels are relatively short, then suppressed individuals, who obtain a small amount of resources as the result of competition, can compensate for this disadvantage using energy from their reserves. This probably makes size distributions less dependent on population density or environmental conditions than in the case of organisms that use assimilated energy immediately.

Competition for Different Kinds of Resources and Stochastic Growth Models

All models of competition between animals presented earlier in this chapter assume that resources are homogeneous and continuously distributed in space. Because of this, only the deterministic version of the model describing growth of individuals has been presented up to this time. However, experimental evidences showing influences of the form of resources on the size distribution of competing individuals in some specific cases should be included in the model. Nakamura and Kasahara (1977a) performed experiments on one group of carp fry fed on *Cladocera* larvae and on another (with the same number of individuals) fed on a mixture of *Cladocera* and mosquito larvae. After 18 days, skewness coefficients of body length distribution were positive in both populations, but in the first, it was seven times smaller than in the second group with the wider spectrum of food size particles. Similar results were obtained in experiments with perch, *Perca fluviatilis* (Dgebuadze, 2001).

The possibility of finding food items with different energetic values can lead to changes in size hierarchy in the population. Wethey (1983) investigated a sessile organism, the intertidal barnacle *Balanus balanoides*. The rank of sizes was not strictly conserved during growth. There were reversals in the size orders—smaller individuals in the initial stages of growth become greater later and vice versa. Also, the positive skewness of size distributions at the beginning decreased over the growth season. The explanation, according to the author, may lie in the fact that the food—suspended particles—for barnacle feeding by filtration may be of a different size. Therefore, the deterministic process of filtration and random (independent of animal size) events of finding of large particles by small animals should be considered together.

Stochastic models seem to describe this situation much better than deterministic models. For only one type of food particles, such a stochastic model can be formulated in the following way (extension of the model in the case of more food particles is obvious). Let us assume that the environment consists of food particles. Each particle, when consumed, increases individual body weight by g. There is some probability (P) of catching a food particle and increasing the weight. It depends on the total number of food particles in the environment, which is equal to V. An individual that caught more food particles in the past has a greater probability of catching the next particle than an individual that caught a smaller number of particles in the past. These assumptions are expressed:

$$P = P(w, V). \tag{19}$$

Let us assume that the respiration in a unit of time is not related to the success or failure in catching of food at this time but depends on the actual weight of an individual, according to the power function in equation (4). Thus, the individual with weight $w(t)$ at time t will at time $t + 1$ weigh

$$w(t+1) = w(t) + g - a_2 w(t)^{b_2}, \tag{20}$$

with probability $P(w, V)$, or weight

$$w(t+1) = w(t) - a_2 w(t)^{b_2}, \tag{21}$$

with probability $1 - P(w, V)$ under the condition in which the number of food particles in the environment is V.

The stochastic model presented here can be used to describe the growth of predators competing for their prey—food particles, in this case. Probability P as the function of w and V should fulfill the same conditions formulated earlier for function of resource partitioning in the deterministic version of the model (see section on growth of competing individuals), in order to obtain positive skewness of weight distribution for competing predators.

Fractal Models of Growth

Mathematical models of individual growth usually only describe changes in body weight. But an individuals's growth and development are characterized by changes not only in body weight but also in the structure of the organism. There are two types of organisms in general. Growth and development of unitary organisms is realized in one step. Modular organismic growth is an iterative process in which step-by-step modules are added to the existing structure. The growth of a unitary individual comes to an end when the organism reaches its final form. Growth of a modular organism never ends—it has no final form. A modular organism (genet) can exist in a form of many parts (ramets) composing complicated, most often 3-D structure. Ramets can be more or less isolated from each other. Different ramets of this same genet experience different growth conditions. Numerous species exist in the form of modular organisms. Plants and colonial animals are modular organisms. For Harper (1967, 1977), the idea that a plant is a set of shoots with meristems (modules) is the basis for understanding biology of plant populations. White (1979) even used the term *metapopulation* to describe a single plant.

A modular way of living also means a high degree of plasticity of growth. Plasticity of individual growth is assumed in many individual-based forest dynamic models (Kellomäki and Strandman, 1995; Gavrikov and Sekretenko, 1996; Perttunen et al., 1998) in which a tree is presented as a three-dimensional structure able to change its form in reaction to the presence of competitors. Sorrensen-Cothern et al. (1993) argued that the results of forest dynamic models are sensitive to this assumption. Dynamics of forests differ, depending on whether we assume or neglect plasticity of growth.

Models presented earlier in this chapter described unitary organisms or treated modular organisms as unitary ones. However, a new modeling technique has emerged that is very useful for description of modular organisms. Because modular organisms are often self-similar, some researchers suspect that they represent some approximations of fractals. Prusinkiewicz and Lindenmayer (1990) were the first to propose some elements of fractal modeling to describe changes in geometric structure of a growing modular organism. All tree-like geometric structures can be described by defining a geometric interpretation from some set of characters called L-systems. Each character means drawing a directed section on the computer display. A string, which is a prescription of drawing an organism, is the product of the iterative application of some given rules to an initial set of characters. On each iteration step, we obtain a structure that is similar to structures obtained in previous steps of iterations and is self-similar—all its parts are similar to the whole structure.

L-systems are not the only technique of fractal modeling. There are also other methods applied most often to modular animals, but all of them are based on the iteration of some geometric structures. Kaandorp (1994) used these approaches in the models of growth of marine sessile organisms (e.g., sponges and corals) in a 3-D space filled with food-particles falling down from above. The model describes not only the increase of successive structures but also the spatial distribution of food particles. It is possible to follow on the computer monitor the growth of an individual under different habitat conditions involving the direction of food inflow, angle of incident light in the case of plants, ambient temperature, and so on.

Fractal modeling is a complex task. At present, due to technical constraints (primarily the long duration of computer calculations), it is only possible to make simulations and computer visualization of a single isolated organism. However, if fractal modeling is to be of any interest to ecologists, it has to describe not only the growth of an isolated individual but also many competing individuals. Méch and Prusinkiewicz (1996) presented the first attempt at fractal modeling of the competition for light between two trees.

Resource partitioning between competing individuals is one of the main problems of individual-based approach in ecology (see section on competition). The question is very difficult from the empirical perspective, because most often it is posed relative to unitary animals with complex behavior. It seems much easier if posed relative to modular organisms (Harper, 1986). Competition between modular organisms resembles the possession of some part of common space with resources. In many cases, we need to apply only some geometrical or physical rules to calculate resources belonging to each competitor.

Single Population Dynamics: Why Individual Variability is Important

The Model

What is the purpose of analyzing the growth of competing individuals and their size distributions in even-age populations? Weight is highly correlated with many other features of an individual (Peters, 1983), and these relationships are not only statistical but in many cases also causal. If individuals differ in weight, they differ also in progeny production and survival probability. This influences the number of individuals in the next generation. Therefore, by studying interactions between individuals and the individual variability they produce, we can also analyze population dynamics.

A simple, proposed model of single-population dynamics with nonoverlapping generations uses all terms introduced earlier (Uchmański, 2000a). Let us assume that the generations do not overlap. The number of individuals in generation t is equal to N_t. The number of individuals in the next generation $t + 1$ is given by the following equation:

$$N_{t+1} = \sum_{i=1}^{N_t} z_i, \quad (22)$$

where z_i is the number of offspring produced by the i-th individual from the generation t. The number of offspring can be different for different individuals; it can even equal zero for some of them. Lets us assume additionally that growth of all individuals from a generation takes place simultaneously—they start and end it in the same time. Individuals reproduce after growth ends. There is no mortality during growth. All individuals die at the end of the generation. Earlier, they compete globally, and resources are unevenly partitioned among them [different models for K can be used here—later, the result of application of functions of resource partitioning is presented (see section on global competition)]. As the effect of this, individuals reach different final weights [see equation (11)]. According to Matessi (1982), the number of progeny produced by the i-th individual is proportional to the difference between its final weight and some threshold weight w_{thr}:

$$z_i = Round[c(w_{i,end} - w_{thr})], \quad (23)$$

where c is constant. Function *Round* rounds a real number to the nearest integer. Individuals with final weight lower than w_{thr} die without leaving progeny. To ensure some variability in initial weights of individuals, let us assume that initial weights of progeny of a given parental individual are chosen from a normal distribution.

The between-generation dynamics of resources V can be described by the following equation:

$$V_{t+1} = V_t + g - \sum_{i=1}^{N_t} C_i^{cum}, \quad (24)$$

where C_i^{cum} is cumulative consumption of resources by the i-th individual during its life, and g is the inflow of resources into the system. The sum describes the cumulative consumption of resources by all individuals in the generation t.

Equal Individuals

Let us analyze by means of the model described in the previous section a population consisting of equal individuals. This null model describes the dynamics of a lineage that started with one individual in the first generation, then describe its progeny in subsequent generations. All individuals are equal: They have the same initial weight, obtain the same amount of resources, and their growth rates and final weights are equal. This null model produces only solutions with relatively short population extinction times. One type of solution is observed: The exponential growth of the population followed by exhaustion of resource and extinction of the population.

The key factor shaping population dynamics in this case is the assumption that each individual can produce only 0, 1, 2, ... offspring [see equation (23)]. At the beginning, the population increases exponentially. In this stage, number of progeny is greater than 1. Sooner or later, resources decrease due to consumption by an increasing number of individuals. This leads to the situation in which each individual can produce only one offspring. Now, the number of individuals in the population is constant for several generations. However, resources are still consumed. As a result of this, the number of progeny can only be equal to zero, and the population becomes extinct.

Variable Individuals

The population's extinction time is longer if the population consists of several lineages that differ with respect to initial weight of progeny. Each lineage increases exponentially. However, the resulting decrease in the amount of resources leads to the subsequent extinction of the lineages. Elimination of a lineage temporarily produces better food conditions for the rest of the population, but a further decrease in resources leads to extinction of the next lineage. This process continues until only the last lineage remains, but it will also become extinct soon, due to the mechanisms described earlier.

In the case described, the number of lineages in the population can decrease, but the number of individuals in each lineage can only increase, be constant, or be equal to zero, because all individuals are equal in each lineage. The longest extinction times of the population, and population dynamics other than exponential increase, were observed only if initial weights of progeny differed from each other. This assumption means that the number of individuals in each lineage can also decrease, due to the selective reproduction and mortality of variable individuals.

If a population consists of identical individuals, it tends to grow exponentially. In the end, the population becomes extinct when resources are not sufficient even for producing one progeny. A population consisting of variable individuals behaves in a different way. Most often, it will not become extinct after reaching the first maximum in population numbers, because even in deteriorating resource condition there are now individuals that will reproduce. So after the first maximum population numbers decrease, then increase once again. The second increase phase is possible because, during the preceding decrease in population numbers, resources were exploited by a relatively small number of individuals that started to increase again before population numbers reached their first minimum. This sequence of oscillations can eventually last a long time. But in each population minimum, there is some probability that the population will become extinct. This can occur just in the first minimum, but it may also not occur for a long time, producing a whole distribution of extinction times as the result of repeated simulations.

What Promotes Population Persistence?

A population made up of identical individuals with the same initial weights and characterized by deterministic growth is not regulated (see previous section). This population can only grow exponentially, then drop to zero as resources decline.

Let us assume that individuals are subject to random mortality over their lifecycle. Only some of them, randomly selected, have a chance to survive until the age of reproduction. The extinction time of a population comprising identical individuals subject to random mortality is longer when compared with the results of the model without random mortality (Uchmański, 1999). Random mortality reduces population size, thereby increasing the amount of resources for individuals that survive to the period of reproduction. Occasionally, solutions appear showing that it is also possible to get a second maximum of numbers. Also, a very low chance of more maxima cannot be excluded. The final result, however, is always the same. The population becomes extinct most often after the first population maximum.

Random mortality incorporated into this above model is a form of individual variation. However, it divides the population into two categories only: individuals that will reproduce, and individuals that will not survive to the period of reproduction. It leads to a very weak regulation of the population, because it produces in each generation a group of still-identical individuals that survive and reproduce. Individual variation of a different type is needed for effective regulation of the population. The solution is to assume differences in individuals arising from asymmetric competition. This can be done by assuming that growth of competing individuals is deterministic and initial weights take random values, or that the initial weights are the same for all individuals and only growth is stochastic. This type of variation in growth, observed when individuals compete asymmetrically with each other, has a property that did not occur earlier: Even under good resource conditions, there are individuals that do not reproduce, and vice versa, even under poor resource conditions there are individuals that will reproduce.

Functions of resource partitioning were used in these models presented (see section on global competition). These functions can be constructed in different ways, assuming different ways of determining the rank of individuals in the population (Uchmański, 1999). Competition is *soft* if the rank of an individual is determined relative to other individuals competing with it in the current generation. If the rank of an individual is determined relative to some absolute scale in the same way in each generation, we call the competition *hard*. The soft type of establishing hierarchy among competing individuals seems much more natural, but only in contrast to the hard method of resource partitioning that its consequences for population dynamics become evident. Because the soft method of hierarchy establishment produces greater differences in assimilation rates between individuals, it leads also to greater persistence of population, measured by its extinction time compared to hard competition.

Different methods of resource partitioning can be assumed in the model: Competition may be a scramble, a contest, or it can lead to monopolization of resources (see section on global competition). Persistence of the population depends on the way resources are partitioned among the individuals (Uchmański, 2000b). In practical terms, there are no large, qualitative diffrences in the outcome between the models assuming scramble and contest competition. Both types of resource partitioning yield similar population extinction times. The characteristic property of the models with contest and scramble competition is the wide range of parameters that produce a high persistence of the population. There is also a range of parameters at which resource monopolization yields a population extinction time similar to those in the models with contest and scramble competition. But the width of the range of the most important model parameter, such as that determining the number of individuals monopolizing resources, is very narrow. Resource monopolization most often resembles equal resource partitioning among the best individuals. As we know, a population consisting of equal individuals increases exponentially. This is why in the models with resource monopolization that longer extinction times can be obtained for only a very small number of individuals monopolizing the resources.

The extinction time of the population depends also on the number of times an individual competes during its life (Grimm and Uchmański, 1994a, 1994b; Uchmański, 1999). It is much longer if an individual competes many times during its life, in comparison to situations in which it competes only once.

Some General Questions

Classical theoretical ecology uses state variable models. This approach to modeling ecological systems assumes that their dynamics can adequately be described by developing appropriate differential or difference equations. In this case, densities of the component populations of a system represent the state variable, and its rate of change in the state variable is a function of the state variable itself. Due to this approach, theoretical ecology uses the achievements of the mathematical theory of dynamic systems, including, most recently, the theory of deterministic chaos.

Tools used to construct models of the dynamics of ecological systems, namely, the mathematical theory of difference and differential equations, have a very strong influence on the way we see ecological systems. Analysis of difference and differential equations means most often looking for stability conditions of their solutions (according to various formulations of this term) (Grimm et al., 1992). This is why classical ecology is dominated by the stability concepts. Nonequilibrium approaches in ecology are rare. Ecologists simply believe that the world they observe is in a state of equilibrium, or is approaching it after a disturbance. Individual-based models of population dynamics presented in this chapter describe the population as a system without equilibrium behavior. A population of globally competing individuals exhibits oscillations and will sooner or later be subject to extinction. However, such a population is also regulated, due to individual variability. It is not the perfect, fine-tuned regulation possible in classical ecological models. We have to call the regulation observed in individual-based models rather imperfect (Uchmański, 2000). These inherently nonequilibrium properties (Huston, 1994) of individual-based models emerge because a population consists of individuals. Their number is 0, 1, 2, ... (an integer) and it cannot be a real number, as is the case with population density in classical models.

As we can see, the application of an individual-based approach in ecology leads to a fundamental controversy about the nature of ecological systems. Much depends, of course, on the way we define individual-based models (see earlier section on definition). There are authors who believe there is no such controversy and that individual-based modeling is simply a continuous enlargement of the classical approach (Baveco, 1997). However, if such controversy exists, it yields another fundamental question: How it is possible that life can exist in nonequilibrium forms of ecological systems? Constructing an individual-based metapopulation model is one way of answering this question.

Classical theoretical ecology did not lead to an essential turning-point in the understanding of the dynamics of ecological systems. We would even say that it caused some crisis in ecological theory, reflected in the fact that classical ecological theory cannot explain well-documented facts, such as population cycles, and yields contradicting hypotheses in community ecology. I believe that the individual-based approach may be a remedy for this crisis in theoretical ecology.

Applications

My intention in this chapter is to present a general scheme for constructing individual-based models, the main problems we face before reaching the final form of the model, and some general controversies it yields. However, it appears that individual-based models are not built to provide answers to some important general questions from a theoretical point of view. As Grimm (1999) pointed out in his review of individual-based models published in the last 10 years, most were constructed to answer some specific, applied question. Here, only chosen examples of application of individual-based modeling are presented [for more complete reference list, see Grimm (1999) or Grimm et al. (1999)].

Progress has been made in the use of the individual-based approach to describe specified ecological systems and to solve problems connected with nature conservation.

Baveco (1998) used this approach to solve problems concerned with protection of tree frogs and root voles in the heterogeneous agricultural landscape of the Netherlands. There is a mosaic of suitable and unsuitable habitats for these species. The model describes the lifecycle of individuals, their movements in this complex habitat, and answers the question concerning factors that promote population persistence of these species, as well as predicting their future survival in the face of different scenarios of habitat transformation.

Ulbrich and Henschel (1999) used a very similar approach presented earlier in this chapter in modeling the population dynamics of the social spider, *Stegodyphus dumicola*. They show that contest competition for food leads to the regulation of the colony size. However, at low food levels, the probability of colony extinction is high, because only few individuals survive and reproduce.

For a long time, the individual-based approach was traditionally used to describe forest dynamics [Shugart et al., 1992; for a recent example, see Liu and Ashton (1998)]. PawŁowski (2000) constructed an individual-based model of changes in the number of trees during 1000 years in a 1-ha spruce forest located in the mountains of southwestern Poland. It describes the competition between trees of different age in a natural stand. This is why competition is strongly asymmetric. An old tree falling down is an elementary event in this population. It opens space for seedlings located in the neighborhood. Trees differentiate and population is dominated by several big trees—winners of competition, if the competition is strong and asymmetric. When their life ends, numerous seedlings that were strongly suppressed by dominant trees start the next generation. But soon they also differentiate, and a small number of big trees once again dominate the population, yielding population dynamics characterized by more or less regular population piks (with numerous young trees, appearing after an old tree falls) and long periods of low population number. The population dynamics will be different if the competition is weak and more symmetric, with no differentiation of trees in this case. Uniform trees growing simultaneously dominate the generations and produce the population dynamics characterized by regular oscillations of the number of trees in the population.

One of the biggest programs for restoration ecology concerns the Everglades region of south Florida. This program uses a mathematical model simulating the fates of animal species most important to the Florida ecosystems at different hydrological regimes. The model, which is called ATLSS (Across Trophic Level System Simulation; DeAngelis et al., 1998), is a complex simulation model composed of many submodels describing biotic and abiotic components of the environment and all trophic levels of ecosystems in the Everglades region. The model uses data from different sources, ranging from direct observations of animals to satellite photographs and GISs. It is also a mixed model with respect to the techniques of mathematical modeling. Some submodels are constructed using classical mathematical ecology, whereas others that describe main animal species (Florida panther, white-tailed deer, alligators, crocodiles, and different bird species) are individual-based models.

The ATLSS model gave rise to discussions about some fundamental problems concerning the individual-based approach in ecology. Its authors have emphasized that one of the most important arguments for applying individual-based models is that individuals making up a population do not live in a uniform habitat. If so, we cannot use classical models that assume identical environmental conditions for all individuals of a local population. There are, however, unanswered questions about the possibility of verification of such complex models. On the one side, attempts are made to construct a unified and relatively easily available computer program [e.g., see Lorek and Sonneschein

(1999), who discuss this problem, or Baveco and Smeulders (1994), who presented an object-oriented simulation tool for individual-based modeling]. On the other hand, the authors of individual-based models prefer to construct models on their own, from the beginning, and to verify them using their own experience and knowledge of the object of the model.

Conclusions

Complexity and lack of generality are often described as the main limitations of individual-based modeling in ecology. Intense application of computer simulations and visualization techniques to present results are quoted as other faults of those who use individual-based models. These arguments seem especially pertinent for people accustomed to classical ecological models. Difference and differential equations with population density as a state variable are relatively simple, usually have analytical solutions and, what is more important, leave us with the feeling that we have found a general solution to some important problem. However, it is neither the kind of mathematical or computer technique we use nor the degree of problem generality that is important. Correct assumptions of the model and the possibility of falsification of the hypotheses are important. One cannot falsify hypotheses about the competitive exclusion principle built on the basis of the classical Lotka–Volterra competition model simply because there is no species in the world that fulfills the assumptions of this model. Life on an ecological level seems to be complex and diverse, making attempts to find general models futile. Nevertheless, many typical patterns can be distinguished in nature. Such pattern-oriented modeling (Grimm et al., 1996) is the future of individual-based modeling in ecology. We believe that it will produce results on a level of generality that will enable us to understand and predict the behavior of ecological systems.

There is great deal of data concerning ecophysiology and the behavior of individuals among numerous plant and animal species. However, most are taken from experiments on isolated individuals, or are the results of averaging individual variability. Interactions between individuals are the most important and also the most difficult part of individual-based modeling in ecology. There is still much to do if we want to understand the mechanisms and predict the results of interactions between individuals. As I have shown in this chapter, interactions between individuals produce variability that usually cannot be neglected or averaged, and is another main part of all individual-based models in ecology. One of the most important lessons we have learned so far from individual-based modeling is that the result of the activity of individuals measured on the population level is not equal to the average activity of individuals multiplied by number of individuals but is probably equal to the sum of activities of all individuals in the population.

References

Attchinson, J., and Brown, J. A. C. (1957). *The lognormal distribution*. Cambridge, UK: Cambridge University Press.

Bart, J. (1995). Acceptance criteria for using individual-based models to make management decisions. *Ecological Applications* 5: 411–420.

Baveco, J. M. (1997). *Population dynamics in object-oriented and individual-based models*. DLO Institute for Forestry and Nature Research (IBN-DLO) Scientific Contribution No. 6, Wageningen.
Baveco, J. M. (1998). Population dynamics in complex heterogeneous landscapes. *INTECOL Congress Abstracts*. A. Farina, J. Kennedy, V. Bossu (Eds.). Proceeding of the VII International Congress of Ecology. INTECOL, Florence, 36.
Baveco, J. M., and Smeulders, A. W. M. (1994). Objects for simulations: Smalltalk and ecology. *Simulation* 62: 42–57.
Begon, M., Firbank, L., and Wall, R. (1986). Is there a self-thinning rule for animal populations? *Oikos* 46: 122–124.
Benjamin, L. R., and Hardwick, R. C. (1986). Sources of variation and measures of variability in even-aged stands of plants. *Annals of Botany* 58: 757–778.
Bolker, B. M., Deutschman, D. H., Hartvigsen, G., and Smith D. L. (1997). Individual-based modelling: What is the difference? *Trends in Ecology and Evolution* 12: 111.
Bonan, G. B. (1988). The size structure of theoretical plant population: Spatial pattern and neighbourhood effects. *Ecology* 69: 1721–1730.
Bonan, G. B. (1991). Density effects on size structure of annual plant populations: An indication of neighbourhood competition. *Annals of Botany* 68: 341–347.
Brand, D. G., and Magnussen, S. (1988). Asymmetric, two-sided competition in even-aged monocultures of red pine. *Canadian Journal of Forest Research* 18: 901–910.
Cianciara, S. (1980). Biometric and bioenergetic characterization of the development of *Cloëon dipterum* (L.). *Polskie Archiwum Hydrobiologii* 27: 377–406.
Collins, J. P. (1979). Intrapopulation variation in the body size at metamorphosis and timing of metamorphosis in the bull-frog, *Rana catesbeiana*. *Ecology* 60: 738–749.
Czárán, T. (1998). *Spatiotemporal models of population and community dynamics*. London: Chapman & Hall.
DeAngelis, D. L., and Gross, L. J. (Eds.). (1992). *Individual-based model and approaches in ecology: Populations, communities and ecosystems*. New York: Chapman & Hall.
DeAngelis, D. L., Gross, L. J., Huston, M. A., Wolff, W. F., Fleming, D. M., Comiskey, E. J., and Sylvester, S. M. (1998). Landscape modelling for Everglades ecosystem restoration. *Ecosystems* 1: 64–75.
DeAngelis D. L., Rose, K. A., and Huston, M. A. (1994). Individual-oriented approaches to modelling ecological populations and communities. In: S. A. Levin (Ed.), *Frontiers in mathematical biology* (pp. 390–410). Berlin: Springer-Verlag.
Dgebuadze, Yu., Yu. (2001). Ekologiceskije zakonomernosti rosta ryb. Moskow: Nauka [in Russian].
Duncan, A., and Klekowski, R. Z. (1975). Parameters of energy budget. In: W. Grodziński, R. Z. Klekowski and A. Duncan (Eds.), *Methods for ecological bioenergetics* (pp. 97–147). Oxford, UK: Blackwell Scientific.
Dunning, J. B., Stewart, D. J., Danielson, B. J. N. R., Noon, B.R., Root, T. L., Lamberson, R. H., and Stevens, E. E. (1995). Spatially explicit population models: Current forms and future uses. *Ecological Applications* 5: 3–11.
Ebenman, B., and Persson, L. (Eds.). (1988). *Size-structured populations: Ecology and evolution*. Berlin: Springer-Verlag.
Firbank, L. G., and Watkinson, A. R. (1985). A model of interference within plant monocultures. *Journal of Theoretical Biology* 116: 291–311.
Ford, D. E. (1975). Competition and stand structure in some even-aged plant monocultures. *Journal of Ecology* 63: 311–333.
Ford, E. D., and Diggle, P. J. (1981). Competition for light in a plant monoculture modeled as a spatial stochastic process. *Annals of Botany* 48: 481–500.
Ford, E. D., and Nowbould, P. J. (1970). Stand structure and dry weight production through the sweet chestnut (*Castanea sativa* Mill.) copice cycle. *Journal of Ecology* 58: 275–296.
Ford, E. D., and Nowbould, P. J. (1971). The leaf canopy of coppiced deciduous woodland: I. Development of structure. *Journal of Ecology* 59: 843–862.
Fromm, J., and Rasmussen G. (1984). A growth model, gastric evacuation, and body composition in rainbow trout, *Salmo gairdneri* Richardson, 1936. *Dana* 3: 61–139.
Gates, D. J., McMurtrie, R., and Borough, C. J. (1983). Skewness reversal of distribution of stem diameter in plantations of *Pinus radiata*. *Australian Forest Research* 13: 267–270.
Gavrikov, V. L., and Sekretenko, O. P. (1996). Shoot-based three-dimensional model of young Scots pine growth. *Ecological Modelling* 88: 183–193.
Geber, M. A. (1989). Interplay of morphology and development on size inequality: A *Polygonum* greenhouse study. *Ecological Monographs* 59: 267–288.

Grimm, V. (1999). Ten years of individual-based modelling in ecology: What have we learned and what could we learn in the future. *Ecological Modelling* 115: 129–148.

Grimm, V., Frank, K., Jeltsch, F., Brandl, R., Uchmański, J., and Wissel, C. (1996). Pattern-oriented modelling in population ecology. *Science of the Total Environment* 183: 151–166.

Grimm, V., Schmidt, E., and Wissel, C. (1992). On the application of stability concepts in ecology. *Ecological Modelling* 63: 143–161.

Grimm, V., and Uchmański, J. (1994a). Ecological systems are not dynamic systems: Some consequences of individual variability. In: J. Grasman and G. van Straten (Eds.), *Predictability and nonlinear modelling in natural sciences and economics* (pp. 248–259). Dordrecht: Kluwer.

Grimm, V., and Uchmański, J. (1994b). Dichteabhängigkeit und asymmetrische Konkurrenz zwischen Individuen: Ein individuenbasiertes Modell. *Verhandlungen der Gesellschaft für Ökologie* 23: 311–319. [In German]

Grimm, V., Wyszomirski, T., Aikman, D., and Uchmański, J. (1999). Individual-based modelling and ecological theory: Synthesis of a workshop. *Ecological Modelling* 115: 275–282.

Groom, M. J., and Pascual, M. A. (1998). The analysis of population persistence: An outlook on the practice of viability analysis. In: P. L. Fiedler and P. M. Kareiva (Eds.), *Conservation biology: For the coming decade* (pp. 4–26). New York: Chapman & Hall.

Hara, T. (1984a). A stochastic model and the moment dynamics of the growth and size distribution in plant population. *Journal of Theoretical Biology* 109: 173–190.

Hara, T. (1984b). Dynamics of the stand structure in plant monocultures. *Journal of Theoretical Biology* 110: 223–239.

Hara, T. (1986). Growth of individuals in plant monocultures. *Annals of Botany* 57: 55–68.

Hara, T. (1988). Dynamics of size structure in plant populations. *Trends in Ecology and Evolution* 3: 129–133.

Hara, T. (1994). Growth and competition in clonal plants—persistence of shoot populations and species diversity. *Folia Geobot. Phytotax. Praha* 29: 181–201.

Hara, T., and Wyszomirski, W. (1994). Competitive asymmetry reduces spatial effects on size–structure dynamics in plant populations. *Annals of Botany* 73: 285–297.

Harper, J. L. (1967). A Darwinian approach to plant ecology. *Journal of Ecology* 55: 247–270.

Harper, J. L. (1977). *Population biology of plants*. London: Academic Press.

Harper, J. L. (1986). Biologia populacyjna i ewolucja organizmów klonalnych. Moduty i rozgałęzienia a pobieranie skladników pokarmowych. *Wiadomości Ekologiczne* 32: 327–359. [In Polish]

Huston, M. A. (1986). Size bimodality in plant populations: An alternative hypothesis. *Ecology* 67: 265–269.

Huston, M. A. (1994). *Biological diversity*. Cambridge, UK: Cambridge University Press.

Huston, M. A., and DeAngelis, D. L. (1987). Size bimodality in monospecific populations: A critical review of potential mechanism. *American Naturalist* 129: 678–707.

Huston, M., DeAngelis, D. L., and Post, W. (1988). New computer models unify ecological theory. *BioScience* 38: 682–691.

Ivlev, V. S. (1961). *Experimental ecology of the feeding of fishes*. New Haven, CT: Yale University Press.

Jobling, M. (1993). Bioenergetics: Feed intake and energy partitioning. In: J. C. Rankin and F. B. Jensen (Eds), *Fish ecophysiology* (pp. 1–44). New York: Chapman and Hall.

Judson, O. P. (1994). The rise of the individual-based model in ecology. *Trends in Ecology and Evolution* 9: 9–14.

Kaandorp, J. A. (1994). *Fractal modelling: Growth and form in biology*. Berlin: Springer-Verlag.

Kaiser, H. (1979). The dynamics of populations as a result of the properties of individual animals. In: U. Holbach and J. Jacobs (Eds.), *Population ecology* (pp. 109–136). Fortschr. Zool. 25, Stuttgart: Gustav Fischer. Verlag.

Kellomaki, S., and Strandman, H. (1995). A model for the structural growth of young Scots pine crowns based on light interception by shoots. *Ecological Modelling* 80: 237–250.

Klekowski, R. Z., Prus, T., and Zyromska-Rudzka, H. (1967). Elements of energy budget of *Trybolium castaneum* (Hbst) in its development cycle. In: K. Petrusewicz (Ed.), *Secondary productivity of terrestrial ecosystems (principles and methods)* (pp. 859–879). Warszawa: Państwowe Wydawnictwo Naukowe.

Knox, R. G., Peet, R. K., and Christensen, N. L. (1989). Population dynamics in loblolly pine stands: Changes in skewness and size inequality. *Ecology* 70: 1153–1166.

Koehl, M. A. R. (1989). Discussion: From individuals to populations. In: J. Roughgarden, R. M. May and A. Levin (Eds.), *Perspectives in ecological theory* (pp. 39–53). Princeton: Princeton University Press.

Konarzewski, M. (1995). Allocation of energy to growth and respiration in avian postembryonic development. *Ecology* 76: 8–19.

Kooijman, S. A. L. M. (1993). *Dynamic energy budgets in biological systems: Theories and applications in ecotoxicology*. Cambridge, UK: Cambridge University Press.

Kozłowski, J. (1980). Density dependence, the logistic equation, and r- and K-selection: A critique and an alternative approach. *Evolutionary Theory* 5: 89–101.

Koyama, H., and Kira, T. (1956). Intraspecific competition among higher plants: VIII. Frequency distribution of individual plant weight as affected by the interactions between plants. *Journal of Biology*, Osaka City University 7: 73–94.

Latto. J. (1994). Evidence for self-thinning rules in animals. *Oikos* 69: 531–534.

Liu, J., and Ashton. P. S. (1998). FORMOSAIC: An individual-based spatially explicit model for simulating forest dynamics in landscape mosaics. *Ecological Modelling* 106: 177–200.

Łomnicki, A. (1978). Individual differences between animals and the natural regulation of their number. *Journal of Animal Ecology* 47: 461–475.

Łomnicki, A. (1980). Regulation of population density due to individual differences and patchy environment. *Oikos* 35: 185–193.

Łomnicki, A. (1988). *Population ecology of individuals*. Princeton, NJ: Princeton University Press.

Łomnicki, A. (1992). Population ecology from the individual perspective. In: D. L. DeAngelis and L. J. Gross (Eds.), *Individual-based models and approaches in ecology: Populations, communities and ecosystems* (pp. 3–17). New York: Chapman & Hall.

Łomnicki, A. A. (1999). Individual-based models and the individual-based approach to population ecology. *Ecological Modelling* 115: 191–198.

Lorek, H., and Sonnenschein, M. (1999). Modelling and simulation software to support individual-based modelling. *Ecological Modelling* 115: 199–216.

Magnuson, J. J. (1962). An analysis of aggressive behaviour, growth and competition for food and space in medaka [*Oryzias letipas* (Pisces; Cyprinodontidae)]. *Canadian Journal of Zoology* 40: 313–363.

Majkowski, J., and Uchmański, J. (1980). Theoretical foundations of individual growth equation in animals. *Polish Ecological Studies* 6: 7–31.

Matessi, C. (1982). A theoretical approach to dynamics of single population. In: C. Barigozzi (Ed.), *Vito Volterra symposium on mathematical models in biology* (pp. 222–244). Lecture Notes in Biomathematics No. 39. Berlin: Springer-Verlag.

Měch, R., and Prusinkiewicz, P. (1996). Visual models of plants interacting with their environment. *Computer Graphics: Proceedings of Annual Conference Series*, 1996: 397–410.

Metcalfe, N. B., Taylor, A. C., and Thorpe, J. E. (1995). Metabolic rate, social status and life-history strategies in Atlantic salmon. *Animal Behaviour* 49: 431–436.

Metz, J. A. J., and Diekmann, O. (Eds.). (1986). *The dynamics of physiologically structured populations*. Lecture Notes in Biomathematics No. 68. Berlin: Springer-Verlag.

Miller, T. E., and Weiner, J. (1989). Local density variation may mimic effects of asymmetric competition on plant size variability. *Ecology* 70: 1188–1191.

Mithen, R., Harper, J. L., and Weiner, J. (1984). Growth and mortality of individual plants as a function of "available area". *Oecologia* 62: 57–60.

Mohler, C. L., Marks, P. L., and Sprugel, D. G. (1978). Stand structure and allometry of trees during self-thinning of pure stands. *Journal of Ecology* 66: 599–614.

Murray, J. D. (1989). *Mathematical biology*. Berlin: Springer-Verlag.

Nagashima, H., Terashima, I., and Katoh, S. (1995). Effects of plant density on frequency distributions of plant height in *Chenopodium album* stands: Analysis based on continuous monitoring of height-growth of individual plants. *Annals of Botany* 75: 173–180.

Nakamura, N., and Kasahara, S. (1977a). A study on the phenomenon of the tobi-koi or shoot carp: I. On the earliest stage at which the shoot carp appears. *Bamidgeh* 29: 41–44.

Nakamura, N., and Kasahara, S. (1977b). A study on the phenomenon of the tobi-koi or shoot carp: II. On the effect of particle size and quality and quantity of the food. *Bamidgeh* 29: 44–47.

Nakamura, N., and Kasahara, S. (1977c). A study on the phenomenon of the tobi-koi or shoot carp: III. On the results of culturing the modal group and the growth of carp fry reared individually. *Bamidgeh* 29: 48–52.

Nakamura, N., and Kasahara, S. (1977d). A study on the phenomenon of the tobi-koi or shoot carp: IV. Effects of adding a small number of large individuals to the experimental batches of carp fry and of culture density upon the occurrence of shoot carp. *Bamidgeh* 29: 53–56.

Nakano, S. (1995). Individual differences in resource use, growth and emigration under the influence of a dominance hierarchy in fluvial red-spotted masu salmon in a natural habitat. *Journal of Animal Ecology* 64: 75–84.

Obeid, M., Machin, D., and Harper, J. L. (1967). Influence of density on plant to plant variation in fiber flax, *Linum usitatissimum* L. *Crop Science* 7: 471–473.

Ogden, J. (1970). Plant population structure. *Proceedings of New Zealand Ecological Society* 17: 1–9.
Pawlowski, W. J. (2000). *Dynamika liczebności populacji roślin: Model osobniczy*. Doctoral thesis, Institute of Ecology, Polish Academy of Sciences, Dziekanów Leśny.
Perttunen, J., Sievänen, R., and Nikinmaa, E. (1998). LIGNUM: A model combining the structure and the functioning of trees. *Ecological Modelling* 108: 189–198.
Peters, R. H. (1983). *The ecological implications of body size*. Cambridge, UK: Cambridge University Press.
Prusinkiewicz, P., and Lindenmayer, A. (1990). *The algorithmic beauty of plants*. New York: Springer-Verlag.
Rabinowitz, D. (1979). Biomodal distribution of seedling weight in relation to density of *Festuca paradoxa* Desv. *Nature* 277: 297–298.
Schmitt, J., Ehrhardt, D. W., and Cheo, M. (1986). Light-dependent dominance and suppression in experimental radish populations. *Ecology* 67: 1502–1507.
Schwinning, S., and Weiner, J. (1998). Mechanism determining the degree of size-asymmetry in competition among plants. *Oecologia* 113: 447–455.
Shugart, H. H., Smith, T., and Post, W. M. (1992). The potential for application of individual-based simulation models for assessing the effects of global changes. *Annual Review of Ecology and Systematics* 23: 15–38.
Sorrensen-Cothern, K. A., Ford, E. D., and Sprugel, D. G. (1993). A model of competition incorporating plasticity through modular foliage and crown development. *Ecological Monographs* 63: 277–304.
Stearns, S. C. (1992). *The evolution of life histories*. Oxford, UK: Oxford University Press.
Stephan, T., and Wissel, C. (1999). The extinction risk of a population exploiting a resource. *Ecological Modelling* 115: 217–225.
Sushchenya, L. M. (1975). *Kolicestvennyje zakonomernosti pitanija rakoobraznych*. Minsk: Nauka i Technika. [In Russian]
Sushchenya, L. M., and Khmeleva, N. N. (1967). Potreblenie piscy kak funkcja vesa tela u rakoobraznych. *Doklady Akademi Nauk SSSR* 76: 1428–1431. [In Russian]
Symons, P. L. (1968). Increase in aggression and in strength of the social hierarchy among juvenile Atlantic salmon deprived of food. *Journal of Fisheries Research Board of Canada* 25: 2387–2401.
Thomas, S. C., and Weiner, J. (1989). Including competitive asymmetry in measures of local interference in plant populations. *Oecologia* 80: 349–355.
Turner, M. D., and Rabinowitz, D. (1983). Factors affecting frequency distribution of plant mass: The absence of dominance and suppression in competing monocultures of *Festuca paradoxa*. *Ecology* 64: 469–475.
Uchmański, J. (1985). Differentiation and frequency distributions of body weights in plants and animals. *Philosophical Transactions of Royal Society London, B* 310: 1–75.
Uchmański, J. (1987). Resource partitioning among unequal competitors. *Ekologia Polska* 35: 71–87.
Uchmański, J. (1999). What promotes persistence of a single population: An individual-based model. *Ecological Modelling* 115: 227–241.
Uchmański, J. (2000a). Individual variability and population regulation: An individual-based model. *Oikos* 90: 539–548.
Uchmański, J. (2000b). Resource partitioning among competing individuals and population persistence: An individual-based model. *Ecological Modelling* 131: 21–32.
Uchmański, J., Aikman, D., Wyszomirski, T., and Grimm, V. (Eds.). (1999). Individual-based modelling in ecology. *Ecological Modelling* 115(2, 3). [Special issue]
Uchmański, J., and Dgebuadze, Yu., Yu. (1990). Factors affecting skewness of weight distributions in even-aged populations: a numerical example. *Polish Ecological Studies* 16: 297–311.
Uchmański, J., and Grimm, V. (1996). Individual-based modelling in ecology: What makes the difference? *Trends in Ecology and Evolution* 11: 437–441.
Uchmański, J., Grimm, V., and Wyszomirski, T. (1999). Individual- based approach in ecology. In: A. Farina (Ed.), *Perspectives in ecology: A glance from the VII International Congress of Ecology* (pp. 187–195). Leiden: Backhuys.
Ulbrich, K., and Henschel, J. R. (1999). Intraspecific competition in a social spider. *Ecological Modelling* 115: 243–251.
Weiner, J. (1990). Asymmetric competition in plant populations. *Trends in Ecology and Evolution* 5: 360–364.
Westoby, M. (1984). The self-thinning rule. *Advances in Ecological Researches* 6: 167–225.
Wethey, D. S. (1983). Intrapopulation variation in growth of sessile organisms: Natural populations of intertidial barnacle *Balanus balanoides*. *Oikos* 40: 14–23.
White, J. (1979). The plant as a metapopulation. *Annual Review of Ecology and Systematics* 10: 109–145.
White, J., and Harper, J. L. (1970). Correlated changes in plant and number in plant populations. *Journal of Ecology* 58: 467–485.

Wilbur, H. M., and Collins, J. P. (1973). Ecological aspects of amphibian metamorphosis. *Sciences* 182: 73–92.

Wissel, C., Stephan, T., and Zaschke, S.-H. (1994). Modelling extinction and survival of small populations. In: H. Remmert (Ed.), *Minimum animal populations* (pp. 67–103). Berlin: Springer-Verlag.

Wyszomirski, T. (1983). A simulation model of the growth of competing individuals of a plant population. *Ekologia Polska* 31: 73–92.

Wyszomirski, T. (1986). Growth, competition and skewness in a population of one-dimensional individuals. *Ekologia Polska* 34: 615–641.

Wyszomirski, T. (1992). Detecting and displaying size bimodality: Kurtosis, skewness and bimodalizable distributions. *Journal of Theoretical Biology* 158: 109–128.

Wyszomirski, T., Wyszomirska, I., and Jarzyna, I. (1999). Simple mechanisms of size distribution dynamics in crowded and uncrowded virtual monocultures. *Ecological Modelling* 115: 253–273.

Yamagishi, H. (1962). Growth relation in some small experimental populations of rainbow trout fry, *Salmo gairdneri* Richardson, with special reference to social relations among individuals. *Japanese Journal of Ecology* 12: 43–53.

Yamagishi, H. (1969). Postembryonal growth and its variability on the three marine fishes with special reference to the mechanism of growth variation in fishes. *Researches on Population Ecology* 11: 14–33.

Yamagishi, H., Maruyama, T., and Mashiko, K. (1974). Social relation in a small experimental population of *Odontobutis obscurus* (Temminck et Schlager) as related to individual growth and food intake. *Oecologia* 17: 187–202.

15

Concepts and Application of Nonlinear Complex Systems Theory to Ecological Succession

Gian-Reto Walther, Jörg Petersen, and Richard Pott

Introduction

Ecologists have long noted and studied the tendency of species communities to change through time. This process, know as succession, is probably one of the oldest principals in ecological theories. Since its first formalization in the early 1900s, hundreds of descriptions, commentaries, and interpretations have been published, and extended, inconclusive controversies have been generated. Although much is known about succession, change in species assemblages through time remains a complex process, still far from being totally understood or predicted. However, there is increasing awareness among ecologists that nonlinear dynamics are the rule rather than the exception in ecological systems. Thus, ecological applications dealing with time-series data require methods that can incorporate nonlinear relations. In this regard, cellular automata models are a promising approach to yield insights into nonlinear dynamics of ecological succession and improve the application thereof to conservation and land management issues.

The Application of Ecological Succession

Dealing with succession, different approaches have been suggested (see also Wildi, 2001). Clements (1916) strongly promoted a monoclimax theory, suggesting that under a given climate, the development will follow one direction only and end up in one specific state. As to systems theory, such a system is said to have a point attractor (e.g., Hastings et

Gian-Reto Walther, Jörg Petersen, and Richard Pott • Institute of Geobotany, University of Hannover, Nienburger Street 17, D-30167 Hannover, Germany al., 1993, and references therein). The concept of Watt (1947) extends the previous one, suggesting as final stage a cycle of permanently altering states (i.e., a cyclic attractor), and Whittaker (1953) introduced the continuity of climax types that vary with environmental conditions. Finally, Whittaker (1967) assumes that the final state depends on the initial state, in other terms, leading to a strange attractor, which raises the question of whether it is possible at all to predict a final state of a system following ecological succession (cf. Lorenz, 1963).

The term "*ecological succession*" has increasingly been used in a wide variety of biological research applications in recent years. With regard to microbes, the concept has been applied to study the seasonal variation of the activity of phytoplankton (Giraudet et al., 1999) or changes in microbial populations (e.g., Simpson et al., 1999; Krause et al., 1999). Pinca and Dallot (1997) use the term with regard to zooplankton community structure in the western Mediterranean Sea, and Gherardi and Bosence (1999) for encrusting organisms in reefs of the South Atlantic in Brazil. An example of how to understand ecological succession of invertebrates in agricultural landscapes is presented by Keesing and Wratten (1998), whereas Spitzer et al. (1999) focus on insect biodiversity and its variability.

Botanical case studies with regard to the aforementioned concept include the work of Akande (1996) and Nakatsubo (1997) on the role of bryophytes in ecological succession. Different strategies of members of the Poaceae family and their importance in terms of ecological succession in meadows are discussed by Bajon et al. (1996) and Karl et al. (1999), with an example of exotic grasses in a northern mixed-grass prairie. Further examples for ecological succession applications include spatial patterns of the distribution of mires and their association to changes in the hydrological conditions in Poland (Okruszko, 1998) as well as temporal changes of understory vegetation in Canadian pine forests (Bruelisauer et al., 1996). In agricultural systems of northern Italy, Zanin et al. (1997) interpret the dynamics of the weed flora under different tillage management on the basis of ecological succession. Aspects of ecological succession in animal–plant interactions are provided by Anderson (1999) for aquatic ecosystems, with the effects of gastropod grazers on the intertidal algae assemblages; by M'Closkey (1997) for terrestrial ecosystems, with the response in the distribution of a lizard species to changes in the plant community in the Sonoran desert, or Taksdal (1997), with the role of the heather beetle on the development of *Calluna vulgaris* heathlands in Norway. Tassin and Riviere (1998) present a model of ecological succession incorporating birds in the evolution of planted forest stands on the island of Reunion. In the same context of ecological succession, Kangmin (1999) studied the impact of environmental changes on Lake Taihus fish communities in China. Mageau et al. (1998) developed a more general simulation model to assess ecosystem response to anthropogenic stress whereas Bruelisauer et al. (1997) interpreted ecological succession based on applications of nonequilibrium thermodynamics to living systems.

The term *ecological succession* has also been used in the context of island biogeography by Whittaker (2000), and in palaeoecological studies by Brett (1998), Kieft et al. (1998), and Patterson et al. (1997).

Finally, studies on ecological succession have also been carried out on human disease vectors such as the *Anopheles* mosquitoes in India (Mishra and Singh, 1997; Sharma, 1996; Shukla et al., 1996). Reviews on ecological succession are presented by McCook (1994), with regard to vegetation dynamics, by Szwagrzyk (1995), and by Wali (1999) with respect to the rehabilitation of disturbed terrestrial ecosystems. This list of examples must not be considered exhaustive, but it illustrates the wide application of the term *ecological succession* in biological research.

The Science of Nonlinear Complex Systems Theory

Since the work of Lorenz (1963), the science of nonlinear complex systems theory has increasingly attracted attention. The very complex behavior of nonlinear systems can make any predictability of ecological succession in the longer term extremely difficult, if not impossible. Previously, irregular fluctuations in real data sets of ecological succession series were interpreted as a consequence of random environmental fluctuations or sampling errors (May and Oster, 1976). But that such fluctuations may arise in deterministic models, in which all parameters are exactly known, has been quite remarkable (Begon et al., 1996). Thus, the classical theories became realistic only in special cases (Anand and Orlóci, 1997). Considering ecological systems, nonlinear dynamics are likely to be the rule rather than the exception (Hastings et al., 1993). Instead of breaking up natural systems into ever finer structures, with the objective to reduce complexity and facilitate the mathematical computability with tools of the linear quantum theory, complex systems theory offers simple but nonlinear mathematical models that might be helpful to describe a major part of the dynamics of ecological succession.

From the early 1970s, even simple models on population dynamics were seen to display chaotic fluctuations (May, 1974, 1976; May and Oster, 1976) Since that time, the science of nonlinear dynamical systems (i.e., chaos theory) has generated much interest and is growing at an ever-accelerating speed (Czárán, 1998), providing a promising framework in which to pose questions about ecological succession, and allowing for much more flexibility in modeling (Anand, 1997).

One main approach in the search for chaos in real populations is the development of techniques for distinguishing chaos from random noise in the analysis of time series. Whereas in a system that is merely "noisy" the forecasting accuracy is roughly independent of the prediction interval, in chaotic systems, the accuracy of forecasts falls off with prediction time (Begon et al., 1996). Two main types of mathematical models are offered, namely, differential models and interacting particle systems (IPSs) (cf. Czárán, 1998). We focus on an important subclass of IPSs (i.e., cellular automata). For ecological investigations, cellular automata have been applied for a wide range of phenomena (cf. Czárán, 1998; Balzter et al., 1998), because they are discontinuous in space and time, implementing state transition rules that contain no stochastic elements (probabilities). Therefore, they are suited to translate the knowledge of field ecologists into mathematical forms rather than to generate differential equations.

Principles of Cellular Automata

We outline the principles of cellular automata based on the model presented by Gassmann et al. (2000). An introduction to the simulation with cellular automata is also provided by Weimar (1997).

The principles of cellular automata are based on nonlinear relations (rules) and feedback effects. With an $N \times N$ interaction matrix, the time evolution of individuals on a test area given by $N \times N$ cells can be described (see later discussion). The model system contains N different actors, and every cell represents exactly one actor (e.g., species representative) at a time. The values within the matrix are derived from field ecologists' familiarity with the investigation area and the observed behavior of the considered actors (e.g., species). The diagonal elements of the matrix describe the intraspecific preference; all the others describe the degree of interspecific preferences.

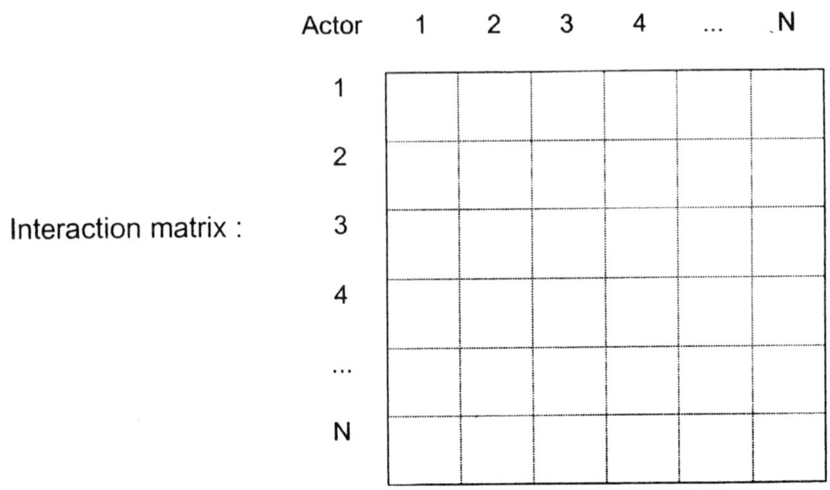

A persistence factor for each particular actor can be attributed to the model and described by the so-called "site advantage," balancing the advantage of already occupying one cell for the following step:

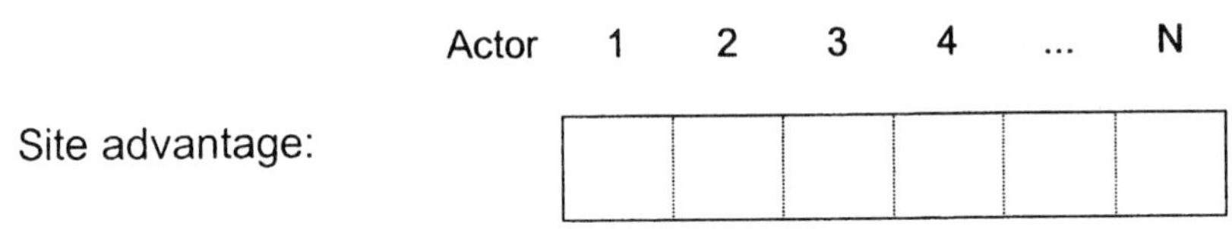

A group of sensitivity vectors may be included in the model, describing the effects of external factors such as environmental parameters or management practices. Fluctuations of the external factors can be simulated on the basis of normalized Gaussian distributions but also offer the possibility to include upward or downward trends.

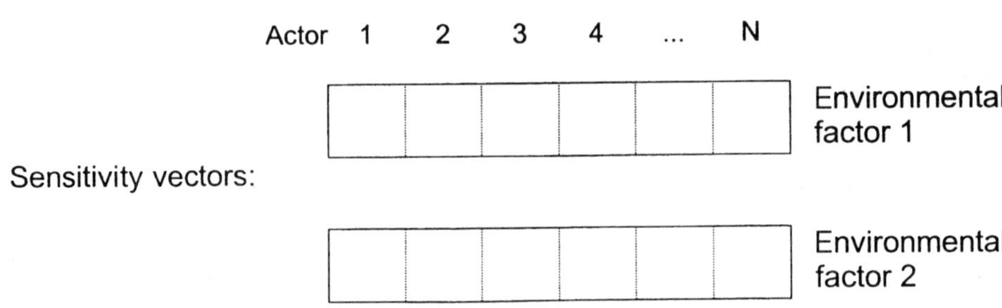

At the beginning, the initial conditions can either be specific (i.e., constructed) or follow a random distribution of actors, with every cell attributed to a different actor. The evolution dynamics defining the modification of a given pattern within 1 year are described by the following interactions:

- The nearest neighbors of each cell.
- The considered cell itself.
- The site advantage.
- The sensitivity to the external parameters.

The time is calculated in iteration steps representing, for example, 1 year. With every step, the state of every cell is reassessed as determined by the aforementioned set of rules. The state of each cell at the time t depends on the state of its adjacent neighbors at time $t - 1$ and is further modified by the site advantage and sensitivity effects. To suppress boundary effects, periodic boundary conditions can be chosen; that is, the calculated rectangle is considered as one period of a repeating infinite pattern. A more detailed description of the model used for the simulation of observed types of plants and plant communities is provided by Gassmann et al. (2000). Further information on the use of IPSs (also called "cellular automata") in biological studies are provided by Silvertown et al. (1992) and Belde and Richter (1997). We concentrate on spatiotemporal dynamics of plant communities of terrestrial ecosystems and present two examples of the application of cellular automata.

The Application of Cellular Automata Models

Simulation of Observed Types of Dynamics of Species

From long-term monitoring series of ecological succession, the following features of dynamic behavior can be distinguished: Species can show constant numbers, with only small fluctuations. They can fluctuate considerably in number but appear permanently in the considered area. Species can grow in number or appear from a former, unknown cryptic state. Or they can decay or disappear. Finally, species can display intermittent behavior by appearing irregularly on a site (cf. Klötzli, 1995).

Gassmann et al. (2000) designed an IPS model to simulate these basic observed traits. Although a quite simple model with relatively few parameters was chosen, all the five types of dynamics—constant, fluctuating, growing, decaying, and intermittent—were reflected by the model (Figure 15.1).

Different model runs with the same intra- and interspecific interrelationships showed a strong dependence on the development from the initial conditions and external variables. Even a slight change of one pixel in the initial conditions could initiate a completely different pattern in further development (for details, see Gassmann et al., 2000; Walther et al., 1999). This sensitivity, often called the "butterfly effect" (cf. Lorenz, 1963), limits the predictability for future development of ecological succession in complex systems. In different runs, the same actor may display different types of dynamics, depending on the initial conditions and the development of the external factors. Again, a variety of alternative developments imply further uncertainties for predictability in the longer term. Constant external factors lead to an earlier "extinction" of particular actors, reducing the diversity of the system, and ending in an impoverished pattern of only one (monoculture) or two remaining actors. For particular behavior types, the pattern of the external factors is crucial for their existence. It has been shown, that fluctuating external factors may offer opportunities for the appearance of particular species, reflecting the feature of an intermittent behavior (Figure 15.2).

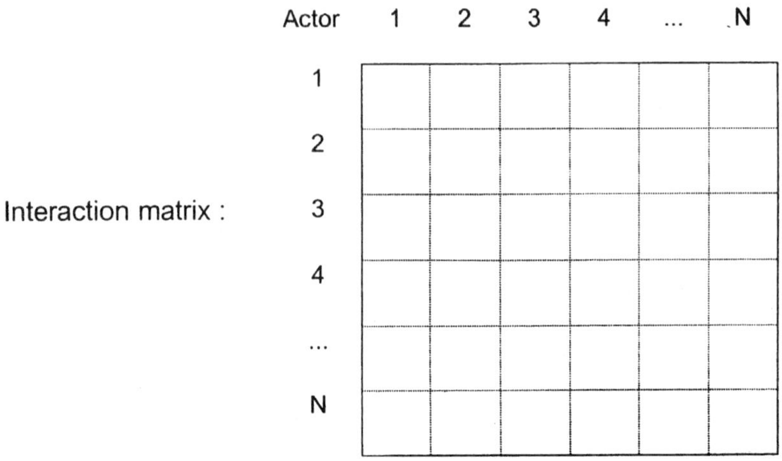

Figure 15.1 Model run over 20 years with five actors representing all five types of dynamics (from Walther et al., 1999, modified).

The aforementioned example is an attempt to show that models permit replication of basic traits of observed types of dynamics. The result of this example underlines the importance of the details in the initial conditions. Minor changes may result in substantial long-range effects. Similar instabilities must be expected to arise from small changes of the external factors.

A Cellular Automaton Model to Predict Vegetation Dynamics in Dunes and Dune Slacks of the Wadden Sea Islands in Germany

In the model of dune and dune-slack areas, the vegetation is classified as vegetation units that contains plant communities based on the Braun-Blanquet method (Braun-Blanquet, 1964). Lammerts and Grootjans (1998) demonstrated that in the Dutch islands, soil pH, lowest water table, and the percentage of organic matter in the topsoil are key factors for the differentiation of dune-slack vegetation. Therefore, and according to the research of Petersen (2000), the site is characterized by the five important factors altering the invasion of these areas: wetness, acidity, organic matter, salinity, and human impact (Figure 15.3). These environmental variates are implemented as integer values. The wetness content is expressed as moisture-(F)-value from 1 (dry) to 12 (wet), based on the classification of Ellenberg et al. (1992). The pH-value (from 3 to 8) represents the acidity. Organic matter content is divided into four classes: less than 4% , 4–8%, 8–15%, and more than 15%. For the Salt-(S)-value, there are three classes from 0 (no salt) to 2

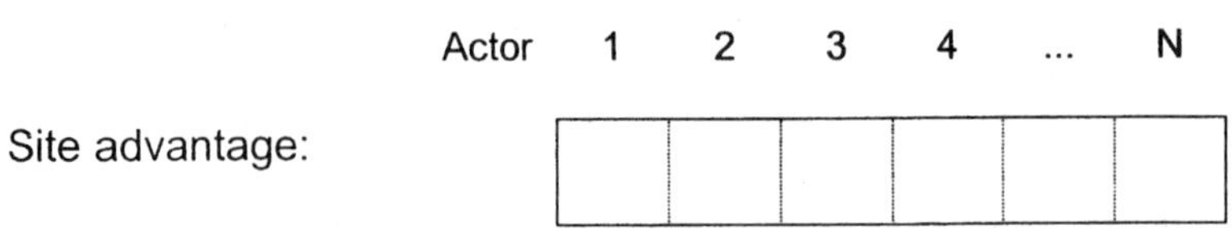

Figure 15.2 Different runs with the same actor but different random initial distributions and environmental fluctuations (from Walther et al., 1999, modified; for details see Gassmann et al., 2000; Walther et al., 1999)

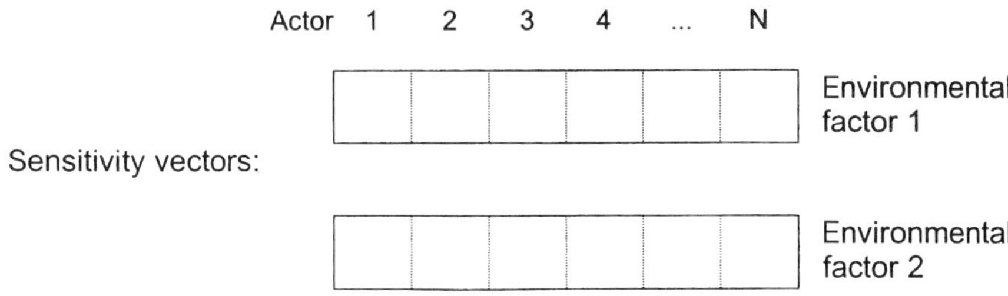

Figure 15.3 Cellular automation model for predicting vegetation dynamics in dunes and dune slacks of the Wadden Sea islands (from Petersen et al., 2001, modified)

(salt)—also based on the classification of Ellenberg et al. (1992). A simple switch (0, 1) decides whether there is human interference or not (see also Petersen et al., 2001). The site is designed as a state space that is spanned by the five environmental variates, so each local state of the environment can be understood as a point in this five-dimensional space.

Around the optimal realization, each vegetation unit occupies a certain area inside the state space described by a probabilistic multivariate distribution. Regarding the environmental factors, the distance from this optimum is determined by the tolerance of the vegetation. The probability of occurrence of a vegetation unit is determined by a multivariate probability density function (here, a Gaussian function). For transferring the ecological conditions and the vegetation into a spatial distribution, and to involve neighborhood relationships, a cellular automaton is used. The model allows simulation of the colonization and vegetation dynamics on dunes and dune slacks. In addition, feedback effects from the vegetation on site factors can be parameterized with the model for long-term studies. The calculated configurations can directly be imported in a geographic information system (GIS) for visualization. Coupled with the results of groundwater table simulations, the effects of changes in the groundwater table on the development of the vegetation in wet dune slacks can be predicted. The knowledge derived from these results serves as a basis for decision-making in resource management and nature conservation.

The Variety of Applications of Cellular Automata Models in Ecology

Cellular automata models are increasingly being used to examine plant competition and vegetation dynamics (Grist, 1999). For instance, Dunkerley (1999) uses cellular automata to model the responses of chenopod shrublands to interannual rainfall variability in arid Australia, and Jeltsch et al. (1996), to simulate the spatial arrangement of trees in semiarid savannas. Banded vegetation patterns have been investigated by Thiery et al. (1995) in arid and semiarid zones of Niger, and Puigdefabregas et al. (1999) in an Argentinean *Nothofagus* forest. In general, cellular automata have been widely used to model forest dynamics in order to analyze the large-scale characteristics of succession (e.g., Lett et al., 1999; Elmoznino and Lobry, 1997), with examples of rain forest dynamics (Alonso and Sole, 2000) or dieback phenomena (Jeltsch and Wissel, 1994).

On the landscape level, Hiebeler (2000) modeled the population response to varying habitat heterogeneity. Colasanti and Grime (1993) used a cellular automaton model to simulate secondary succession and the spatial pattern of vegetation development on gradients in resource concentration and disturbance intensity, whereas Herben (1996) presented an example using the cellular automaton to analyze spatiotemporal patterns in permanent plots. On a smaller scale, Bedward (1995) simulated the patch dynamics of a bryophyte community on a rock face, applying the same methodology. Balzter et al. (1998) provided a short review of the implementation of cellular automata models in ecology and presented a model of the population dynamics of three plant species. A single-species experiment on the effect of nutrient supply and spacing on neighbor relations was presented by Molofsky (1999) with *Cardamine pensylvanica*, and Carey (1996) used a cellular automaton to predict the distribution of an orchid species (*Himantoglossum hircinum*) in a changed climate.

Along with the examples in botany, there are also a number of applications of cellular automata in zoology, for instance, simulation of population dynamics of termites (Miramontes and De Souza, 1996), ants (Britton et al., 1996; Cole and Cheshire, 1996), and ticks (Kiszewski et al., 1997), as well as the spatial dynamics of gypsy moth outbreaks (Zhou and Liebhold, 1995). A model for the influence of the greenhouse effect on the geographical distribution and population dynamics of insects and microorganisms was provided by Karafyllidis (1998), whereas Maree et al. (1999a) used cellular automata to model phototaxis, and migration and thermotaxis (Maree et al., 1999b) of an amoeba (*Dictyostelium discoideum*). Burrows and Hawkins (1998) presented a model of the patch dynamics in the limpet barnacle mosaic on rocky shores. Vabo and Nottestad (1997) simulated the reaction of herring shoals, and Bahr and Bekoff (1999) the behavior of birds. Models on interactions between two species were presented by Huxel and Hastings (1998) and Sherratt et al. (1997).

In human medicine, cellular automata are used, for example, to model cardiac excitation kinetics (Freudenberg et al., 2000), pulmonary gas exchange (Abrams et al., 1999), repertoire dynamics of the immune system (Bernardes and Zorzenon dos Santos, 1997), and dynamics of tumor patterns (Chaplain and Anderson, 1997; Smolle et al., 1994) and HIV (Pandey, 1998).

Apart from these examples with biotic actors included in the model, others model biochemical reactions (e.g., Wurthner et al., 2000) or the dynamics of abiotic factors such as fire (Karafyllidis and Thanailakis, 1997), water release and absorption in soils (Aitkenhead et al., 1999), oil slick movement and spreading (Karafyllidis, 1997), or urban development (Wu, 1998). We conclude the list of examples with reviews on self-replication models (Sipper, 1998; Reggia et al., 1998; cf. also Machal, 1998).

Conclusions

Hastings et al. (1993) consider most natural systems and the interrelationships therein too complex to be easily comprehensible. Wilhelm and Brüggemann (2000) argue that "infinite complexity of nature requires (···) also a kind of model with infinite complexity" (p. 232) and consider cellular automata the most promising approach possessing this property. Cellular automata have proven their usefulness for a better understanding of many natural phenomen (Elmoznino and Lobry, 1997) and implementation in applied issues of resource management and nature conservation. The aforementioned examples reflect the great variety of applications.

Relevance for field ecologists is indicated by the fact that such models may provide information for short-term development of ecological succession, even though definite limits are set for the predictability in the long run (cf. also Oreskes et al., 1994). In other words: "Dynamics is predictable in the short term, but unpredictable in the long term" (Anand, 1997, p. 61). Numerical models can provide insights for a better understanding of the dynamical and structural instability of ecological systems and yield more confidence in the interpretation of observed dynamics in ecological succession. However, dealing with complex ecosystems always implies an arrangement fraught with uncertainties.

References

Abrams, J. H., Slovut, D. P., Lim, J. P. K., and Bugedo, G. (1999). Modelling inert pulmonary gas exchange using cellular automata. *Complexity* 5(2): 36–45.

Aitkenhead, M. J., Foster, A. R., Fitz-Patrick, E. A., and Townend, J. (1999). Modelling water release and absorption in soils using cellular automata. *Journal of Hydrology* 220(1–2): 104–112.

Akande, A. O. (1996). Ecological succession in the bryophyte communities of the warm spring forest in Ikogosi, Nigeria. *Nigerian Journal of Botany* 9–10: 27–34.

Alonso, D., and Sole, R. V. (2000). The DivGame Simulator: A stochastic cellular automata model of rainforest dynamics. *Ecological Modelling* 133(1–2): 131–141.

Anand, M. (1997). The fundamental nature of vegetation dynamics—a chaotic synthesis. *Coenoses* 12(2–3): 55–62.

Anand, M., and Orlóci, L. (1997). Chaotic dynamics in a multispecies community. *Environmental and Ecological Statistics* 4(4): 337–344.

Anderson, M. J. (1999). Distinguishing direct from indirect effects of grazers in intertidal estuarine assemblages. *Journal of Experimental Marine Biology and Ecology* 234(2): 199–218.

Bahr, D. B., and Bekoff, M. (1999). Predicting flock vigilance from simple passerine interactions: Modelling with cellular automata. *Animal Behaviour* 58(4): 831–839.

Bajon, R., Richard, M., and Briane, J. P. (1996). Intraspecific biodiversity in meadows: II. Two examples in Poaceae: The *Koeleria cristata* (L.) Pers, s.l. and the *Holcus lanatus* L. *Holcus mollis* L. s.l. complex. *Acta Botanica Gallica* 143(4–5): 251–262.

Balzter, H., Braun, P. W., and Koehler, W. (1998). Cellular automata models for vegetation dynamics. *Ecological Modelling* 107(2–3): 113–125.

Bedward, M. (1995). Simple models of pattern of process. *Proceedings of the Linnean Society of New South Wales* 115: 17–23.

Begon, M., Mortimer, M., and Thompson, D. J. (1996). *Population ecology: An unified study of animals and plants* (3rd ed.). Oxford, UK: Blackwell.

Belde, M., and Richter, O. (1997). Zellulärer Automat zur Simulation der Sukzession auf Niedermoorstandorten bei unterschiedlichen Managementmassnahmen. *Verhandlungen der Gesellschaft für Ökologie* 27: 189–198.

Bernardes, A. T., and Zorzenon dos Santos, R. M. (1997). Immune network at the edge of chaos. *Journal of Biology* 186(2): 173–187.

Braun-Blanquet, J. (1964). *Pflanzensoziologie*. Wien: Springer.

Brett, C. E. (1998). Sequence stratigraphy, paleoecology and evolution: Biotic clues and responses to sea-level fluctuations. *Palaios* 13(3): 241–262.

Britton, N. F., Partridge, L. W., and Franks, N. R. (1996). A mathematical model for the population dynamics of army ants. *Bulletin Mathematical Biology* 58(3): 471–492.

Bruelisauer, A. R., Bradfield, G. E., and Maze, J. (1996). Quantifying organizational change after fire in lodgepole pine forest understorey. *Canadian Journal of Botany* 74(11): 1773–1782.

Bruelisauer, A. R., Bradfield, G. E., and Maze, J. (1997). A thermodynamic interpretation of succession. *Coenoses* 11(2): 73–76.

Burrows, M. T., and Hawkins, S. J. (1998). Modelling patch dynamics on rocky shores using deterministic cellular automata. *Marine Ecology Progress Series* 167: 1–13.

Carey, P. D. (1996). DISPERSE: A cellular automaton for predicting the distribution of species in a changed climate. *Global Ecology and Biogeography Letters* 5(4–5): 217–226.

Chaplain, M. A. J., and Anderson, A. R. A. (1997). Mathematical modelling, simulation and prediction of tumour-induced angiogenesis. *Invasion Metastasis* 16(4–5): 222–234.

Clements, F. E. (1916). *Plant succession: An analysis of the development of vegetation*. Washington, DC: Carnegie Institute.

Colasanti, R. L., and Grime, J. P. (1993). Resource dynamics and vegetation processes: A deterministic model using two-dimensional cellular automata. *Functional Ecology* 7(2): 169–176.

Cole, B. J., and Cheshire, D. (1996). Mobile cellular automata models of ant behavior: Movement activity of *Leptothorax allardycei*. *American Naturalist* 148: 1–15.

Czárán, T. (1998). *Spatiotemporal models of population and community dynamics*. London: Chapman & Hall.

Dunkerley, D. L. (1999). Banded chenopod shrublands of arid Australia: Modelling responses to interannual rainfall variability with cellular automata. *Ecological Modelling* 121(2–3): 127–138.

Ellenberg, H., Weber, H., Düll, R., Wirth, V., Werner, W., and Paulissen, D. (1992). Zeigerwerte von Pflanzen in Mitteleuropa. *Scripta Geobotanica* 18: 1-248.

Elmoznino, H., and Lobry, C. (1997). Cellular automata and modelization of forest dynamics. *Ecologie Brunoy* 28(4): 307–321.

Freudenberg, J., Schiemann, T., Tiede, U., and Höhne, K. H. (2000). Simulation of cardiac excitation patterns in a three-dimensional anatomical heart atlas. *Computers in Biology and Medicine* 30(4): 191–205.

Gassmann, F., Klötzli, F., and Walther, G.-R. (2000). Simulation of observed behaviour types of plants and plant communities. *Journal of Vegetation Science* 11: 397–408.

Gherardi, D. F. M., and Bosence, D. W. J. (1999). Modelling of the ecological succession of encrusting organisms in recent coralline-algal frameworks from Atol das Rocas, Brazil. *Palaois* 14(2): 145–158.

Giraudet, H., Abrial, D., Berthon, J. L., and Buisson, B. (1999). Seasonal variation of the alkaline phosphatase activity of phytoplankton in the hypereutrophic Grangent reservoir (Loire). *Annales de Limnologie* 35(4): 213–221.

Grist, E. P. M. (1999). The significance of spatio-temporal neighbourhood on plant competition for light and space. *Ecological Modelling* 121(1): 63–78.

Hastings, A., Hom, C. L., Ellner, S., Turchin, P., and Godfray, H. C. J. (1993). Chaos in ecology: Is mother nature a strange attractor? *Annual Review of Ecological Systems* 24: 1–33.

Herben, T. (1996). Permanent plots as tools for plant community ecology. *Journal of Vegetation Science* 7(2): 195–202.

Hiebeler, D. (2000). Populations on fragmented landscapes with spatially structured heterogeneities: Landscape generation and local dispersal. *Ecology* 81(6): 1629–1641.

Huxel, G. R., and Hastings, A. (1998). Population size dependence, competitive coexistence and habitat destruction. *Journal of Animal Ecology* 67(3): 446–453.

Jeltsch, F., Milton, S. J., Dean, W. R. J., and van Rooyen, N. (1996). Tree spacing and coexistence in semiarid savannas. *Journal of Ecology* 84(4): 583–595.

Jeltsch, F., and Wissel, C. (1994). Modelling dieback phenomena in natural forests. *Ecological Modelling* 75–76: 111–121.

Kangmin, L. (1999). Management and restoration of fish communities in Lake Taihu, China. *Fisheries Management Ecology* 6(1): 71–81.

Karafyllidis, I. (1997). A model for the prediction of oil slick movement and spreading using cellular automata. *Environment International* 23(6): 839–850.

Karafyllidis, I. (1998). A model for the influence of the greenhouse effect on insect and microorganism geographical distribution and population dynamics. *BioSystems* 45: 1–10.

Karafyllidis, I., and Thanailakis, A. (1997). A model for predicting forest fire spreading using cellular automata. *Ecological Modelling* 99: 87–97.

Karl, M. G., Heitschmidt, R. K., and Haferkamp, M. R. (1999). Vegetation biomass dynamics and patterns of sexual reproduction in a northern mixed-grass prairie. *American Midland Naturalist* 141(2): 227–237.

Keesing, V., and Wratten, S. D. (1998). Indigenous invertebrate components in ecological restoration in agricultural landscapes. *New Zealand Journal of Ecology* 22(1): 99–104.

Kieft, T. L., Murphy, E. M., Haldeman, D. L., Amy, P. S., Bjornstad, B. N., McDonald, E. V., Ringelberg, D. B., White, D. C., Stair, J., Gsell, R. P., Griffiths, T. C., Holben, W. E., and Boone D. R. (1998). Microbial transport survival, and succession in a sequence of buried sediments. *Microbial Ecology* 36(3): 336–348.

Kieszewski, A. E., Awerbuch, T., Sandberg, S., and Spielman, A. (1997). A cellular automata model for exploring the relation between sexual behaviour and fertility in ticks. *Journal of Biological Systems* 5(2): 203–213.

Klötzli, F. (1994). Grundsätze ökologischen Handelns. *DVGW Schriftenreihe Wasser* 78: 9–24.

Klötzli, F. (1995). Projected and chaotic changes in forest and grassland plant communities: Preliminary notes and theses. *Annali di Botanica* (Roma) 53: 225–231.

Krause, D. O., Smith, W. J. M., Ryan, F. M. E., Mackie, R. I., and McSweeney, C. S. (1999). Use of 16S-rRNA based techniques to investigate the ecological succession of microbial populations in the immature lamb rumen: Tracking of a specific strain of inoculated *Ruminococcus* and interactions with other microbial populations *in vivo*. *Microbial Ecology* 38(4): 365–376.

Lammerts, E., and Grootjans, A. (1998). Endangered dune slack plants, gastronomers in need of mineral water. In: I. A. Salman; H. Berends, and M. Bonazountas (Eds.), *Coastal management and habitat conservation* (pp. 355–359). Leiden: EUCC Publications.

Lett, C., Silber, C., and Barret, N. (1999). Comparison of a cellular automata network and an individual-based model for the simulation of forest dynamics. *Ecological Modelling* 121(2–3): 277–293.

Lorenz, E. N. (1963). Deterministic non-periodic flow. *Journal of Atmospheric Science* 20: 130–141.

Machal, P. (1998). John von Neumann: The founding father of artificial life. *Artificial Life* 4(3): 229–235.

Mageau, M. T., Costanza, R., and Ulanowicz, R. E. (1998). Quantifying the trends expected in developing ecosystems. *Ecological Modelling* 112(1): 1–22.

Maree, A. F. M., Panfilov, A. V., and Hogeweg, P. (1999a). Phototaxis during the slug stage of *Dictyostelium discoideum*: A model study. *Proceedings of the Royal Society of London B: Biological Sciences* 266(1426): 1351–1360.

Maree, A. F. M., Panfilov, A. V., and Hogeweg, P. (1999b). Migration and thermotaxis of *Dictyostelium discoideum* slugs, a model study. *Journal of Theoretical Biology* 199(3): 297–309.

May, R. M. (1974). Biological populations with non-overlapping generations: Stable points, stable cycles and chaos. *Science* 186: 645–647.

May, R. M. (1976). Simple mathematical models with very complicated dynamics. *Nature* 261: 459–467.

May, R. M., and Oster, G. F. (1976). Bifurcations and dynamic complexity in simple ecological models. *American Naturalist* 110: 573–599.

M'Closkey, R. T. (1997). Response of a lizard species to the senescence of a dominant plant in the Sonoran desert. *Ecoscience* 4(1): 43–47.

McCook, L. J. (1994). Understanding ecological community succession: Causal models and theories, a review. *Vegetatio* 110(2): 115–147.

Miramontes, O., and De Souza, O. (1996). The nonlinear dynamics of survival and social facilitation in termites. *Journal of Theoretical Biology* 181(4): 373–380.

Mishra, A. K., and Singh, N. (1997). Observations on mosquito breeding in rice fields in two ecological terrains of District Jabalpur, Madhya Pradesh. *Indian Journal of Malariology* 34(4): 197–203.

Molofsky, J. (1999). The effect of nutrients and spacing on neighbor relations in *Cardamine pensylvanica*. *Oikos* 84(3): 506–514.

Nakatsubo, T. (1997). The role of bryophytes in terrestrial ecosystems with special reference to forests and volcanic deserts. *Japan Journal of Ecology* 47(1): 43–54.

Okruszko, H. (1998). Hydrologic conditions of ecological succession of Biebrza and Upper Narew Valleys. *Zeitschrift für Kulturtechnik und Landentwicklung* 39(1): 1–5.

Oreskes, N., Shrader-Frechett, K., and Belitz, K. (1994). Verification, validation and confirmation of numerical models in the Earth sciences. *Science* 263: 641–646.

Pandey, R. B. (1998). A stochastic cellular automata approach to cellular dynamics for HIV: Effect of viral mutation. *Theoretical Bioscience* 117(1): 32–41.

Patterson, R. T., McKillop, W. B., Kroker, S., Nielsen, E., and Reinhardt, E. G. (1997). Evidence for rapid avian-mediated foraminiferal colonization of Lake Winnipegosis, Manitoba, during the Holocene Hypsithermal. *Journal of Paleolimnology* 18(2): 131–143.

Petersen, J. (2000). *Die Dünentalvegetation der Wattenmeer-Inseln in der südlichen Nordsee*. Husum: Husum Verlag.

Petersen, J., Pott, R., and Richter, O. (2001). Dünentäler—Ein gefährdeter Lebensraum im Interessenkonflikt zwischen Naturschutz und Grundwasserbewirtschaftung. *Zentralblatt für Geologie und Paläontologie*, Teil 1, (1/2): 93–110.

Pinca, S., and Dallot, S. (1997). Zooplankton community structure in the western Mediterranean Sea related to mesoscale hydrodynamics. *Hydrobiologia* 356: 127–142.

Puigdefabregas, J., Gallart, F., Biaciotto, O., Allogia, M., and del Barrio, G. (1999). Banded vegetation patterning in a subantarctic forest of Tierra del Fuego, as an outcome of the interaction between wind and tree growth. *Acta Oecologica* 20(3): 135–146.

Reggia, J. A., Lohn, J. D., and Chou, H. H. (1998). Self-replicating structures: Evolution, emergence and computation. *Artificial Life* 4(3): 283–302.

Sharma, V. P. (1996). Ecological changes and vector-borne disease. *Tropical Ecology* 37(1): 57–65.

Sherratt, J. A., Eagan, B. T., and Lewis, M. A. (1997). Oscillations and chaos behind predator–prey invasion: Mathematical artifact or ecological reality? *Philosophical transactions of the Royal Society of London. Series B: Biological Sciences* 352(1349): 21–38.

Shukla, R. P., Kohli, V. K., Pandey, A. C., Ojha, V. P., and Pathak, P. K. (1996). Ecology of immature mosquitoes in paddy fields of district Nainital, UP, India. *Mosquito Borne Diseases Bulletin* 13(3–4): 31–36.

Silvertown, J., Holtier, S., Johnson, J., and Dale, P. (1992). Cellular automaton models of interspecific competition for space—the effect of pattern on process. *Journal of Ecology* 80: 527–534.

Simpson, J. M., McCracken, V. J., White, B. A., Gaskins, H. R., and Mackie, R. I. (1999). Application of denaturant gradient gel electrophoresis for the analysis of the porcine gastrointestinal microbiota. *Journal of Microbiological Methods* 36(3): 167–179.

Sipper, M. (1998). Fifty years of research on self-replication: An overview. *Artificial Life* 4(3): 237–257.

Smolle, J., Hofmann-Wellenhof, R., and Kerl, H. (1994). Pattern interpretation by cellular automata (PICA)—Evaluation of tumour cell adhesion in human melanomas. *Analytical Cellular Pathology* 7(2): 91–106.

Spitzer, K., Bezdek, A., and Jaros, J. (1999). Ecological succession of a relict central European peat bog and variability of its insect biodiversity. *Journal of Insect Conservation* 3(2): 97–106.

Szwagrzyk, J. (1995). Theory of ecological succession compared with results of recent studies: Examples from the forests of North America. *Wiadomosci Botaniczne* 39(1–2): 71–80.

Taksdal, G. (1997) . Heathlands and the heather beetle (*Lochmaea suturalis*). *Fauna-Oslo* 50(1): 20–28.

Tassin, J., and Riviere, J. N. (1998). A method of assessing the impact of forest plantations on avifauna: Application to the Reunion coast. *Bois et Forets des Tropiques* 258: 37–46.

Thiery, J. M., D'Herbes, J. M., and Valentin, C. (1995). A model simulating the genesis of banded vegetation patterns in Niger. *Journal of Ecology* 83(3): 497–507.

Vabo, R., and Nottestad, L. (1997). An individual based model of fish school reactions: Predicting antipredator behaviour as observed in nature. *Fisheries Oceanography* 6(3): 155–171.

Wali, M. K. (1999). Ecological succession and the rehabilitation of disturbed terrestrial ecosystems. *Plant and Soil* 213(1–2): 195–220.

Walther, G.-R., Klötzli, F., and Gassmann, F. (1999). On the application of chaos theory to analyse fluctuations in wet meadows: Preliminary results. *Annali di Botanica* (Roma) 57: 31–40.

Watt, A. S. (1947). Pattern and process in the plant community. *Journal of Ecology* 35: 1–22.

Weimar, J. R. (1997). *Simulation with cellular automata*. Berlin: Logos.

Whittaker, R. H. (1953). A consideration of climax theory: The climax as a population and pattern. *Ecological Modelling* 23: 41–78.

Whittaker, R. H. (1967). Gradient analysis of vegetation. *Biological Reviews* 42: 207–264.

Whittaker, R. J. (2000). Scale, succession and complexity in island biogeography: Are we asking the right questions? *Global Ecology and Biogeography* 9(1): 75–85.

Wildi, O. (2001). Statistical design and analysis in long-term vegetation monitoring. In: C. A. Burga and A. Kratochwil (Eds.), *Biomonitoring: General and applied aspects on regional and global scales* (pp. 17–39). London: Kluwer Academic.

Wilhelm, T., and Brüggemann, R. (2000). Goal functions for the development of natural systems. *Ecological Modelling* 132(3): 231–246.

Wu, F. (1998). Simulating urban encroachment on rural land with fuzzy- logic-controlled cellular automata in a geographical information system. *Journal of Environmental Management* 53(4): 293–308.

Wurthner, J. U., Mukhopadhyay, A. K., and Peimann, C. J. (2000). A cellular automaton model of cellular signal transduction. *Computers in Biology and Medicine* 30: 1–21.

Zanin, G., Otto, S., Riello, L., and Borin, M. (1997). Ecological interpretation of weed flora dynamics under different tillage systems. *Agriculture, Ecosystems and Environment* 66(3): 177–188.

Zhou, G., and Liebhold, A. (1995). Forecasting the spatial dynamics of gypsy moth outbreaks using cellular transition models. *Landscape Ecology* 10(3): 177–189.

16

Soil Organisms and Litter Decomposition

Grizelle González

. . . The increasingly structured organization of contemporary science has left research students as the only large category of scientists who will habitually look at organisms. It seems important to re-open the windows on nature which, though not the source of hypotheses, is the source of experience on which intuitive faculties can operate.—J.E. Satchell (1976)

Introduction

The processes associated with the physical breakdown of litter, the transfer of organic matter and nutrients to the soil, and the release of carbon dioxide to the atmosphere are all components of decomposition. Resource quality, decomposer organisms, and physicochemical conditions influence the decomposition of plant residues (Figure 16.1). Resource quality is defined by the chemical composition of the plant residues such as the C : N ratio, lignin, and polyphenol contents (Melillo et al., 1982; Palm and Sanchez, 1991; Tian et al., 1997). Physicochemical conditions include both climate and soil parent material, and help determine abiotic soil characteristics that in turn influence litter quality and, ultimately, the activity and composition of microbial and invertebrate communities (Wardle and Lavelle, 1997). Although decomposition is mainly the result of microbial activities, soil fauna are important in conditioning the litter and in stimulating microbial actions (Coleman and Crossley, 1996).

This trilateral model of the controls of plant litter decomposition *sensu* Swift et al. (1979) remains conceptually the most robust model of factors affecting decomposition. It is also the zeal of current experimental studies that try to tease apart the independent

Grizelle González • International Institute of Tropical Forestry, USDA Forest Service, P.O. Box 25000, San Juan, PR 00928- 5000 USA; E-mail: ggonzalez@fs.fed.us

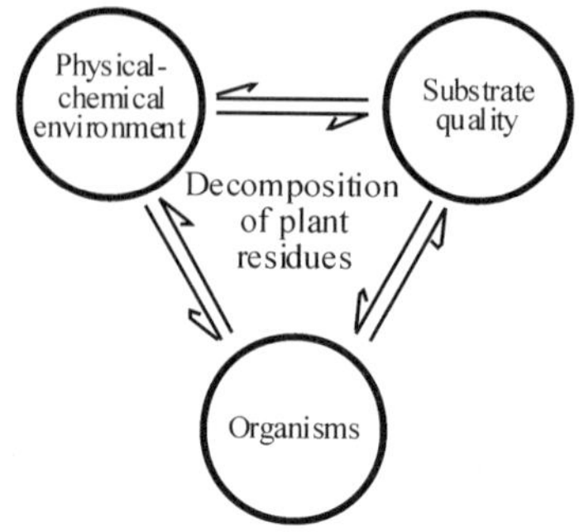

Figure 16.1. The regulation of decomposition by three interacting factors, the physico-chemical environment, and resource quality acting through decomposer organisms. Modified from Swift et al. (1979).

effects of these regulating factors. This is not an easy task, because the inference that soil organisms are an important regulating factor of ecosystem processes appears to the general scientific community as opaque (Seastedt, 2000) and intuitive. The function of soil arthropods may be considered opaque because the attributes of the fauna are in general terms captured into abiotic variables that are easier to measure (Seastedt, 2000) and incorporated to ecosystem models (e.g., CENTURY and GEM; González and Seastedt, 2001). Soil fauna effects on mineralization processes often seem intuitive because their importance is seldom properly quantified and put in the context of the other regulating factors. Most soil biologists encounter complex organismal interactions that vary among ecosystems and organism(s) of interest, as well as among the total or independent effects of the functionality, abundance, and diversity of the fauna. This tangle of interactions is imbedded within the organismal part of the simple and well-accepted three-factor model. Therefore, by no means should the effects of soil organisms be downplayed, but rather should be brought to light.

In this chapter, I discuss where the study of soil organisms currently stands in respect to theoretical setting presented in the literature by various models of plant litter decomposition and the contribution of soil microorganisms, soil fauna, and their interactions to decomposition and nutrient mineralization processes. At the end of the chapter, I present some concluding remarks to spark interest in the future directions of soil biology.

Modeling Decomposition

Several equations describing plant litter decomposition have been proposed in order to predict its rate (Heal et al., 1997). Working at continental and global scales, Meentemeyer (1978) found that the actual evapotranspiration (AET), which is a measure of the concurrent availability of energy and moisture to an ecosystem (a climatic variable), and lignin (a litter quality index) were good predictors of decomposition rates (Figure 16.2). Since then, most studies have concentrated on substrate quality, and climate effects on decomposition and have excluded explicit recognition of the decomposer flora and fauna in the process (González and Seastedt, 2001). However, two limitations can be found in this model: (1) AET is not an adequate predictor of mass loss rate in some biomes (e.g., subarctic and semiarid regions, Coûteaux et al., 1995), and (2) it assumes that the biota, the organisms responsible for the decomposition, are totally predicted by climate and substrate quality characteristics. In addition, the results from Meentemeyer (1978) indicate that the relative control of lignin on decomposition is not the same in different climates (Figure 16.2), suggesting the reduced control by lignin in cool climates is the result of reduced fragmentation ability by litter animals. Clearly, if the decomposer fauna are important in the decomposition process and major shifts occur in their composition that are not captured by the climatic and substrate quality effects, this model should fail. If the biota are linearly controlled by the physicochemical composition of the environment, this approach is justified. However, the global pattern observed in the composition and abundance of the decomposer fauna indicate that this is not the case.

The abundance of various soil fauna changes with latitude (Figure 16.3; Swift et al., 1979). The soil microfauna are relatively more abundant in the temperate regions than in the tropics, whereas soil macrofauna are more common in tropical regions than in temperate zones (González and Seastedt, 2000). In addition, the hierarchy of the determinants of decomposition might not be the same across latitudinal gradients (e.g., Heneghan et al., 1999; González and Seastedt, 2001; González et al., 2001). The relative importance of the three regulating factors varies in a predictable way within and between sites (Heal et al., 1997). Lavelle et al. (1993) suggested that the hierarchy of the determinants of decomposition is different in temperate and tropical ecosystems (Figure 16.4). In tropical ecosystems, climatic determinants are likely to be less important than the biological regulation by soil macrofauna (Figure 16.4). Strong seasonal differences in climate are most likely the dominant control of plant litter decomposition in temperate areas. Coleman and Crossley (1996) stated that fauna are potentially more important in tropical areas, for example, "breakdown and decomposition are viewed as the result of biota acting on substrate quality within the constraints of climate" (p. 111).

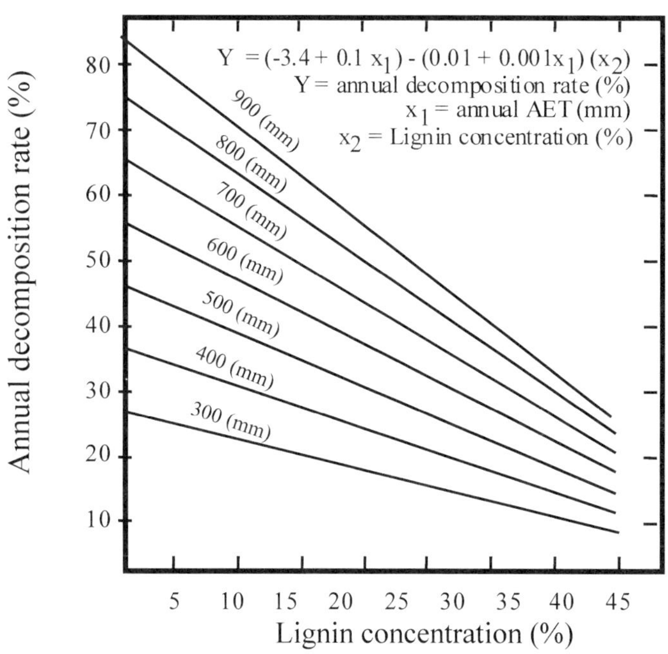

Figure 16.2. Generalized model of the changes in the slope and intercept of the relationship between initial lignin concentration (%) and annual weight loss (%) with climatic actual evapotranspiration. From Meentemeyer (1978).

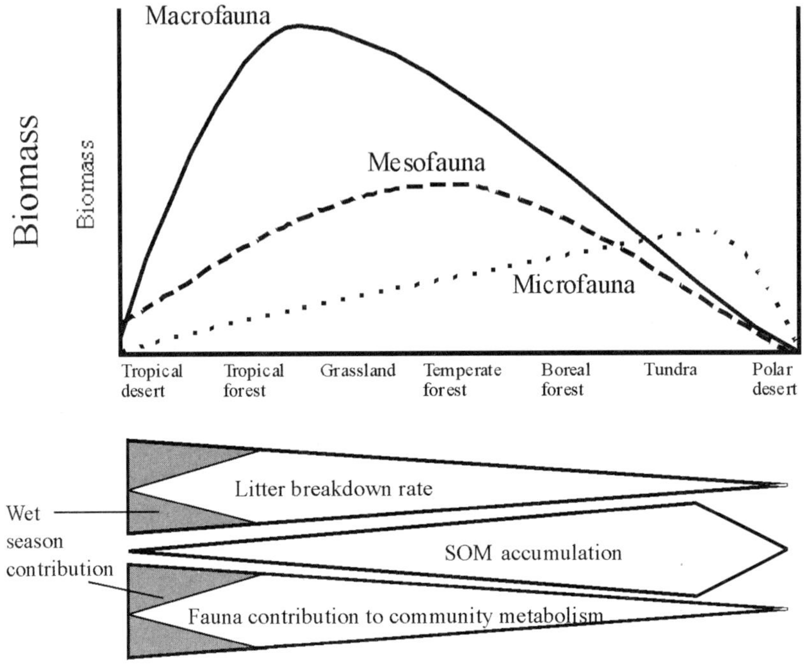

Figure 16.3. Hypothetical patterns of latitudinal variation in the contribution of the macro-, meso-, and microfauna to total soil fauna biomass. The effects on litter breakdown rates of changes in the relative importance of the three fauna size groups are represented as a gradient, together with the fauna contribution to soil community metabolism. The favorability of the soil environment for microbial decomposition is represented by the cline of soil organic matter (SOM) accumulation from the poles to the equator; SOM accumulation is promoted by low temperatures and waterlogging where microbial activity is impeded. From Swift et al. (1979).

The relative control of climate and substrate quality on decomposition processes across latitudinal gradients has been pondered in the literature. Coûteaux et al. (1995) suggested that (1) climate is the dominant factor in areas subject to unfavorable (dry and cold) weather conditions, whereas (2) litter quality is the dominant factor under favorable (wet and warm) conditions. Berg et al. (1998) suggested that if climate and site conditions are constant, then the chemical composition and physical structure of the organic matter regulate decomposition rates. Both the Coûteaux et al. (1995) and Berg et al. (1998) interpretations of the controls of decomposition are concordant with the Meentemeyer (1978) model, because both identify the same causal mechanisms for decay (González and Seastedt, 2001).

If modeling efforts of decomposition rates pragmatically assume that the biota are completely constrained by the physicochemical environment and substrate quality characteristics then they will be limited particularly when applied to lower hierarchical levels (from global to local scales). For example, González and Seastedt (2001) found the lowest density of soil fauna in a tropical dry forest, a site with an intermediate value of AET, when compared to a tropical wet and subalpine forests. Whitford et al. (1981) and Schaefer et al. (1985) showed that termites are capable of improving the microclimate and fragmentation of the litter in arid ecosystems via incorporation of the material into the soils, resulting in faster decomposition rates than those predicted from the previous models. The loss of this group, together with the general decline in abundance of macroinvertebrate decomposers,

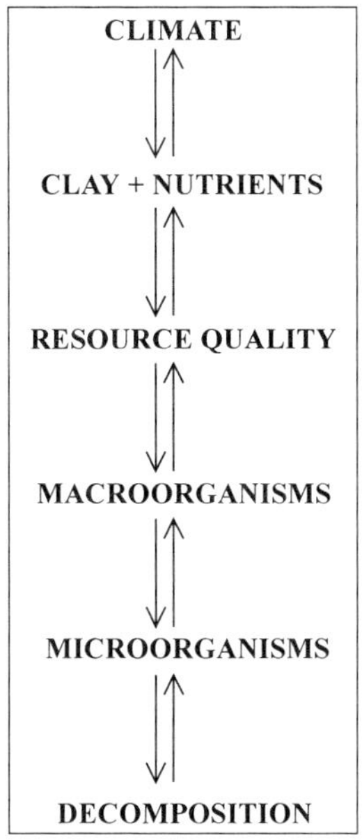

Figure 16.4. An hierarchical model of the factors determining decomposition rates in terrestrial ecosystems. Modified from Lavelle et al. (1993).

is hypothesized to reduce the ability of the fauna to uniformly affect decomposition rates in colder regions as well.

Fortunately, several process- and organism-oriented models have incorporated soil biological data in their approach to predict the functioning of ecosystems particularly due to climate change (see Smith et al., 1998). However, it is complicated to incorporate the effects of soil animals into mathematical models because (1) the independent effects of soil fauna abundance, diversity, and activity on ecosystem processes are unknown, and (2) "animals are rate-limiting or rate-regulating factors in the ecosystem, and (as such) their importance may exceed their direct role in energy release" (MacLean, 1974, p. 200). For example, MacLean recognized soil fauna as direct and indirect consumers of dead organic matter and tried to account for their respiration (Figure 16.5). He found that in most terrestrial ecosystems, animals contribute little to the total community respiration, and that the soil fauna contribution was overcome by error terms associated to the estimation of primary production and microbial decomposition, providing a reason for the exclusion of soil fauna in the analysis. Yet if we consider that models with only an implicit description of soil organisms cannot be used predictively with full confidence (Smith et al., 1998), then modeling efforts in which functional or taxonomic groups of soil organisms are explicitly described should be developed especially, if we are to accurately predict ecosystem processes at multiple hierarchical scales.

Microorganisms

Soil microorganisms usually occupy less than 1% of the soil volume, yet their numbers are high (Schinner, 1996) and their role important in regulating ecosystem processes. A high microbial biomass usually reflects the abundance of degradable organic

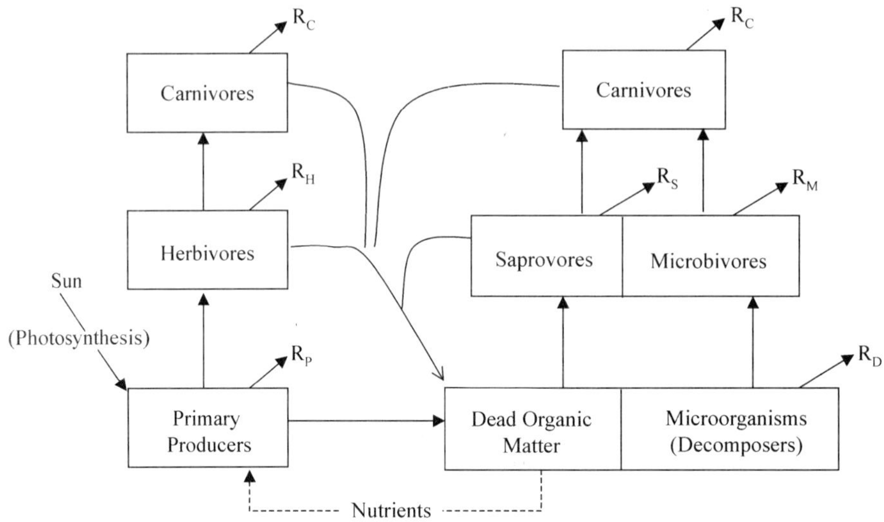

Figure 16.5. The intervention of animals in ecosystems. From MacLean (1974).

compounds (Yang and Insam, 1991), because microbes colonize and chemically break down the organic matter via utilization of enzymatic activities that release nutrients from complex organic compounds and the mineral fractions of the soil (Sharma et al., 1995; Schinner, 1996). The microorganisms ultimately make nutrients available for plant growth. Singh et al. (1989) showed a decline in microbial biomass and nutrient pools during a wet period coupled with rapid plant growth and an increase of nutrients in the microbial biomass during the dry period. Most of the aboveground net primary production of terrestrial ecosystems is returned to the soil system, and the recycling of those plant materials drives elaborate belowground food webs, where the chemical form of the nutrient elements become modified (Coleman and Crossley, 1996; Wardle and Lavelle, 1997). The microbial biomass reflects the status of accumulation and conservation of nutrients in the soil (e.g., Singh et al., 1989).

The microorganisms associated with the decomposition of organic material are extremely diverse. Fungi, bacteria, actinomycetes, and algae are soil microorganisms that have different functional capabilities. For example, soil enzymes are mostly of bacterial and fungal origin (Schinner, 1996). It has been shown that actinomycetes can comprise up to 70% of the microflora in the gut content and the casts of earthworms (e.g., Contreras, 1980). The decomposition of plant remains is facilitated by actinomycetes, because many groups synthesize cellulase and lignin-degrading enzymes (McCarthy, 1987; Crawford, 1988, as cited in Willington and Toth, 1994). Actinomycetes are also known sources of antibiotics that could have important implications in the determination of the composition of soil microbial communities (Krištůfek et al., 1993). Microalgae contribute to the fertility of the soils as they fix nitrogen and produce organic matter production via photosynthesis. The functional role and the metabolic activities of the different components of the microbial communities are associated with the rates of decomposition, nutrient mineralization processes and soil fertility (Swift, 1976). Additionally, differences in the functionality of the microorganisms result in varying interactions with the soil fauna, the resource quality, and the physicochemical environment (Sharma et al., 1995).

Much is known about the factors that control total microbial biomass in the soil, but less information is available on (1) the relative proportions of functional groups to the total microbial biomass and (2) how their relative abundance affects nutrient mineralization process. We need to understand population dynamics of microorganisms and overcome methodological constraints in order to characterize specific functions of microorganisms (Colores et al., 1996). In this sense, physiological and kinetic approaches are the most commonly used methods for estimating the biomass of specific metabolic functional groups. The substrate-induced respiration (SIR) technique relies on the respiratory response of an active population of microbes that is stimulated by the addition of a particular carbon substrate over a given period of time (Anderson and Domsch, 1978). This induced respiration is converted to biomass using a coefficient that is related to the efficiency of the mineralization process (Carter et al., 1999) that varies from soil to soil. Therefore, the SIR method might be inappropriate for comparative studies of the same functional group in different soils if a constant conversion coefficient is used. The substrate-induced growth response (SIGR) method uses growth kinetics instead of conversion factors to obtain the biomass and maximum growth rate of microorganisms. These characteristics are a reflection of metabolic activity and are independent of soil type (e.g., Schmidt, 1992; Colores et al., 1996; Lipson et al., 1999). SIGR models used to describe the growth of the microbial populations must take into account all the physiological changes induced in the microbial biomass after the substrate amendment (Panikov and Sizoka, 1996).

The SIR method has been used with selective inhibitors that allow for the separation of prokaryotic and eukaryotic contributions to the total respiratory response (Beare et al., 1990). This technique identifies the metabolically active component of the microbial community as related to the fungal and bacterial contributions. The bacterial : fungal ratio has been associated with different stages of the decomposition process and the quality of the decomposing substrate. For example, bacterial biomass dominates sites where there are decomposing residues of low carbon : nitrogen ratio and base-rich, high pH soils (Parkinson and McLean, 1998). Fungal biomass is often associated with the slow metabolism of recalcitrant organic residues with high lignin or polyphenol contents (Doube and Brown, 1998). The SIR has been evaluated mostly in temperate forest or agricultural soils (e.g., Anderson and Domsch, 1975, 1978; West and Sparling, 1986), in agricultural soils of different water contents (West and Sparling, 1986; Wardle and Parkinson, 1990), and in decaying plant residues (e.g., Beare et al., 1990; Neely et al., 1991). It has not been used extensively in tropical soils, because its suitability needs to be cautiously assessed. Both Yang and Insam (1991) and Feigl et al. (1995) tested the SIR protocol in tropical soils and showed that microbial C was underestimated when the standard conversion factor was applied (*sensu* Sparling et al., 1990). Lodge (1996) argued for a high incidence of streptomycin resistance in soils from a tropical rain forest in Puerto Rico. Antibiotic resistance may indicate exposure due to resource competition in the microbial community. A high antibiotic incidence might be related to a high abundance of soil macrofauna, for example, earthworms that support actinomycete dominance in their gut flora and surrounding soil (e.g., Contreras, 1980; Krištůfek et al., 1993), but whether there is a pattern of antibiotic resistance in different forest types is unknown. Limited information is available on the roles played by inhibitory microorganisms in soil ecology due to failures to detect antibiotics in the soil (Goodfellow and Williams, 1983; Krištůfek et al., 1993).

Soil Fauna

There are two courses by which soil fauna can affect plant litter decomposition and the rates of mineralization and humification of soil organic matter: directly, by physically modifying the substrate and soil environments, and indirectly, through interactions with the microbial community (Figure 16.6; e.g., Seastedt, 1984; Brown, 1995; Lavelle et al., 1997). The discussion in this section focuses on the direct effects by which soil fauna affect plant litter decomposition and nutrient mineralization processes. The indirect effects of soil fauna on decomposition are evaluated in the upcoming soil fauna–microbial interaction section of this chapter.

Soil fauna modify the soil environment by mixing organic and mineral particles, and changing the water infiltration and aeration regimens. Tillering by soil fauna directly alters soil physical, chemical, and biological properties. The effects of the substrate modification by the soil fauna to the decomposition process are diverse. The breakdown of litter increases the surface area for microbial use and leaching of nutrients. Soil fauna can also augment the nutrient pool in soil solution by adding nitrogenous compounds present in their excreta and dead tissue (González and Zou, 1999). However, it has been argued that

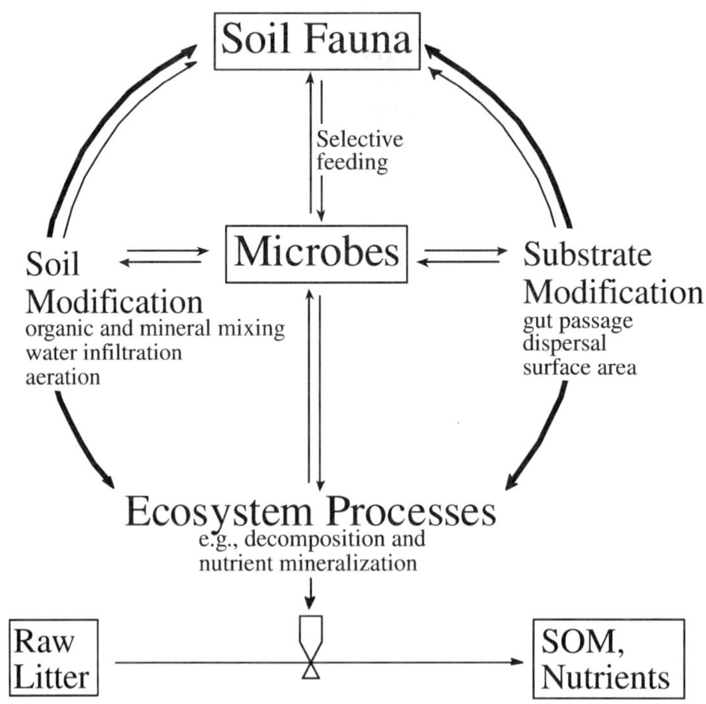

Figure 16.6. Conceptual model indicating direct and indirect paths by which soil fauna affect ecosystem processes (e.g., decomposition and mineralization) and the interaction with microorganisms. Modified from González et al. (2001).

arthropod feces might reduce the surface : volume ratio of organic materials, increasing their persistence in the soil and nutrient unavailability to plants (Webb, 1977).

The quantification of soil fauna effects is feasible by recognizing abiotic, microbial, and soil fauna components of the decomposition constant (Seastedt, 1984). There are three main methods for measuring the mass loss of a given substrate due to the effects of the biota: (1) litterbag studies, (2) chemical treatments, and (3) radioisotope techniques. Litterbag studies exclude invertebrates of a particular size by varying the mesh size of the bags. This method could fail to predict accurately faunal contributions to the percentage of mass remaining, because small mesh-sized bags can change the microclimate of the confined litter, and fragments of litter from bigger mesh-sized bags can escape (Seastedt, 1984). Naphthalene is a chemical treatment commonly used to repel litter microarthropods from the soil surface. However, naphthalene application can stimulate microbial activities (González et al., 2001). Therefore, manipulative experiments based on faunal exclusions using biocides need to be judiciously interpreted. Carbon isotope techniques represent a practical tool for measuring the turnover rate of organic matter when used in combination with plants of different isotopic origin. This method promises to help quantify the effects of soil fauna on the labile and recalcitrant portions of the soil organic carbon pool over the long term.

The range of values for the contribution of soil fauna activities to litter decomposition varies widely along latitudinal gradients, because it is dependent on confounding effects of the size, abundance, diversity, and functionality of the fauna (e.g., Hansen, 1999; Heneghan et al., 1999; Irmler, 2000; González and Seastedt, 2001). Anderson (1977) argued that small animal species tend to have large geographic distribution patterns as a result of the size and ubiquity of the microsites they occupy, whereas larger ones show greater discontinuity in distribution (Figure 16.7). There is also evidence of trophic separation of *Cryptostigmata* according to mean body size. For example, Anderson (1977) analyzed the gut content of 12 species and found that the largest mite generally fed on leaf litter and the smallest species were mycophagous (Figure 16.8). Heneghan et al. (1999) showed that the decomposition of oak litter proceeds faster in Puerto Rican and Costa Rican forests than in a temperate forest in North Carolina. In this study, microarthropods had little effect on the decomposition in the temperate forest, whereas their influence was pronounced at the tropical sites, in spite of similar or even lower species diversity.

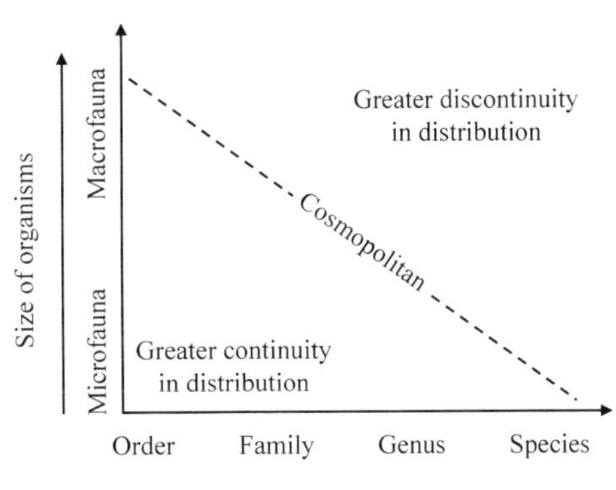

Figure 16.7. Biogeographical distribution of soil animals in relation to body size and taxonomic level. Modified from Anderson (1977).

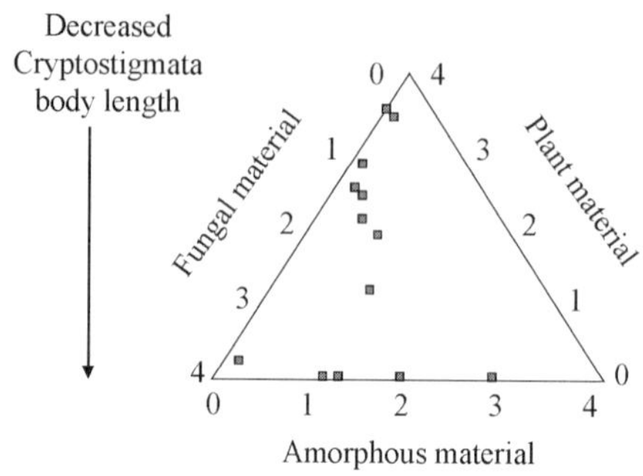

Figure 16.8. Cumulative mean proportions of the major food items selected by the most abundant *Cryptostigmata* species. Modified from Anderson (1977).

Therefore, the results from Heneghan et al. (1999) point to the importance of the faunal assemblage structure as a major determinant of decomposition rates in the tropics.

Swift et al. (1979) hypothesized that soil macrofauna were more abundant in the tropics than in temperate ecosystems. Macrofauna eat larger amounts of litter than microfauna. In fact, faunal effects on litter breakdown can be up to 66% in a tropical wet forest, a site of high abundance of macrofauna and diversity of functional groups (González and Seastedt, 2001). Anderson (1977) theorized that most classes and orders of soil animal communities are cosmopolitan, families are more restricted in distribution, and genera and species show a high degree of endemism (Figure 16.7). It is because of the cosmopolitan distribution of microarthropods that they are often represented as a constant in most field studies on litter decomposition (Heneghan et al., 1998). Still if we deem the tremendous lack of taxonomic expertise and comparative studies of soil organisms and ecosystem processes across geographical zones, we realize the extent of uncertainty in our understanding of this aspect of the global carbon cycling.

Soil Fauna–Microbial Interactions

Soil fauna can affect plant litter decomposition through interactions with microbial communities (Figure 16.6). These interactions are multiform, because they are greatly influenced by the direction and magnitude of the activities of the soil organisms involved. Indirect effects of soil fauna on microbial communities can be generally considered positive because soil fauna increase the surface area for microbial use, stimulate microbial populations in their gut, and increase the dispersal of spores on their integument (Anderson, 1987). However, soil fauna can feed on the microbial biomass and alter their turnover rates. The net effect of soil fauna and microbial interactions on nutrient mineralization would be, for the most part, dependent on the feeding behavior of the soil fauna on the microorganisms. For example, oribatid mites (*Cryptostigmata*) may stimulate fungal growth by grazing on senescent hyphae (Griffiths and Bargett, 1997), but, depending on the foraging intensity, net microarthropod effects on microbial activity can also be negative (González et al., 2001). Hanlon and Anderson (1979) reported inhibition of soil microbial respiration as the number of grazing collembolans was increased beyond an optimal number.

In addition to feeding methods, faunal effects on microbes can be considered in terms of relative body size (Visser, 1985; Anderson, 1987). Anderson states that the "indirect physical effect of animals on microbial populations and activities increase with body size." Concordingly, Visser (1985) argued that microarthropods have a high chance of selective feeding on the microbial population. González et al. (2001) found that exclusion of soil fauna resulted in a higher microbial biomass and lower maximum growth rate of microbes in subalpine forests. Given the dominance and abundance of microarthropods in the subalpine forest, soil fauna could affect ecosystem processes primarily through direct interactions with the microorganisms in this site. In contrast with the results found in the subalpine forests, González et al. (2001) found that exclusion of soil fauna in two tropical forests (dry and wet) did not affect the biomass and maximum growth rates of microorganisms; suggesting a larger net regulatory effect of soil fauna on microbes in the tropics via indirect mechanisms.

It is known that the main energy sources for bacteria are fresh aboveground litter, root-derived carbon, and small organic matter molecules that have been enzymatically processed by fungi (Clarholm, 1994). Yang and Insam (1991) found a higher fraction of bacterial biomass in an AB horizon that in the A horizon in soils from a tropical rain forest in China, supporting the idea that fungal biomass is more abundant in the initial stages of decomposition than in the later stages. Soil fauna can selectively feed on the microbial population, and we need to account for the superimposed effects of grazing by the soil fauna on the nutrient release from organic particles. Therefore, the intimacy of the fauna–microbial association can be related not only to body size but also to the formation of biogenic structures and the digestive process (Figure 16.9; Lavelle, 1997).

Concluding Remarks

More than 20 years have passed since Swift et al. (1979) presented a conceptual trilateral model of plant litter decomposition. It remains the most robust and widely

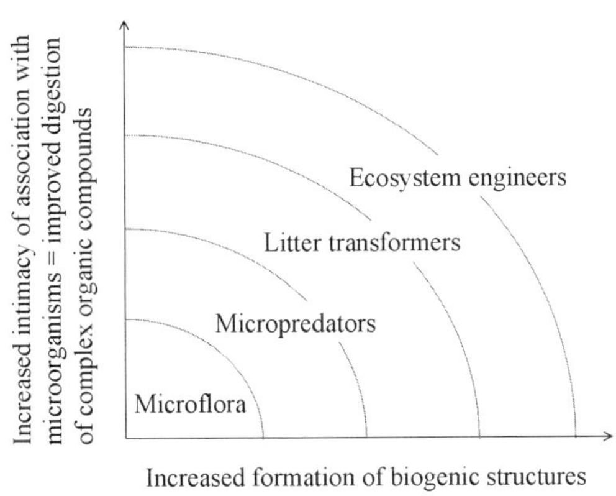

Figure 16.9. Interactions among micro- and macro-organisms in soils. As the size of organisms increases, their relationships with microflora gradually shift from predation to external and internal mutualism, and they produce biogenic structures of increasing strength. When present, larger organisms tend to be more competitive than smaller ones, but their activity is more frequently limited by low temperature and moisture conditions. Modified from Lavelle (1997).

accepted model, as demonstrated by its citation in most introductory paragraphs of studies and reviews dealing with the topic. Yet quantitative efforts of modeling keep treating biotic effects on decomposition like a black box. Soil biologists have long recognized that a tangle of confounding effects is imbedded within the overall organismal portrait. Soil fauna, microorganisms, and their interactions are important regulating factors of decomposition, and their relative hierarchical control varies spatially and temporally.

In recent years, there has been a tremendous advancement in understanding the particular roles and effects of soil organisms on ecosystem processes driven by micro- and mesocosms studies. There is still the need for comprehensive and manipulative field studies that try to tease apart the distinct effect of fauna and microorganisms. This is not an easy task, for the interactions of the abundance, diversity, activity, and functionality of soil organisms (fauna and microorganisms) are at play.

Kinetic methods have been proven useful in determining soil microbial communities and functional groups. They can be used for assessing inhibitory effects of soil fauna on microorganisms or as tool for determining competition for resources within the microbial compartment. Methods of soil fauna exclusion in the field should be carefully evaluated. The application of biocides to litter and soil surfaces is a popular technique to quantify soil fauna effects on plant litter decomposition and mineralization rates. However, this technique is likely to have a nontarget impact on the microbial community that might mask soil fauna effects.

The recent advancement in the study of soil microbial–faunal interactions has opened a new dimension in the understanding of soil biology and ecosystem ecology. Many scientists are currently puzzling with biodiversity issues that relate to ecosystem process and function. The field of soil biology is not exempt from this trend in ecology. The implications of soil biodiversity to ecosystem functioning should be carefully examined. The biocomplexity of the activities and interactions of soil organisms can have profound effects on the functioning of systems and the globe.

Acknowledgments

I am grateful to Timothy R. Seastedt for insights and discussions about soil fauna and decomposition, and to William A. Gould, Ariel E. Lugo, Frederick N. Scatena, and Timothy R. Seastedt for commenting on an earlier version of this manuscript.

References

Anderson, J. M. (1977). The organization of soil animal communities. In: U. Lohm and T. Persson (Eds.), *Organisms as components of ecosystems: Proceedings of the VI International Soil Zoology Colloquium of the International Society of Soil Science (ISSS)* (pp. 15–23). Stockholm: Swedish Natural Science Research Council.

Anderson, J. M. (1987). Interactions between invertebrates and microorganisms: Noise or necessity for soil processes? In: M. Fletcher, T. R. G. Gray, and J. G. Jones (Eds.), *Ecology of microbial communities* (pp. 125–145). Cambridge, UK: Cambridge University Press.

Anderson, J. M., and Domsch, K. H. (1975). Measurement of bacterial and fungal contributions to respiration of selected agricultural and forest soils. *Canadian Journal of Microbiology* 21: 314–322.

Anderson, J. M., and Domsch, K. H. (1978). A physiological method for the quantitative measurement of microbial biomass in soils. *Soil Biology and Biochemistry* 10: 215–221.

Beare, M. H., Neely, C. L., Coleman, D. C., and Hargrove, W. L. (1990). A substrate-induced respiration (SIR) method for measurement of fungal and bacterial biomass on plant residues. *Soil Biology and Biochemistry* 22: 585–594.

Berg, B., Kniese, P. P., Roomer, R., and Verhoef, H. A. (1998). Long-term decomposition of successive organic substrate in a nitrogen-saturated Scots pine forest soil. *Forest Ecology and Management* 107: 159–172.

Brown, G. G. (1995). How do earthworms affect microfloral and faunal community diversity? *Plant Soil* 170: 209–231.

Carter, M. R., Gregorich, E. G., Angers, D. A., Beare, M. H., Sparling, G. P., Wardle, D. A., and Voroney, R. P. (1999). Interpretation of microbial biomass measurements for soil quality assessment in humid temperate regions. *Canadian Journal of Soil Science* 79: 507–520.

Clarholm, M. (1994). The microbial loop. In: K. Ritz, J. Dighton, and K. E. Giller (Eds.), *Beyond the Biomass* (pp. 221–230). New York: Wiley.

Coleman, D. C., and Crossley, D. A., Jr. (1996). *Fundamentals of soil ecology.* New York: Academic Press.

Colores, G. M., Schmidt, S. K., and Fisk, M. C. (1996). Estimating the biomass of microbial functional groups using rates of growth-related soil respiration. *Soil Biology and Biochemistry* 28: 1569–1577.

Contreras, E. (1980). Studies on the intestinal actinomycete flora of *Eisenia lucens* (Annelida, Oligochaeta). *Pedobiologia* 20: 411–416.

Coûteaux, M.-M., Bottner, P., and Berg, B. (1995). Litter decomposition, climate and litter quality. *Trends in Ecology and Evolution* 10: 63–66.

Crawford, D. L. (1988). Biodegradation of agricultural and rural wastes. In: M. Goodfellow, S. T. Williams, and M. Mordarski (Eds.), *Actinomycetes in biotechnology* (pp. 433–439). London: Academic Press.

Doube, B. M., and Brown, G. G. (1998). Life in complex community: Functional interactions between earthworms, organic matter, microorganisms, and plants. In: C. A. Edwards (Ed.), *Earthworm ecology* (pp. 179–212). Boca Raton, FL: St. Lucie Press.

Feigl, B. J., Sparling, G. P., Ross, D. J., and Cerri, C. C. (1995). Soil microbial biomass in Amazonian soils: Evaluation of methods and estimates of pool sizes. *Soil Biology and Biochemistry* 27: 1467–1472.

González, G., Ley, R. E., Schmidt, S. K., Zou, X., and Seastedt, T. R. (2001). Soil ecological interactions: Comparisons between tropical and subalpine forests. *Oecologia* 128: 549–556.

González, G., and Seastedt, T. R. (2000). Comparison of the abundance and composition of litter fauna in tropical and subalpine forests. *Pedobiologia* 44: 545–555.

González, G., and Seastedt, T. R. (2001). Soil fauna and plant litter decomposition in tropical and subalpine forests. *Ecology* 82: 955–964.

González, G., and Zou, X. (1999). Earthworm influence on N availability and the growth of *Cecropia scheberiana* in tropical pasture and forest soils. *Pedobiologia* 43: 824–829.

Goodfellow, M., and Williams, S. T. (1983). Ecology of actinomycetes. *Annual Review of Microbiology* 37: 189–216.

Griffiths, B. S., and Bargett, R. D. (1997). Interactions between microbes-feeding invertebrates and soil microorganisms. In: J. D. Elsas, J. T. Trevors, and E. M. H. Wellington (Eds.), *Modern soil microbiology* (pp. 165–182). New York: Marcel Dekker.

Hanlon, R. D. G., and Anderson, J. M. (1979). The effects of collembola grazing on microbial activity in decomposing leaf litter. *Oecologia* 38: 93–99.

Hansen, R. A. (1999). Red oak litter promotes a microarthropod functional group that accelerates its decomposition. *Plant and Soil* 209: 37–45.

Heal, O. W., Anderson, J. M., and Swift, M. J. (1997). Plant litter quality and decomposition: An historical overview. In: G. Cadish and K. E. Giller (Eds.), *Driven by nature: plant litter quality and decomposition* (pp. 3–30). Wallingford, UK: CAB International.

Heneghan, L., Coleman, D. C., Zou, X., Crossley, D. A., Jr., and Haines, B. L. (1998). Soil microarthropod community structure and litter decomposition dynamics: A study of tropical and temperate sites. *Applied Soil Ecology* 9: 33–38.

Heneghan, L., Coleman, D. C., Zou, X., Crossley, D. A., Jr., and Haines, B. L. (1999). Soil microarthropod contributions to decomposition dynamics: Tropical–temperate comparisons of a single substrate. *Ecology* 80: 1873–1882.

Irmler, U. (2000). Changes in the fauna and its contribution to mass loss and N release during leaf litter decomposition in two deciduous forests. *Pedobiologia* 44: 105–118.

Krištůfek, V., Ravasz, K., and Pižl, V. (1993). Actinomycete communities in earthworm guts and surrounding soils. *Pedobiologia* 37: 379–384.

Lavelle, P. (1997). Faunal activities and soil processes: Adaptative strategies that determine ecosystem function. *Advances in Ecological Research* 27: 93–132.

Lavelle, P., Bignell, D., Lepage, M., Wolters, V., Roger, P., Ineson, P., Heal, O. W., and Dhillion, S. (1997). Soil function in a changing world: The role of invertebrate ecosystem engineers. *European Journal of Soil Biology* 33:159–193.

Lavelle, P., Blanchart, E., Martin, A., and Martin, S. (1993). A hierarchical model for decomposition in terrestrial ecosystems: Application to soils of the humid tropics. *Biotropica* 25: 130–150.

Lipson, D. A., Schmidt, S. K., and Monson, R. K. (1999). Links between microbial population dynamics and nitrogen availability in an alpine ecosystem. *Ecology* 80: 1623–1631.

Lodge, D. J. (1996). Microorganisms. In: D. P. Reagan and R. B. Waide (Eds.), *The food web of a tropical rain forest* (pp. 53–108). Chicago: The University Chicago Press.

MacLean, S. F., Jr. (1974). Primary production, decomposition and the activity of soil invertebrates in tundra ecosystems: A hypothesis. In: A. J. Holding, O. W. Heal, S. F. MacLean, Jr., and P. W. Flanagan (Eds.), *Soil organisms and decomposition in tundra* (pp. 197–200). Stockholm, Sweden Fairbanks, Alaska: Tundra Biome Steering Committee.

McCarthy, A. J. (1987). Lignocellulose-degrading actinomycetes. *Federation of European Microbiological Societies (FEMS) Microbiology Reviews* 46: 145–163.

Meentemeyer, V. (1978). Macroclimate and lignin control of litter decomposition rates. *Ecology* 59: 465–472.

Melillo, J. M., Aber, J. D., and Muratore, J. F. (1982). Nitrogen and lignin control of hardwood leaf litter decomposition dynamics. *Ecology* 63: 621–626.

Neely, C. L., Beare, M. H., Hargrove, W. L., and Coleman, D. C. (1991). Relationships between fungal and bacterial substrate-induced respiration, biomass and plant residue decomposition. *Soil Biology and Biochemistry* 23: 947–954.

Palm, C. A., and Sánchez, P. A. (1991). Nitrogen release from the leaves of some tropical legumes as affected by their lignin and polyphenolic contents. *Soil Biology and Biochemistry* 23: 83–88.

Panikov, N. S., and Sizoka, M. V. (1996). A kinetic method for estimating the biomass of microbial functional groups in soils. *Journal of Microbiological Methods* 24: 219–230.

Parkinson, D., and McLean, M. A. (1998). Impacts of earthworms on the community structure of other biota in forest soils. In: C. A. Edwards (Ed.), *Earthworm ecology* (pp. 213–228). Boca Raton, FL: St. Lucie Press.

Schaefer, D., Steinberger, Y., and Whitford, W. G. (1985). The failure of nitrogen and lignin control of decomposition in a North American desert. *Oecologia* 65: 382–386.

Schinner, F. (1996). Introduction. In: F. Schinner, R. Öhlinger, E. Kandeler, and R. Margesin (Eds.), *Methods in soil biology* (pp. 3–6). Berlin: Springer-Verlag.

Schmidt, S. K. (1992). A substrate-induced growth-respiration method for estimating the biomass of microbial functional groups in soil and aquatic systems. *FEMS Microbiology Reviews* 101: 197–206.

Seastedt, T. R. (1984). The role of microarthropods in decomposition and mineralization processes. *Annual Review of Entomology* 29: 25–46.

Seastedt, T. R. (2000). Soil fauna and controls of carbon dynamics: Comparisons of rangelands and forests across latitudinal gradients. In: C. Coleman and P. F. Hendrix (Eds.), *Invertebrates as webmasters in ecosystems* (pp. 293–311). London: CAB International.

Sharma, G. D., Mishra, R. R., and Kshattriya, S. (1995). Fungi and litter decomposition in the tropics. In: M. V. Reddy (Ed.), *Soil organisms and litter decomposition in the tropics* (pp. 39–57). IBH. Pub. Co. PVT. LTD., Janpah, New Delhi: Oxford.

Singh, J. S., Raghubanshi, A. S., Singh, R. S., and Srivastava, S. C. (1989). Microbial biomass acts as a source of plant nutrients in dry tropical forest and savanna. *Nature* 338: 499–500.

Smith, P., Andrén, O., Brussaard, L., Dangerfield, M., Ekschmitt, K., Lavelle, P., and Tate, K. (1998). Soil biota and global change at the ecosystem level: Describing soil biota in mathematical models. *Global Change Biology* 4: 773–784.

Sparling, G. P., Feltham, C. W., Reynolds, J., West, A. W., and Singleton, P. (1990). Estimation of soil microbial C by a fumigation- extraction method: Use on soils of high organic matter content, and a reassessment of the k_{EC}-factor. *Soil Biology and Biochemistry* 20: 743–747.

Swift, M. J. (1976). Species diversity and the structure of microbial communities in terrestrial habitats. In: J. M. Anderson and A. Macfadyen (Eds.), *The role of terrestrial and aquatic organisms in decomposition processes* (pp. 185–222). London: Blackwell Scientific.

Swift, M. J., Heal, O. W., and Anderson, J. M. (1979). *Decomposition in terrestrial ecosystems*. London: Blackwell Scientific.

Tian, G., Brussaard, L., Kang, B. T., and Swift, M. J. (1997). Soil fauna-mediated decomposition of plant residues under constrained environmental and residue quality conditions. In: G. Cadisch and K. E. Giller (Eds.), *Driven by nature: Plant litter quality and decomposition* (pp. 125–134). London: CAB International.

Visser, S. (1985). Role of the soil invertebrates in determining the composition of soil microbial communities. In: A. H. Fitter, D. Atkinson, D. J. Read, and M. B. Usher (Eds.), *Ecological interactions in soil: Plants, microbes and animals* (pp. 297–317). Oxford, London: Blackwell Scientific.

Wardle, D. A., and Lavelle, P. (1997). Linkages between soil biota, plant litter quality and decomposition. In: G. Cadisch and K. E. Giller (Eds.), *Driven by nature: Plant litter quality and decomposition* (pp. 107–124). London: CAB International.

Wardle, D. A., and Parkinson, D. (1990). Response of the soil microbial biomass to glucose, and selective inhibitors, across a soil moisture gradient. *Soil Biology and Biochemistry* 22: 825–834.

Webb, D. P. (1977). Regulation of deciduous forest litter decomposition by soil arthropod feces. In: W. J. Matson (Ed.), *The role arthropods in forest ecosystems* (pp. 57–69). New York: Springer-Verlag.

West, A. W., and Sparling, G. P. (1986). Modifications to the substrate-induced respiration method to permit measurement of microbial biomass in soils of differing water contents. *Journal of Microbiological Methods* 5: 177–189.

Whitford, W. G., Meentemeyer, V., Seastedt, T. R., Cromack, K., Jr., Crossley, D. A., Jr., Santos, P., Todd, R. L., and Waide, J. B. (1981). Exceptions to the AET model: Deserts and clear-cut forests. *Ecology* 62: 275–277.

Willington, E. M. H., and Toth, I. K. (1994). Actinomycetes. In: R. W. Weaver, J. S. Angle and P. S. Bottomley (Eds.), *Methods in soil analysis: Part 2. Microbiological and biochemical properties* (pp. 269–290). Soil Science Society of America Series No. 5 (Madison, Wisconsin) WI.

Yang, J. C., and Insam, H. (1991). Microbial biomass and relative contributions of bacteria and fungi in soil beneath tropical rain forests, Hainan Island, China. *Journal of Tropical Ecology* 7: 385–393.

17

Applied Ecology of Succession in Pine Forests of western Japan

Michiro Fujihara, Yoshio Hada, and Gentaro Toyohara

Introduction

One of the most representative vegetation types in western Japan is *Pinus densiflora* forest, given that most primary evergreen, broad-leaved forests have been largely destroyed by human activities. Pine trees regenerate and persist at sites where forest cutting, land clearing, and fire have occurred, because *P. densiflora* requires full sunlight and bare mineral soil for seed germination. For a long time, pine trees have been used for timber and fuel; shrubs for fuel; and litter, for compost. However, after the 1960s, petrochemicals such as propane gas and chemical fertilizer became widely used in Japan. Since then, pine forests have been abandoned and have gradually been replaced by broad-leaved forests. In addition, pine forests have suffered from mass dieback due to pine wilt disease, which is caused by epidemic attacks of the nematode *Bursaphelenchus xylophilus* (Steiner et Buhrer) Nickel, vectored by the beetle *Monochamus alternatus* Hope (Kiyohara and Tokushige, 1971; Morimoto and Iwasaki, 1972). Many pine forests in Japan suffer from pine wilt disease except in northern areas such as Hokkaido and Aomori Prefecture (Kishi, 1995; Rutherford et al., 1990). One-fourth of pine forests in Japan have been infested by the disease (Mamiya, 1983).

Disturbances influence the stand structure through modification of the physical environment (White and Pickett, 1985). Abrams and Scott (1989) documented a hypothetical example of disturbance-mediated accelerated succession. On the one hand, disturbance by pine wilt disease influences the forest ecosystem and succession. On the other hand, the severity of the disturbance seems to be affected by the successional stage of pine forests (Fujihara et al., 1992; Fujihara, 1996).

Michiro Fujihara • Himeji Institute of Technology, Awaji Landscape Planning and Horticulture Academy, 945-2 Nojimatokiwa, Hokudan-cho, Tsuna-gun, Hyogo 656-1726, Japan. E-mail: fujihara@awaji.ac.jp **Yoshio Hada** • Faculty of Informatics, Department of Biosphere-Geosphere System Science, Okayama University of Science, 1-1 Ridai-cho, Okayama 700-005, Japan. E-mail: hada@rsch.tuis.ac.jp **Gentaro Toyohara** • Miyajima Natural Botanical Garden, Department of Biological Sciences, Graduate School of Science, Hiroshima University, Mitsumaruko-yama, Miyajima-cho, Saiku-gun, Hiroshima 739-0543, Japan. E-mail gtoycohar@hiroshima-u.ac.jp

Toyohara (1984) used phytosociological methods to describe vegetation types of pine forests based on species composition. Pine forests, which cover about 70% of the forest area in Hiroshima Prefecture, were identified with the order Pinetalia densiflorae, and were classified into three alliances and 10 associations, accompanied by several subassociations (Toyohara, 1984). Differences in alliances and associations of pine forests represent the potential natural vegetation (Toyohara, 1984). In each association, differences in vegetation types seem to represent the successional sere, depending on human disturbance, topography, and land use history. The successional sere and distribution of each type of pine forest were clarified in Hiroshima and Okayama Prefectures, western Japan (Toyohara, 1984; Fujihara et al., 1989, 1992; Hada et al., 1994; Nishimoto and Hada, 1994; Kishida et al., 1996, 1997; Kojima et al., 1996; Takama et al., 1996; Miyamoto et al., 1997).

In this chapter, we try to clarify the development of secondary *P. densiflora* forests in western Japan after stand-level dieback through the use of phytosociological analysis, vegetation mapping, estimation of forest structure before dieback, and size–structure and tree ring analysis. We discuss proper forest management from the viewpoint of succession.

Study Areas and Methods

The studies were undertaken in two rural areas in the northern coastal region of the Inland Sea (Sanyo district), western Japan. Low mountains and acidic rocks such as granite and rhyolite occur over most of the area, and sedimentary rocks are scattered locally. Funakoshi, Hiroshima Prefecture, is located at 34°28′N, 132°28′E, and Washu, Okayama Prefecture, is located at 34°22′N, 133°52′E (Figure 17.1). The mean annual temperature and precipitation are 15.4°C and 1,608 mm at Hiroshima, the nearest meterological station to Funakoshi, and 15.2°C and 1,202 mm at Kurashiki, the nearest station to Washu (Anonymous, 1982). According to Kira's warmth and coldness indices, these study areas belong to the evergreen broad-leaved forest zone of *Camellietea japonicae* (Nakanishi and Hattori, 1979). The original natural vegetation is considered to be *Castanopsis cuspidata–Pasania glabra* community (Toyohara, 1984). At present, however, natural forests have

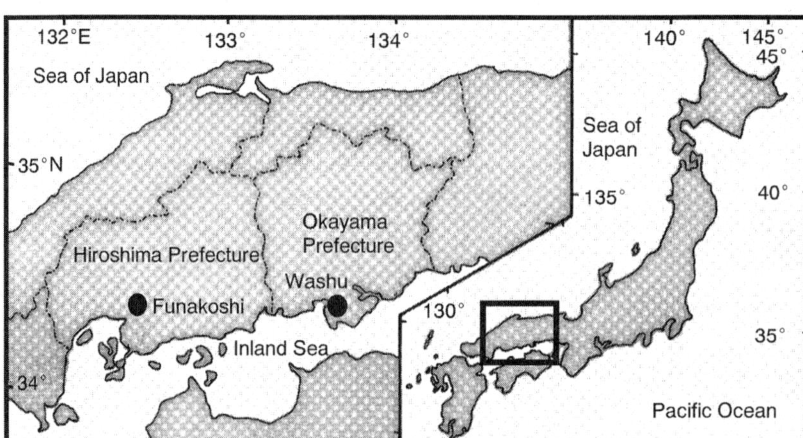

Figure 17.1. Study sites.

largely been destroyed and replaced by secondary growth communities such as secondary *P. densiflora* forest (Itow, 1983; Toyohara, 1984). Pine wilt disease had spread widely throughout most of the study areas.

Study plots of 40–100 m^2 were set in each area. Cover abundance and sociability (Braun-Blanquet, 1964) for all vascular plants were recorded in each plot. Species name, tree height (h), and trunk diameter at breast height (*dbh*) and at ground level (D_0) were recorded for all trees taller than 1.3 m in each plot. Trunk diameter at D_0 was measured for dead fallen pine trees.

Vegetation types were classified by phytosociological methods (Braun-Blanquet, 1964; Mueller-Dombois and Ellenberg, 1974). Vegetation maps were made in 1985 and 1993 with the use of field surveys based on floristic criteria, establishing whether diagnostic species of each vegetation type were present or absent in the field (Mueller-Dombois and Ellenberg, 1974).

Surviving pine forests were recognized as living pine trees having more than 25% coverage. Distribution maps of surviving pine forests were drawn after field surveys in 1985 and 1993.

For fallen dead pine trees, *dbh* was estimated from the diameter of the remaining trunk at D_0 with the following equation:

$$dbh = a \times D_0 + b,$$

where *a* and *b* are coefficients for species in Japanese red pine forests. These coefficients were calculated from standing trees in each plot by the least squares method (Fujihara, 1995). The height of fallen dead pine trees was estimated from *dbh* with the following equation:

$$1/h = 1/(A \times dbh) + 1/h^*,$$

where *A* and *h** are coefficients specific to *P. densiflora* forests; they were calculated from standing trees in each plot by the least squares method (Fujihara, 1995).

The mortality of pine trees was calculated with the following equation: Mortality of pine trees = (number of dead pine trees/total number of dead and live pine trees) × 100.

To estimate annual radial growth, we analyzed the tree rings of *Quercus serrata* in each plot at Washu. Discs and cores were obtained from sample trees, and annual ring widths were measured under a binocular microscope at an accuracy of 0.1 mm.

Successional Stages of Japanese Red Pine Forests

Figure 17.2 shows the vegetation types and diagnostic species of the secondary pine forests in the study areas. Four forest communities were recognized: (I) Querco glaucae–Pinetum densiflorae (Japanese red pine stage), (II) *Quercus serrata–Quercus glauca* community (deciduous and evergreen mixed oak stage), (III) *Quercus glauca* community (evergreen oak stage), and (IV) *Gleichenia japonica* community (fern stage). Querco glaucae–Pinetum densiflorae (Japanese red pine stage) is characterized by the occurrence of light-demanding species, such as *Juniperus rigida* and *Miscanthus sinensis*. This association was subdivided into five vegetational substages: (A) *Eurya japonica* substage, (B) *Q. glauca* substage, (C) *Ardisia japonica* substage, (D) *Ardisia crenata* substage, and (E) *Dryopteris erythrosora* substage. From the *E. japonica* substage (I-A) of the Japanese red pine stage to the evergreen oak stage (III), the number of components of evergreen

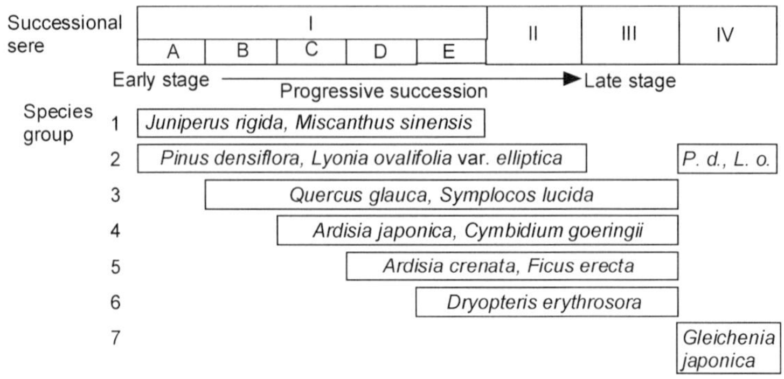

Figure 17.2. Schema of the vegetation types of pine forests.

broad-leaved forest and shade-tolerant species increased, but the number of light-demanding species decreased. This order of communities represents a successional sere from early to late stage (Fujihara, 1995; Fujihara et al., 1989, 1992; Toyohara and Fujihara, 1998). The *G. japonica* community is characterized by the dominance of a large-leaved fern (*G. japonica* Spr.) on the forest floor. The number of species was quite low in this community.

Change in Distribution of Each Type of Pine Forest Affected by Pine Wilt Disease

Vegetation changes before and after the mass dieback of pine trees were studied at Funakoshi, Hiroshima Prefecture. Figure 17.3 shows the distributions of each vegetation type in 1985 and 1993. Pine wilt disease progressed during these 8 years. The Querco glaucae–Pinetum densiflorae covered more than 80% of the study area in 1985 (Figure 17.3A). The *E. japonica* and *Q. glauca* substages of this association were found on ridges and the upper parts of slopes. The *A. japonica* substage was found on the middle parts of slopes, and the *A. crenata* substage was found on the lower parts of slopes and on concave slopes.

After the pine wilt disease progressed, the vegetation of the area changed (Figure 17.3B). The *E. japonica* and *Q. glauca* substages were found on the ridges and the upper parts of slopes, as in 1985. The *A. crenata* substage was found on the lower parts of slopes and on concave slopes. The *Q. serrata–Q. glauca* community (deciduous and evergreen mixed *Quercus* stage) appeared on the lower parts of slopes in 1993. The *Q. glauca* community (evergreen *Quercus* stage) also appeared on the lower parts of slopes in 1993. Small patches of *G. japonica* community appeared frequently on the middle parts of slopes. The area of Querco glaucae–Pinetum densiflorae was reduced from 9.9 ha (80.7%) to 7.8 ha (63.2%). The combined area of the *E. japonica*, *Q. glauca*, and *A. japonica* substages decreased from 4.7 ha (35.5%) to 1.6 ha (13.2%), whereas that of the *A. crenata* and *D. erythrosora* substages increased from 5.2 ha (42.2%) to 6.2 ha (50.0%) (Toyohara and Fujihara, 1998). Fujihara et al. (1991) and Fujihara (1996) also showed that deciduous oak (*Q. serrata*) and evergreen oak (*Q. glauca*) forests became widely distributed after mass dieback of pine trees.

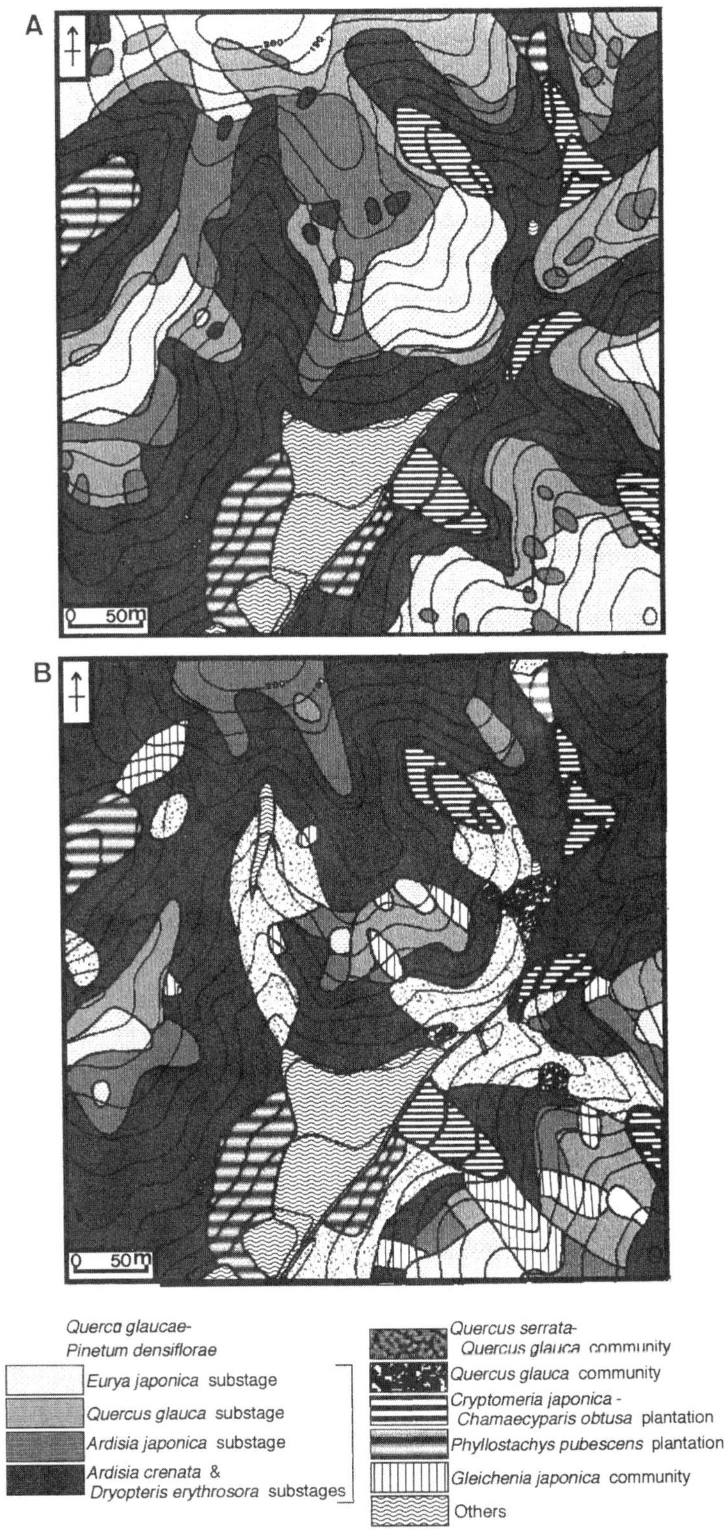

Figure 17.3. Vegetation maps showing the distribution of each vegetation type in 1985 and 1993 at Funakoshi. Modified from Toyohara and Fujihara (1998).

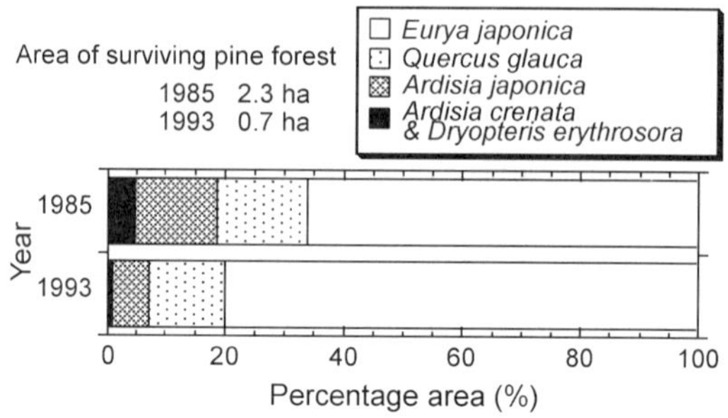

Figure 17.4. The percentage area of surviving pine forests in 1985 and 1993 at Funakoshi.

As the disease progressed, the surviving forests became still more restricted to the ridge and upper parts of the slopes. The area of surviving pine forests decreased from 2.3 ha to 0.7 ha (Toyohara and Fujihara, 1998). Most of the surviving pine forests belonged to the *E. japonica* and *Q. glauca* substages (Figure 17.4).

Vegetation change was related to forest management after the mass dieback of pine trees (Sakamoto et al., 1995). The area of the early successional stage increased after mass dieback on Miyajima Island, Hiroshima Prefecture; this change was due mainly to disturbance by selective cutting and transportation of dead pine trees (Toyohara et al., 1986). On the other hand, progressive succession occurred in pine forests where no management had been undertaken after dieback (Toyohara et al., 1986). No forest management was conducted in our study areas; progressive succession occurred in most of the area.

Hada et al. (1994) and Kishida et al. (1996) suggested that the recovery rate of laurel forests was strongly related to geology, and that the rate of succession on a Mesozoic formation area was more rapid than that on a granite area.

Stand Structure of Pine Forests

Figure 17.5 shows the height frequencies of trees in four vegetation types of Querco glaucae–Pinetum densiflorae at Funakoshi. The height of the tallest tree increased from the *E. japonica* substage to the *A. crenata* substage. The total tree density decreased in the same order. The heights of dead pine trees were different in each vegetation substage. In the *E. japonica* substage, small, dead, standing pine trees were found, indicating that the mortality (32.6%) was the result of thinning of suppressed trees within a decade rather than of canopy tree mortality caused by pine wilt disease, but many pine trees still dominated in most height classes.

The mortality was 67.7% in the *Q. glauca* substage, 75.9% in the *A. japonica* substage, and 94.3% in the *A. crenata* substage. In the late successional stages of pine forests nearly all canopy pine trees were dead and fallen. All living and dead pine trees were taller than 2 m, which suggests that pine trees had not regenerated recently in the late successional stages. In the *A. japonica* and *A. crenata* substages, evergreen broad-leaved trees such as *Q. glauca*, *Symplocos lucida* Sieb. et Zucc., and *Dendropanax trifidus* (Thunb.) Makino existed in the shrub layer.

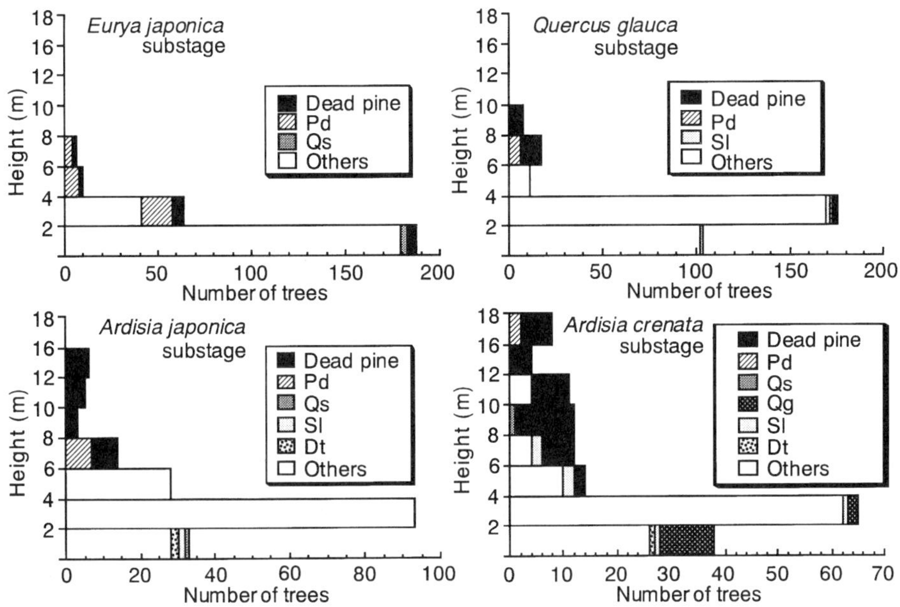

Figure 17.5. Height frequencies of each vegetation type in pine forests at Funakoshi. Pd: *Pinus densiflora*, Qs: *Quercus serrata*, Sl: *Symplocos lucida*, Dt: *Dendropanax trifidus*, Qg: *Quercus glauca*.

Growth of Individual Trees after the Mass Dieback

Figure 17.6 shows the growth rates before 1970 and after 1971 of *Q. serrata* in the severely disturbed pine forests at Washu, Okayama Prefecture. Pine wilt disease arrived here around 1970 (Fujihara, 1996). The radial growths of all sample trees increased after the disappearance of canopy pine trees. After the disease struck, the growth of suppressed oak trees accelerated, and the stand structure of forests changed dramatically. The dominant species changed from *P. densiflora* to *Q. serrata*. The change in the growth rates of individual trees led to a change in stand structure. The growth of surviving trees located near the canopy openings created by the death of the canopy pine trees began to increase, and these trees became dominant (Fujihara and Toyohara, 1996, Fujihara et al., 1991).

Landscape Structure of Pine-Dominated Rural Regions

Some pine forests change to evergreen broad-leaved forests after the canopy pines disappear, as in Funakoshi, but the forests at Washu were dominated by deciduous broad-leaved trees (*Q. serrata*). This dominance was due to the landscape configuration, dispersal agents, and land use history. In the coastal area of the Inland Sea, we found five types of pine forests as a successional sere from early to late stages—(A) *E. japonica* (*Wikstroemia sikokiana*), (B) *Q. glauca*, (C) *A. japonica*, (D) *A. crenata*, and (E) *D. erythrosora*—but we did not find the latter three at Washu (Fujihara et al., 2000). Fewer individuals and species of natural evergreen broad-leaved forests grew at Washu than in species-rich pine forest. At

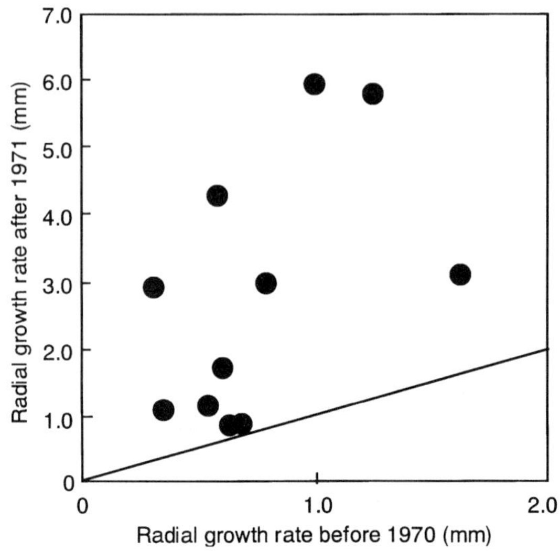

Figure 17.6. Radial growth rates before 1970 and after 1971 of *Quercus serrata* in the forests affected by pine wilt disease at Washu.

Washu, human disturbance, such as clear-cutting, land clearance, and fire, was severe, and the precipitation is the lowest in Japan. The dominant species of evergreen broad-leaved forests have disappeared from this area. These dominant species bear acorns and their dispersal ability is low. Therefore, development into evergreen broad-leaved forest takes a long time. In species-rich forests, the dominant species change from pine to evergreen broad-leaved types directly after pine wilt disease (Fujihara, 1995, 1997). In degraded areas, for succession from deciduous oak to evergreen oak forest to take place, the establishment of evergreen broad-leaved trees, such as *Q. glauca* and *Castanopsis cuspidata* (Thunb. ex Murray) Schottky, is important.

Stand-Level Dieback and Proper Management of Pine Forests

From the viewpoint of phytosociology, pine wilt disease was severe in the late successional stages located on the lower part of slopes. The abandoned pine forests, where succession is taking place, have a feeble constitution and are vulnerable to attack by pine wilt disease. Changes in the ectomycorrhizal flora are largely dependent on changes in the forest soil (Blasius and Oberwinkler, 1989; Tyler, 1989). Mycorrhizae tend to increase the resistance of *P. densiflora* seedlings to pine wilt disease (Kikuchi et al., 1991). The successional stage of mycorrhizal flora might affect the susceptibility of a pine stand to pine wilt disease (Fukuda et al., 1999).

Pinus densiflora is a typical pioneer tree. Pine seedlings and saplings could not establish and regenerate in most of the stands that experienced severe canopy dieback of pines, because *Q. serrata*, *Q. variabilis*, or *Q. glauca* already occurred in the shrub or herb layers before the dieback and became dominant thereafter. After the disappearance of the canopy pine trees, the growth of previously suppressed oak trees was accelerated (Fujihara, 1996). The pine wilt disease shifted the dominance of the secondary forests from Japanese red pine (*P. densiflora*) to deciduous oak (*Q. serrata*) or evergreen oak (*Q. glauca*). In the early successional stages located on ridges and the upper part of slopes, pine wilt disease was not as severe, and pine forests remained or regenerated. The survival of pine forests was restricted to the early successional stages.

The disappearance of canopy pine trees has affected ecosystems dramatically. This phenomenon is known as stand-level dieback (Mueller-Dombois 1987, 1993; Akashi and Mueller-Dombois, 1995) and can be divided into three types: replacement dieback, displacement dieback, and stand-reduction dieback (Mueller-Dombois, 1991). In the *E. japonica* substage, pine trees survive or regenerate. This is regarded as replacement dieback (Figure 17.7). In the *Q. glauca*, *A. japonica*, *A. crenata*, and *D. erythrosora* substages, numerous canopy pines die and oak trees become dominant, because the later successional species in these severely disturbed pine forests are already established before the dieback. This is regarded as displacement dieback. In the *G. japonica* type of pine forest, the tree and lower tree layers are open, because *G. japonica* prevents trees from establishing both before and after mass dieback of pines. The coverage of *G. japonica* increases after mass dieback. This stand-reduction dieback leads to a *G. japonica* community.

Ikegami and Nakagoshi (1999) conducted a risk assessment for pine death and reported that the site condition of surviving pine forest was similar to the typical condition of a pine forest. These results suggest that secondary forest vegetation is being replaced by potential natural vegetation. Based on ecological research, we propose a future landscape

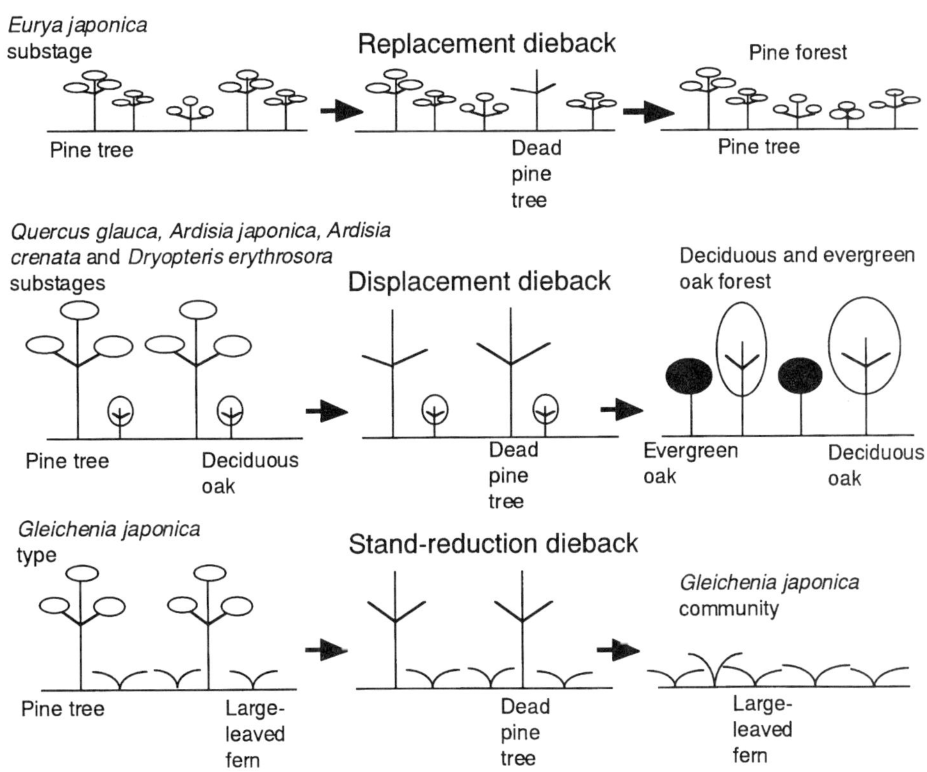

Figure 17.7. Schema of stand-level dieback of pine forests.

of secondary forests affected by pine wilt disease. We suggest that the late successional stages should be managed by accelerating their progressive succession to deciduous and evergreen broad-leaved forests. Early successional stages of pine forest should be managed as a seed source for pine forests. Secondary forests are resources for the local and global environment. Both rural and urban people should manage the secondary pine forests for many purposes, such as recreational use.

References

Abrams, S. D., and Scott, M. L. (1989). Disturbance-mediated accelerated succession in two Michigan forest types. *Forest Science* 35: 42–49.

Akashi, Y., and Mueller-Dombois, D. (1995). A landscape perspective of the Hawaiian rain forest dieback. *Journal of Vegatation Science* 6: 449–464.

Anonymous. (1982). The monthly normals of temperature and precipitation at climatological stations in Japan (1951–1978). Technical Data Series No. 46. Tokyo: Japan Meteorological Agency.

Blasius, D., and Oberwinkler, F. (1989). Succession of mycorrhizae: A matter of tree and stand age. *Annales des Sciences Forestieres* 46: 758–761.

Braun-Blanquet, J. (1964). *Pflanzensoziologie* (Vol. 3). Wien: Springer-Verlag.

Fujihara, M. (1995). Succession of secondary pine forests after pine wilt disease in San-yô district, western Japan. *Natural History Research* 3: 161–171.

Fujihara, M. (1996). Development of secondary pine forests after pine wilt disease, western Japan. *Journal of Vegetation Science* 7: 497–504.

Fujihara, M. (1997). Structure of *Pinus luchuensis* forests affected by pine wilt disease in northern Taiwan. *Natural History Research Special Issue* 4: 113–126.

Fujihara, M., Hada, Y., and Toyohara, G. (2000). Stand structure of *Pinus* forests affected by pine wilt disease. In: P. S. White, L. Mucina, and J. Leps (Eds.), *Proceedings International Association for Vegetation Science Symposium* (pp. 90–94). Uppsala: Opulus Press.

Fujihara, M., Toyohara, G., Hada, Y., and Iwatsuki, Z. (1989). Growth of *Pinus densiflora* of secondary pine forests in Hiroshima Prefecture, western Japan. *Hikobia* 10: 339–349.

Fujihara, M., Toyohara, G., Hada, Y., and Iwatsuki, Z. (1992). Successional stages and degree of damage of secondary pine forests in Hiroshima City, western Japan. *Japanese Journal of Ecology* 42: 71–79.

Fujihara, M., Yoshino, Y., and Toyohara, G. (1991). Change of vegetation after catastrophic withering of pine forests by pine wilt disease on Mt. Ushita, Hiroshima City, western Japan. *Hikobia* 11: 85–91.

Fukuda, K., Ichihara, Y., Ugawa, S., and Suzuki, K. (1999). Mycorrhizal fungal flora in pine forests infested with pine wilt disease. In: K. Futai, K. Togashi, and T. Ikeda (Eds.), *Sustainability of pine forests in relation to pine wilt and decline* (pp. 274–279). Tokyo: Shokado.

Hada, Y., Koshin, C., Fukuzawa, Y., and Nishimoto, T. (1994). Secondary forest of coastal area in Okayama Prefecture, regarding geological and geographical factors. *Bulletin of the Okayama Prefecture Nature Conservation Center* 1: 11–28.

Ikegami, Y., and Nakagoshi, N. (1999). Risk assessment for pine death by analyzing site condition. In: K. Futai, K. Togashi, and T. Ikeda (Eds.), *Sustainability of pine forests in relation to pine wilt and decline* (pp. 234–238). Tokyo: Shokado.

Itow, S. (1983). Secondary forests and coppices in southern Japan. In: W. Holzner, M. J. A. Werger, and I. Ikushima (Eds.), *Man's impact on vegetation* (pp. 317–326). The Hague: Junk.

Kikuchi, J., Tsuno, N., and Futai, K. (1991). The effect of mycorrhizae as a resistance factor of pine trees to the pinewood nematode. *Journal of the Japanese Forest Society* 73: 216–218.

Kishi, Y. (1995). *The pine wood nematode and Japanese pine swayer.* Tokyo: Thomas.

Kishida, S., Toyohara, G., and Deguchi, H. (1996). A phytosociological study on the secondary forests in the coastal regions of the Sanyo district, western Japan. *Hikobia* 12: 85–94.

Kishida, S., Toyohara, G., and Deguchi, H. (1997). Successional processes related to the geology of secondary forests in the Sanyo district, western Japan (I): The successional sere of the inland region. *Hikobia* 12: 257–266.

Kiyohara, T., and. Tokushige, Y. (1971). Inoculation experiments on a nematode, *Bursaphelenchus* sp., onto pine trees. *Journal of the Japanese Forest Society* 53: 210–218.

Kojima, Y., Toyohara, G., and Deguchi, H. (1996). A phytosociological study on the forest vegetation of the Sera Platform, Hiroshima Prefecture. *Hikobia* 12: 123–132.

Mamiya, Y. (1983). Pathology of the pine wilt disease caused by *Barsaphelenchus xylophilus*. *Annual Review of Phytopathology* 221: 201–220.

Miyamoto, K., Toyohara, G., and Deguchi, H. (1997). On the boundary between the *Querco glaucae–Pinetum densiflorae* and the *Querco myrsinaefoliae–Pinetum densiflorae* in Hiroshima Prefecture. *Hikobia* 12: 241–255.

Morimoto, K., and Iwasaki, A. (1972). The role of *Monochamus alternatus* (Coleoptera: Cerambycidae). *Journal of the Japanese Forest Society* 54: 177–183.

Mueller-Dombois, D. (1987). Natural dieback in forests. *Bioscience* 37: 575–583.

Mueller-Dombois, D. (1991). The mosaic theory and the spatial dynamics of natural dieback and regeneration in Pacific forests. In: H. Remmert (Ed.), *The mosaic-cycle concept of ecosystems* (pp. 46–60). Berlin: Springer-Verlag.

Mueller-Dombois, D. (1993). Forest decline in the Hawaiian Islands: A brief summary. In: R. F. Huettl and D. Mueller-Dombois (Eds.), *Forest decline* (pp. 229–235). Berlin: Springer-Verlag.

Mueller-Dombois, D., and Ellenberg, H. (1974). *Aims and methods of vegetation ecology.* New York: Wiley.

Nakanishi, S., and Hattori, T. (1979). A Castanopsis type association of the Setouchi district in southwestern Japan. In: A. Miyawaki (Ed.), *Vegetation und Landschaft Japans. Bulletin of the Yokohama Phytosociological Society Japan* 16: 113–140.

Nishimoto, T., and Hada, Y. (1994). Forest vegetation in the Okayama Prefectural Nature Conservation Center: 1. Species composition and community structure. *Bulletin of the Okayama Prefecture Nature Conservation Center* 2: 13–24.

Rutherford, T. A., Mamiya, Y., and Webster, J. M. (1990). Nematode-infested pine wilt disease: Factors influencing its occurrence and distribution. *Forest Science* 36: 145–155.

Sakamoto, K., Fujikawa, N., Chiba, K., and Yoshikawa, K. (1995). Stand structure and dynamics of *Pinus densiflora* forests damaged by pine wilt disease in the southern part of Okayama Prefecture. *Journal of the Japanese Society of Revegetation Technology* 21: 18–28.

Takama, H., Toyohara, G., and Deguchi, H. (1996). Succession of secondary forests on the Sera Platform, Hiroshima Prefecture. *Hikobia* 12: 111–121.

Toyohara, G. (1984). A phytosociological study and a tentative draft on vegetation mapping of the secondary forests in Hiroshima Prefecture with special reference to pine forests. *Journal of Science of the Hiroshima University* 19: 131–170.

Toyohara, G., and Fujihara, M. (1998). Succession of secondary forests affected by pine wilt disease in western Japan followed on vegetation maps. *Applied Vegetation Science* 1: 259–266.

Toyohara, G., Okuda, T., Fukushima, A., and Nishiura, H. (1986). Change of the vegetation after pine damage in Miyajima Island, Hiroshima. *Japanese Journal of Ecology* 35: 609–619.

Tyler, G. (1989). Edaphical distribution patterns of macrofungal species in deciduous forest in south Sweden. *Acta Ecologia* 10: 309–326.

White, P. S., and Pickett, S. T. A. (1985). Natural disturbance and patch dynamics: An introduction. In: S. T. A. Pickett and P. S. White (Eds.), *The ecology of natural disturbance and patch dynamics* (pp. 3–13). Orlando, FL: Academic Press.

18

Ethnobotany and Biodiversity Conservation

Rainer W. Bussmann

Introduction

During the last decades, species extinction has increased at a frightening rate. Covering only 10% of the earth's land surface, tropical forests are believed to harbor more than 50% of the planet's species (Wilson, 1985). The need for the integration of local indigenous knowledge for a sustainable management and conservation of natural resources receives more and more recognition (Posey, 1992; Pimbert and Pretty, 1995). Moreover, increased emphasis is being placed on possible economic benefits, especially of the medicinal use of tropical forest products, instead of pure timber harvesting—an approach particularly appealing to countries with difficult economic conditions. Most research efforts, due to lack of manpower, time, and resources, focus only on either "biodiversity assessments" or ethnobotanical inventories, or try to implement management and use measures without having a sound scientific base to do so. Often, the needs of the local populations (e.g., their dependence on plant resources for health care) are entirely ignored.

During the 1970s, the World Health Organization (WHO) was very productive in advocating the integration of traditional medicine into public health programs in Third World countries. The Alma Ata Declaration proclaimed "health for all in the year 2000." At the beginning of the new millennium, we have to ask: Where are we with regard to this ideal?

Unfortunately, not only are we a long way from its fulfillment in the Third World, but, as managed care issues in the United States clearly indicate, contemporary health delivery systems are also not meeting the needs of the postmodern world, especially with regard to mental health.

In an era of booming research in receptor pharmacology, molecular biology, and computerized drug design, it is easy to forget that medicinal plants still continue to play a

Rainer W. Bussmann • Department of Plant Physiology, University of Bayreuth, 95440 Bayreuth, Germany; e-mail: ceja.andina@t-online.de dominant role in the primary health care of about 80% of the world's population living in developing nations. But natural products, and medications derived from them, also remain an essential feature of the health care systems of the remaining 20% of the population residing is developed countries. Studies reveal that more than 50% of all medications currently used in developing nations have a natural product origin. Of the 25 best-selling pharmaceutical agents, 12 are natural product–derived (Baker et al., 1995; Farnsworth et al., 1985). Also, the American herbal industry has grown exponentially in the last 25 years, from a few mail-order suppliers of crude bulk herbs to an industry estimated at $4 billion retail today (Brevoort, 1998). In Europe in 1994, it had a value of $6.5 billion (Breevort, 1994). Traditional lore has proven to be an important source of therapeutic drugs. Many antibiotics, tranquilizers, sedatives, anesthetics, pain relievers, and laxatives have come from this source. Examples are digitoxin and digoxin (for heart failure), egotamine (for migraine), salicin (for pain and inflammation), morphine (for pain), reserpine (for hypertension), quinine from Loja (for malaria), tubocurarine (for surgery), and a host of others (Maybury-Lewis, 1992). One-fourth of all prescriptions sold every year in the United States have plant chemicals as active ingredients. Consumers spend more than $6 billion a year on medicines derived from tropical plants (Plotkin, 1993).

Many drugs are waiting to be "discovered" by modern science, particularly in the tropics. Harvard anthropologist David Maybury-Lewis (1992) described this situation and the role that traditional medicine can play in the discovery process:

> The tropical rain forest is home to approximately 70 percent of the million or so species of higher plants that are believed to inhabit the earth. Scientists have named only about 250,000 of these species, which are becoming extinct at a rate faster than we can name them. Of those plants that have been named, only a tiny fraction has been scientifically studied. Among these are many of our most important drugs. Tribal peoples, on the other hand, use thousands of species of higher plants in their medical practices. Many of these have not yet been botanically named, let alone studied by scientists. We are just beginning to appreciate how much more tribal peoples know about plants and their properties than we do.
>
> The conflagration of the tropical rain forest threatens not only countless species of plants but also the cultures and individuals who know their properties and use them in their daily lives. What we are witnessing makes the burning of the library of ancient Alexandria look insignificant by comparison. It is as if the greatest medical library in the world is burning faster than we can read its contents, which we have just begun to catalog.

Pharmaceutical research up to the 1950s relied heavily on natural product development for the discovery of prototype drugs (i.e., substances having unique chemical structures and eliciting effects by new mechanisms). However, the synthetic chemistry revolution brought massive random drug testing in the 1970s and 1980s, and the advent of very specific bioassays. By the mid-1980s, most pharmaceutical companies avoided exploring plant-derived substances in their search for new drugs, pinning their hopes on the potential of receptor- and enzyme-targeted drug designs. This trend appeared to be reversing in the decade of the 90s, due to recent discoveries by a small but growing group of medical anthropologists, ethnobotanists, and ethnopharmacologists—researchers who study the relationships between plants and people. Recently, statistical analyses by some researchers at the Bayer pharmaceutical company have recognized that natural compounds cannot be replaced by synthetic products (Henkel et al., 1999). The advent of alternative medical practices in industrialized nations shows a tendency to correlate with this trend. Herbalism, acupuncture, chiropractic, osteopathy, and so on, are used by millions of people worldwide as health resources (Trachtman, 1994). According to a recent survey, as many as one-third of all Americans use alternative medicine annually (Eisenberg et al., 1993).

Ethnobotanist Mark Plotkin (1993, p.) describes the ethical issues surrounding a major aspect of this problem, specifically the search for plant-derived therapeutic drugs:

> When I began my work, no pharmaceutical firms had any interest in ethnobotany. But given the increasing concern and publicity devoted to the rain forest issue, I was confident the situation would change—and I was right. In February 1989, an article describing my field research appeared in *Smithsonian* magazine.
>
> Within two weeks of the article's appearance, I was besieged by venture capitalists who saw ethnobotany as a fail-safe route to quick riches. "We'll raise some capital," one would-be tycoon said, "set up a lab, find some cures, and synthesize the compounds. Then we'll sell it all off to a big drug company and pocket a ton of money." My most serious objection to their schemes stemmed from the take-the-money-and-run approach; they had virtually no interest in the people who were teaching me about the plants or about the fate of the forest in which these plants were found.

The United Nations Convention on Biological Diversity and Intellectual Property Rights—signed by more than 157 countries in Rio de Janeiro in June 1992—reflects a fundamental change in how the international scientific community perceives the environment and issues in human ecology. However, individual nations are still left with the task of developing viable policies that effectively promote "bioprospecting" and "sustainable development," while protecting the rights—and the cultures—of local communities (Reid, 1993–1994; Posey and Dutfield, 1995; Jacoby and Weiss, 1996; Iwu, 1996).

An approach to bioprospecting in Costa Rica has been taken by Merck & Company, Ltd. In 1991, the company entered into a 2-year, $1.1 million contract with the National Biodiversity Institute (INBio). Under this agreement INBio provides chemical extracts to Merck in exchange for a sampling budget, technical assistance, and training, as well as royalties on any resulting chemical products. A similar agreement involving indigenous medicinal knowledge was recently negotiated between Conservation International, Bristol-Myers Squibb, Suriname, and the National Institutes of Health (United States) (Conservation International Fact Sheet). Also, the Smithsonian Institution (National Museum of Natural History (NMNH/SI) has contracted with Pharma-Cognetics, Inc. for 3 years of botanical work in Panama, Guyana, and Ecuador (NMNW/SI, 1996). Mooney (1993) criticizes M–INBio and describes other pharmaceutical projects in Latin America.

Another innovation that has emerged in recent years is the Shaman's Apprentice Program developed by ethnobotanist Mark Plotkin while a vice-president at Conservation International in Washington, D.C. This program is designed to help indigenous peoples preserve vital shamanic lore that is endangered by acculturation and/or massive migration away from tribal and peasant communities to the teeming urban areas of the developing world. In effect, there is either no one left to whom shamans can transmit traditional lore or the knowledge itself is no longer valued by community members. In response to this situation, Plotkin (1993) makes a strong case for grassroots efforts:

> If we are to safeguard the rights of the indigenous peoples, protect endangered species, find new foods to feed the hungry and new medicines to cure the sick, now is the time to act. We must develop a proactive, holistic approach to the environmental problems we face, realizing that previously overlooked or even ridiculed world views like shamanic wisdom can help us find answers to some of the questions we face. If we don't our children and grandchildren will inherit a world infinitely less diverse biologically and culturally than the one into which we were born.

Medical anthropologist Joe Bastien (1992) makes an equally strong case for the ecological viability of Andean ethnomedicine:

> In this age of "Save the Earth," ethnomedical practitioners use adaptive strategies and folk traditions that are living and dynamic systems subject to change in response to environment and community, providing examples of biocultural integration. . . . Andean medical systems provide a myriad of adaptive strategies to some of the most variable environmental zones of the world, from the tropical Amazon to the high Sierras.

Although Conservation International is developing the Shaman's Apprentice Program among 40 indigenous peoples in 19 countries, there is currently very little published documentation of this important initiative. However, Plotkin (1993) provides some initial guidelines. He relates how, in 1988, he presented the chief of a Tirio village in Suriname with a 200-page manuscript on his work with medicinal plants. In response, the chief assigned two young men to serve as apprentices under the tutelage of the Jaguar Shaman. In addition, the apprentices were to work with Plotkin in fact-checking data, adding new information, translating into Tirio, writing a handbook, and teaching younger Indians about the healing plants in the village schoolhouse. Once established, the program was administered by the indigenous people themselves.

The key element in this educational dialogue has been one of restoring respect for traditional ways. As Plotkin (1993) puts it:

> I feel strongly that this effort has helped validate their culture in the eyes of the Indians. . . . This research constitutes a true partnership between Western and Indian cultures; both share in any potential material benefits, but more important, this approach to ethnobotany helps the indigenous peoples understand the potential global importance of a fundamental aspect of their culture.

On November 30, 1999, *The New York Times* reported that Plotkin had given up on bioprospecting, which had become extremely controversial among indigenous peoples. Currently, he is focusing exclusively on the preservation of shamanic healing methods. He has also severed his ties with Shaman Pharmaceuticals, which abandoned bioprospecting for the marketing of dietary supplements in February 1999.

Ethnobotany and Conservation: A Case Study from Southern Ecuador

Ethnobotanical Studies and Intellectual Property Rights in Latin America

Traditional medicine or ethnomedicine, a set of empirical practices embedded in the knowledge of a social group, is often transmitted orally from generation to generation, with the intent to solve health problems. An alternative to Western medicine, it is strongly linked to religious beliefs and practices of indigenous cultures. Medicinal plant lore or herbal medicine is a major component of traditional medicine.

In Latin American countries, herbal medicine is deeply rooted, practiced extensively by indigenous groups, and frequently used by a broad cross-section of the larger society. Knowledge is transmitted from one generation to the next by traditional healers, shamans, or *curanderos*, and has survived the rigors of the Spanish conquest and extensive *mestizaje*, or racial intermixing. Herbal medicine is, however, a dynamic phenomenon in constant evolution, and additional knowledge has been acquired by natural selection over the centuries.

Although one of the smallest countries in Latin America, Ecuador hosts some of the most diverse flora on Earth. The frontier region of Ecuador and Peru in the southern part of the country, with its still-pristine mountain forests (Figure 18.1, upper left) is known as one of the areas with the highest biodiversity worldwide. Low passes in the Andean chain permit migratory movements from the Amazon Basin to the coast and vice versa. Additionally, a rapid transition of the humid Northern Páramos to the drier central Andes and the Northern Peruvian lowlands occurs in the area. The tropical montane forests of Ecuador are thus one of the "hotspots" of species diversity. The traditional use of nontimber products, particularly of medicinal plants, has deep roots not only in indigenous communities, but is also practiced in a wide section of Latin American society. The use of medicinal herbs is often an economically inevitable alternative to expensive Western medicine. The base knowledge of this traditional use is passed from one generation to the next. Particularly, the medical use, however, represents a highly dynamic, always evolving process, in which new knowledge is constantly being obtained and linked to traditional practices. However, the southern part of the country is relatively unexplored, and little work has been done on vegetation structure, ecology, and ethnobotany. Also, despite the fact that this area is located at the northern end of what Peruvian anthropologist Lupe Camino (1992) calls the "health axis" of Central Andean ethnomedicine, no anthropological research has been published on the rich shamanic lore found here. Despite their huge importance as water catchments, erosion barriers and source of a wide array of timber and non- timber products the destruction of forest resources is accelerating. Huge tracts of land are cleared for agriculture and pastoral use (Figure 18.1, upper right), leading to widespread erosion (Figure 18.1, lower left). Especially in Southern Ecuador, gold-mining (Figure 18.1, lower right) is an additional threat, not only destroying the original vegetation cover, but also leading to heavy mercury pollution downstream.

In September 1995, the knowledge and expertise of three researchers was put to work in a unique ethnobiological team effort. The team was coordinated by Douglas Sharon, a medical anthropologist with years of experience researching traditional medicine in northern Peru (Sharon, 1980, 1994; Joralemon and Sharon, 1993). It included Ezra Béjar, an ethnopharmacologist with multidisciplinary experience in the area of phyto-pharmacodynamics in the United States (Béjar, 1997; Béjar et al., 2000) and ethnobotany in Mexico (Instituto Mexicano del Jeguno Social, 1982; Aguilar et al., 1984), and Rainer Bussmann, a plant ecologist from Germany who had conducted ecological surveys of montane and lowland tropical forest vegetation in East Africa and Ecuador. Working with a local healer, Sr. Cruz Roa, the research team studied medicinal plants in an area around the town of San Pedro de Vilcabamba, establishing an interdisciplinary field methodology to identify species, possible curative properties, and therapeutic applications of medicinal plants found in the area. Because only a limited amount of time in the field was available, an intensive focus on one specialist was the most productive method to follow. Girault's (1987) documentation of 600 plants with a Bolivian healer justifies such an approach.

Two databases of medicinal plants—one in English, the other in Spanish—have been created. The purpose of the databases was to create a book in Spanish for local dissemination (Béjar et al., 2001). Andean antecedents for the book already existed (Kothari, 1993; Ruiz, 1993, for the northern highlands of Ecuador; Quezada et al., 1992, for the central highlands; Franquemont et al., 1990; Roersch, 1994, for southern highland Peru; Duke and Vásquez, 1994, for the Peruvian Amazon; Bastien, 1987, for highland Bolivia). In Mexico, Alcorn's work (1984) on Huastec Mayan ethnobotany analyzed the dynamics of plant–human interrelationships and documented 550 medicinal plants.

Figure 18.1. Upper left: Pristine montane cloud forest in Southern Ecuador; upper right: recent forest clearing for agricultural use; lower left: large-scale erosion on exposed mountainsides; lower right: gold mining area in southern Ecuador.

There are also many positive initiatives at the present time. For example, in Peru, Fernando Cabieses (personal communication, 1997), director of the National Institute of Traditional Medicine, housed within the Ministry of Health in Lima, reports that there is a network of some 16 botanical gardens in the country that are focusing on plants used in traditional medicine. Two hundred plants are being grown in the Lima garden.

In Bolivia, a dialogue between shamans and doctors was the outcome of workshops for biomedical and ethnomedicinal practitioners, as described by Bastien (1992):

> Doctors soon realized that ethnomedical practitioners used adaptive and popular strategies for curing which peasants readily accepted. Ethnomedical practitioners became aware of public health measures and medicines that could help their patients and members of the community. [It is] an educational resource that can teach doctors about synergetic healing and communication with patients. At the least, it represents an alternative healing system and provides a balancing mechanism to restore the pendulum's from mechanistic healing to humanistic and psychosocial healing.

In Mexico, Abigail Aguilar directs the herbarium of the Mexican Social Security Institute where, since the 1970s, over 12,600 voucher specimens of medicinal plants, with their popular names and uses, have been identified and catalogued. They correspond to 3,000 species. Also, the Ethnobotanical Garden of the National Institute of Anthropology and History, founded in 1979 at Cuernavaca, has published a *Catalog of Medicinal Plants* containing 236 of 330 plants grown at the garden (Avilés and Suárez, 1994). In addition, Paul Hersch heads a program to grow these plants in rural communities and to coordinate collaborative efforts with international phytotherapists.

In Ecuador, since our 1995 project, there have been some relevant innovations. In 1996, the Royal Danish Academy of Sciences and Letters published a comprehensive work on the ethnobotanical use of plants by three indigenous peoples of coastal Ecuador (Barfod and Kvist, 1996). In May 1997, Douglas Sharon—following a precedent established with the help of a Peruvian *curandero*, Eduardo Calderón, at the end of the 1970s—was invited to join a group of Ecuadorian scholars in teaching a course on traditional medicine at the Faculty of Medical Sciences of the University of Loja. As of August 10, 1998, Article 44 of the new Constitution of Ecuador stated that the Republic will "recognize, respect, and promote the development of traditional and alternative medicine."

Béjar et al. (2001) will be the primary vehicle for dissemination in Ecuador. The principal target audience for this textbook for daily use and in educational programs, will be the citizens of San Pedro de Vilcabamba. Additional audiences in-country will be the academic community and general public. At a grassroots level, use of the book in a Curandero Apprentice Program could enhance transmission of herbal lore from one generation to the next. The database will be the medium for disseminating information to Ecuadorian government agencies and the scientific community. However, dissemination of project results—whether in printed or digital format—poses a serious ethical dilemma with regard to indigenous IPRs (intellectual property rights). Thomas Greaves (1995) aptly describes this problem:

> Many people believe, of course, that indigenous societies should financially benefit from any economic benefit that comes from commercializing products identified by their knowledge, and there is a minor industry of people seeking ways to cut native societies into the royalty stream. Unfortunately, however, most companies now mining for ethnobotanical leads are doing it in libraries, screening the information that anthropologists, ethnobotanists, and others have gathered, published and placed in the public domain. Because the information is in the

> public domain, these companies' legal obligation to the societies that sourced the cultural information is zero....
>
> The royalties-from-drugs example... is misleading, however, because it leaves us with the impression that, for indigenous peoples, IPR is really a question of how native peoples can profit from the commercialization of their medical knowledge. It is not. A far more common reaction is that commercializing the knowledge of ones' ancestors is repulsive. The efficacy of healing substances is ascribed to a cosmology of forces and spiritual relationships, not simply a biochemical reaction. Ripping a substance out of its context is an egregious act.

Medicinal Plants of Southern Ecuador: Ethnobotany and Conservation Applications

Documentation of Plant Uses

To date, 142 plant species with medicinal properties, belonging to 65 plant families were identified, and their ecological requirements and use were documented (Table 18.1; Béjar et al., 2001). The most widely used family was Asteraceae, with 15 species employed, followed by Lamiaceae (7). Thirty-nine families were represented by only one species, and 14, by two. Interestingly, only little more than two-thirds of these species are indigenous to southern Ecuador, whereas more than 30% represent introduced species of mainly Asian and European origin. This depicts the described dynamics in plant use in a striking way: Interesting neophytes are constantly being tested for their potential use and added to the traditional repertoire.

For most treatments, the green parts of the plants were employed. Roots, and particularly seeds or bark, were used to a much smaller extent. Mostly, the collected material was cooked, and the resulting broth was either swallowed or applied topically. The central focus of medicinal care is the treatment of "magical" illnesses, like mal aire ("bad air"), fright, attacks by the devil, and cleansing from evil spirits. Fifty species, or 35% of all plants, were to some extent used for this category. Rheumatic illnesses and problems of the respiratory system (including colds, flu, cough), as well as urinary tract problems, mostly caused by damp and chilly living conditions during the wet season, and the widespread use of open fires in the houses, followed next. The large category "others" contained a wide array of illnesses such as nerves, ailments, toothache, eye ailments, blood pressure problems, cancerous systems, diabetes, headache, inflammations, asthma, allergies, depression, high cholesterol, ulcers, and so on. Many species encountered had a variety of uses, and in most treatments a mixture of different plants was used.

Most treatments are performed in the houses of the individual healers (Figure 18.2, upper left), who normally have their *mesas* (healing altars) ready to be set up. In most cases in southern Ecuador, a "western" altar, without many power objects is employed (Figure 18.2, upper right), in contrast to northern Peru, where normally a "traditional" mesa is set up. This difference is rooted in the fact that traditional healing is still illegal in Ecuador. Additionally, traditional cures are often performed outdoors, either close to holy lagoons or waterfalls, or at special ceremonial sites. A curing ceremony normally involves purifications of the patient by spitting or spraying blessed and enchanted herbal extracts on the whole body to fend off evil spirits (Figure 18.2, upper left).

The pure ethnobotanical documentation of plant use does not contribute much to conservation efforts. For this reason, the main focus is placed on application of such results for sustainable development activities. The dissemination of traditional knowledge (Béjar et al., 2001) is a first important step toward the long-term use and conservation of local forest resources. Linking basic ecological research, ethnobotany, and conservation,

Table 18.1. Medicinal Plants of Southern Ecuador and Their Uses

Vernacular Name	Use	Scientific Name
Ajinco	Critical illnesses involving body and stomach pains	*Artemisia absinthium* L., *A. vulgare* L.
Ajo Caspa; Ajo Macho	Protection against sorcery	*Allium sativum* L.
Albaca	During childbirth	*Ocimum basilicum* L.
Alcanfora	Fright (children), Mal Aire, skin irritations	*Ocotea* spp.
Anís	Chills, colds	*Pimpinella anisum* L.
Apio Negro	To reduce swellings	*Apium graveolens* L.
Arco Iris	Yellow diarrhea of children, fright	*Streptosolen jamesonii* (Benth.) Miers
Balsa	Very light timber	*Ochroma pyramidale* (Cav. ex Lam.) Urb.
Berro	Lungs	*Nasturtium officinale* R. Br.
Bledo	Food	*Amaranthus spinosus* L.
Buenas Tardes	Purgative	*Mirabilis jalapa* L.
Cabuyo	Rheumatism, for cleansing to defend against sorcerers during sessions, to curse somebody	*Furcraea andina* Trel.
Cadillo	Kidneys	*Triumfetta mollissima* Kunth.
Calahuala	Kidneys	*Polypodium* spp.
Camalonga	Against evil eye, fever, heart problems	*Thevetia peruviana* (Pers.) K. Schum.
Camotillo	Inflammation of arms and feet, rheumatism	*Salvia microphylla* Kunth.
Caña Agria	Kidneys, diabetes	*Costus glabratus* SW., *C. laevis* Ruiz & Pav.
Caña Caña	Kidneys	*Costus* spp.
Canchalagua	Colds, flu, cough	*Centaurium erythraea* L.
Cardo Santo	Lungs, sore eyes, fright	*Argemone mexicana* L.
Cascarilla	Reduce fever, flu, aires, aching bones, colds, cough, malaria	*Cinchona pubescens* Vahl.
Cebada	Diarrhea, tiredness	*Hordeum vulgare* L.
Cebolla Blanca	Cough	*Allium cepa* L.
Cerraja	Amoebas, liver	*Sonchus asper* (L.) Hill.
Chamana	Muscular pain, sore bones, rheumatism	*Dodonaea viscosa* Jacq.
Chamico	Asthma, rheumatism	*Datura stramonium* L.
Chanca Piedra	Stomachaches	*Phyllanthus niruri* L.
Chicoria	Shivering fits, fever, malaria, high blood pressure	*Pseudelephantopus funckii* (Turcz.) Philipson
Chila Chilo	Fright, mal aire	*Eupatorium dendroides* Spreng.
Chilca Redonda	Vomiting, mal aire	*Baccharis obtusifolia* HBK
Chin Chil	Fright, mal aire, skin irritations	*Tagetes minuta* L.
Chirimoya	Colds, helps women during childbirth	*Annona cherimoya* Miller
Cholo Valiente	Injury, pain in the bones, cold feet, rheumatism	*Centropogon* spp.
Chonta	Construction of houses and roofs, ceremonial staffs	*Bactris* spp.
Chuhuasi	Kidneys, inflammations	*Maytenus laevis* Reiscek
Chupaya	Tumors, infections	*Puya* spp.

(*continued*)

Table 18.1. (*continued*)

Vernacular Name	Use	Scientific Name
Chuquiragua	Colds, flu, coughs, malaria, sore bones	*Chuquiragua jussieui* J.F. Gmel.
Cicuta	Animal wounds	*Polygonum hydropiperoides* Michx.
Cimora toro	Good luck, against sorcery	*Sanchezia oxysepala* Mildbr.
Clavel	Nervous system, nose, nasal passage, heart	*Dianthus caryophyllus* L.
Cola de Caballo	Kidneys	*Equisetum bogotense* H.B.K.
Cominillo	Diarrhea, stomachaches	*Apium leptophyllum* (Pers.) F. Mueller ex Benth.
Condorango	Kidneys	*Marsdenia cundurango* Rchb. f.
Congona	Heart, nerves	*Peperomia congona* Ruiz. & Pav.
Cordoncillo	Wounds	*Piper aduncum* L.
Cucharillo	Diabetes	*Adiantum concinnum* Willd. ex HBK
Culantrillo	Kidneys	*Eryngium foetidum* L.
Cutuyuyo	Against fever	*Portulaca oleracea* L.
Díctamo del Campo	Allergies	*Cyperus* spp.
Diente de León	Liver, high blood pressure	*Taraxacum officinale* Weber
Doradilla	Sore bones	*Selaginella arthritica* Alston
Escancel	Infections, liver, headaches, depression	*Aerva lanata* (L.) Juss., *Alternanthera mexicana* (Schltdl.) Hieron
Eucalipto	Sore throat, bronchitis	*Eucalyptus globulus* Labill.
Flor de Christo	Nerves	*Epidendrum calanthum* Rchb. f., *E. catillus* Warcz., *E. cochlidium* Lindl.
Fresmo	Nervous system, menstrual cramps	*Tecoma stans* (L.) Juss. ex Kunth.
Gorondilla	Urinary system to remove obstructions	*Opuntia pubescens* Wendl. ex Pfeiffer
Grama Dulce	Kidneys	*Cynodon dactylon* (L.) Pers.
Grenadilla	Infections, menstrual cramps	*Passiflora ligularis* Juss.
Guayavilla	Diarrhea	*Psidium guineese* Sw.
Guayusa	Diuretic, sorcery, anemia	*Ilex guayusa* Loes.
Hierba de la Justicia	Fright, anxiety, psychological tension, heart, pneumonia Astonishment in children	*Justicia* spp.
Hierba del Coche	Fright	*Cuphea racemosa* (L. f.) Spreng.
Hierba del Oso	Rheumatism	*Saurauia* spp.
Hierba del Perro	Diarrhea with a cold	*Holcus lanatus* L.
Hierba Gallinazo	Fright, mal aire	*Cassia occidentalis* L.
Hierba Mora	To cure sorcery	*Solanum nigrum* L.
Hierbabuena	Stomachaches	*Mentha viridis* (L.) L., *Mentha x piperita* Stokes
Hierbas del Toro	Leucorrhea	*Cuphea loxensis* Kunth.
Higuerilla	Sprains, fractures	*Ricinus communis* L.
Hinojo	Food, infections	*Foeniculum vulgare* Mill.
Huachuma	Ulcers, purgative	*Echinopsis pachanoi* (Britton & Rose) Friedrich & G.D. Rowley
Huichinge	Liver, infections, diarrhea	*Bidens pilosa* L.
Ishpingo	Fright, mal aire, skin irritations	*Ocotea floribunda* (Sw.) Mez.

Table 18.1. (*continued*)

Vernacular Name	Use	Scientific Name
La Chichera	Fright, birth and post partum protection	*Lepidium* spp.
Laritaco	Espanto, boils, rheumatism, "bad wind"	*Vernonia patens* Kunth.
Lechuguilla	Diarrhea	*Onoseris spinosa* Kunth.
Llantén	Liver, kidneys	*Plantago australis* Lam., P. major L.
Lutuyuyo	Diarrhea	*Basella obovata* Kunth.
Malva Blanca	Laxative	*Malva erecta* J. Presl. & C. Presl., *M. sylvestris* L.
Malva Olorosa	Hemostatic, diarrhea	*Pelargonium odoratissimum* (L.) L'Her. ex Ait.
Manzanilla	Stomach pains, fever-related colds, diarrhea	*Matricaria chamomilla* L.
Marapico	Toothache, nerves, tumors	*Iochroma loxense* (Kunth.) Miers
Margarita	Cleansing from bad air	*Chrysanthemum leucanthemum* L.
Matico	Kidneys	*Piper angustifolium* Lam.
Menta	Expel intestinal worms, colds, rheumatism	*Mentha viridis* L.
Molle	Seasoning for food, rheumatism	*Schinus molle* L.
Monte de la Huanchaca	Fright	*Brickellia diffusa* (Vahl.) A. Gray
Moradilla	Kidneys, lungs	*Alternanthera porrigens* (jacq.) Kuntze
Mortiño	Hangover, shame, kidneys, nose, vomiting	*Solanum nigrum* L.
Mosquera	Stomach pains, pneumonia, rheumatism, tooth extraction	*Croton wagneri* Müll. Arg.
Naranja Agria	Stomachache, high blood pressure, deodorant	*Citrus aurantium* L.
Níspero	Liver, to reduce cholesterol, diabetes	*Eriobotrya japonica* (Thunb.) Lindl.
Nuez Moscada	Mal aire, sorcery	*Myristica fragrans* Houtt.
Ortiga Blanca	Muscular pains	*Urtica urens* L.
Ortiga de Caballo	Muscular pains	*Urera baccifera* (L.) Gaudich ex Wedd.
Ortiga de Leones del Oriente	Muscular pains	*Urera* spp.
Ortiga Negra	Muscular pains	*Urtica leptophylla* Kunth
Pacunga	Stomach	*Siegesbeckia orientalis* L.
Paico	Parasites	*Chenopodium ambrosoides* L.
Palitalia	Internal bleeding caused by a blow	*Parietaria debilis* G. Forst.
Palo Santo	Anemia, incense, deodorant, ears	*Bursera greveolens* (Kunth.) Triana & Planch.
Palta	Toothache, relieves women's abdominal pain	*Persea americana* Miller
Pedromera Blanca/Rosada	Gastrointestinal problems	*Ageratum conyzioides* L.
Pega Pega	Lesions	*Desmodium triflorum* (L.) DC.
Penca de Caballo	Rheumatism	*Agave americana* L.
Pepiso	Rheumatism	*Cantua quercifolia* Juss.
Perdromera Blanca	Gastrointestinal problems	*Ageratum hirsutum* Lam., *A. hirtum* Lam., *A. latifolium* Cav.

(*continued*)

Table 18.1. (*continued*)

Vernacular Name	Use	Scientific Name
Perejil	Food condiment, nosebleed, infections	*Petroselinum crispum* (Mill.) Nyman
Pico Pico	Swollen stomach, vomiting, purgative	*Acnistus arborescens* (L.) Schltdl.
Pimpinella de Bejuco	Mal aire	*Sanguisorba officinalis* L.
Piñon	Wounds, strong purgative	*Jatropa curcas* L.
Pitayo	Sorcery, liver	*Hylocereus polyrhizus* (F.A.C. Weber) Britten & Rose
Poleo	Cold, flu	*Minthostachys mollis* (Kunth.) Giesb.
Preñadilla	Diuretic, venereal diseases, female fertility, lungs	*Pilea microphylla* (L.) Liebm.
Romero	Mal aire, nerves, sore eyes	*Rosmarinus officinale* L.
Ruda	Mal aire, sore red eyes caused by mal aire, after childbirth	*Ruta graveolens* L.
Salvia	Mal aire, eyesight, headache	*Buddleja americana* L.,
San Agustin Lluyo	Purgative	*Lantana comunis* L., *L. trifolia* (L.) Kuntze
San Antonio	Wounds	*Cuphea carthagenensis* (Jacq.) J.F. Macbr.
Sangre de Drago	Wounds, infections (particularly in women), kidneys	*Croton lechleri* Müll. Arg.
Santa Maria	Fright, sorcery, against the devil	*Tanacetum parthenium* (L.) Sch. Bip.
Sauce	Fever, for children after having lost a tooth, to prevent infection	*Salix humboldtiana* Willd.
Sauco Común	Diarrhea, scurvy	*Cestrum auriculatum* L'Herit
Savia Reál	Rheumatism	*Salvia ramicifolia* H.B.K.
Sávila Penca	Liver, internal infections, to protect the home	*Aloe vera* (L.) Burm. f.
Suche	Hernia	*Plumeria rubra* L.
Suelda con Suelda	Stomach pains, fright, wounds	*Pseudelephantopsus spicatus* (B. Juss. ex Aubl.) C.F. Baker
Tabaco	Against external parasites, insecticide, fungicide	*Nicotiana rustica* L.
Tabaco	Against external parasites, insecticide, fungicide	*Nicotiana tabacum* L.
Tiatina	Stomachache, physical blows against the body	*Scoparia dulcis* L.
Toronjíl	Nerves, nervous system	*Melissa officinalis* L.
Trébol	Diarrhea, scurvy	*Oxalis* spp.
Trensilla	Kidneys	*Loricaria thyrsoidea* (Cuatrec.) Dillon & Sogast
Triguillo	Women's problems, uterine and urinary tract infections	*Bromus pitensis* Kunth.
Tululuche	Rheumatism	*Solanum albidum* Dunal
Uña de Gato	Cancerous systems	*Mimosa acantholoba* (Humb. & Bonpl. ex Willd.) Poir
Valeriana	Nervous system, nerves, heart	*Valeriana decussata* Ruiz & Pav.
Verdalonga	Liver	*Portulaca oleracea* L.
Vira Vira	Mal aire	*Gnaphalium americanum* Mill.
Yanacaspe	Toothache, broken teeth	*Heliotropium* spp.

Figure 18.2. Upper left: Traditional healing ceremony, with curandero spraying patient to cleanse off evil spirits; upper right: curing Altar; lower left: medicinal herb market; lower right: standardized medicinal herbs for local markets.

however, needs work in a variety of different fields. Apart from scientists and involvement of the local population, local nongovernmental organizations play a major role as nexus between these two players. In southern Ecuador, the Nature and Culture International (NCI) is one of the most successful examples of the conservation of biological and cultural diversity, by linking scientific investigations to its application in sustainable development and education.

Interdisciplinary Investigation

Located between the provincial capitals Loja and Zamora, the main research area of NCI, around Estación Científica San Francisco, the Reserva Biológica San Francisco, covers 1000 ha of the northern slopes of Cordillera de Consuelo in Zamora–Chinchipe Province, Ecuador, adjacent to Podocarpus National Park. Ranging from 1,800 to 3,150 m, it contains a complete pristine transect of the montane forest of the region, as well as comparable anthropogenically disturbed areas in the close vicinity. Further research areas in other ecozones of southern Ecuador complement the research network.

Since 1997, most public and private Ecuadorian universities and research institutions, in close cooperation with 20 departments of nine German universities, as well as U.S. colleagues (Missouri Botanical Garden, Chicago Fields Museum, University of Arizona) are engaged in investigating the basic ecological key parameters in the tropical mountain forest ecosystem and applied approaches in areas of reforestation and agroecology, to produce a sound database for sustainable projects and education. The scientific work is beginning to shift from basic to applied ecology, incorporating, particularly, projects dealing with nutrient turnover in agroecosystems, propagation of and reforestation with indigenous timber species, and silvopastoral systems, as well as conservation and management of indigenous fruit and medicinal plant species with agricultural potential. To allow a broad-scale participation of Ecuadorian students, NCI runs a stipend program, financing these projects of young local scientists under supervision of local and international senior researchers. Apart from investigation in its own reserves, NCI scientists are involved in assessments for the extension of Podocarpus National Park within the frame of the Peru–Ecuador peace agreement and helping the indigenous Saraguro community of Yacuambi to install its own conservation area.

Environmental Education

Building an understanding of environmental issues as early as possible is of paramount importance. To achieve this, a broad environmental education program has been established in southern Ecuador. Frequent public lectures about ecological topics, as well as local development needs and a popular weekly environmental newspaper page, attract mainly the adult population.

A local Natural and Cultural History Museum, focusing on the main ecosystems of the region, its different peoples, and their traditions, will provide a focus for the whole zone. Schoolchildren are addressed by the formation of ecological clubs that provide early possibilities to be involved in environmental issues, to adapt to urban city greening, waste management), as well as rural settings (sustainable agricultural techniques, forest resources). During an annual "Ecological Week," the schools of the region are invited to improve their local environment, and every year, a binational science fair, with the participation of the schools in northern Peru and southern Ecuador, strengthens environmental consciousness and fortifies the peace process between the two neighbors. The production of educative materials about different topics of cloud- and dry-forest ecology, as well as different plant and animal groups, fits into the school curricula. The establishment of organic native fruit and medicinal plant gardens in schools provides the children with theoretical and practical experience.

Sustainable and Community Development

The improvement of the lives and health of the local population, together with careful resource use, are equally important. Events such as the "First Congress of Ecuadorian Nationalities in Southern Ecuador" provide a forum for indigenous and local communities. Donations of vital field equipment for National Park guards help to improve the conservation status of the ecosystem.

Sustainable development projects try to focus again on both urban and rural populations. In the context of urban development, activities range from more environmentally friendly garbage disposal and humus production out of biodegradable waste, to the establishment of small health centers.

Most important, however, are approaches to find possible income alternatives linking scientific research results closely with their application for the needs of a growing population, a large percentage of which depends directly on resources from the surrounding forest areas, or has a high demand for forest products.

The ethnobotanic–socioeconomic survey indicated clearly that indigenous tree species are highly appreciated as building timber, and for furniture and traditional uses such as beehives (Figure 18.3, upper left, example from Kenya). However, the supply of timber is dwindling rapidly, and shortages are a common phenomenon already, with prices rising. Introduced exotics such as *Eucalyptus globulus* or *Pinus radiata* and *P. patula*—which would be readily available from plantations—are not used at all, if possible. High demand leads to extraction of immature trees and poor logging practices, such as using power saws to produce planks directly in the forest, lead to a loss of up to 50% of the base material. Overaged machinery in sawmills and carpentry leads to additional waste. Reforestation with indigenous species is almost entirely nonexistent, because there is no information about phenology, reproductive biology, and germination of these species. Here, ethnobotanical research, together with investigations on basic ecology, has a direct impact on conservation by fostering reforestation with native species instead of exotics, thus diminishing the pressure on the forest as a natural timber source, and providing farmers who plant valuable trees with additional income. Another direct implication is the use of "useless" exotic timbers in the production of traditional bricks and pottery, lowering the waste of high-quality indigenous wood. Additionally, new marketing strategies for traditional ceramics provide better sales and livelihood.

The poorer population especially relies to a large extent on medicinal plants for health care, because Western medicines are much too expensive to be affordable, and herbal markets are common (Figure 18.2, lower left). Some traditional medicinal products are highly esteemed among the whole population and lead inevitably to overharvesting. Again, in this context, ethnobotany can contribute directly to conservation. Documenting the exact use of medicinal plants and their ecology, schemes for sustainable harvesting, or even agricultural production can be elaborated. The most popular herbal tea of southern Ecuador, *Horchata*, consists of more than 30 medicinal herbs. Instead of collecting the base material, the ingredients are now grown by local farmers in small, organically

Figure 18.3. Upper left: traditional beehive (Kenyan Mountain Forest); upper right: production of traditional pottery; lower left: garden landscape in Ecuador; lower right: medicinal organic tea from southern Ecuador.

managed gardens (Figure 18.3, lower left). The harvest is delivered to a central production unit, where the plants are carefully dried, and the clean, standardized final product (Figure 18.3, lower right) can easily be sold in a wider market. This gives the producers a much better, more reliable income than pure collection and helps protect the surrounding forests as a seed source for production; the tea sold contains a certified and mold-free herbal mixture that is much safer for the consumer. Similar approaches pertain to production of standardized medicinal products for the local market (Figure 18.2, lower right). Recently, the growth of organic coffee under multi-indigenous species canopies, with local vegetables and medicinal herbs in the undergrowth, has been implemented. Future planned projects include reforestation with indigenous species in dry forest areas to capture mist for irrigation and drinking water purposes, pasture management, and production of organic sugar.

Conclusions

Attempts to conserve and manage biological diversity need to integrate the views, needs, and traditional knowledge of the local communities involved. The linkage of scientists, nongovernmental organizations, and local people is inevitable in this process; every link needs specifically to contribute to its success. On the scientific level, pure ethnobotanical studies have to be included in more widespread interdisciplinary research, uniting many disciplines of natural and social science to provide a sound base for project implementation. The main results of such research have to be made available in the local languages to be useful. For the long term, only management and conservation measures based on such an integrated approach are viable.

Acknowledgments

I gratefully acknowledge the continuing support by Deutsche Forschungsgemeinschaft (DFG) and Nature and Culture International (NCI). Special thanks are due to all my Ecuadorian, American, Mexican, and German colleagues.

References

Aguilar, A., Béjar, E., Camacho, J., Chávez, M., Delgado, J., Ibañez, R., Lozoya, X., Mechéz-Lozoya, M., Mellado, V., Romero, M., Silva, A., Zolla, C., and Zurita, M. (1984). *Medicina tradicional y herbolaria*. México: Instituto Mexicano de Seguro Social.

Alcorn, J. (1984). *Huastec Mayan ethnobotany*. Austin: University of Texas Press.

Avilés, M., and Suárez, G. (1994). *Catálogo de plantas medicinales*. Morelos, México: Jardín Etnobotánico, Centro I.N.A.H.

Baker, J., Borris, R., Carté, B., Cordell, G., Soejarto, D., Cragg, G., Gupta, M., Iwo, M., Madulid, D., and Tyler, V. (1995). Natural product discovery and development: New perspectives on international collaboration. *Journal of Natural Products* 58(9): 1325–1357.

Barfod, B. B. A., and Kvist, L. (Eds.). (1996). *Comparative ethnobotanical studies of the Amerindian groups in coastal Ecuador*. Copenhagen: The Royal Danish Academy of Sciences and Letters.

Bastien, J. (1987). *Healers of the Andes: Kallawaya herbalists and their medicinal plants*. Salt Lake City: University of Utah Press.

Bastien, J. (1992). Drum and stethoscope: Integrating ethnomedicine and biomedicine in Bolivia. Salt Lake City: University of Utah Press.

Béjar, E. (1997). Pharmacological search for active ingredients in medicinal plants of Latin America. In: T. Shuman, M. Garrett, and L. Wozniak (Eds.), *International Symposium on Herbal Medicine: A holistic approach* (pp. 63–81). San Diego: San Diego State University International Institute for Human Resources Development.

Béjar, E., Bussmann, R. W., Roa, C., and Sharon, D. (2001). Herbs of southern Ecuador—hierbas del sur ecuatoriano. Cincinnati: Muska and Lipman.

Béjar, E., Reyes-Chilpa, R., and Jiménez-Estrada, M. (2000). Bioactive phytochemicals from selected plants used in the XVI century Mexican medicine. In: Ur-Rahman (Ed.), *Studies in natural products chemistry* (Vol. 24, pp. 1–50). Amsterdam: Elsevier Scientific.

Brevoort, P. (1994). The economics of botanicals: The U.S. experience. In: *Botanicals: A role in U.S. health care?* Washington, DC: National Institute of Health/OAM Conference.

Breevort, P. (1998). The booming U.S. botanical market: A new overview. *HerbalGram* 44: 33–46.

Camino, L. (1992). *Cerros, plantas y lagunas poderosas: La medicina al norte del Perú*. Lima: Lluvia Editores.

Cox, P., and Balick, M. (1994). The ethnobotanical approach to drug discovery. *Scientific American* 270(6): 82–87.

Duke, J., and Vázquez, J. (1994). *Amazonian ethnobotanical dictionary*. Boca Raton, FL: CRC Press.

Eisenberg, D., Kessler, R., Foster, C., Norlock, F., Calkins, D., and Delbanco, D. (1993). Conventional medicine in the United States: Prevalence, costs, and patterns of use. *New England Journal of Medicine* 328(4): 246–252.

Farnsworth, N., Akerele, O., Bingel, A., Soejarto, D., and Guo, Z. (1985). Medicinal plants in therapy. *Bulletin of the World Health Organization* 63(6): 965–981.

Franquemont, C., Plowman, T., Franquemont, E., Niezgoda, C., King, S., Sperling, C., and Davis, W. (1990). The ethnobotany of Chinchero, an Andean community in southern Peru. *Fieldiana*, Botany, new series 24.

Girault, L. (1987). Kallawaya: Curanderos itinerantes de los Andes. La-Paz: UNICEF-OPS-OMS.

Greaves, T. (1995). Cultural rights and ethnography. *General Anthropology* 1(1): 3–6.

Henkel, T., Brunne, R., and Reichel, F. (1999). Statistical investigation into the structural complementarity of natural products and synthetic compounds. *Angewandte Chemie* 38(5): 647–649.

Instituto Mexicano del Seguro Social. (1982). *La medicina tradicional en Chiapas* (4 vols). Mexico City: Author.

Iwu, M. (1996). Implementing the Biodiversity Treaty: How to make international cooperative agreements work. *Trends in Biotechnology* 14–3(146): 67–107.

Jacoby, C., and Weiss, C. (1996). Recognizing property rights in traditional biocultural contribution (unpublished manuscript).

Joralemon, D., and Sharon, D. (1993). *Sorcery and shamanism: Curanderos and clients in northern Peru*. Salt Lake City: University of Utah Press.

Kothari, B. (1993). Nucanchic panpa janpicuna: Plantas medicinales del Campo. Quito: Abya-Yala.

Maybury-Lewis, D. (1992). *Millennium: Tribal wisdom and the modern world*. New York: Viking.

Mooney, P. (1993). Aprovechando la diversidad: Una nota sobre la diversidad biológica y el conocimiento indígena. *América Indígena* 3: 41–55.

National Museum of Natural History, Smithsonian Institution. (1996). Agreement may lead to new medicines from tropical plants. *Quest Annual Reports 1995*: 7.

Pimbert, M., and Pretty, J. (1995). *Parks, people and professionals: Putting "participation" into protected area management*. UNRISD Discussion Paper No. 57, Geneva:

Plotkin, M. (1993). *Tales of a shaman's apprentice: An ethnobotanist searches for new medicines in the Amazon rain forest*. New York: Viking.

Posey, D. (1992). Traditional knowledge, conservation and "the rain forest harvest". In: M. Plotkin and L. Famolare (Eds.), *Sustainable harvest and marketing of rain forest products* (pp. 46–50). Washington, DC: Island Press.

Posey, D., and Dutfield, G. (1995). Beyond intellectual property rights: Toward traditional resource rights for indigenous and local communities (unpublished manuscript).

Quezada, A., Hermida, C., Vega, G., Canizares, E., and Hermida, J. (1992). *La práctica médica tradicional en las provincias del Azuay, Cañar, el Oro y Morona Santiago*. Cuenca: IDICSA.

Reid, W. (1993–1994). The economic realities of biodiversity. *Issues in Science and Technology* 10(2): 48–55.

Roersch, C. (1994). *Plantas medicinales en el Sur Andino del Perú*. Königstein: Koeltz Scientific Books.

Ruiz, E. (1993). *Plantas medicinales Andinas: Uso práctico*. Quito: Abya-Yala.

Sharon, D. (1980). El chamán de los cuatro vientos. Siglo veintiuno editores, México.

Sharon, D. (1994). Tuno y sus colegas: Notas comparativas. In: L. Millones and M. Lemlij (Eds.), *En el nombre del señor: Shamanes, demonios y curanderos del norte del Perú* (pp. 128–147). Lima: Australis S.A.

Trachtman, P. (1994). NIH looks at the implausible and the inexplicable. *Smithsonian* 25(6): 110–112, 114–123.

Wilson, E. O. (1985). The biological diversity crisis. *Bioscience* 55(11): 700–706.

INDEX